AF669294

The Most Successful Way of JEE Preparation

DPP

Daily Practice Problems

JEE Main & Advanced

Straight Line & Circle

with Point & Pair of Straight Line

MATHEMATICS Vol-4

Amit M Agarwal

arihant
ARIHANT PRAKASHAN (SERIES), MEERUT

DPP

Daily Practice Problems

arihant

ARIHANT PRAKASHAN (SERIES), MEERUT

Administrative & Production Offices

Corporate Office 'Ramchhaya' 4577/15, Agarwal Road, Darya Ganj, New Delhi -110002
Tele: 011- 47630600, 43518550; Fax: 011- 23280316

Head Office Kalindi, TP Nagar, Meerut (UP) - 250002
Tel: 0121-2401479, 2512970, 4004199; Fax: 0121-2401648

Sales & Support Offices

Agra, Ahmedabad, Bengaluru, Bareilly, Chennai, Delhi, Guwahati, Hyderabad, Jaipur, Jhansi, Kolkata, Lucknow, Meerut, Nagpur & Pune

ISBN : 978-93-13193-52-4

Published by Arihant Publications (I) Ltd.

Production Team

Publishing Managers	Mahendra Singh Rawat & Keshav Mohan	*Inner Designer*	Mazher Chaudhary
Project Head	Sorabh Chaudhary	*Page Layouting*	Sunil Sharma, Hemant
Cover Designer	Shanu Mahad	*Proof Reading*	Himanshu

For further information about the products from Arihant, log on to www.arihantbooks.com or email to info@arihantbooks.com

/arihantpub /@arihantpub Arihant Publications /arihantpub

ABOUT DPP

Daily Practice Problems

Indian Institutes of Technology better known as IITs have always been the dream destination of each science group students who want to pursue their career as Engineers. These are the institutes which have got very high value not even in India but in the whole world. To get the admission in such institutes one need to face entrance Examinations; JEE Main & JEE Advanced, out of which JEE Main is considered as the preliminary exam & JEE Advanced is the final exam to get into the IITs. Approximately per year 12 lakh students enroll themselves for JEE Main and only the top 2.25 lakh in JEE Main merit can attempt for JEE Advanced, the fact clearly express the degree of toughness of JEE Main & JEE Advanced.

To have success in such an exam one needs to be very planned & systematic in its preparation. A very popular wise saying 'Rome was not built in a day' suggests that to achieve a big target like IITs we need to work continuously. For continuous work, one need to have proper planning that how much work he/she has to do on regular basis, i.e., on daily basis. Each student preparing for JEE has many books but unfortunately none of them gives the students a plan for daily studies, which is the most effective success key for JEE.

Keeping the daily practice needs of the students in mind we have come up with DPP Series (Daily Practice Problems) having approximately 26 books. The primary focus of this series is on achieving success through practice & proper planning. For that we have divided the whole syllabus of JEE Main & JEE Advanced; Physics, Chemistry & Mathematics, into different volumes (around 8 volumes of Mathematics & 9 volumes each of Physics & Chemistry), each volume has Daily Practice Problems Sheets having specific questions on various Topics of the individual chapter of the volume ensuring the complete Practice & Assessment of the topic.

DPP
Daily Practice Problems

In this way we have tried to give you an effective daily planning for JEE keeping in mind the daily hours that a student can devote to his practice, you are just required to practice one sheet of each subject on daily basis. This habit of daily practice will keep you away from exam stress and thoughts like; how much I have prepared/am I preparing properly or not/ do I need to study more on daily basis or what I am doing is ok, etc., which regularly tease your mind and can give you panic at any stage of your preparation.

Some special features that we have tried to incorporate in our Daily Practice Problems (DPP) series are discussed below

1. Micro Level coverage of all the topics of a Chapter/ Unit given in JEE Main & Advanced Syllabus.
2. Each DPP has various types of questions; Subjective Questions, Single/ Multiple Correct Options, Statements type, Comprehension based, Integer Value Answer type, Matching, etc., making you practice & ready for all the question formats of JEE.
3. New & innovative questions in each DPP challenging & improving your problem solving skills.
4. Along with Topical Coverage, Revisal DPPs for JEE Main & JEE Advanced with each Chapter, covering multidisciplinary questions proving comprehensive practice.
5. JEE Main & JEE Advanced Archive (collection of last 13 years' exams questions) with each chapter.
6. Complete Solutions for each DPP.

We hope that DPP series will prove to be very beneficial for the students preparing for JEE, by giving them daily practice in a systematic & planned manner. The suggestions/criticism from the students, teachers & parents about DPP series will be welcomed whole heartedly and we promise to update this series accordingly from time to time.

Daily Practice Problems

DAILY TASKS...

1 POINT

2 STRAIGHT LINE

3 PAIR OF STRAIGHT LINE

DPP

Daily Practice Problems

4 CIRCLE

1. Point

DPP-1 Distance Formula

Subjective Question

Working Space

Direction *(Q. No. 1) This question is subjective in nature, need to be solved completely on notebook.*

1 Find the distance between the following pairs of points.

i. $(a, 0)$ and $(0, b)$

ii. $(b + c, c + a)$ and $(c + a, a + b)$

iii. $(a \cos \alpha, a \sin \alpha)$ and $(a \cos \beta, a \sin \beta)$

iv. $(am_1^2, 2am_1)$ and $(am_2^2, 2am_2)$

Only One Option Correct Type

Direction *(Q. Nos. 2-15) This section contains 14 multiple choice questions. Each question has four choices (a), (b), (c) and (d), out of which ONLY ONE option is correct.*

2 Find the value of x_i, if the distance between the points $(x_i, 2)$ and $(3, 4)$ is 8.

a. $3 \pm 2\sqrt{15}$ b. $3 \pm \sqrt{15}$
c. $2 \pm 3\sqrt{15}$ d. $2 \pm \sqrt{15}$

3 The length of a line segment AB is 10 units. If the coordinates of one extremity are $(2, -3)$ and the abscissa of the other extremity is 10, then the sum of all possible values of the ordinate of the other extremity is

a. 3 b. -4
c. 12 d. -6

4 A particle begins at the origin and moves successively in the following manner as shown 1 unit to the right, $\frac{1}{2}$ unit up, $\frac{1}{4}$ unit to the right, $\frac{1}{8}$ unit down, $\frac{1}{16}$ unit to the right etc.

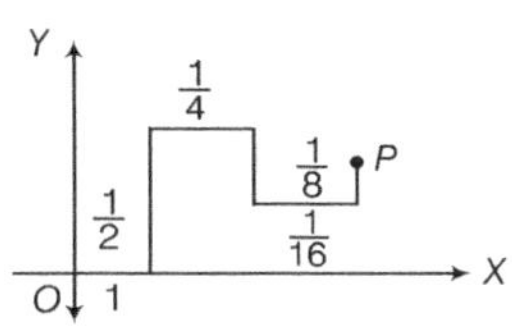

The length of each move is half of the previous move and movement continues in the zig-zag manner infinitely. The coordinates of the point (P) to which the zig-zag converges is

a. $\left(\frac{4}{3}, \frac{2}{3}\right)$ b. $\left(\frac{4}{3}, \frac{2}{5}\right)$
c. $\left(\frac{3}{2}, \frac{2}{3}\right)$ d. $\left(2, \frac{2}{5}\right)$

5 The triangle with vertices $(1, 5)$, $(-3, 1)$ and $(3, -5)$, is

a. isosceles b. equilateral
c. right angled d. None of these

Working Space

6 If the point (4, – 4)(–4, 4) and (x, y) form an equilateral triangle, then

a. $x = -4\sqrt{3}, y = 4\sqrt{3}$ b. $x = 4\sqrt{3}, y = -4\sqrt{3}$

c. $x = 4\sqrt{3}, y = 4\sqrt{3}$ - d. None of these

7 Distance between foot of perpendicular drawn from a point (– 3, 4) on both axes, is

a. 5 units b. 2 units

c. 4 units d. 1 unit

8 If the point (1, 1) (– 1, – 1) and $(-\sqrt{3}, k)$ are vertices of an equilateral triangle, then the value of k will be

a. – 1 b. 1

c. $\sqrt{3}$ d. $-\sqrt{3}$

9 If the points (0, 0), $(2, 2\sqrt{3})$ and (a, b) are the vertices of an equilateral triangle, then (a, b) is

a. (0, – 4) b. (0, 4)

c. (4, 0) d. (– 4, 0)

10 The point on Y-axis equidistant from the points (3, 2) and (– 1, 3), is

a. (0, – 3) b. $\left(0, -\frac{3}{2}\right)$

c. $\left(0, \frac{3}{2}\right)$ d. (0, 3)

11 ABC is an isosceles triangle. If the coordinates of the base are B (1, 3) and C (– 2, 7), the coordinates of vertex A, is

a. $\left(-\frac{1}{2}, -5\right)$ b. (1, 6)

c. $\left(\frac{5}{6}, 6\right)$ d. None of these

12 If the line segment joining the points A (a, b) and B (c, d) subtends an angle θ at the origin, then $\cos\theta$ is equal to

a. $\dfrac{ab + cd}{\sqrt{(a^2 + b^2)(c^2 + d^2)}}$ b. $\dfrac{ac + bd}{\sqrt{(a^2 + b^2)(c^2 + d^2)}}$

c. $\dfrac{ac - bd}{\sqrt{(a^2 + b^2)(c^2 + d^2)}}$ d. None of these

13 If (a, a), $(-a, -a)$, $(-a\sqrt{3}, a\sqrt{3})$ are vertices of a triangle, then it is

a. a right angled triangle b. an isosceles

c. an equilateral triangle d. None of these

14 The locus of $P(x, y)$ such that $\sqrt{x^2 + y^2 + 8y + 16} - \sqrt{x^2 + y^2 - 6x + 9} = 5$ is

a. straight line b. finite line segment

c. infinite ray d. None of these

15 The number of integer points exactly in the interior of the triangle with vertices (0, 0), (0, 10), (10, 0) is k, then total number of combinations of x and y is

a. 32 b. 34

c. 36 d. None of these

DPP-2 Section Formula

Only One Option Correct Type

Working Space

Direction (Q. Nos. 1-11) *This section contains 11 multiple choice questions. Each question has four choices (a), (b), (c) and (d), out of which ONLY ONE option is correct.*

1 If P (1, 2), Q (4, 6), R (5, 7) and S (a, b) are the vertices of a parallelogram $PQRS$, then

a. $a = 2, b = 4$ b. $a = 3, b = 4$

c. $a = 2, b = 3$ d. $a = 3, b = 5$

2 If A and B are the points (– 3, 4) and (2, 1), then the coordinates of the point C on AB produced such that $AC = 2BC$ are

a. (2, 4)   b. (3, 7)

c. (7, – 2)   d. $\left(-\frac{1}{2}, \frac{5}{2}\right)$

3 The line joining the points (1, – 2) and (– 3, 4) is trisected, then the coordinates of the points of the trisection are

a. $\left(-\frac{1}{3}, 0\right), \left(-\frac{5}{3}, 2\right)$ b. $\left(\frac{1}{3}, 0\right), \left(\frac{5}{3}, 2\right)$

c. $\left(0, \frac{1}{3}\right), \left(2, -\frac{5}{3}\right)$ d. None of these

4 ΔABC lies in the plane with A (0, 0), B (0, 1) and C (1, 0). Points M and N are chosen on AB and AC, respectively such that MN is parallel to BC and MN divides the area of ΔABC in half, then the coordinates of M, are

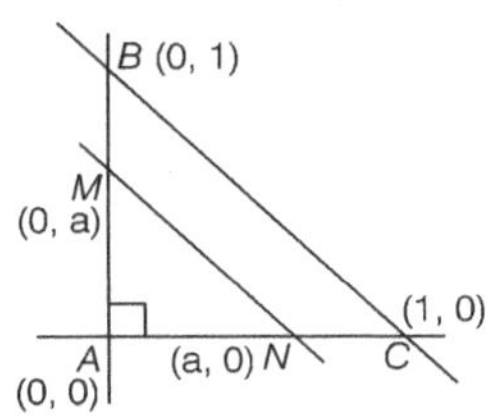

a. $\left(0, \frac{1}{\sqrt{2}}\right)$ b. $\left(0, \frac{1}{2}\right)$

c. $\left(0, \frac{1}{3}\right)$ d. $\left(0, \frac{1}{4}\right)$

5 The line segment joining the points (3, 4) and (7, 8) is divided by the line $2x + 3y + 7 = 0$ in the ratio

a. $\frac{1}{5} : 9$ externally

b. $\frac{1}{5} : \frac{1}{9}$ externally

c. $5 : \frac{1}{9}$ externally

d. $5 : 9$ externally

6 If P, Q, R are collinear points such that $P\ (7, 7)$, $Q\ (3, 4)$ and $PR = 10$, then R is

a. (1, 1)
b. (1, − 1)
c. (− 1, 1)
d. (− 1, − 1)

7 The ratio in which X-axis divides the join of the points (2, − 3) and (5, 6), is

a. 2 : − 1
b. 1 : 2
c. 2 : 1
d. None of these

8 If (8, 5), (− 7, − 5) and (− 5, 5) are three vertices of a parallelogram, then its fourth vertex is

a. (10, 15)
b. (15, 10)
c. (10, 5)
d. (5, 10)

9 Which pair of points lie on the same side of $3x - 8y - 7 = 0$?

a. (0, − 1) and (0, 0)
b. (4, − 3) and (0, 1)
c. (− 3, − 4) and (1, 2)
d. (− 1, − 1) and (3, 7)

10 The set of values of b for which the origin and the point (1, 1) lie on the same side of the straight line, $a^2x + aby + 1 = 0$ for all $a \in R$, $b > 0$ are

a. $b \in (2, 4)$
b. $b \in (0, 2)$
c. $b \in [0, 2]$
d. $(2, \infty)$

11 The harmonic conjugates of the point $R\ (5, 1)$ with respect to the points $P\ (2, 10)$ and $Q\ (6, - 2)$, is

a. (8, − 8)
b. (− 8, 8)
c. (8, 8)
d. (− 8, − 8)

Comprehension Type

Direction (Q. Nos. 12-14) This section contains a paragraph, each describing theory, experiments, data, etc. Three questions related to the paragraph have been given. Each question has only one correct answer among the four given options (a), (b), (c) and (d).

Passage

Let $A\ (0, 3)$, $B\ (-2, 0)$, $C\ (6, 1)$ be the vertices of a triangle and $M\ (\beta, \beta + 1)$ be a moving point, then

12 M lies on the curve

a. $y = x + 1$
b. $y = x^2$
c. $x = y + 1$
d. None of these

13 If M and A lie on same side of BC, then

a. $\beta > 2$
b. $\beta < 2$
c. $\beta > -\frac{6}{7}$
d. $\beta < \frac{3}{4}$

14 M lies within ΔABC, if

a. $-\frac{6}{7} < \beta < 4$
b. $-4 < \beta < -\frac{6}{7}$
c. $-\frac{6}{7} < \beta < \frac{3}{2}$
d. None of the above

Matching List Type

***Direction** (Q. No. 15) Choices for the correct combination of elements from Column I and Column II are given as options (a), (b), (c) and (d), out of which one is correct.*

15 Match the given information in Column I with Column II. Select the correct option from the codes given below.

	Column I		Column II
i.	The ratio in which the point $\left(\frac{10}{7}, \frac{33}{7}\right)$ divides the line segment joining the points (1, 3) and (2, 7) is $p:q$, then $(p+q)$ is (p and q are coprime)	p.	$\frac{1}{5}$
ii.	The ratio in which the point (– 2, – 9) divides the line segment joining the points (1, 3) and (2, 7) is $p:q$, then $(p+q)$ is (p and q are coprime)	q.	– 1
iii.	If (p, q) divides internally the line joining (– 1, 2) and (4, – 5) in the ratio 2 : 3, then $(p+q)$ is	r.	5
iv.	If (p, q) divides externally the line joining (– 1, 2) and (4, – 5) in the ratio 2 : 3, then $(p+q)$ is	s.	7

Codes

	i	ii	iii	iv
a.	s	s	p	r
b.	p	q	r	s
c.	s	q	p	r
d.	s	r	p	p

1 POINT

DPP-3 Area

Subjective Questions

Working Space

Direction (Q. Nos. 1-3) *These questions are subjective in nature, need to be solved completely on notebook.*

1 The area of the triangle the coordinates of whose angular points are respectively

i. (0, 4) (3, 6) and (− 8, − 2)

ii. $(a, b+c)$, $(a, b-c)$ and $(-a, c)$

iii. $(a\cos\phi_1, b\sin\phi_1)$, $(a\cos\phi_2, b\sin\phi_2)$ and $(a\cos\phi_3, b\sin\phi_3)$

iv. $(am_1^2, 2am_1)$, $(am_2^2, 2am_2)$ and $(am_3^2, 2am_3)$

v. $\left\{am_1, \frac{a}{m_1}\right\}, \left\{am_2, \frac{a}{m_2}\right\}$ and $\left\{am_3, \frac{a}{m_3}\right\}$

2 Find the area of quadrilaterals, the coordinates of whose angular points, taken in order, are

i. (1, 1), (3, 4), (5, − 2) and (4, − 7)

ii. (− 1, 6), (− 3, − 9), (5, − 8) and (3, 9)

3 Find the area of pentagon, whose vertices taken in order are (0, 4)(3, 0)(6, 1)(7, 5) and (4, 9).

Only One Option Correct Type

Direction (Q. Nos. 4-26) *This section contains 23 multiple choice questions. Each question has four choices (a), (b), (c) and (d), out of which ONLY ONE option is correct.*

4 For what value of k, the points $(k, 2-2k)(1-k, 2k)$ and $(-4-k, 6-2k)$ are collinear?

a. $1, -\frac{1}{2}$ b. $1, \frac{1}{2}$

c. $-1, \frac{1}{2}$ d. $-1, -\frac{1}{2}$

5 The conditions that three points $(a, 0)$, $(0, b)$ and $(1, 1)$ are collinear, is

a. $\frac{1}{a}+\frac{1}{b}=2$ b. $\frac{1}{a}+\frac{1}{b}=1$

c. $\frac{1}{a}+\frac{1}{b}=0$ d. $\frac{1}{a}=\frac{1}{b}$

6 If D, E, F are mid-points of the sides AB, BC and CA of a triangle formed by the points $A(5, -1)$, $B(-7, 6)$ and $C(1, 3)$, then area of ΔDEF is

a. $\frac{2}{5}$ sq units b. $\frac{5}{2}$ sq units

c. 5 sq units d. 10 sq units

Working Space

7 If the area of the triangle with vertices (1, 5), (4, a) and (2, 3) is 3, then a is equal to

a. 7
b. -7
c. 4
d. -4

8 Point P divides the line segment joining $A(-5, 1)$ and $B(3, 5)$ internally in the ratio $\lambda : 1$, if $Q(1, 5)$ and $R(7, 2)$ and area of $\Delta PQR = 2$, then λ is equal to

a. $\frac{23}{3}$
b. $\frac{29}{5}$
c. $\frac{31}{9}$
d. None of these

9 The area (in sq units) common to triangle formed by (0, 0), (0, $2p$), ($2q$, 0) and (0, 0), ($2q$, 0), ($2q$, $2p$), is

a. $2pq$
b. $4pq$
c. pq
d. $\frac{pq}{2}$

10 The points (1, 1), $(0, \sec^2\theta)$, $(\text{cosec}^2\theta, 0)$ are collinear for

a. $\theta = n\pi$
b. $\theta \neq \frac{n\pi}{2}$
c. $\theta = \frac{n\pi}{2}$
d. None of these

11 The area of a quadrilateral constructed by lines $|x| + |y| = 1$, is

a. 4 sq units
b. 3 sq units
c. 2 sq units
d. 1 sq unit

12 $A(6, 3)$, $B(-3, 5)$, $C(4, -2)$ and $D(x, 3x)$ are four points. If the area of ΔDBC and ΔABC are in the ratio 1 : 2, then x is equal to

a. $\frac{11}{8}$
b. $\frac{8}{11}$
c. 3
d. None of these

13 Three points are $A(6, 3)$, $B(-3, 5)$, $C(4, -2)$ and $P(x, y)$ is a point, then the ratio of area of ΔPBC and ΔABC, is

a. $\left|\frac{x + y - 2}{7}\right|$
b. $\left|\frac{x - y + 2}{2}\right|$
c. $\left|\frac{x - y - 2}{7}\right|$
d. None of these

14 If the vertices of a triangle have integral coordinates, then the triangle is

a. isosceles
b. never equilateral
c. equilateral
d. None of the above

15 If points $(a - 2, a - 4)(a, a + 1)(a + 4, 16)$ are collinear , then a is equal to

a. 5
b. -5
c. 7
d. -7

16 Area of a triangle is 5 sq units and two of its vertices are (2, 1) and (3, -2). If its third vertex is on the line $y = x + 3$, then it is

a. (1, 4)
b. (10, 13)
c. $\left(\frac{7}{2}, \frac{13}{2}\right)$
d. $\left(\frac{13}{2}, \frac{19}{2}\right)$

Working Space

17 Area of $\Delta ABC = 20$ units and its vertices A and B are $(-5, 0)$ and $(3, 0)$, respectively. If its vertex C lies on the line $x - y = 2$, then C is

a. $(3, 5)$ b. $(-3, -5)$
c. $(-5, 7)$ d. None of these

18 If x_1, x_2, x_3 and y_1, y_2, y_3 are in GP's whose common ratios are equal, then points $(x_1, y_1), (x_2, y_2)$ and (x_3, y_3) are

a. collinear b. on a circle
c. vertices of a triangle d. None of these

19 Area of the parallelogram (in sq units) formed by the lines $y = mx$, $y = mx + 1$, $y = nx$ and $y = nx + 1$ equals **[IIT JEE 2001]**

a. $\dfrac{|m+n|}{(m-n)^2}$ b. $\dfrac{2}{|m+n|}$
c. $\dfrac{1}{|m+n|}$ d. $\dfrac{1}{|m-n|}$

20 The area of parallelogram formed by the lines $4y - 3x = 1$, $4y - 3x = 3$, $3y - 4x + 1 = 0$, $3y - 4x + 2 = 0$, is

a. $\frac{3}{8}$ sq units b. $\frac{2}{7}$ sq units
c. $\frac{1}{6}$ sq units d. None of these

21 The area enclosed by $2|x| + 3|y| \le 6$, is

a. 3 sq units b. 12 sq units
c. 9 sq units d. 24 sq units

22 The area of the figure (in sq units) formed by $a|x| + b|y| + c = 0$, is

a. $\dfrac{c^2}{|ab|}$ b. $\dfrac{2c^2}{ab}$, if $ab > 0$
c. $\dfrac{c^2}{2|ab|}$ d. None of these

23 If t_1, t_2 and t_3 are distinct, the points $(t_1, 2at_1 + at_1^3)$, $(t_2, 2at_2 + at_2^3)$ and $(t_3, 2at_3 + at_3^3)$ are collinear, if

a. $t_1 t_2 t_3 = 1$ b. $t_1 + t_2 + t_3 = t_1 t_2 t_3$
c. $t_1 + t_2 + t_3 = 0$ d. $t_1 + t_2 + t_3 = -1$

24 If $a > 0$, $b > 0$, then maximum area of the triangle formed by the points $O(0, 0)$, $A(a\cos\theta, b\sin\theta)$ and $B(a\cos\theta, -b\sin\theta)$ is (in sq units)

a. $\dfrac{3ab}{4}$, when $\theta = \dfrac{\pi}{4}$
b. $\dfrac{ab}{2}$, when $\theta = \dfrac{\pi}{4}$
c. $\dfrac{ab}{2}$, when $\theta = \dfrac{\pi}{2}$
d. a^2b^2

25 Let $A(1, 2)$, $B(3, 4)$ be two points and $C(x, y)$ be a point such that $(x-1)(x-3) + (y-2)(y-4) = 0$. If area of ΔABC is 1 sq unit, then the maximum number of positions of C in the XY-plane, is

a. 2 b. 4
c. 8 d. None of these

Working Space

26 If the area of parallelogram formed by the lines $2x - 3y + a = 0$, $3x - 2y - a = 0$, $2x - 3y + 3a = 0$ and $3x - 2y - 2a = 0$ is 10 sq units, then a is equal to

a. 5 b. 10

c. -8 d. -10

One Integer Value Correct Type

Direction (Q. Nos. 27-30) This section contains 4 questions. When worked out will result in an integer from 0 to 9 (both inclusive).

27 If the straight lines $7x - 2y + 10 = 0$, $7x + 2y - 10 = 0$ and $y + 2 = 0$ form an isosceles triangle, of area $2k$, then the value of k is

28 Let P be the point (3, 2). Let Q be the reflection of P about the X-axis. Let R be the reflection of Q about the $y = -x$ and let S be the reflection of R through the origin. $PQRS$ is a convex quadrilateral. If the area of $PQRS$ is $5k$, then k is equal to

29 Consider three lines $L_1 : ax + ky - al - a^3m = 0$, $L_2 : bx + ky - bl - b^3m = 0$, $L_3 : cx + ky - lc - c^3m = 0$ in XY-plane. Lines are concurrent, then the value of $a + b + c$ is

30 A point P is selected inside an equilateral triangle. If sum of lengths of perpendicular dropped on to sides from P is 2015, then $\left(\frac{\text{length of altitude of triangle}}{2015}\right)$, is equal to

DPP-4 Centroid, Circumcentre, Orthocentre and Incentre of a Triangle

Only One Option Correct Type

Working Space

Direction *(Q. Nos. 1-20) This section contains 20 multiple choice questions. Each question has four choices (a), (b), (c) and (d), out of which ONLY ONE option is correct.*

1 If $(-4, 6), (2, 3)$ and $(-2, -5)$ are vertices of a triangle, then its incentre is

a. $(-1, 2)$ b. $(2, -1)$
c. $(1, 2)$ d. $(2, 1)$

2 Orthocentre of a triangle whose vertex are $(8, -2), (2, -2)$ and $(8, 6)$, is

a. $(8, -2)$ b. $(8, 6)$
c. $\left(\frac{18}{3}, \frac{2}{3}\right)$ d. $(0, 0)$

3 If the vertices of a triangle are $(a, b-c)$, $(b, c-a)$ and $(c, a-b)$, then the centroid of the triangle lies

a. on Y-axis b. on X-axis
c. at origin d. None of these

4 The distance of orthocentre of the triangle $(2, 3)$, $(4, 5)$ and $(-1, 10)$ from $(2, 3)$, is

a. $2\sqrt{2}$ b. $\sqrt{2}$ c. 4 d. 2

5 If $A(4, -3)$, $B(3, -2)$ and $C(2, 8)$ are vertices of a triangle, then the distance of its centroid from Y-axis, is

a. $\frac{9}{2}$ b. 1 c. 3 d. $\frac{1}{2}$

6 An equilateral triangle whose orthocentre is $(3, -2)$ and one side is X-axis, then vertex of triangle which is not on X-axis, is

a. $(3, -6)$ b. $(1, -2)$ c. $(9, -2)$ d. $(3, -3)$

7 An equilateral triangle whose circumcentre is $(-2, 5)$ and one side is on Y-axis, then length of side of triangle, is

a. 6 b. $2\sqrt{3}$
c. $4\sqrt{3}$ d. 4

8 The circumcentre of the triangle formed by the points $(a\cos\alpha, a\sin\alpha)$, $(a\cos\beta, a\sin\beta)$, $(a\cos\gamma, a\sin\gamma)$, is

a. $(0, 0)$

b. $\left[\frac{a}{3}(\cos\alpha + \cos\beta + \cos\gamma), \frac{a}{3}(\sin\alpha + \sin\beta + \sin\gamma)\right]$

c. $(a, 0)$

d. None of the above

Working Space

9 The mid-points of the sides of a triangle are (5, 0), (5, 12) and (0, 12). The orthocentre of this triangle, is

a. (0, 0) b. (0, 24) c. (10, 0) d. $\left(\frac{13}{3}, 8\right)$

10 The orthocentre of triangle with vertices $\left(2, \frac{\sqrt{3}-1}{2}\right), \left(\frac{1}{2}, -\frac{1}{2}\right)$ and $\left(2, -\frac{1}{2}\right)$, is

a. $\left(\frac{3}{2}, \frac{3-\sqrt{3}}{6}\right)$ b. $\left(2, -\frac{1}{2}\right)$ c. $\left(-\frac{1}{2}, \frac{1}{2}\right)$ d. $\left(\frac{5}{4}, \frac{\sqrt{3}-2}{4}\right)$

11 The incentre of triangle with vertices $(1, \sqrt{3})$, (0, 0) and (2, 0), is **[AIEEE 2002]**

a. $\left(1, \frac{\sqrt{3}}{2}\right)$ b. $\left(\frac{2}{3}, \frac{1}{\sqrt{3}}\right)$ c. $\left(\frac{2}{3}, \frac{\sqrt{3}}{2}\right)$ d. $\left(1, \frac{1}{\sqrt{3}}\right)$

12 Orthocentre of triangle with vertices (0, 0), (3, 4) and (4, 0), is **[IIT JEE 2003]**

a. $\left(3, \frac{5}{4}\right)$ b. (3, 12) c. $\left(3, \frac{3}{4}\right)$ d. (3, 9)

13 If a vertex of a triangle is (1, 1) and the mid-points of two sides through this vertex are (– 1, 2) and (3, 2), then the centroid of the triangle, is **[AIEEE 2005]**

a. $\left(\frac{1}{3}, \frac{7}{3}\right)$ b. $\left(1, \frac{7}{3}\right)$ c. $\left(-\frac{1}{3}, \frac{7}{3}\right)$ d. $\left(-1, \frac{7}{3}\right)$

14 Let $A(h, k)$, $B(1, 1)$ and $C(2, 1)$ be the vertices of a right angled triangle with AC as its hypotenuse. If the area of the triangle is 1 sq unit, then the set of values which k can take is given by **[AIEEE 2007]**

a. {– 1, 3} b. {– 3, – 2} c. {1, 3} d. {0, 2}

15 Coordinates of the vertices of ΔABC are (12, 8), (– 2, 6) and (6, 0), then the correct statement, is

a. triangle is right but not isosceles
b. triangle is isosceles but not right angle
c. triangle is obtuse
d. the product of the abscissae of the centroid, orthocentre and circumcentre is 160

16 The orthocentre of the ΔABC is B and the circumcentre is $S(a, b)$. If A is the origin, then the coordinates of C are

a. $(2a, 2b)$ b. $\left(\frac{a}{2}, \frac{b}{2}\right)$ c. $(\sqrt{a^2+b^2}, 0)$ d. None of these

17 The x-coordinate of the incentre of the triangle that has the coordinates of mid-point of its sides as (0, 1) (1, 1) and (1, 0) is **[JEE Main 2013]**

a. $2+\sqrt{2}$ b. $2-\sqrt{2}$ c. $1+\sqrt{2}$ d. $1-\sqrt{2}$

18 The medians of the triangle meet at (0, – 3) and its two vertices are at (– 1, 4) and (5, 2). Then, the third vertex is at

a. (4, 15) b. (– 4, – 15) c. (– 4, 15) d. (4, – 15)

19 Consider the points $P(2, -4)$, $Q(4, -2)$ and $R(7, 1)$. The points P, Q, R

a. form an equilateral triangle
b. form a right angled triangle
c. form an isosceles triangle which is not equilateral
d. are collinear

Working Space

20 AB is the diameter of semi-circle k, C is an arbitrary point on the semi-circle (other than A or B) and S is the centre of the circle inscribed into ΔABC, then measure of

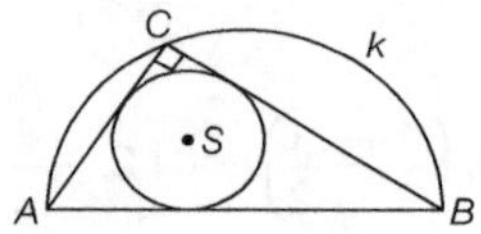

a. $\angle ASB$ changes as C moves on k
b. $\angle ASB$ is the same for all positions of C but it cannot be determined without knowing the radius
c. $\angle ASB = 135°$, for all C
d. $\angle ASB = 150°$, for all C

Matching List Type

Direction *(Q. No. 21) Choices for the correct combination of elements from Column I and Column II are given as options (a), (b), (c) and (d), out of which one is correct.*

21 Match the Column I with Column II and mark the correct option from the options given below.

	Column I		Column II
i.	The points $(2, -2)$, $(-2, 1)$ and $(5, 2)$	p.	are the vertices of a right angled triangle
ii.	The points $(1, -2)$, $(-3, 0)$ and $(5, 6)$	q.	are the vertices of a right angled isosceles triangle
iii.	The points $(3, 7)$, $(6, 5)$ and $(15, -1)$	r.	are the vertices of an equilateral triangle
iv.	The points $(2, 2)$, $(-2, -2)$ and $(-2\sqrt{3}, 2\sqrt{3})$	s.	do not form a triangle

Codes

	i	ii	iii	iv
a.	p	q	r	s
b.	q	p	s	r
c.	q	s	p	r
d.	p	r	s	q,r

One Integer Value Correct Type

Direction *(Q. Nos. 22-25) This section contains 4 questions. When worked out will result in one integer from 0 to 9 (both inclusive).*

22 If the two vertices of a triangle are $(7, 2)$ and $(1, 6)$ and its centroid is $(4, 6)$, then the coordinate of the third vertex is (a, b). The value of $(a + b - 10)$, is

23 If a and b are real numbers between 0 and 1 such that the point $(a, 1)$, $(1, b)$ and $(0, 0)$ form an equilateral triangle, then $(a - b)$ is

24 Consider the triangle with vertices $A(-2, 4)$, $B(10, -2)$, $C(-2, -8)$. If G is centroid of the triangle and the area of ΔBGC is Δ_1, then $\frac{\Delta_1}{4}$ is

25 A ΔABC is formed by three lines $x + y + 2 = 0$, $x - 2y + 5 = 0$ and $7x + y - 10 = 0$. P is a point inside the ΔABC such that areas of ΔPAB, ΔPBC and ΔPCA are equal. If the coordinates of the point P are (a, b) and the area of the ΔABC is δ, then $(a + b + \delta - 10)$, is

DPP-5 Shifting of Origin, Rotation of Axes and Locus

Only One Option Correct Type

Working Space

Direction *(Q. Nos. 1-20) This section contains 20 multiple choice questions. Each question has four choices (a), (b), (c) and (d), out of which ONLY ONE option is correct.*

1 If the axes are rotated through an angle of 30° in the anti-clockwise direction, then coordinates of point $(4, -2\sqrt{3})$ with respect to new axes are

a. $(2, \sqrt{3})$ b. $(\sqrt{3}, -5)$
c. $(2, 3)$ d. $(\sqrt{3}, 2)$

2 Reflecting the point $(2, -1)$ about Y-axis, coordinate axes are rotated at 45° angle in negative direction without shifting the origin. The new coordinates of the points are

a. $\left(-\frac{1}{\sqrt{2}}, -\frac{3}{\sqrt{2}}\right)$ b. $\left(\frac{1}{\sqrt{2}}, \frac{3}{\sqrt{2}}\right)$
c. $\left(-\frac{3}{\sqrt{2}}, \frac{1}{\sqrt{2}}\right)$ d. None of these

3 Without changing the direction of coordinate axes, to which point should origin be transferred, so that the equation $x^2 + y^2 - 4x + 6y - 7 = 0$ is changed to an equation which contains no term of first degree?

a. $(3, 2)$ b. $(2, -3)$
c. $(-2, 3)$ d. None of these

4 Keeping coordinate axes parallel to the origin, is shifted to a point $(1, -2)$, then transformed equation of $x^2 + y^2 = 2$, is

a. $x^2 + y^2 + 2x - 4y + 3 = 0$ b. $x^2 + y^2 + 2x + 4y + 3 = 0$
c. $x^2 + y^2 - 2x - 4y + 3 = 0$ d. $x^2 + y^2 - 2x + 4y + 3 = 0$

5 To remove xy term from the second degree equation $5x^2 + 8xy + 5y^2 + 3x + 2y + 5 = 0$, the coordinate axes are rotated through an angle θ, then θ equals

a. $\frac{\pi}{2}$ b. $\frac{\pi}{4}$
c. $\frac{3\pi}{8}$ d. $\frac{\pi}{8}$

6 The locus of a point, which moves in such a way that its distance from the origin $(0, 0)$ is thrice the distance from X-axis, is

a. $x^2 - 8y^2 = 0$ b. $4x^2 - y^2 = 0$
c. $x^2 + 8y^2 = 0$ d. $x^2 - 4y^2 = 0$

Working Space

7 If sum of squares of distances of a point from axes is 4, then its locus is

a. $x + y = 2$ b. $x^2 + y^2 = 16$

c. $x + y = 4$ d. $x^2 + y^2 = 4$

8 Let A (2, 3) and B (– 4, 5) be two fixed points, a point P moves in such a way that $\Delta PAB = 12$ sq units, then its locus is

a. $x^2 + 6xy + 9y^2 + 22x + 66y - 23 = 0$

b. $x^2 + 6xy + 9y^2 + 22x + 66y + 23 = 0$

c. $x^2 + y^2 + 2x + 4y + 1 = 0$

d. $x^2 + 6xy + 9y^2 - 22x - 66y - 23 = 0$

9 The locus of moving point P, such that $2PA = 3PB$, where A (0, 0) and B (4, – 3), is

a. $5x^2 + 5y^2 + 72x + 54y + 225 = 0$

b. $5x^2 - 5y^2 + 72x + 54y + 225 = 0$

c. $5x^2 + 5y^2 - 72x + 54y + 225 = 0$

d. None of the above

10 Two fixed points A (a, 0) and B (– a, 0). If $\angle A - \angle B = \theta$, then the locus of point C of ΔABC will be

a. $x^2 + y^2 + 2xy \cot\theta = a^2$ b. $x^2 - y^2 + 2xy \tan\theta = a^2$

c. $x^2 + y^2 + 2xy \tan\theta = a^2$ d. $x^2 - y^2 + 2xy \cot\theta = a^2$

11 The extremities of diagonal of a right angled triangle are (2, 0) and (0, 2), then locus of its third vertex is

a. $x^2 + y^2 - 2x - 2y = 0$ b. $x^2 + y^2 + 2x - 2y = 0$

c. $x^2 + y^2 - 2x + 2y = 0$ d. $x^2 + y^2 + 2x + 2y = 0$

12 If the sum of the distances of a point from the origin and from the line $x = 2$ is always equal to 4, then locus of this point is a

a. straight line b. circle

c. parabola d. None of these

13 If A (a, 0) and B (– a, 0) are two fixed points and a point P moves such that $\angle APB = 90°$, then locus of P is

a. $x^2 + y^2 = 2a^2$ b. $x^2 + y^2 = a^2$

c. $x^2 + y^2 + 2a^2 = 0$ d. None of these

14 If P (1, 0), Q (– 1, 0) and R (2, 0) are three given points, then locus of a point S satisfying the relation $SQ^2 + SR^2 = 2SP^2$, is

a. a straight line parallel to X-axis

b. a straight line parallel to Y-axis

c. a circle through origin

d. a circle with centre at origin

15 Locus of centroid of the triangle whose vertices are ($a \cos t$, $a \sin t$), ($b \sin t$, $- b \cos t$) and (1, 0), where t is a parameter is **[AIEEE 2003]**

a. $(3x + 1)^2 + (3y)^2 = (a^2 - b^2)$

b. $(3x - 1)^2 + (3y)^2 = a^2 - b^2$

c. $(3x - 1)^2 + (3y)^2 = a^2 + b^2$

d. $(3x + 1)^2 + (3y)^2 = a^2 + b^2$

Working Space

16 If the equation of the locus of a point equidistant from points (a_1, b_1) and (a_2, b_2) is $(a_1 - a_2)x + (b_1 - b_2)y + c = 0$, then c is equal to **[AIEEE 2004]**

a. $\sqrt{a_1^2 + b_1^2 - a_2^2 - b_2^2}$
b. $\frac{1}{2}(a_2^2 + a_1^2 - b_1^2 - b_2^2)$
c. $\frac{1}{2}(a_2^2 + b_2^2 - a_1^2 - b_1^2)$
d. None of these

17 $A(2, -3)$ and $B(-2, 1)$ are vertices of ΔABC. If the centroid of the triangle lies on the line $2x + 3y = 1$, then locus of vertex C is

a. $3x + 2y = 5$
b. $2x - 3y = 7$
c. $2x + 3y = 9$
d. $3x - 2y = 4$

18 A point $P(x, y)$ moves so that the sum of the distances from P to the coordinate axes is equal to the distance from P to the point $A(1, 1)$. The equation of the locus of P in the first quadrant, is

a. $(x + 1)(y + 1) = 1$
b. $(x + 1)(y + 1) = 2$
c. $(x - 1)(y - 1) = 1$
d. $(x - 1)(y - 1) = 2$

19 A stick of length 10 units rests against the floor and a wall of a room. If the stick begins to slide on the floor, then the locus of its middle point is

a. $x^2 + y^2 = 2.5$
b. $x^2 + y^2 = 25$
c. $x^2 + y^2 = 100$
d. None of these

20 A variable line passes through $P(2, 3)$ and cuts the coordinates axes at A and B. If the parallelogram $OABC$ (where, O is the origin) is completed, then the number of ordered pairs (x, y) of integers which lie on the locus of point C, is

a. 5
b. 6
c. 7
d. 8

Revisal Problems

Only One Option Correct Type

Working Space

Direction *(Q. Nos. 1-8) This section contains 8 multiple choice questions. Each question has four choices (a), (b), (c) and (d), out of which ONLY ONE option is correct.*

1 If the centroid and circumcentre of a triangle are (3, 3) and (6, 2) respectively, then the orthocentre is

a. (−3, 5) b. (−3, 1) c. (3, −1) d. (9, 5)

2 If by shifting the origin to the point (1, 2) without rotating the coordinate axes, the equation $y^2 - 8x - 4y + 12 = 0$ changes to $y^2 = 4ax$, then a is equal to

a. 1 b. 2 c. −2 d. −1

3 If two vertices A and B of ΔABC have the coordinates (3, −2) and (5, 4) respectively and the orthocentre is at the origin O, then the coordinates of the orthocentre of ΔOAC are

a. (5, 4) b. (3, −2)
c. (0, 0) d. None of these

4 If the point $[x_1 + t(x_2 - x_1), y_1 + t(y_2 - y_1)]$ divides the join of (x_1, y_1) and (x_2, y_2) internally, then

a. $t < 0$ b. $t = 2$
c. $0 < t < 1$ d. None of these

5 A and B are fixed points. The vertex C of ΔABC moves such that $\cot A + \cot B =$ constant. The locus of C is a straight line is

a. perpendicular to AB b. parallel to AB
c. inclined at an angle $(A - B)$ to AB d. None of these

6 If two vertices of a triangle are (5, −1) and (−2, 3) and if its orthocentre lies at the origin, then the coordinates of the third vertex are

a. (4, 7) b. (−4, −7)
c. (2, −3) d. (5, −1)

7 Let ABC be an isosceles triangle with $AB = BC$. If base BC is parallel to X-axis and m_1 and m_2 are the slopes of medians drawn through the angular points B and C, then

a. $m_1 m_2 = -1$ b. $m_1 + m_2 = 0$
c. $m_1 m_2 = 2$ d. None of these

8 The number of integral points (integral point means both the coordinates should be integer) exactly in the interior of the triangle with the vertices (0, 0), (0, 21) and (21, 0) is

a. 133 b. 190
c. 233 d. 105

Comprehension Type

Direction *(Q. Nos. 9-17) This section contains 3 paragraphs, each describing theory, experiments, data, etc. Nine questions related to the paragraphs have been given. Each question has only one correct answer among the four given options (a), (b), (c) and (d).*

Passage I

ABC is a triangle right angled at A with vertices A, B, C in the anti-clockwise sense in that order. $A(1, 2)$, $B(-3, 1)$ and vertex C lies on the X-axis. $BCEF$ is a square with vertices B, C, E, F in the clockwise sense in that order. ACD is an equilateral triangle with vertices A, C, D in the anti-clockwise sense in that order.

9 Slope of AF is

a. 7/10 b. 7/9
c. 9/10 d. 11/10

10 The abscissa of centroid of ΔBCE is

a. -1 b. $-1/2$
c. $-1/3$ d. $-2/3$

11 If $D = (\alpha, \beta)$, then $(4\beta - 4)^2$ is

a. 2 b. 3
c. 4 d. 5

Passage II

Let $A(0, \beta)$, $B(-2, 0)$ and $C(1, 1)$ be the vertices of a triangle, then

12 $\angle A$ of the ΔABC will be obtuse, if β lies in

a. $(-1, 2)$ b. $(2, 5/2)$
c. $(-1, 2/3) \cup (2/3, 2)$ d. None of these

13 If I_1 is the internal value of β for which A is obtuse and I_2 is the internal value of β for which A is largest angle of ΔABC, then

a. $I_1 = I_2$ b. I_1 is subset of I_2
c. I_2 is subset of I_1 d. None of these

14 All the values of β for which $\angle A$ of the ΔABC is largest lie in interval

a. $(-2, 1)$ b. $(-2, 2/3) \cup (2/3, 1)$
c. $(-2, 2/3) \cup (2/3, \sqrt{6})$ d. None of these

Passage III

One vertex of a triangle is a fix point from where these lines passes

$$(a-b)x - (b-c)y + (c-a) = 0$$
$$(b-c)x - (c-a)y + (a-b) = 0$$
$$(c-a)x - (a-b)y + (b-c) = 0$$
$$x\cos^2\theta - y\sin^2\theta = 1$$

another vertex is (3, 4) third vertex lie on X-axis, let at $(\lambda, 0)$.

15 Condition on λ such that it represent triangle is

a. $\lambda \neq 7/5$ b. $\lambda \neq 2/3$
c. $\lambda = 7/5$ d. $\lambda = 2/3$

16 Locus of centroid of Δ, is

a. $y = 1$ b. $x = 1$
c. $x + y = 1$ d. None of these

17 Condition on a and k such that $x^2 = 4a(y - k)$ does not intersect locus of centroid of Δ, is

a. $a > 0, k < 1$ b. $a < 0, k < 1$
c. $a < 0, k > 1$ d. None of these

One Integer Value Correct Type

Direction *(Q. Nos. 18-20) This section contains 3 questions. When worked out will result in an integer from 0 to 9 (both inclusive).*

18 If the lines $x + ay + a = 0$, $bx + y + b = 0$ and $cx + cy + 1 = 0$ (a, b, c being distinct $\neq 1$) are concurrent. Then, the value of $\frac{a}{a-1} + \frac{b}{b-1} + \frac{c}{c-1}$, is

19 A triangle is formed by the line $x + y = 0$, $x - y = 0$ and $px + qy = 1$. If $p^2 + q^2 = 1$, $(p^2 - q^2 \neq 0)$ and locus of its circumcentre is $(x^2 + y^2) = k(x^2 - y^2)^2$, then $(k - 1)$ is equal to

20 $A_1, A_2, A_3, \ldots, A_n$ are points on the line $y = x$ lying in the positive quadrant such that $OA_n = n \cdot OA_{n-1}$, 'O' being the origin. If $OA_1 = 1$ and the coordinates of A_n are $[2520\sqrt{2}, 2520\sqrt{2}]$, then the value of n is

Working Space

JEE Main & AIEEE Archive

(Compilation of Last 13 Years Questions)

1 The x-coordinate of the incentre of the triangle that has the coordinates of mid-points of its sides as (0, 1), (1, 1) and (1, 0) is **(JEE Main 2013)**

a. $2+\sqrt{2}$ b. $2-\sqrt{2}$
c. $1+\sqrt{2}$ d. $1-\sqrt{2}$

2 If the line $2x+y=k$ passes through the point which divides the line segment joining the points (1, 1) and (2, 4) in the ratio 3 : 2, then k is equal to **(AIEEE 2012)**

a. $\frac{29}{5}$ b. 5
c. 6 d. $\frac{11}{5}$

3 If $A(2,-3)$ and $B(-2,1)$ are two vertices of a triangle and third vertex moves on the line $2x+3y=9$, then the locus of the centroid of the triangle is **(AIEEE 2011)**

a. $2x-3y=1$
b. $x-y=1$
c. $2x+3y=1$
d. $2x+3y=3$

4 Three distinct points A, B and C given in the 2-dimensional coordinate plane such that the ratio of the distance of any one of them from the point (1, 0) to the distance from the point (– 1, 0) is equal to $\frac{1}{3}$. Then, the circumcentre of the ΔABC is at the point **(AIEEE 2009)**

a. $\left(\frac{5}{4}, 0\right)$ b. $\left(\frac{5}{2}, 0\right)$
c. $\left(\frac{5}{3}, 0\right)$ d. (0, 0)

5 Let $A(h, k)$, $B(1, 1)$ and $C(2, 1)$ be the vertices of a right angled triangle with AC as its hypotenuse. If the area of the triangle is 1, then the set of values which k can take is given by **(AIEEE 2007)**

a. {1, 3} b. {0, 2}
c. {–1, 3} d. {–3, –2}

6 If a vertex of a triangle is (1, 1) and the mid-points of two sides through this vertex are (–1, 2) and (3, 2), then the centroid of the triangle is **(AIEEE 2005)**

a. $\left(\frac{1}{3}, \frac{7}{3}\right)$ b. $\left(1, \frac{7}{3}\right)$
c. $\left(-\frac{1}{3}, \frac{7}{3}\right)$ d. $\left(-1, \frac{7}{3}\right)$

7 If the equation of the locus of a point equidistant from the points (a_1, b_1) and (a_2, b_2) is $(a_1-a_2)x+(b_1-b_2)y+c=0$, then the value of c is **(AIEEE 2003)**

a. $\frac{1}{2}(a_2^2+b_2^2-a_1^2-b_1^2)$
b. $a_1^2-a_2^2+b_1^2-b_2^2$
c. $\frac{1}{2}(a_1^2+a_2^2+b_1^2+b_2^2)$
d. $\sqrt{a_1^2+b_1^2-a_2^2-b_2^2}$

8 Locus of centroid of the triangle whose vertices are $(a\cos t, a\sin t)$, $(b\sin t, -b\cos t)$ and (1, 0), where t is a parameter, is **(AIEEE 2003)**

a. $(3x-1)^2+(3y)^2=a^2-b^2$
b. $(3x-1)^2+(3y)^2=a^2+b^2$
c. $(3x+1)^2+(3y)^2=a^2+b^2$
d. $(3x+1)^2+(3y)^2=a^2-b^2$

9 A triangle with vertices (4, 0), (–1, –1), (3, 5) is **(AIEEE 2002)**

a. isosceles and right angled
b. isosceles but not right angled
c. right angled but not isosceles
d. neither right angled nor isosceles

10 The incentre of the triangle with vertices $(1, \sqrt{3})$, (0, 0) and (2, 0) is **(AIEEE 2002)**

a. $\left(1, \frac{\sqrt{3}}{2}\right)$ b. $\left(\frac{2}{3}, \frac{1}{\sqrt{3}}\right)$
c. $\left(\frac{2}{3}, \frac{\sqrt{3}}{2}\right)$ d. $\left(1, \frac{1}{\sqrt{3}}\right)$

2. Straight Line

DPP-1 Slopes and Types of Straight Line

Only One Option Correct Type

Working Space

Direction (*Q. Nos.* 1-30) *This section contains* 30 *multiple choice questions. Each question has four choices (a), (b), (c) and (d), out of which ONLY ONE option is correct.*

1 The equation of a line passing through $(4, -2)$ and $(-1, 3)$, is

a. $x + y = 0$ b. $x - y = 2$
c. $x + y + 2 = 0$ d. $x + y = 2$

2 In the equation $(y - y_1) = m(x - x_1)$, if m and x_1 are fixed and different lines are drawn for different values of y_1, then (where, $m \neq \infty$)

a. there will be one line only
b. there will be a set of parallel lines
c. the lines will pass through the single point
d. None of the above

3 If the lines $ax - y - b = 0$ and $x - 2by + a = 0$ meet at the point (1, 2), then

a. $a = 1, b = 3$ b. $a = 3, b = -1$
c. $a = 3, b = 1$ d. $a = -3, b = 1$

4 The equation of line passing through the points $(at_1, a/t_1)$ and $(at_2, a/t_2)$, is

a. $t_1 t_2 y + x = a(t_1 + t_2)$ b. $t_1 t_2 y + x = 2a(t_1 + t_2)$
c. $t_1 t_2 x + y = a(t_1 + t_2)$ d. None of these

5 The equation of the line passing through the point $(-3, 2)$ and parallel to X-axis, is

a. $x - 3 = 0$ b. $x + 3 = 0$ c. $y - 2 = 0$ d. $y + 2 = 0$

6 $A(1, 3)$ and $C(5, 1)$ are two opposite vertices of a rectangle. If the slope of the line on which the other two vertices B and D lie is 2, then equation of BD, is

a. $2x - y = 4$ b. $2x - y = 1$
c. $2x + y = 4$ d. $2x + y = 1$

7 A straight line through the point $A(3, 4)$ is such that its intercept between the axes is bisected at A. Its equation is **[AIEEE 2006]**

a. $3x - 4y + 7 = 0$ b. $4x + 3y = 24$
c. $3x + 4y = 25$ d. $x + y = 7$

8 The intercept made by a line on Y-axis is double to the intercept made by it on X-axis. It is passes through (1, 2), then its equation is

a. $2x + y = 4$ b. $2x + y + 4 = 0$
c. $2x - y = 4$ d. $2x - y + 4 = 0$

9 If the point $(3, -4)$ divides the intercept of a line between the coordinate axes in the ratio 2 : 3. Then, its equation is

a. $2x + 3y + 1 = 0$ b. $3x - 4y + 1 = 0$
c. $2x - 3y = 5$ d. $2x - y = 10$

Working Space

10 The sum of intercepts made by a line passing through point (4, 3) on coordinate axes is −1. Its equation is **[AIEEE 2004]**

a. $\frac{x}{2}+\frac{y}{3}=1$ or $\frac{x}{2}+\frac{y}{1}=1$ b. $\frac{x}{2}-\frac{y}{3}=-1$ or $\frac{x}{-2}+\frac{y}{1}=-1$

c. $\frac{x}{2}+\frac{y}{3}=-1$ or $\frac{x}{-2}+\frac{y}{1}=-1$ d. $\frac{x}{2}-\frac{y}{3}=1$ or $\frac{x}{-2}+\frac{y}{1}=1$

11 A line makes an angle 45° with X-axis and at a distance of $\sqrt{2}$ from the origin. Its equation is

a. $x+y\sqrt{2}=1$ b. $\sqrt{2}x+y=1$ c. $y-x=2$ d. $y-x=\sqrt{2}$

12 The equation of lines on which the perpendicular from the origin makes 30° angle with X-axis and which form a triangle of area $\frac{50}{\sqrt{3}}$ with axes, are

a. $x\pm\sqrt{3}y-10=0$ b. $\sqrt{3}x+y\pm 10=0$

c. $x+\sqrt{3}y\pm 10=0$ d. None of these

13 A line passes through the point (1, 2) and makes 60° angle with X-axis. A point on this line at a distance 3 from the point (1, 2), is

a. $\left(\frac{-5}{2}, 2-\frac{3\sqrt{3}}{2}\right)$ b. $\left(\frac{3}{2}, 2+\frac{\sqrt{3}}{2}\right)$ c. $\left(\frac{5}{2}, 2+\frac{3\sqrt{3}}{2}\right)$ d. None of these

14 If a line passes through the point $P(1, 2)$ makes an angle of 45° with the X-axis and meets the line $x+2y-7=0$ at Q, then PQ equals

a. $\frac{2\sqrt{2}}{3}$ b. $\frac{3\sqrt{2}}{3}$ c. $\sqrt{3}$ d. $\sqrt{2}$

15 The line PQ whose equation $x-y=2$ cuts the X-axis at P and Q is (4, 2). The line PQ is rotated about P through 45° in anti-clockwise direction. The equation of the line PQ in the new position is

a. $y=-\sqrt{2}$ b. $y=2$ c. $x=2$ d. $x=-2$

16 A straight line through the point (2, 2) intersects the lines $\sqrt{3}x+y=0$ and $\sqrt{3}x-y=0$ at the points A and B. The equation to the line AB so that the ΔOAB is equilateral is **[AIEEE 2002]**

a. $x-2=0$ b. $y-2=0$ c. $x+y-4=0$ d. None of these

17 The graph of the function $\cos x\cos(x+2)-\cos^2(x+1)$, is **[IIT JEE 2001]**

a. a straight line passing through $(0, -\sin^2 1)$ with slope 2

b. a straight line passing through (0, 0)

c. a parabola with vertex $(1-\sin^2 1)$

d. a straight line passing through the point $(\pi/2, -\sin^2 1)$ and parallel to X-axis

18 The number of integer value of m, for which the x-coordinate of the point of intersection of the lines $3x+4y=9$, $y=mx+1$ is also an integer is **[IIT JEE 2001]**

a. 2 b. 0 c. 4 d. 1

19 All points inside the triangle formed by the points (1, 3), (5, 0) and (−1, 2) satisfy

a. $3x+2y\geq 0$ b. $2x+y-13\geq 0$

c. $2x-3y+12\leq 0$ d. None of these

20 A line passes through (2, 2) and cuts a triangle of area 9 sq units from the first quadrant. The sum of all possible values for the slope of a line is

a. −2.5 b. −2 c. −1.5 d. −1

Working Space

21 The diagonals of a parallelogram $PQRS$ are along the lines $x + 3y = 4$ and $6x - 2y = 7$. Then, $PQRS$ must be a

a. rectangle
b. square
c. cyclic quadrilateral
d. rhombus

22 A and B are any two points on the positive X and Y-axes, respectively satisfying $2(OA) + 3(OB) = 10$. If P is the middle point of AB, then the locus of P is

a. $2x + 3y = 5$
b. $2x + 3y = 10$
c. $3x + 2y = 10$
d. $3x + 2y = 10$

23 The sides of a ΔABC lie on the lines $3x + 4y = 0$, $4x + 3y = 0$ and $x = 3$. Let (h, k) be the centre of the circle inscribed in ΔABC. Then, value of $(h + k)$, is

a. 0
b. $\frac{1}{4}$
c. $\frac{-1}{4}$
d. $\frac{1}{2}$

24 If m and b are real numbers and $mb > 0$, then the line whose equation $y = mx + b$ cannot contain point

a. (0, 2009)
b. (2009, 0)
c. (0, – 2009)
d. (20, – 100)

25 If the x-intercept of the line $y = mx + 2$ is greater than $\frac{1}{2}$, then the gradient of the line lies in the interval

a. $(-1, 0)$
b. $\left(\frac{-1}{4}, 0\right)$
c. $(-\infty, -4)$
d. $(-4, 0)$

26 The greatest slope along the graph represented by $4x^2 - y^2 + 2y - 1 = 0$, is

a. –3
b. –2
c. 2
d. 3

27 In a ΔABC, side AB has the equation $2x + 3y = 29$ and the side AC has the equation, $x + 2y = 6$. If the mid-point of BC is (5, 6), then the equation of BC is

a. $x - y = -1$
b. $5x - 2y = 13$
c. $21x + 31y = 291$
d. $3x - 4y = -9$

28 A variable rectangle $PQRS$ has its sides parallel to fixed directions. Q and S lie, respectively on the lines $x = a$, $x = -a$ and P lies on X-axis. Then, the locus of R is

a. a straight line
b. a circle
c. a parabola
d. pair of straight lines

29 A rectangular billiard table has vertices at $P(0, 0)$, $Q(0, 7)$, $R(10, 7)$ and $S(10, 0)$. A small billiard ball starts at $M(3, 4)$ and moves in a straight line to the top of the table, bounces to the right side of the table, then comes to rest at $N(7, 1)$. The coordinates of the point, where it hits the right side, are

a. 3.7
b. 3.8
c. 3.9
d. 4

30 Line AB passes through point (2, 3) and intersects the positive X and Y-axes at $A(a, 0)$ and $B(0, b)$, respectively. If the area of ΔAOB is 11, the numerical value of $4b^2 + 9a^2$, is

a. 220
b. 240
c. 248
d. 284

Working Space

One or More than One Options Correct Type

Direction (Q. Nos. 31-35) This section contains 5 multiple choice questions. Each question has four choices (a), (b), (c) and (d), out of which ONE or MORE THAN ONE are correct.

31 A and B are two fixed points whose coordinates are (3, 2) and (5, 4), respectively. The coordinates of a point P, if ABP is an equilateral triangle, is/are

a. $(4-\sqrt{3}, 3+\sqrt{3})$ b. $(4+\sqrt{3}, 3-\sqrt{3})$ c. $(3-\sqrt{3}, 4+\sqrt{3})$ d. $(3+\sqrt{3}, 4-\sqrt{3})$

32 Straight lines $2x + y = 5$ and $x - 2y = 3$ intersect at the point A. Points B and C are chosen on these two lines such that $AB = AC$. Then, the equation of a line BC passing through the point (2, 3), is

a. $3x - y - 3 = 0$ b. $x + 3y - 11 = 0$ c. $3x + y - 9 = 0$ d. $x - 3y + 7 = 0$

33 The straight lines $x + y = 0$, $3x + y - 4 = 0$ and $x + 3y - 4 = 0$ form a triangle which is

a. isosceles b. right angled c. obtuse angled d. equilateral

34 The x-coordinates of the vertices of a square of unit area are the roots of the equation $x^2 - 3|x| + 2 = 0$ and the y-coordinates of the vertices are the roots of the equation $y^2 - 3y + 2 = 0$. Then, the possible vertices of the square is/are

a. (1, 1), (2, 1), (2, 2), (1, 2) b. (−1, 1), (−2, 1), (−2, 2), (−1, 2)
c. (2, 1), (1, − 1), (1, 2), (2, 2) d. (−2, 1), (−1, − 1), (−1, 2), (−2, 2)

35 The area of the ΔABC is 20 sq units. The coordinates of vertex A are (−5, 0) and B are (3, 0). The vertex C lies on the line, $x - y = 2$. The coordinates of C are

a. (5, 3) b. (−3, − 5) c. (−5, − 7) d. (7, 5)

Comprehension Type

Direction (Q. Nos. 36-38) This section contains a paragraph, describing theory, experiments, data, etc. Three questions related to the paragraph have been given. Each question has only one correct answer among the four given options (a), (b), (c) and (d).

Passage

Let $ABCD$ is a square with sides of unit length. Points E and F are taken on sides AB and AD, respectively so that $AE = AF$. Let P be a point inside the square $ABCD$.

36 The maximum possible area of quadrilateral $CDFE$, is

a. $\frac{1}{8}$ b. $\frac{1}{4}$ c. $\frac{5}{8}$ d. $\frac{3}{8}$

37 The value of $(PA)^2 - (PB)^2 + (PC)^2 - (PD)^2$ is equal to

a. 3 b. 2 c. 1 d. 0

38 Let a line passing through point A divides the square $ABCD$ into two parts so that area of one portion is double the other, then the length of portion of line inside the square is

a. $\frac{\sqrt{10}}{3}$ b. $\frac{\sqrt{13}}{3}$ c. $\frac{\sqrt{11}}{3}$ d. $\frac{2}{\sqrt{3}}$

One Integer Value Correct Type

Direction (Q. Nos. 39 and 40) This section contains 2 questions. When worked out will result in an integer from 0 to 9 (both inclusive).

39 The orthocentre of the triangle formed by the lines $2x + 3y - 1 = 0$, $x + 2y - 1 = 0$, $ax + by - 1 = 0$ is at origin, then $(a + b)$ is

40 If the point $(1, \lambda)$ always remains in the interior of the triangle formed by the lines $y = x$, $y = 0$ and $x + y = 4$, where $\lambda \in (a, b)$. Then, $a + b$ is

DPP-2 Angle between Straight Lines, Image and Foot of Perpendicular

Subjective Questions

Direction (Q. Nos. 1-3) *These questions are subjective in nature, need to be solved completely on notebook.*

Working Space

1 Find the angles between the pairs of straight lines.

i. $x - y\sqrt{3} = 5$ and $\sqrt{3}x + y = 7$

ii. $x - 4y = 3$ and $6x - y = 11$

iii. $y = (2 - \sqrt{3})x + 5$ and $y = (2 + \sqrt{3})x - 7$

iv. $y = 3x + 7$ and $3y - x = 8$

v. $(m^2 - mn)y = (mn + n^2)x + n^3$ and $(mn + m^2)y = (mn - n^2)x + m^3$

2 i. Find the sum of the abscissae of all the points on the line $x + y = 4$ that lie at a unit distance from the line $4x + 3y - 10 = 0$.

ii. Find the equations of all the sides of a square whose each side is of length 4 units and centre is (1, 1). Given that one pair of sides is parallel to $3x - 4y = 0$.

3 Find the equations to the straight lines which divide internally and externally, the line joining $(-3, 7)$ to $(5, -4)$ in the ratio of 4 : 7 and which are perpendicular to this line.

Only One Option Correct Type

Direction (Q. Nos. 4-30) *This section contains 27 multiple choice questions. Each question has four choices (a), (b), (c) and (d), out of which ONLY ONE option is correct.*

4 If the coordinates of the points A, B and C are $(-1, 5)$, $(0, 0)$ and $(2, 2)$, respectively and D is the mid-point of BC, then the equation of the perpendicular drawn from B to the line AD, is

a. $2x + y = 0$ b. $x + 2y = 0$

c. $x - 2y = 0$ d. $2x - y = 0$

5 The equation of a line perpendicular to the line $5x - 2y + 7 = 0$ and passing through the point of intersection of lines $y = x + 7$ and $x + 2y + 1 = 0$, is

a. $2x + 5y = 0$ b. $2x + 5y = 20$

c. $2x + 5y = 10$ d. None of these

6 The equation of the line which passes through $(a\cos^3\theta, a\sin^3\theta)$ and perpendicular to the line $x\sec\theta + y\,\text{cosec}\,\theta = a$, is

a. $x\cos\theta + y\sin\theta = 2a\cos 2\theta$

b. $x\sin\theta - y\cos\theta = 2a\sin 2\theta$

c. $x\sin\theta + y\cos\theta = 2a\cos 2\theta$

d. $x\cos\theta - y\sin\theta = a\cos 2\theta$

Working Space

7 The ends of the base of an isosceles trianlge are at $(2a, 0)$ and $(0, a)$ and one side is parallel to Y-axis. The equation of the other side is

a. $x + 2y - a = 0$
b. $x + 2y = 2a$
c. $3x + 4y - 4a = 0$
d. $3x - 4y + 4a = 0$

8 If α, β and γ are three different values, then equation $(\beta - \gamma)x + (\gamma - \alpha)y + (\alpha - \beta) = 0$ and $(\beta^3 - \gamma^3)x + (\gamma^3 - \alpha^3)y + (\alpha^3 - \beta^3) = 0$ represents same line, if

a. $\alpha + \beta + \gamma = 0$
b. $\alpha + \beta + \gamma = 1$
c. $\alpha = \beta + \gamma$
d. None of these

9 A straight line parallel to the lines $3x - y - 3 = 0$ and $3x - y - 5 = 0$ and lies between them. If its distance from these lines are in the ratio 3 : 5, then its equation will be

a. $3x - y + 1 = 0$
b. $3x - y + 2 = 0$
c. $3x - y = 0$
d. None of these

10 The equation of the line perpendicular to the line $\frac{x}{a} - \frac{y}{b} = 1$ and passing through the point at which it cuts X-axis, is

a. $\frac{x}{a} + \frac{y}{b} + \frac{a}{b} = 0$
b. $\frac{x}{b} + \frac{y}{a} = \frac{b}{a}$
c. $\frac{x}{b} + \frac{y}{a} = 0$
d. $\frac{x}{b} + \frac{y}{a} = \frac{a}{b}$

11 The perpendicular bisector of the line segment joining $P(1, 4)$ and $Q(k, 3)$ has y-intercept -4. Then, a possible value of k, is **[AIEEE 2008]**

a. 1
b. 2
c. -2
d. -4

12 The line $p(p^2 + 1)x - y + q = 0$ and $(p^2 + 1)^2 x + (p^2 + 1)y + 2q = 0$ are perpendicular to a common line for **[AIEEE 2009]**

a. exactly one value of p
b. exactly two values of p
c. more than two values of p
d. No values of p

13 The line parallel to the X-axis and passing through the intersection of the lines $ax + 2by + 3b = 0$ and $bx - 2ay - 3a = 0$, is (where $(a, b) \neq (0, 0)$) **[AIEEE 2005]**

a. below the X-axis at a distance of $\frac{3}{2}$ from it
b. below the X-axis at a distance of $\frac{2}{3}$ from it
c. above the X-axis at a distance of $\frac{3}{2}$ from it
d. above the X-axis at a distance of $\frac{2}{3}$ from it

14 The line L given by $\frac{x}{5} + \frac{y}{b} = 1$ passes through the point (13, 32). The line K is parallel to L and has the equation $\frac{x}{c} + \frac{y}{3} = 1$, Then, the distance between L and K, is **[AIEEE 2010]**

a. $\frac{22}{\sqrt{15}}$
b. $\sqrt{17}$
c. $\frac{17}{\sqrt{15}}$
d. $\frac{23}{\sqrt{17}}$

Working Space

15 The lines $x + y = |a|$ and $ax - y = 1$ intersect each other in the first quadrant. Then, the set of all possible values of a is the interval. **[AIEEE 2011]**

a. $(-1, 1]$ b. $(0, \infty)$
c. $[1, \infty)$ d. $(-1, \infty)$

16 A line is drawn through the point (1, 2) to meet the coordinate axes at P and Q such that it forms a ΔOPQ, where O is the origin. If the area of ΔOPQ is least, then the slope of the line PQ is **[AIEEE 2012]**

a. $\frac{-1}{2}$ b. $\frac{-1}{4}$
c. -4 d. -2

17 If $x - 3y = p$, $ax + 2y = q$ and $ax + y = r$ form a right angled triangle, then **[IIT JEE 2013]**

a. $a^2 - 6a - 12 = 0$ b. $a^2 - 9a + 12 = 0$
c. $a^2 - 9a + 18 = 0$ d. $a^2 - 6a - 18 = 0$

18 The coordinates of foot of a perpendicular drawn on line $3x - 4y - 5 = 0$ to the point (0, 5), is

a. (1, 3) b. (2, 3)
c. (3, 2) d. (3, 1)

19 Consider the lines $2x + 3y = 7$, $2x + 3y = 12$ and point $A(3, -5)$. Then,

a. A lies between the lines
b. sum of perpendicular distance from A to the line is $\frac{5}{\sqrt{13}}$
c. distance between lines is $\frac{19}{\sqrt{13}}$
d. None of the above

20 The ratio in which the line $3x + 4y + 2 = 0$ divides the distance between $3x + 4y + 5 = 0$ and $3x + 4y - 5 = 0$, is

a. 7 : 3 b. 3 : 7
c. 2 : 3 d. None of these

21 The product of perpendiculars drawn from the point $(\pm\sqrt{a^2 - b^2}, 0)$ to the line $\frac{x}{a}\cos\theta + \frac{y}{b}\sin\theta = 1$, is

a. $a^2 - b^2$ b. $a^2 + b^2$ c. a^2 d. b^2

22 The points on the line $x + y = 4$ which lie at a unit distance from the line $4x + 3y = 10$ are

a. (3,1), (−7, 11) b. (−3, 1), (−7, 11)
c. (3, 1), (7, 11) d. (11, 3), (−7, 11)

23 Let line L be $2x + y = 2$. If the coordinate axes be rotates at 45° angle, then the lengths of the intercepts of L on new axes will be

a. $\sqrt{2}, 1$ b. $1, \sqrt{2}$
c. $2\sqrt{2}, 2\sqrt{\frac{2}{3}}$ d. $\frac{2\sqrt{2}}{3}, 2\sqrt{2}$

24 The reflection of the point (3, 8) with respect to the line $x + 3y = 7$, is

a. (5, − 6) b. (−1, − 4)
c. (0, − 1) d. (−9, − 4)

Working Space

25 If p is the length of the perpendicular from the origin on the line $\frac{x}{a}+\frac{y}{b}=1$, then

a. $p^2 = a^2 + b^2$

b. $p^2 = \frac{1}{a^2} + \frac{1}{b^2}$

c. $\frac{1}{p^2} = \frac{1}{a^2} + \frac{1}{b^2}$

d. None of these

26 If the base of an equilateral triangle is $x + y = 2$ and its opposite vertex is (2, 1), then the length of its each side is

a. $\frac{2}{3}$

b. $\frac{3}{2}$

c. $\sqrt{\frac{3}{2}}$

d. $\sqrt{\frac{2}{3}}$

27 A line cuts intercepts a and b on the coordinate axes. If after rotating the axes at a constant angle without changing the origin it makes intercepts p and q on the new axes, then

a. $a^2 + b^2 = p^2 + q^2$

b. $a^{-2} + b^{-2} = p^{-2} + q^{-2}$

c. $a^2 + p^2 = b^2 + q^2$

d. $a^{-2} + p^{-2} = b^{-2} + q^{-2}$

28 A straight line through the origin O meets the parallel lines $4x + 2y = 9$ and $2x + y = -6$ at points P and Q, respectively. Then, the point O divides the segment PQ in the ratio **[IIT JEE 2002]**

a. 1 : 2

b. 3 : 4

c. 2 : 1

d. 4 : 3

29 The vertex of right angled triangle lies on the straight line $2x - y - 10 = 0$ and the two other vertices, at points (2, – 3) and (4, 1). Then, the area of triangle in sq units is

a. $\sqrt{10}$

b. 3

c. $\frac{33}{5}$

d. 11

30 Point P lies on the line $l\{(x, y) \mid 3x + 5y = 15\}$. If '$P$' is also equidistant from the coordinate axes, then P can be located in which of the four quadrants?

a. Only I

b. Only II

c. I or II

d. Only IV

Matching List Type

Direction *(Q. Nos. 31 and 32) Choices for the correct combination of elements from Column I and Column II are given as options (a), (b), (c) and (d), out of which one is correct.*

31 Consider the line $Ax + By + C = 0$.

Match the Column I with Column II and mark the correct option code.

	Column I		Column II
i.	x-intercept is finite and y-intercept is infinite.	p.	$A = 0$; $B, C \neq 0$
ii.	x-intercept is infinite and y-intercept is finite.	q.	$C = 0$; $A, B \neq 0$
iii.	Both x and y intercepts are zero.	r.	$A, B = 0$ $C \neq 0$
iv.	Both x and y intercepts are infinite.	s.	$B = 0$; $A, C \neq 0$

Codes

	i	ii	iii	iv
a.	s	p	q	r
b.	p	q	r	s
c.	s	r	q	p
d.	p	r	s	q

32 Match the Column I with Column II and select the correct option from the given codes.

Working Space

	Column I		Column II
i.	Passing through the point (2, 3) and perpendicular to $4x - 3y = 10$.	p.	$4y + 11x = 10$
ii.	Passing through the point (−6, 10) and perpendicular to $7x + 8y = 5$.	q.	$4y + 3x = 18$
iii.	Passing through the point (2, − 3) and perpendicular to the line joining (5, 7) and (−6, 3).	r.	$x + 4y + 16 = 0$
iv.	Passing through the point (−4, − 3) and perpendicular to the line joining (1, 3) and (2, 7).	s.	$7y - 8x = 118$

Codes

	i	ii	iii	iv		i	ii	iii	iv
a.	p	q	r	s	b.	q	s	p	r
c.	s	r	q	p	d.	r	s	p	q

Comprehension Type

Direction (Q. Nos. 33-35) *This section contains a paragraph, describing theory, experiments, data, etc. Three questions related to the paragraph have been given. Each question has only one correct answer among the four given options (a), (b), (c) and (d).*

Passage

In the diagram, a line is drawn through the points $A(0, 16)$ and $B(8, 0)$. Point P is chosen in the first quadrant on the line passing through A and B. Points C and D are chosen on the X and Y-axes respectively, so that $PDOC$ is a rectangle. Then,

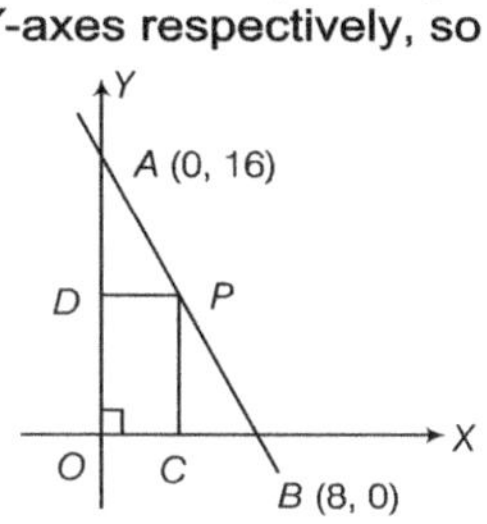

33 Perpendicular distance of the line AB from the point (2, 2), is

a. $\sqrt{4}$ b. $\sqrt{10}$
c. $\sqrt{20}$ d. $\sqrt{50}$

34 Sum of the coordinates of P if $PDOC$ is a square, is

a. $\frac{32}{3}$ b. $\frac{16}{3}$
c. 16 d. 11

35 Number of possible ordered pairs(s) of all positions of the point P on AB, so that the area of the rectangle $PDOC$ is 30 sq units, is

a. 3 b. 2
c. 1 d. 0

DPP-3 Equation of Straight Line making an Angle 'α' with Given Straight Line

Only One Option Correct Type

Working Space

Direction (Q. Nos. 1-9) *This section contains 9 multiple choice questions. Each question has four choices (a), (b), (c) and (d), out of which ONLY ONE option is correct.*

1 A vertex of an equilateral triangle is (2, 3) and the equation of the opposite side is $x + y = 2$, then the equation of other two sides of the triangle, are

a. $(y-3)=(2\pm\sqrt{3})(x-2)$
b. $(y-3)=(4\pm\sqrt{3})(x+2)$
c. $(y-3)=(2\pm\sqrt{3})(x+4)$
d. $(y-3)=(6\pm\sqrt{3})(x-2)$

2 (1, 2) is vertex of a square whose one diagonal is along the X-axis. The equations of sides passing through the given vertex are

a. $2x-y=0,\ x+2y+5=0$
b. $x-2y+3=0,\ 2x+y-4=0$
c. $x-y+1=0,\ x+y-3=0$
d. None of these

3 The equation of the lines which pass through the origin and are inclined at an angle $\tan^{-1} m$ to the line $y = mx + c$, are

a. $y=0,\ 2mx+(1-m^2)y=0$
b. $y=0,\ 2mx-(m^2-1)y=0$
c. $x=0,\ 2mx+(m^2-1)y=0$
d. None of these

4 If one of the diagonals of a square is along $x = 2y$ and one of its vertices is (3, 0), then its sides through this vertex are given by the equations

a. $y-3x+9=0,\ 3y+x-3=0$
b. $y+3x-9=0,\ 3y+x-3=0$
c. $y-3x+9=0,\ 3y-x+3=0$
d. $y-3x+3=0,\ 3y+x+9=0$

5 A straight line L through the point $(3,-2)$ is inclined at an angle of 60° to the line $\sqrt{3}x + y = 1$. If L also intersects the X-axis, then the locus of L is **[IIT JEE 2011]**

a. $y+\sqrt{3}x-2-3\sqrt{3}=0$
b. $y-\sqrt{3}x+2+3\sqrt{3}=0$
c. $\sqrt{3}y-x+3+2\sqrt{3}=0$
d. $\sqrt{3}y+x-3+2\sqrt{3}=0$

6 A square of side a lies above X-axis and has one vertex at the origin. The side passing through the origin makes an angle $\alpha\ (0<\alpha<\pi/4)$ with the positive direction of X-axis. The equation of its diagonal not passing through the origin is

a. $y(\cos\alpha+\sin\alpha)+x(\cos\alpha-\sin\alpha)=a$
b. $y(\cos\alpha-\sin\alpha)+x(\sin\alpha-\cos\alpha)=a$
c. $y(\cos\alpha+\sin\alpha)+x(\sin\alpha-\cos\alpha)=a$
d. $y(\cos\alpha+\sin\alpha)+x(\sin\alpha+\cos\alpha)=a$

Working Space

7 Let $0 < \alpha < \pi/2$ be fixed angle. If $P(\cos\theta, \sin\theta)$ and $Q[\cos(\alpha - \theta), \sin(\alpha - \theta)]$, then Q is obtained from P by

a. clockwise rotation around origin through an angle α
b. anti-clockwise rotation around origin through an angle α
c. reflection in the line through origin with slope $\tan\alpha$
d. reflection in the line through origin with slope $\tan(\alpha/2)$

8 If the lines $y = 3x + 1$ and $2y = x + 3$ are equally inclined to the line $y = mx + 4$, then m is

a. $\frac{1-3\sqrt{2}}{7}$
b. $\frac{1+3\sqrt{2}}{7}$
c. $\frac{1\pm 3\sqrt{2}}{7}$
d. $\frac{1\pm 5\sqrt{2}}{7}$

9 The diagonal of the parallelogram whose sides are $lx + my + n = 0$, $lx + my + n' = 0$, $mx + ly + n = 0$, $mx + ly + n' = 0$ include an angle

a. $\tan^{-1}\left(\frac{2lm}{l^2 + m^2}\right)$
b. $\tan^{-1}\left(\frac{l^2 - m^2}{l^2 + m^2}\right)$
c. $\frac{\pi}{2}$
d. $\frac{\pi}{3}$

Comprehension Type

Direction (Q. Nos. 10-14) *This section contains 2 paragraphs, each describing theory, experiments, data, etc. Five questions related to the paragraphs have been given. Each question has only one correct answer among the four given options (a), (b), (c) and (d).*

Passage I

Given two straight lines AB and AC whose equations are $3x + 4y = 5$ and $4x - 3y = 5$, respectively. Then, the possible equation of line BC through (1,2), such that ΔABC is isosceles, is $l_1 = 0$ or $l_2 = 0$, then answer the following questions :

10 If $l_1: ax + by + c = 0$ and $l_2 : dx + ey + f = 0$, where $a, b, c, d, e, f \in I$ and $a, d > 0$, then $c + f$, is

a. 1
b. 2
c. 3
d. 4

11 A straight line through $P(2, c + f - 1)$, inclined at an angle of 60° with positive Y-axis. The coordinates of one of the points on it at a distance $(c + f)$ units from point P is (c, f obtained from previous equation).

a. $(2 - 2\sqrt{3}, 5)$
b. $(3 + 2\sqrt{3}, 3)$
c. $(2 + 3\sqrt{2}, 4)$
d. $(2 + 3\sqrt{2}, 3)$

12 If (a, b) are the coordinates of the point obtained in previous question, then find the equation of line which is at the distance $(b - 2a - 1)$ units from origin and make equal intercept on coordinate axes in first quadrant.

a. $x + y + 4\sqrt{6} = 0$
b. $x + y + 2\sqrt{6} = 0$
c. $x + y - 4\sqrt{6} = 0$
d. $x + y - 10\sqrt{2} = 0$

Working Space

Passage II

An equilateral ΔABC has its centroid at the origin and the base BC lies along the line $x + y = 1$.

13 Area of equilateral ΔABC, is (in sq units)

a. $\frac{3\sqrt{3}}{2}$ b. $\frac{3\sqrt{3}}{4}$ c. $\frac{3\sqrt{2}}{2}$ d. $\frac{2\sqrt{3}}{4}$

14 Gradient of the other two lines, are

a. $\sqrt{3}, \sqrt{2}$ b. $\sqrt{3}, \frac{1}{\sqrt{3}}$
c. $\sqrt{2}+1, \sqrt{2}-1$ d. $2+\sqrt{3}, 2-\sqrt{3}$

One or More than One Options Correct Type

Direction (Q. Nos. 15-18) This section contains 4 multiple choice questions. Each question has four choices (a), (b), (c) and (d), out of which ONE or MORE THAN ONE are correct.

15 A line passes through the origin and makes an angle of $\pi/4$ with the line $x - y + 1 = 0$, then

a. equation of the line is $x = 0$
b. equation of the line is $y = 0$
c. the point of intersection of the line with the given line is $(-1, 0)$
d. the point of intersection of the line with the given line is $(0, 1)$

16 Equation of a straight line passing through the point $(2, 3)$ and inclined at an angle of arc $\tan\frac{1}{2}$ with the line $y + 2x = 5$, is

a. $y = 3$ b. $x = 2$
c. $3x + 4y - 18 = 0$ d. $4x + 3y - 17 = 0$

17 If one vertex of an equilateral triangle of side a, lies at the origin and the other lies on the line $x - \sqrt{3}y = 0$. The coordinates of the third vertex are

a. $(0, a)$ b. $\left(\frac{\sqrt{3}a}{2}, \frac{-a}{2}\right)$ c. $(0, -a)$ d. $\left(\frac{-\sqrt{3}a}{2}, \frac{a}{2}\right)$

18 If $A(1, 3)$ and $C(7, 5)$ are opposite vertices of a square, then the equation of a side through A, is

a. $x + 2y - 7 = 0$ b. $x - 2y + 5 = 0$
c. $2x + y - 5 = 0$ d. $2x - y + 1 = 0$

One Integer Value Correct Type

Direction (Q. Nos. 19 and 20) This section contains 2 questions. When worked out will result in an integer from 0 to 9 (both inclusive).

19 Number of lines that can be drawn through the point $(4, -5)$ so that its distance from $(-2, 3)$ will be equal to 12, is equal to

20 Two mutually perpendicular straight lines through the origin form an isosceles triangle with the line $2x + y = 5$. Then, the area of the triangle, is

DPP-4 Angle Bisector and Family of Concurrent Lines

Only One Option Correct Type

Working Space

Direction (*Q. Nos. 1-23*) *This section contains 23 multiple choice questions. Each question has four choices (a), (b), (c) and (d), out of which ONLY ONE option is correct.*

1 The equation of the bisector of the acute angle between the lines $3x - 4y + 7 = 0$ and $12x + 5y - 2 = 0$, is

a. $11x - 3y - 9 = 0$
b. $11x - 3y + 9 = 0$
c. $21x + 77y - 101 = 0$
d. None of these

2 The equation of the line which bisects the obtuse angle between the lines $x - 2y + 4 = 0$ and $4x - 3y + 2 = 0$, is

a. $(4 - \sqrt{5})x - (3 - 2\sqrt{5})y + (2 - 4\sqrt{5}) = 0$
b. $(4 + \sqrt{5})x + (3 + 2\sqrt{5})y + (2 + 4\sqrt{5}) = 0$
c. $(4 + \sqrt{5})x - (3 + 2\sqrt{5})y + (2 + 4\sqrt{5}) = 0$
d. None of the above

3 The equation of the bisector of the angle between the lines $3x - 4y + 7 = 0$ and $12x - 5y - 8 = 0$, is

a. $99x - 77y + 51 = 0$, $21x + 27y - 131 = 0$
b. $99x - 77y + 51 = 0$, $21x + 27y + 131 = 0$
c. $99x - 77y + 131 = 0$, $21x + 27y - 51 = 0$
d. None of the above

4 The equation of a straight line passing through the point (4, 5) and equally inclined to the lines $3x = 4y + 7$ and $5y = 12x + 6$, is

a. $9x - 7y = 1$, $7x + 9y = 73$
b. $8x + 7y = 1$ and $7x - 9y = 75$
c. $6x - 7y = 1$ and $6x + 9y = 73$
d. None of these

5 If (a, a^2) falls inside the angle made by the lines $y = \frac{x}{2}$, $x > 0$, and $y = 3x$, $x > 0$ then a belongs to **[AIEEE 2006]**

a. $(3, \infty)$
b. $\left(\frac{1}{2}, 3\right)$
c. $\left(-3, -\frac{1}{2}\right)$
d. $\left(0, \frac{1}{2}\right)$

6 Let $P(-1, 0)$, $Q(0, 0)$, $R(3, 3\sqrt{3})$ be three points. The equation of the bisector of the $\angle PQR$, is **[AIEEE 2007]**

a. $\frac{\sqrt{3}}{2}x + y = 0$
b. $x + \sqrt{3}y = 0$
c. $\sqrt{3}x + y = 0$
d. $x + \frac{\sqrt{3}}{2}y = 0$

Working Space

7 The equation of the line passing through the point of intersection of the line $4x - 3y - 1 = 0$ and $5x - 2y - 3 = 0$ and parallel to the line $2x - 3y + 2 = 0$, is

a. $x - 3y = 1$ b. $3x - 2y = 1$
c. $2x - 3y + 1 = 0$ d. $2x - y = 1$

8 The fix point through which the line $(2\cos\theta + 3\sin\theta)x + (3\cos\theta - 5\sin\theta)y - (5\cos\theta - 2\sin\theta) = 0$ passes for all values of θ is

a. (0, 0) b. (1, 1)
c. (2, 1) d. None of these

9 The straight line $ax + by + c = 0$, where $3a + 2b + 4c = 0$ are concurrent at the point

a. $\left(\frac{1}{2}, \frac{3}{4}\right)$ b. $\left(\frac{3}{4}, \frac{1}{2}\right)$ c. $\left(-\frac{3}{4}, -\frac{1}{2}\right)$ d. $\left(-\frac{3}{4}, \frac{1}{2}\right)$

10 The straight line through the point of intersection of $ax + by + c = 0$ and $a'x + b'y + c' = 0$ are parallel to the Y-axis has the equation

a. $x(ab' - a'b) + (cb' - c'b) = 0$
b. $x(ab' + a'b) + (cb' + c'b) = 0$
c. $y(a'b - ab') + (a'c - ac') = 0$
d. None of the above

11 The line $(3x - y + 5) + \lambda(2x - 3y - 4) = 0$ will be parallel to Y-axis, if λ is

a. $\frac{1}{3}$ b. $-\frac{1}{3}$ c. $\frac{3}{2}$ d. $-\frac{3}{2}$

12 For the straight lines given by the equation $(2 + k)x + (1 + k)y = 5 + 7k$, for different values of k. Which of the following statements is true?

a. Lines pass through the point (2, – 9)
b. Lines pass through the point (–2, 9)
c. Lines are parallel
d. None of the above

13 If a, b and c are in an AP, then $ax + by + c = 0$ will always pass through a fixed point whose coordinates are

a. (1, – 2) b. (–1, 2) c. (1, 2) d. (–1, – 2)

14 The fix point through which the line $x(a + 2b) + y(a + 3b) = a + b$ always passes for all values of a and b is

a. (2, 1) b. (1, 2)
c. (2, – 1) d. (1, – 2)

15 In the equation $y - m_1x - c_1 + \lambda(y - m_2x - c_2) = 0$, λ is a variable quantity which may have any value, then the straight line represented by this equation passes through the fixed point is

a. $\left(\frac{c_1 + c_2}{m_1 + m_2}, \frac{m_1c_2 + m_2c_1}{m_1 + m_2}\right)$ b. $\left(\frac{c_2 - c_1}{m_1 - m_2}, \frac{m_1c_2 - m_2c_1}{m_1 - m_2}\right)$
c. $\left(\frac{c_1 - c_2}{m_1 + m_2}, \frac{m_1c_2 - m_2c_1}{m_1 + m_2}\right)$ d. None of these

16 If the algebraic sum of the perpendicular distance from the points (2, 0), (0, 2) and (1, 1) to a variable straight line is zero, then find the coordinates through which the line passes.

a. (1, 1) b. (0, 0) c. (3, 2) d. (2, 3)

Working Space

17 A is a point on either of two lines $y + \sqrt{3}\,|x| = 2$ at a distance of $\frac{4}{\sqrt{3}}$ units from their point of intersection. The coordinates of the foot of perpendicular from A on the bisector of the angle between them are

a. $\left(\frac{-2}{\sqrt{3}}, 2\right)$ b. $(0, 0)$ c. $\left(\frac{2}{\sqrt{3}}, 2\right)$ d. $(0, 4)$

18 If non-zero numbers a, b and c are in HP, then the straight line $\frac{x}{a} + \frac{y}{b} + \frac{1}{c} = 0$ always passes through a fixed point, that point is **[AIEEE 2005]**

a. $\left(1, -\frac{1}{2}\right)$ b. $(1, -2)$ c. $(-1, -2)$ d. $(-1, 2)$

19 If the lines $x + y + 1 = 0$, $4x + 3y + 4 = 0$ and $x + \alpha y + \beta = 0$, where $\alpha^2 + \beta^2 = 2$ are concurrent, then

a. $\alpha = 1, \beta = -1$ b. $\alpha = 1, \beta = \pm 1$ c. $\alpha = -1, \beta = \pm 1$ d. $\alpha = \pm 1, \beta = 1$

20 Given that the family of lines, $a(3x + 4y + 6) + b(x + y + 2) = 0$. The line of the family situated at the greatest distance from the point $P(2, 3)$ has equation

a. $4x + 3y + 8 = 0$ b. $5x + 3y + 10 = 0$
c. $15x + 8y + 30 = 0$ d. None of these

21 In a ΔABC, if $A(2, -1)$ and $7x - 10y + 1 = 0$ and $3x - 2y + 5 = 0$ are equations of an altitude and an angle bisector respectively drawn from B, then equation of BC is

a. $x + y + 1 = 0$ b. $5x + y + 17 = 0$
c. $4x + 9y + 30 = 0$ d. $x - 5y - 7 = 0$

22 The line $(k + 1)^2 x + ky - 2k^2 - 2 = 0$ passes through a point regardless of the value of k. Which of the following is the line with slope 2 passing through the point?

a. $y = 2x - 8$ b. $y = 2x - 5$ c. $y = 2x - 4$ d. $y = 2x + 8$

23 The line $2x + 3y = 6$, $2x + 3y = 8$ cut the X-axis at A and B. A line l is drawn through the point $(2, 2)$ meets the X-axis at C in such a way that the abscissa of A, B and C are in GP. The equation of line l is

a. $3x - 5y = 16$ b. $3x + 5y = 16$ c. $5x + 3y = 16$ d. $5x - 3y = 16$

Statement Type

Direction (Q. Nos. 24 and 25) *This section is based on Statement I and Statement II. Select the correct answer from the codes given below :*

a. Both Statement I and Statement II are correct and Statement II is the correct explanation of Statement I
b. Both Statement I and Statement II are correct and Statement II is not the correct explanation of Statement I
c. Statement I is correct but Statement II is incorrect
d. Statement II is correct but Statement I is incorrect

24 **Statement I** Centroid of the triangle whose vertices are $A(-1, 11)$, $B(-9, -8)$ and $C(15, -2)$ lies on the internal angle bisector of the vertex A, because

Statement II ΔABC is isosceles with B and C as base angles.

25 Consider the line $L : 3x + y + 4 = 0$ and the points $A(-5, 6)$ and $B(3, 2)$.

Statement I There is exactly one point on the line L which is equidistant from the points A and B, because

Statement II The points A and B are on the different sides of the line.

One or More than One Options Correct Type

Working Space

Direction *(Q. Nos. 26-34) This section contains 9 multiple choice questions. Each question has four choices (a), (b), (c) and (d), out of which ONE or MORE THAN ONE are correct.*

26 The equations of the sides AB, BC and CA of a ΔABC are $2x + y = 0$, $x + py = q$ and $x - y = 3$, respectively and if the centroid is (2, 3), then

a. $p + q = 74$
b. equation of BC is $x + 15y + 59 = 0$
c. equation of median through A is $5x - y - 7 = 0$
d. equation of altitude through A is $11x - y - 13 = 0$

27 Two sides of a rhombus $OABC$ (lying entirely in first quadrant or fourth quadrant) of area equal to 2 sq units, are $y = \frac{x}{\sqrt{3}}$ and $y = \sqrt{3}x$. Then, possible coordinates of B is/are (O being origin)

a. $(1+\sqrt{3}, 1+\sqrt{3})$
b. $(-1-\sqrt{3}, -1-\sqrt{3})$
c. $(\sqrt{3}-1, \sqrt{3}-1)$
d. None of these

28 If the lines $x - 2y - 6 = 0$, $3x + y - 4 = 0$ and $\lambda x + 4y + \lambda^2 = 0$ are concurrent, then

a. $\lambda = 2$
b. $\lambda = -3$
c. $\lambda = 4$
d. $\lambda = -4$

29 If the lines $ax + by + c = 0$, $bx + cy + a = 0$ and $cx + ay + b = 0$ are concurrent $(a + b + c \neq 0)$. Then,

a. $a^3 + b^3 + c^3 - 3abc = 0$
b. $a = b$
c. $a = b = c$
d. $a^2 + b^2 + c^2 - ab - bc - ca = 0$

30 If the point (α, α^2) lies inside the triangle formed by the lines $2x + 3y + 1 = 0$, $x + 2y - 3 = 0$ and $5x - 6y - 1 = 0$, then α can be in the interval

a. $\left(-\infty, -\frac{3}{2}\right)$
b. $\left(-\frac{3}{2}, -1\right)$
c. $(1, \infty)$
d. $\left(\frac{1}{2}, 1\right)$

31 Two lines from the family of lines $(1 + 2\lambda)x - (1 + \lambda)y + 1 = 0$ and the line $x + y = 5$ form an equilateral triangle. The equation of the two lines can be

a. $y - 2 = (2 + \sqrt{3})(x - 1)$
b. $y - 2 = (2 - \sqrt{3})(x - 1)$
c. $y - 2 = (3 - \sqrt{2})(x - 1)$
d. $y - 2 = (3 - \sqrt{2})(x - 2)$

32 The diagonals of rhombus $ABCD$ intersect at point (1, 2) and its sides are parallel to the lines $x - \sqrt{3}y + 2\sqrt{3} = 0$ and $\sqrt{3}x - y + 3 = 0$. If the vertex A lies on X-axis, then possible coordinates of vertex C are

a. (−1, 4)
b. (−1, − 4)
c. (3, 4)
d. None of these

33 The area of a triangle is 5 sq units. Two of its vertices are (2, 1) and (3, −2) the third vertex lies on $y = x + 3$. The coordinates of the third vertex can be

a. $\left(-\frac{3}{2}, \frac{3}{2}\right)$
b. $\left(\frac{3}{4}, -\frac{3}{2}\right)$
c. $\left(\frac{7}{2}, \frac{13}{2}\right)$
d. $\left(-\frac{1}{4}, \frac{11}{4}\right)$

34 Locus of mirror image of (4, 3) about the variable line $(2 - 3m)x + (1 + 4m)y + (2 - m) = 0$ (where, m is variable), is

a. $x^2 + y^2 + \frac{8}{11}x + \frac{8}{11}y = 0$
b. $11x^2 + 11y^2 + 18x + 18y + 96 = 0$
c. $11x^2 + 11y^2 - 18x - 18y - 96 = 0$
d. passing through $\left(-\frac{62}{11}, \frac{-41}{11}\right)$

Comprehension Type

Working Space

Direction (Q. Nos. 35-40) *This section contains 2 paragraphs, each describing theory, experiments, data, etc. Six questions related to the paragraphs have been given. Each question has only one correct answer among the four given options (a), (b), (c) and (d).*

Passage I

The equations of two equal sides *AB* and *AC* of an isosceles ΔABC are $x + y = 5$ and $7x - y = 3$. If the area of ΔABC is 5 units, then answer the following questions :

35 Which of the following cannot represent *BC*?

a. $3y - x = 7$
b. $y + 3x = 12$
c. $y + 3x = 20$
d. $3y + x = -7$

36 If $\left(\frac{7}{2}, \frac{3}{2}\right)$ is one of the possible position of *B*, then which of the following cannot be possible position of *C*?

a. $\left(\frac{1}{2}, \frac{1}{2}\right)$
b. $\left(\frac{3}{2}, \frac{15}{2}\right)$
c. $\left(-\frac{3}{2}, \frac{13}{2}\right)$
d. None of these

37 The number of possible ΔABC must be

a. 3
b. 4
c. 1
d. 2

Passage II

The vertex *A* of ΔABC is $(3, -1)$. The equation of median *BE* and angular bisector *CF* are $6x + 10y - 59 = 0$ and $x - 4y + 10 = 0$. Then,

38 Slope of the side *BC* must be

a. $\frac{1}{9}$
b. $-\frac{2}{9}$
c. $\frac{1}{7}$
d. None of these

39 The equation of *AB* must be

a. $x + y = 2$
b. $x + 4y = 10$
c. $18x + 13y = 41$
d. $23x - y = 70$

40 The length of the side *AC* must be

a. $\sqrt{71}$ units
b. $\sqrt{83}$ units
c. $\sqrt{85}$ units
d. None of these

DPP-5 Locus

Only One Option Correct Type

Working Space

Direction *(Q. Nos. 1-27) This section contains 27 multiple choice questions. Each question has four choices (a), (b), (c) and (d), out of which ONLY ONE option is correct.*

1 For variable lines $\frac{x}{a} + \frac{y}{b} = 1, a + b = 10$, the locus of mid-point of the intercept of this line between coodinate axes is

a. $10x + 5y = 1$
b. $x + y = 10$
c. $x + y = 5$
d. $5x + 10y = 1$

2 A line moves in such a way that the sum of the intercepts made by it on the axes is always c (constant). The locus of the mid-point of its intercept between the axes is

a. $x + y = 2c$
b. $x + y = c$
c. $2(x + y) = c$
d. None of these

3 A variable line passes through a fixed point (a, b) and meets the coordinates axes at A and B. The locus of point of intersection of lines through A, B parallel to coordinate axes is

a. $\frac{x}{a} + \frac{y}{b} = 2$
b. $\frac{a}{x} + \frac{b}{y} = 1$
c. $\frac{x}{a} + \frac{y}{b} = 1$
d. $\frac{x}{a} + \frac{y}{b} = 3$

4 The locus of point of intersection of the lines $x \cos \alpha + y \sin \alpha = a$ and $x \sin \alpha - y \cos \alpha = b$, is

a. $x^2 - y^2 = a^2 - b^2$
b. $x^2 - y^2 = a^2 + b^2$
c. $x^2 + y^2 = a^2 - b^2$
d. $x^2 + y^2 = a^2 + b^2$

5 A point moves so that square of its distance from the point $(3, -2)$ is numerically equal to its distance from the line $5x - 12y = 13$. The equation of the locus of the point is

a. $13x^2 + 13y^2 - 83x + 64y + 182 = 0$
b. $x^2 + y^2 - 11x + 16y + 26 = 0$
c. $x^2 + y^2 - 11x + 16y = 0$
d. None of the above

6 A straight line segment of length l moves with its ends on the coordinate axes. The locus of the point which divides the line in the ratio of 1:2, is

a. $9x^2 + 36y^2 = l^2$
b. $36x^2 + 9y^2 = l^2$
c. $36x^2 + 9y^2 = 4l^2$
d. $9x^2 + 36y^2 = 4l^2$

7 If for the variable line $\frac{x}{a} + \frac{y}{b} = 1$, the relation $1/a^2 + 1/b^2 = 1/c^2$ (c is constant) is satisfied, then the locus of the foot of perpendicular from the origin on this line is

a. $x^2 + y^2 = c^2$
b. $x^2 + y^2 = 2c^2$
c. $2(x^2 + y^2) = c^2$
d. None of these

Working Space

8 The locus of the mid-point of the intercept of the variable line $x \cos \alpha + y \sin \alpha = p$ (p is constant) between the coordinate axes is

a. $\frac{1}{x^2} + \frac{1}{y^2} = \frac{1}{p^2}$
b. $\frac{1}{x^2} + \frac{1}{y^2} = \frac{2}{p^2}$
c. $\frac{1}{x^2} + \frac{1}{y^2} = \frac{4}{p^2}$
d. None of these

9 If $A(2, -3)$ and $B(-2, 1)$ are vertices of a ΔABC and centroid of the triangle lies on the line $2x + 3y = 1$, then locus of the vertex C, is

a. $3x + 2y = 5$
b. $2x - 3y = 7$
c. $2x + 3y = 9$
d. $3x - 2y = 3$

10 If sum of the distance of a point from two perpendicular lines in a plane is 1, then its locus is

a. square
b. circle
c. straight line
d. two intersecting lines

11 The area of the triangle formed by the intersection of a line parallel to X-axis and passing through $P(h, k)$ with the lines $y = x$ and $x + y = 2$ is $4h^2$, then locus of P is

[IIT JEE 2005]

a. $y = \pm 2x + 1$
b. $y = \pm x + 1$
c. $y = \pm 3x + 1$
d. None of these

12 A variable straight line passes through a fixed point (a, b) intersecting the coordinates axes at A and B. If O is the origin, then the locus of the centroid of ΔOAB is

a. $bx + ay - 3xy = 0$
b. $bx + ay - 2xy = 0$
c. $ax + by - 3xy = 0$
d. $ax + by - 2xy = 0$

13 AB is a variable line sliding between the coordinate axes in such a way that A lies on X-axis and B lies on Y-axis. If P is a variable point on AB such that $PA = b$, $PB = a$ and $AB = a + b$, then equation of locus of P is

a. $\frac{x^2}{a^2} + \frac{y^2}{b^2} = 1$
b. $\frac{x^2}{a^2} - \frac{y^2}{b^2} = 1$
c. $x^2 + y^2 = a^2 + b^2$
d. None of these

14 A and B are fixed points. The vertex C of ΔABC moves such that $\cot A + \cot B =$ constant. The locus of C is a straight line

a. a perpendicular to AB
b. parallel to AB
c. inclined at an angle $(A - B)$ to AB
d. None of these

15 Let A be the fixed point $(0, 4)$ and $B(2t, 0)$ be a moving point. Let M be the mid-point of AB and let the perpendicular bisector of AB meets the Y-axis at R. The locus of mid-point P of MR, is

a. the parabola $y = x^2 + 2$
b. is the ellipse having vertex (0, 2)
c. is the parabola having length of latusrectum, 1
d. the parabola $x^2 + y = 2$

16 Vertices of a variable triangle are $A(3, 4)$, $B(5 \cos \theta, 5 \sin \theta)$ and $C(5 \sin \theta, -5 \cos \theta)$. Then, locus of its orthocentre is

a. $(x + y - 1)^2 + (x - y - 7)^2 = 100$
b. $(x + y - 7)^2 + (x - y - 1)^2 = 100$
c. $(x + y - 7)^2 + (x + y - 1)^2 = 100$
d. $(x + y - 7)^2 + (x - y + 1)^2 = 100$

Working Space

17 If the distance of any point (x, y) from the origin is defined as $d(x, y) = \max\{|x|, |y|\}$, then the locus of the point (x, y), where $d(x, y) = 1$, is

a. a square of area 1 sq unit
b. a square of area 4 sq units
c. a triangle
d. None of these

18 If $P(1, 0)$, $Q(-1, 0)$ and $R(2, 0)$ are three given points, then the locus of the point S satisfying the relation, $SQ^2 + SR^2 = 2SP^2$, is

a. a straight line parallel to X-axis
b. a circle passing through the origin
c. a circle with centre at origin
d. a straight line parallel to Y-axis

19 Consider a quadratic equation in Z with parameters x and y as $z^2 - xz + (x - y)^2 = 0$. The parameters x and y are the coordinates of a variables point P w.r.t. an orthonormal coordinate system in a plane. If the quadratic equation has equal roots, then locus of P is

a. a circle
b. a line pair through the origin of coordinates with slope 1/2 and 2/3
c. a line pair through the origin of coordinates with slope 3/2 and 2
d. a line through the origin of coordinates with slope 3/2 and 1/2

20 Given $A(1, 1)$ and AB is any line through it, cutting the X-axis at B. If AC is perpendicular to AB and meets the Y-axis at C, then the equation of locus of mid-point P on BC, is

a. $x + y = 1$
b. $x + y = 2$
c. $x + y = 2xy$
d. $2x + 2y = 1$

21 Given $\frac{x}{a} + \frac{y}{b} = 1$ and $ax + by = 1$ are two variable lines, a and b being the parameters connected by the relation $a^2 + b^2 = ab$. The locus of the point of intersection has the equation

a. $x^2 + y^2 + xy - 1 = 0$
b. $x^2 + y^2 - xy + 1 = 0$
c. $x^2 + y^2 + xy + 1 = 0$
d. $x^2 + y^2 - xy - 1 = 0$

22 Triangle formed by the lines $x + y = 0$, $x - y = 0$ and $lx + my = 1$. If l and m vary subject to the condition $l^2 + m^2 = 1$. Then, the locus of its circumcentre is

a. $(x^2 - y^2)^2 = x^2 + y^2$
b. $(x^2 + y^2)^2 = x^2 - y^2$
c. $x^2 + y^2 = 4x^2y^2$
d. $(x^2 - y^2)^2 = (x^2 + y^2)^2$

23 The distance of the point $P(x_1, y_1)$ from each of the two straight lines through the origin is d. The equation of two straight lines is

a. $(xy_1 - yx_1)^2 = d^2(x^2 + y^2)$
b. $d^2(xy_1 - yx_1)^2 = x^2 + y^2$
c. $d^2(xy_1 + yx_1)^2 = x^2 + y^2$
d. $(xy_1 + yx_1)^2 = d^2(x^2 + y^2)$

24 If the vertices A and B of a ΔABC are given by $(2, 5)$ and $(4, -11)$, respectively and C moves along the line $L = 9x + 7y + 4 = 0$, then the locus of the centroid of the ΔABC, is

a. a circle
b. any straight line
c. a line parallel to L
d. a line perpendicular to L

Working Space

25 A, B and C are points in the xy-plane such that $A(1, 2)$, $B(5, 6)$ and $AC = 3BC$. Then,

a. ABC is a unique triangle
b. there can be only two such triangles
c. No such triangle is possible
d. there can be infinite number of such triangles

26 Each member of the family of parabolas $y = ax^2 + 2x + 3$ has a maximum or a minimum point depending upon the value of a. The equation to the locus of the maxima and minima for all possible values of a, is

a. a straight line with slope 1 and y-intercept 3
b. a straight line with slope 2 and y-intercept 2
c. a straight line with slope 1 and x-intercept 3
d. a circle

27 The area (in sq units) enclosed by the graphs of $|x + y| = 2$ and $|x| = 1$, is

a. 2
b. 4
c. 6
d. 8

Comprehension Type

Direction (Q. Nos. 28-30) *This section contains a paragraph, describing theory, experiments, data, etc. Three questions related to the paragraph have been given. Each question has only one correct answer among the four given options (a), (b), (c) and (d).*

Passage

Let ABC be an acute angled triangle and AD, BE and CF are its medians, where E and F are the points $E(3, 4)$ and $F(1, 2)$, respectively and centroid of ΔABC is $G(3, 2)$, then answer the following questions:

28 The equation of side AB, is

a. $2x + y = 4$
b. $x + y - 3 = 0$
c. $4x - 2y = 0$
d. None of these

29 Coordinates of D are

a. $(7, -4)$
b. $(5, 0)$
c. $(7, 4)$
d. $(-3, 0)$

30 Height of the altitude drawn from the point A is (in units)

a. $4\sqrt{2}$
b. $3\sqrt{2}$
c. $6\sqrt{2}$
d. $2\sqrt{3}$

SET - 1

Revisal Problems
for JEE Main

Only One Option Correct Type

Working Space

Direction (Q. Nos. 1-36) *This section contains 36 multiple choice questions. Each question has four choices (a), (b), (c) and (d), out of which ONLY ONE option is correct.*

1 The line parallel to the X-axis and passing through the intersection of the lines $ax + 2by + 3b = 0$ and $bx - 2ay - 3a = 0$ is, [where, $(a, b) \neq (0, 0)$]

a. below the X-axis at a distance of $\frac{3}{2}$ from it

b. below the X-axis at a distance of $\frac{2}{3}$ from it

c. above the X-axis at a distance of $\frac{3}{2}$ from it

d. above the X-axis at a distance of $\frac{2}{3}$ from it

2 Find the equation to the straight line which passes through the point $(-4, 3)$ and is such that the portion of it between the axes is divided by the point in the ratio $5:3$, is

a. $9x - 20y + 96 = 0$ b. $2x - y + 11 = 0$
c. $2x + y + 5 = 0$ d. $3x - 2y + 7 = 0$

3 The equations of the perpendicular bisector of the sides AB and AC of a ΔABC are $x - y + 5 = 0$ and $x + 2y = 0$, respectively. If the point A is $(1, -2)$, then the equation of the line BC, is

a. $14x + 23y = 40$ b. $14x - 23y = 40$
c. $23x + 14y = 40$ d. $23x - 14y = 40$

4 The distance of the point $(2, 3)$ from the line $2x - 3y + 9 = 0$ measured along a line $x - y + 1 = 0$, is

a. $5\sqrt{3}$ b. $4\sqrt{2}$
c. $3\sqrt{2}$ d. $2\sqrt{2}$

5 Which pair of points lie on the same side of $3x - 8y - 7 = 0$?

a. $(0, -1)$ and $(0, 0)$ b. $(4, -3)$ and $(0, 1)$
c. $(-3, -4)$ and $(1, 2)$ d. $(-1, -1)$ and $(3, 7)$

6 The reflection of the point $(4, -13)$ in the line $5x + y + 6 = 0$, is

a. $(-1, -14)$ b. $(3, 4)$
c. $(1, 2)$ d. $(-4, 13)$

7 The equation of the bisector of the acute angle between the lines $3x - 4y + 7 = 0$ and $12x + 5y - 2 = 0$, is

a. $11x - 3y + 9 = 0$ b. $3x + 11y - 13 = 0$
c. $3x + 11y - 3 = 0$ d. $11x - 3y + 2 = 0$

8 The points on the line of $x + y = 4$ which lie at a unit distance from the line $4x + 3y = 10$, are

a. $(3, 1)$ and $(-7, 11)$ b. $(3, 4)$ and $(-3, -4)$
c. $(8, 6)$ and $(6, 8)$ d. $(4, 0)$ and $(0, 4)$

9 One side of a rectangle lies along the line $4x + 7y + 5 = 0$. Two of its vertices are $(-3, 1)$ and $(1, 1)$. Then, the equations of other sides are

a. $7x - 4y + 25 = 0, 7x + 4y - 3 = 0, 7x + 4y + 25 = 0$
b. $7x + 4y + 25 = 0, 4x + 7y = 11, 7x - 4y + 25 = 0$
c. $7x - 4y + 25 = 0, 7x - 4y - 3 = 0, 4x + 7y = 11$
d. None of the above

Working Space

10 Line $\frac{x}{a} + \frac{y}{b} = 1$ cuts the coordinate axes at $A(a, 0)$ and $B(0, b)$ and the line $\frac{x}{a'} + \frac{y}{b'} = -1$ at $A'(-a', 0)$ and $B'(0, -b')$. If the points A, B, A' and B' are concyclic, then the orthocentre of the $\Delta ABA'$, is

a. $(0, 0)$
b. $(0, b')$
c. $\left(0, \frac{aa'}{bb'}\right)$
d. $\left(a, \frac{bb'}{a}\right)$

11 The area of the parallelogram formed by the lines $3x + 4y = 7a$, $3x + 4y = 7b$, $4x + 3y = 7c$ and $4x + 3y = 7d$, is

a. $\frac{|(a-b)(c-d)|}{7}$
b. $|(a-b)(c-d)|$
c. $\frac{|(a-b)(c-d)|}{49}$
d. $7|(a-b)(c-d)|$

12 If x_1, y_1 are roots of $x^2 + 8x - 20 = 0$, x_2, y_2 are the roots of $4x^2 + 32x - 57 = 0$ and x_3, y_3 are the roots of $9x^2 + 72x - 112 = 0$, then the points (x_1, y_1), (x_2, y_2) and (x_3, y_3)

a. are collinear
b. form an equilateral triangle
c. form a right angled isosceles triangle
d. are concyclic

13 If the lines $\lambda x + (\sin\alpha)y + \cos\alpha = 0$, $x + (\cos\alpha)y + \sin\alpha = 0$, $x - (\sin\alpha)y + \cos\alpha = 0$ pass through the same point, where $\alpha \in R$, then λ lies in the interval

a. $[-1, 1]$
b. $[-\sqrt{2}, \sqrt{2}]$
c. $[-2, 2]$
d. $(-\infty, \infty)$

14 Family of lines represented by the equation $(\cos\theta + \sin\theta)x + (\cos\theta - \sin\theta)y - 3(3\cos\theta + \sin\theta) = 0$ passes through a fixed point M for all real values of θ. The reflection of M in the line $y - x = 0$, is

a. $(6, 3)$
b. $(3, 6)$
c. $(-6, 3)$
d. $(3, -6)$

15 Area of the quadrilateral formed by the lines $|x| + |y| = 2$, is (in sq units)

a. 8
b. 6
c. 4
d. None of these

16 Consider a parallelogram whose sides are represented by the lines $2x + 3y = 0$, $2x + 3y - 5 = 0$, $3x - 4y = 0$ and $3x - 4y = 3$. The equation of the diagonal not passing through the origin, is

a. $21x - 11y + 15 = 0$
b. $9x - 11y + 15 = 0$
c. $21x - 29y - 15 = 0$
d. $21x - 11y - 15 = 0$

17 If the straight lines, $ax + amy + 1 = 0$, $bx + (m+1)by + 1 = 0$ and $cx + (m+2)cy + 1 = 0$, $m \neq 0$ are concurrent, then a, b, c are in

a. AP only for $m = 1$
b. AP for all m
c. GP for all m
d. HP for all m

18 A ΔABC is formed by the lines $2x - 3y - 6 = 0$, $3x - y + 3 = 0$ and $3x + 4y - 12 = 0$. If the points $P(\alpha, 0)$ and $Q(0, \beta)$ always lie on or inside the ΔABC. Then,

a. $\alpha \in (-1, 2)$ and $\beta \in (-2, 3)$
b. $\alpha \in [-1, 3]$ and $\beta \in [-2, 4]$
c. $\alpha \in [-2, 4]$ and $\beta \in [-2, 4]$
d. $\alpha \in [-1, 3]$ and $\beta \in [-2, 3]$

19 The coordinates of a point P on the line $2x - y + 5 = 0$ such that $|PA - PB|$ is maximum, where $A(4, -2)$ and $B(2, -4)$ will be

a. $(11, 27)$
b. $(-11, -17)$
c. $(-11, 17)$
d. $(0, 5)$

20 A ray of light passing through the point $A(1, 2)$ is reflected at a point B on the X-axis and then passes through $(5, 3)$. Then, the equation of AB is

a. $5x + 4y = 13$
b. $5x - 4y = -3$
c. $4x + 5y = 14$
d. None of these

Working Space

21 The point $P(a^2, a+1)$ is a point in the angle between the lines $3x - y + 1 = 0$ and $x + 2y - 5 = 0$ containing the origin, if

a. $a \geq 1$ or $a \leq -3$ b. $a \in (-3, 0) \cup (1/3, 1)$
c. $a \in (0, 1)$ d. None of these

22 Drawn from the origin are two mutually perpendicular straight lines forming an isosceles triangle together with the straight line, $2x + y = a$. Then, the area (in sq units) of the triangle is

a. $\frac{a^2}{2}$ b. $\frac{a^2}{3}$
c. $\frac{a^2}{5}$ d. None of these

23 The image of the point $A(1, 2)$ by the line mirror $y = x$ is the point B and the image of B by the mirror $y = 0$ is the point (α, β), then

a. $\alpha = -1, \beta = -2$ b. $\alpha = 0, \beta = 0$
c. $\alpha = 2, \beta = -1$ d. None of these

24 The line $x + 3y - 2 = 0$ bisects the angle between a pair of straight lines of which one has equation $x - 7y + 5 = 0$. The equation of the other line is

a. $3x + 3y - 1 = 0$ b. $x - 3y + 2 = 0$
c. $5x + 5y - 3 = 0$ d. None of these

25 On the portion of the straight line, $x + 2y = 4$ intercepted between the axes. A square is constructed on the side of the line away from the origin. Then, the point of intersection of its diagonals has coordinates

a. $(2, 3)$ b. $(3, 2)$
c. $(3, 3)$ d. None of these

26 A light beam emanating from the point $A(3, 10)$ reflects from the straight line $2x + y - 6 = 0$ and then passes through the point $B(7, 2)$. The equation of the reflected beam is

a. $3x - y + 1 = 0$ b. $x + 3y - 13 = 0$
c. $3x + y - 15 = 0$ d. $x - 3y + 5 = 0$

27 The equation of the bisector of the angle between the two lines $3x - 4y + 12 = 0$ and $12x - 5y + 7 = 0$ which contains the points $(-1, 4)$, is

a. $21x + 27y - 121 = 0$ b. $21x - 27y + 121 = 0$
c. $21x + 27y + 191 = 0$ d. None of these

28 The equation of bisector of two lines L_1 and L_2 are $2x - 16y - 5 = 0$ and $64x + 8y + 35 = 0$. If the lines L_1 passes through $(-11, 4)$, the equation of acute angle bisector of L_1 and L_2 is

a. $2x - 16y - 5 = 0$
b. $64x + 8y + 35 = 0$
c. data insufficient
d. None of the above

29 If the algebraic sum of perpendicular distances from the points $(2, 0)$, $(0, 2)$ and $(1, 1)$ to a variable straight line is zero, then the line passing through the point

a. $(-1, -1)$ b. $(1, 1)$
c. $(1, -1)$ d. None of these

30 The degree measure of the angle made with X-axis by a straight line drawn through $(1, 2)$, so that it intersects $x + y = 4$ at a distance $\frac{\sqrt{6}}{3}$ from $(1, 2)$, is/are

a. $105°$ b. $75°$
c. $60°$ d. $30°$

31 If a straight line $ax + by + c = 0$ and $x \cos \alpha + y \sin \alpha = c$ enclose an angle $\pi/4$ between them and meets the straight line $x \sin \alpha - y \cos \alpha = 0$ in the same point, then

a. $a^2 + b^2 = c^2$ b. $a^2 + b^2 = 2$
c. $a^2 + b^2 = 2c^2$ d. $a^2 + b^2 = 4$

Working Space

32 Set of values of α for which the point $P(\alpha, \alpha^2 - 2)$ lies inside the triangle formed by the lines $x + y = 1$, $y = x + 1$ and $y = -1$, is

a. $\left(\frac{-1-\sqrt{13}}{2}, 1\right)$

b. $\left(\frac{1-\sqrt{13}}{2}, -1\right) \cup \left(1, \frac{-1+\sqrt{13}}{2}\right)$

c. $[-1, 1]$

d. None of the above

33 If $P\left(1+\frac{t}{\sqrt{2}}, 2+\frac{t}{\sqrt{2}}\right)$ is any point on the line, then the values of t for which the point P lies between parallel lines $x + 2y = 1$ and $x + 2y = 15/2$ may belong to the interval

a. $\left(-\frac{4\sqrt{2}}{5}, \frac{5\sqrt{2}}{6}\right)$
b. $\left(0, \frac{2\sqrt{5}}{6}\right)$
c. $\left(\frac{4\sqrt{2}}{5}, -\frac{5\sqrt{2}}{6}\right)$
d. None of these

34 Point $P(2, 4)$ is translated through a distance $3\sqrt{2}$ units measured parallel to the line $y - x - 1 = 0$, in the direction of decreasing ordinates, to reach at Q. If R is the image of Q with respect to the line $y - x - 1 = 0$, then coordinates of R are

a. $(-1, 1)$
b. $(5, 7)$
c. $(6, 6)$
d. $(0, 0)$

35 If a, b and c are in GP. Then, $ax + by + c = 0$

a. has a fixed direction
b. always passes through a fixed point
c. forms a triangle with the axes whose area is constant
d. None of the above

36 A family of lines is given by $(1 + 2\lambda)x + (1 - \lambda)y + \lambda = 0$, λ being parameter. The line belonging to this family at the maximum distance from the point (1, 4), is

a. $4x - y + 1 = 0$
b. $33x + 12y + 7 = 0$
c. $12x + 33y = 7$
d. None of these

Statement Type

Direction* (Q. Nos. 37-45) *This section is based on Statement I and Statement II. Select the correct answer from the codes given below :

a. Both Statement I and Statement II are correct and Statement II is the correct explanation of Statement I
b. Both Statement I and Statement II are correct and Statement II is not the correct explanation of Statement I
c. Statement I is correct but Statement II is incorrect
d. Statement II is correct but Statement I is incorrect

37 Consider the lines $L_1 : \frac{x}{3} + \frac{y}{4} = 1$, $L_2 : \frac{x}{4} + \frac{y}{3} = 1$, $L_3 : \frac{x}{3} + \frac{y}{4} = 2$ and $L_4 : \frac{x}{4} + \frac{y}{3} = 2$.

Statement I The quadrilateral formed by these four lines is a rhombus.

Statement II If diagonals of a quadrilateral formed by any four lines are unequal and intersect at right angle, then it is a rhombus.

38 **Statement I** Centroid of the triangle whose vertices are $A(-1, 11)$, $B(-9, -8)$ and $C(15, -2)$ lies on the internal angle bisector of the vertex A.

Statement II ΔABC is isosceles with B and C as base angles.

39 Let points A, B and C are represented by $(a \cos \theta_i, a \sin \theta_i)$, where $i = 1, 2, 3$ and $\cos(\theta_1 - \theta_2) + \cos(\theta_2 - \theta_3) + \cos(\theta_3 - \theta_1) = -3/2$.

Statement I Orthocentre of ΔABC is at origin.

Statement II ΔABC is an equilateral triangle.

Working Space

40 **Statement I** The area of the triangle formed by the points $A(20,22)$, $B(21,24)$ and $C(22,23)$ is the same as the area of the triangle formed by the points $P(0, 0)$, $Q(1,2)$ and $R(2,1)$.

Statement II The area of triangle is invarient w.r.t. the translation of the coordinate axes.

41 **Statement I** In centre of the triangle formed by the lines whose sides are $3x + 4y = 0$, $5x - 12y = 0$ and $y - 15 = 0$ is the point P whose coordinates are (1, 8).

Statement II Point P is equidistant form the 3 lines forming the triangle.

42 **Statement I** Point Q is symmetric to point $P(4, -3)$ w.r.t. bisector of first and third quadrant, then the length of PQ is $7\sqrt{2}$.

Statement II Bisector of the first and third quadrant is perpendicular bisector of PQ.

43 **Statement I** The diagonals of the quadrilateral whose sides are $3x + 2y + 1 = 0$, $3x + 2y + 2 = 0$, $2x + 3y + 1 = 0$ and $2x + 3y + 2 = 0$ include an angle $\pi/2$.

Statement II Diagonals of a parallelogram are bisecting each other.

44 **Statement I** Each point on the line $y - x + 12 = 0$ is at same distance from the lines $3x + 4y - 12 = 0$ and $4x + 3y - 12 = 0$.

Statement II Locus of point which is at equal distance from the two given lines is the angle bisectors of the two lines.

45 **Statement I** Area of triangle formed by the line which is passing through the point (5, 6), such that segment of the line between axis is bisected at the point, with coordinate axes is 60 sq units.

Statement II Area of triangle formed by the lines passing through point (α, β) with axis is maximum when point (α, β) is mid-point of segment of line between axes.

SET - 2

Revisal Problems for JEE Advanced

Only One Option Correct Type

Working Space

Direction (Q. Nos. 1-23) *This section contains 23 multiple choice questions. Each question has four choices (a), (b), (c) and (d), out of which ONLY ONE option is correct.*

1 In ΔABC, $B(2, 3)$, $C(-2, 6)$ and perimeter is 14 units. Then, locus of centroid of ΔABC, is

a. a circle
b. a pair of lines
c. a line
d. an ellipse

2 The number of triangles which are obtuse and which have the points (8, 9), (8, 16) and (20, 25) as the foot of perpendiculars drawn from the vertices on the opposite sides, is

a. 0
b. 1
c. 2
d. 4

3 If the roots of the equation $ax^2 + bx + 10 = 0$ are not real and distinct, where $a, b \in R$ and α and β are values of a and b for which $5a + b$ is minimum, then the family of lines $\alpha(4x + 2y + 1) + \beta(x - y + 1) = 0$ are concurrent at

a. $(1, -1)$
b. $(1, 1)$
c. $\left(-\frac{1}{6}, -\frac{7}{6}\right)$
d. $\left(\frac{1}{6}, \frac{7}{6}\right)$

4 The four sides of a quadrilateral are given by the equation $xy(x-2)(y-3) = 0$. The equation of the line parallel to $x - 4y = 0$ that divides the quadrilateral in two equal areas

a. $x - 4y + 5 = 0$
b. $x - 4y - 5 = 0$
c. $4y = x + 1$
d. $4y + 1 = x$

5 Let $A(1, 1)$ and $B(3, 2)$ be two points. If C is a point on X-axis such that $AC + BC$ is minimum, then the coordinates are

a. $\left(\frac{5}{3}, 0\right)$
b. $\left(\frac{1}{3}, 0\right)$
c. $(3, 0)$
d. None of these

6 Let A and B have coordinates (x_1, y_1) and (x_2, y_2), respectively. We define the distance between A and B as $d(A, B) = \max\{|x_2 - x_1|, |y_2 - y_1|\}$. If $d(0, A) = 1$, where 0 is origin, then the locus of A is

a. a square
b. pair of parallel lines
c. a square of area 2 sq units
d. a square of area 8 sq units

7 The coordinate axes are rotated about the origin O in counter-clockwise direction through an angle 60°. If p and q are the intercepts made on the new axes by a straight line whose equation referred to the original axes is $x + y = 1$, then

a. $p^2 + q^2 = 2$
b. $\frac{1}{p^2} + \frac{1}{q^2} = 2$
c. $p^2 + q^2 = p^2q^2$
d. None of these

8 Consider the family of lines $(x - y - 6) + \lambda(2x + y + 3) = 0$

and $(x + 2y - 4) + \mu(3x - 2y - 4) = 0$

If the lines of these 2 families are at right angle to each other, then the locus of their point of intersection, is

a. $x^2 + y^2 - 3x + 4y - 3 = 0$
b. $x^2 + y^2 + 3x - 4y - 3 = 0$
c. $x^2 + y^2 - 3x - 4y - 3 = 0$
d. $x^2 + y^2 + 3x + 4y - 3 = 0$

Working Space

9 If $A(1, p^2)$, $B(0, 1)$ and $C(p, 0)$ are the coordinates of three points, then the value of p for which the area of the ΔABC is minimum, is

a. $\frac{1}{\sqrt{3}}$ b. $-\frac{1}{\sqrt{3}}$ c. $\frac{1}{\sqrt{3}}$ or $-\frac{1}{\sqrt{3}}$ d. None of these

10 m, n are integers with $0 < n < m$. A is the point (m, n) on the cartesian plane. B is the reflection of A in the line $y = x$. C is the reflection of B in the Y-axis, D is the reflection of C in the X-axis and E is the reflection of D in the Y-axis. The area of the pentagon $ABCDE$, is

a. $2m(m+n)$ b. $m(m+3n)$
c. $m(2m+3n)$ d. None of these

11 The distance between the two parallel lines is 1 unit. A point A is chosen to lie between the lines at a distance d from one of them. ΔABC is equilateral with B on one line and C on the other parallel line. Then, length of the side of the equilateral triangle is

a. $\frac{2}{3}\sqrt{d^2+d+1}$ b. $2\sqrt{\frac{d^2-d+1}{3}}$ c. $2\sqrt{d^2+d+1}$ d. $\sqrt{d^2-d+1}$

12 The graph of $(y - x)$ against $(y + x)$ is shown as,

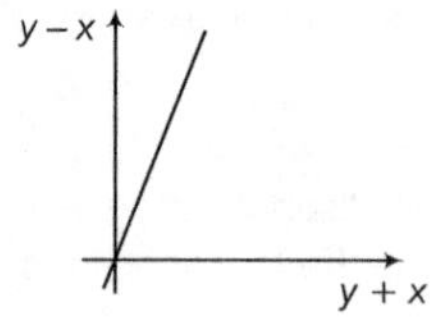

Which one of the following shows the graph of y against x?

a.
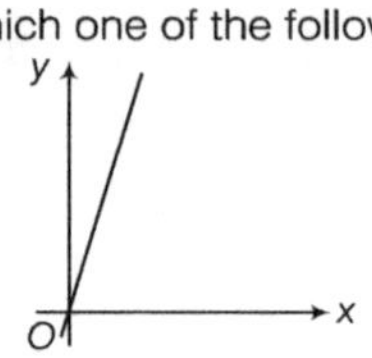

b.
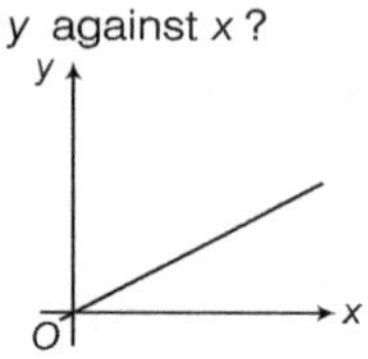

c.
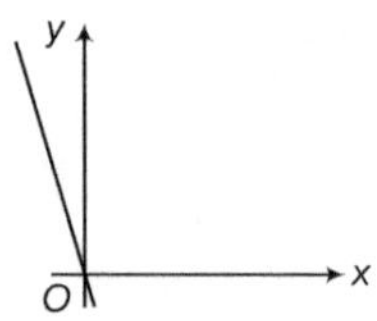

d.
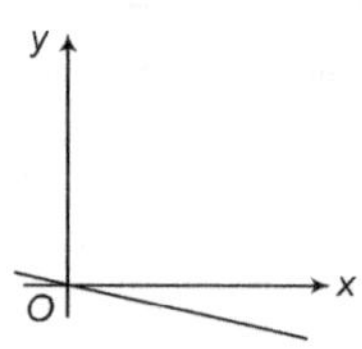

13 P is a point inside the ΔABC. Lines are drawn through P, parallel to the sides of the triangle. The three resulting triangles with the vertex at P have areas 4, 9 and 49 sq units. The area of the ΔABC, is (in sq units)

a. $2\sqrt{3}$ b. 12 c. 24 d. 144

14 A graph is defined in polar coordinates as $r(\theta) = \cos\theta + \frac{1}{2}$. The smallest x-coordinates of any point on the graph is

a. $-\frac{1}{16}$ b. $-\frac{1}{8}$ c. $-\frac{1}{4}$ d. $-\frac{1}{2}$

15 Let P be (5, 3) and a point R on $y = x$ and Q on X-axis is such that $PQ + QR + RP$ is minimum. Then, coordinates of Q are

a. $\left(\frac{17}{4}, 0\right)$ b. (17, 0)
c. $\left(\frac{17}{2}, 0\right)$ d. None of these

16 The equation of circumcircle of a triangle whose sides are $x + ay + 2 = 0$, $ax + y + 2 = 0$ and $x - y = 0$ and has orthocentre (1, 1) is

a. $x^2 + y^2 - 3x - 4y - 1 = 0$ b. $x^2 + y^2 = 1$
c. $x^2 + y^2 - 3x + 4y - 1 = 0$ d. None of these

17 In ΔABC, $a = 4$ and $b = c = 2\sqrt{2}$. A point P moves with the triangle such that the square of its distance from BC is half the rectangle contained by its distance from the other two sides. If D is the centre of locus of P, then

a. locus of P is an ellipse with eccentricity $\sqrt{1/3}$
b. locus of P is a hyperbola with eccentricity $\sqrt{3/2}$
c. area of the quadrilateral $ABCD$ is $\frac{16}{3}$ sq units
d. area of the quadrilateral $ABCD$ is $\frac{32}{3}$ sq units

Working Space

18 The perpendicular bisector of $x + y + 2 = 0$ and $x - y - 1 = 0$ of sides AB and AC of a ΔABC intersect them at $(-1, -1)$ and $(2, 1)$, respectively. If the middle point of side BC is P, then distance of this point from the orthocentre of triangle is

a. $\sqrt{85}$ b. $\sqrt{41}$ c. $\sqrt{13}$ d. None of these

19 The equation of line which is at maximum distance from the point (1, 2) and belonging to family of $(x + 2y + 1) + \lambda(x + y - 2) = 0$, is

a. $4x + 5y = 25$ b. $4x - 5y = 25$ c. $4x - 3y = 25$ d. $4x - 5y = 35$

20 Number of values of λ such that $(\lambda^2 + 1, \lambda)$ is the mirror image of $(\lambda, \lambda + 1)$ w.r.t. $x + 2y - 1 = 0$

a. 0 b. 1 c. 2 d. 3

21 The number of triangles having two vertices as (1, 2) and (6, 2) and incentre (4, 6), is

a. 2 b. 1 c. infinite d. 0

22 If AD, BE and CF are the altitudes of ΔABC whose vertex A is the point $(-4, 5)$. The coordinates of the points E and F are (4, 1) and $(-1, -4)$ respectively. Then, equation of BC is

a. $3x - 4y - 28 = 0$ b. $4x + 3y + 28 = 0$
c. $3x - 4y + 28 = 0$ d. $x + 2y + 7 = 0$

23 The vertices of a triangle are $A(m, n)$, $B(12, 19)$ and $C(23, 20)$, where m and n are integers. If its area is 70 sq units and the slope of the median through A is -5, then the last digit of $(m + n)$ is

a. 3 b. 5 c. 7 d. 8

Statement Type

Direction **(Q. Nos. 24-29)** ***This section is based on Statement I and Statement II. Select the correct answer from the codes given below :***

a. Both Statement I and Statement II are correct and Statement II is the correct explanation of Statement I
b. Both Statement I and Statement II are correct and Statement II is not the correct explanation of Statement I
c. Statement I is correct but Statement II is incorrect
d. Statement II is correct but Statement I is incorrect

24 Let $A(x_1, y_1)$, $B(x_2, y_2)$ and $C(x_3, y_3)$ be the vertices of a ΔABC.

Statement I If $\angle C$ is an obtuse angle, then the quantity $(x_3 - x_1)(x_3 - x_2) + (y_3 - y_1)(y_3 - y_2)$ is negative.

Statement II Diameter of a circle subtends an obtuse angle at any point lying inside the semi-circle.

25 **Statement I** If $(x_1 - x_2)^2 + (y_1 - y_2)^2 = a^2$, $(x_2 - x_3)^2 + (y_2 - y_3)^2 = b^2$ and

$(x_3 - x_1)^2 + (y_3 - y_1)^2 = c^2$

Then, $\begin{vmatrix} x_1 & y_1 & 1 \\ x_2 & y_2 & 1 \\ x_3 & y_3 & 1 \end{vmatrix}^2 = (a + b + c)(b + c - a)(c + a - b)(a + b - c)$.

Statement II Area of ΔABC, where $A(x_1, y_1)$, $B(x_2, y_2)$ and $C(x_3, y_3)$ is $\frac{1}{2}\begin{vmatrix} x_1 & y_1 & 1 \\ x_2 & y_2 & 1 \\ x_3 & y_3 & 1 \end{vmatrix}$.

Working Space

26 **Statement I** Locus of the image of a point $A(0, 0)$ in the line $(2a+b)x+(3a-2b)y+(b-5a)=0$ is $x^2+y^2+2x-2y=0$, (where, a and b are parameters).

Statement II If the lines L is the angle bisector of the line l_1 and l_2, then image of any point on l_1 in the line L lies on L_2.

27 **Statement I** The incentre of triangle formed by the lines

$$L_1: x\cos\left(\frac{\pi}{10}\right)+y\sin\left(\frac{\pi}{10}\right)-2=0$$

$$L_2: x\cos\left(\frac{\pi}{8}\right)-y\sin\left(\frac{\pi}{8}\right)+2=0$$

and $$L_3: x\cos\left(\frac{\pi}{5}\right)+y\sin\left(\frac{\pi}{5}\right)+2=0 \text{ is } (0, 0)$$

Statement II The incentre I of a ΔABC is the point of concurrency of 3 internal angle bisectors of ΔABC.

28 Given a ΔABC whose vertices are $A(x_1, y_1)$, $B(x_2, y_2)$ and $C(x_3, y_3)$. Let there exists a point $P(a, b)$ such that $6a=2x_1+x_2+3x_3$, $6b=2y_1+y_2+3y_3$.

Statement I Area of ΔPBC must be less than the area of ΔABC.

Statement II P lies inside the ΔABC.

29 **Statement I** Let a, b and c be the sides of ΔABC such that $\begin{vmatrix} a & a^2b-1 & a^2+c \\ b & b^3-1 & b^2+c \\ c & c^2b-1 & c^2+c \end{vmatrix}=0$, then triangle is isosceles.

Statement II All the elements of any two rows are equal, then the value of the determinant is equal to zero.

Matching List Type

Direction (Q. No. 30) *Choices for the correct combination of elements from Column I and Column II are given as options (a), (b), (c) and (d), out of which one is correct.*

30 Let the coordinates of the foot of the perpendicular form the vertices of a ΔABC on the opposite sides are $D(20, 25)$, $E(8, 16)$ and $F(8, 9)$.

Match the Column I with Column II and mark the correct option from the given codes.

	Column I		Column II
i.	If the orthocentre of the ΔABC is (p, q), then $p+q$ equals	p.	45
ii.	If the equation of the line BC is $y=-x+C$, then the value of C is	q.	35
iii.	If the centroid of the ΔABC is (a, b), then the sum $(a+b)$ is	r.	25
iv.	If the circumcentre of the ΔABC is (l, m), then $(l+m)$ equals	s.	14

Codes

	i	ii	iii	iv
a.	r	p	q	s
b.	p	r	s	q
c.	s	p	r	q
d.	q	s	p	r

Comprehension Type

Direction *(Q. Nos. 31-54) This section contains 8 paragraphs, each describing theory, experiments, data, etc. Twenty four questions related to the paragraphs have been given. Each question has only one correct answer among the four given options (a), (b), (c) and (d).*

Passage I

The base of an isosceles triangle is equal to 4, the base angle is equal to 45°. A straight line cuts the extension of the base at a point M at the angle θ and bisects the lateral side of the triangle which is nearest to M.

31 The area A of the quadrilateral which the straight line cuts off from the given triangle is

a. $\frac{3+\tan\theta}{1+\tan\theta}$ b. $\frac{3+2\tan\theta}{1+\tan\theta}$

c. $\frac{3+\tan\theta}{1-\tan\theta}$ d. $\frac{3+5\tan\theta}{1+\tan\theta}$

32 The range of values of A for different values of θ, lie in the interval

a. $\left(\frac{5}{2},\frac{7}{2}\right)$ b. $(4,5)$

c. $\left(4,\frac{9}{2}\right)$ d. $(3, 4)$

33 The length of portion of straight line inside the triangle may lie in the range

a. $(2, 4)$ b. $\left(\frac{3}{2},\sqrt{3}\right)$

c. $(\sqrt{2}, 2)$ d. $(\sqrt{2}, \sqrt{3})$

Passage II

Consider a family of lines $(4a+3)x-(a+1)y-(2a+1)=0$, where $a \in R$.

34 The locus of the foot of the perpendicular from the origin on each member of this family, is

a. $(2x-1)^2+4(y+1)^2=5$ b. $(2x-1)^2+(y+1)^2=5$

c. $(2x+1)^2+4(y-1)^2=5$ d. $(2x-1)^2+4(y-1)^2=5$

35 A member of this family with positive gradient making an angle of $\frac{\pi}{4}$ with the line $3x-4y=2$, is

a. $7x-y-5=0$ b. $4x-3y+2=0$

c. $x+7y=15$ d. $5x-3y-4=0$

36 Minimum area of the triangle which a member of this family with negative gradient can make with the positive semi-axes, is

a. 8 b. 6 c. 4 d. 2

Passage III

ΔABC is a right angled triangle at A with vertices A, B, C in the anti-clockwise sense in that order. $A(1,2)$, $B(-3,1)$ and vertex C lies on the X-axis. $BCEF$ is a square with vertices B, C, E, F in the clockwise sense in that order. ΔACD is an equilateral triangle with vertices A, C, D in the anti-clockwise sense in that order.

37 Slope of AF, is

a. $\frac{7}{10}$ b. $\frac{7}{9}$

c. $\frac{9}{10}$ d. $\frac{11}{10}$

38 The abscissa of centroid of ΔBCE, is

a. -1 b. $-\frac{1}{2}$

c. $-\frac{1}{3}$ d. $-\frac{2}{3}$

39 If $D(\alpha, \beta)$, then $(4\beta-4)^2$ is

a. 2 b. 3 c. 4 d. 5

Working Space

Passage IV

A variable line L is drawn through $O(0, 0)$ to meet the line $L_1 : y - x + 10 = 0$, $L_2 : y - x - 20 = 0$ at the points A and B, respectively. A point P is taken on L,

40 Locus of P, if $\frac{2}{OP} = \frac{1}{OA} + \frac{1}{OB}$, is

a. $3x + 3y = 40$ b. $3x + 3y + 40 = 0$
c. $3x - 3y = 40$ d. $3x - 3y + 40 = 0$

41 Locus of P, if $OP^2 = OA \cdot OB$, is

a. $(y - x)^2 = 100$ b. $(y - x)^2 = 50$ c. $(y - x)^2 = 200$ d. $(x - y)^2 = 20$

42 Locus of P if $OP = OA + OB$, is

a. $y - x = 10$ b. $y + x - 20 = 0$
c. $y - x = 30$ d. $x - y - 30 = 0$

Passage V

For points $P(x_1, y_1)$ and $Q(x_2, y_2)$ of the coordinate plane, a new distance $d(P, Q)$ is defined by $d(P, Q) = |x_1 - x_2| + |y_1 - y_2|$. Let $O(0, 0)$, $A(1, 2)$, $B(2, 3)$ and $C(4, 3)$ are four points on XY-plane.

43 Let $R(x, y)$, such that R is equidistant from the points O and A with respect to new distance and if $0 \le x < 1$, $0 \le y < 2$, then R lies on a line segment whose equation is

a. $x + y = 3$ b. $x + 2y = 3$ c. $2x + y = 3$ d. $2x + 2y = 3$

44 Let $S(x, y)$, such that S is equidistant from points O and B with respect to new distance and if $x \ge 2$ and $O \le y < 3$, then locus of S is

a. a line segment of finite length b. a line of the infinite length
c. a ray of finite length d. a ray of infinite length

45 Distance between the circumcentre and orthocentre of ΔABC, is

a. 5 units b. $\sqrt{29}$ units c. 7 units d. $\sqrt{37}$ units

Passage VI

Reflection of a rational point with respect to the line having its equation with rational coefficients given a rational point. Reflection of a rational point under opposite scenario gives irrational points.

46 Number of rational points on a circle with centre $(\sqrt{2}, \sqrt{3})$ and with any radius is

a. 1 b. 2 c. variable d. ∞

47 Number of rational points on a circle with centre $(\sqrt{2}, -\sqrt{2})$ and passing through $(1, A)$ are

a. 1 b. 2 c. variable d. ∞

48 Number of rational points on a circle with centre $(\sqrt{2}, -\sqrt{2})$ and passing through $(1, -1)$ are

a. 1 b. 2 c. variable d. ∞

Passage VII

Let $A(1/2, 0)$, $B(3/2, 0)$, $C(5/2, 0)$ be the given points and P be a point satisfying $\max\{PA + PB, PB + PC\} < 2$.

49 P lies inside

a. a circle and an ellipse b. a circle and a hyperbola
c. a parabola and an ellipse d. None of these

50 The locus of P is symmetric about

a. origin b. the line $y = x$
c. the line $2x - 3 = 0$ d. None of these

51 The area of the region of the point P, is

a. $\sqrt{2}\left(\frac{\pi}{3} - \frac{\sqrt{3}}{4}\right)$ b. $\sqrt{3}\left(\frac{\pi}{3} - \frac{\sqrt{3}}{4}\right)$ c. $2\left(\frac{\pi}{3} - \frac{\sqrt{3}}{4}\right)$ d. None of these

Working Space

Passage VIII

Diagonal AC of rhombus $ABCD$ is a member of both the family of lines $L_1 + \lambda L_2 = 0$ and $L_3 + \mu L_4 = 0$ and vertex B of rhombus is (3, 2). Suppose,

$$L_1 : x + y - 1 = 0$$
$$L_2 : 2x + 3y - 2 = 0$$
$$L_3 : x - y + 2 = 0$$
$$L_4 : 2x - 3y + 5 = 0$$

52 The equation of diagonal AC, is

a. $2x + y + 1 = 0$
b. $x + 2y + 3 = 0$
c. $x + 2y - 1 = 0$
d. $2x + y - 7 = 0$

53 The equation of diagonal BD, is

a. $2x + y + 1 = 0$
b. $2x - y - 4 = 0$
c. $x - 2y + 7 = 0$
d. $x + 2y - 1 = 0$

54 If area of rhombus $ABCD$ is $12\sqrt{5}$ sq units, then coordinates of centroid of ΔACD is

a. $\left(\frac{6}{5}, -\frac{7}{5}\right)$
b. $\left(-\frac{7}{4}, \frac{6}{5}\right)$
c. $\left(\frac{7}{5}, -\frac{6}{5}\right)$
d. $\left(-\frac{6}{5}, \frac{7}{5}\right)$

One Integer Value Correct Type

Direction *(Q. No. 55-60) This section contains 6 questions. When worked out will result in an integer from 0 to 9 (both inclusive).*

55 Consider the system of equations $ax + y = b$, $bx + y = a$, $ax + by = ab$, where $a, b \in \{0, 1, 2, 3, 4\}$, then the number of ordered pairs (a, b) for which system is consistent.

56 Let slope of sides AB and BC of a ΔABC are roots of the equation $x^2 - 4x + 1 = 0$, (ΔABC lying in the first quadrant and B is origin) and in radius of triangle is 1 unit. If coordinates of its incentre are (a, b). Then, find $(a^2 + b^4)$.

57 The Δ ABC, right angled at C has median AD, BE and CF. AD lies along the line $y = x + 3$. BE lies along the line $y = 2x + 4$. If the length of the hypotenuse is 60, the area of ΔABC is abc sq units, then $(a + b + c)$ is

58 Let $x \tan \alpha + y \sin \alpha = \alpha$, $\alpha x \operatorname{cosec} \alpha + y \cos \alpha = 1$. Let P be the point of intersection of the lines in the limiting position when $\alpha \to 0$, if the point P is (h, k), then $|h + k|$ is

59 A ΔABC having vertices of $A(5, 1)$, $B(-1, -7)$ and $C(1, 4)$, respectively. L be the line mirror passing through C and parallel to AB and a light ray emanating from point A goes along the direction of internal bisector of the $\angle A$, which meets the mirror at BC at E and D respectively. If sum of the areas of ΔACE and ΔABC is $(10 + a)$ sq unit. Then, a is equal to

60 If the line $y = \sqrt{3}x$ cuts the curve $x^4 + ax^2y + bxy + cx + dy + 6 = 0$ at A, B, C and D, then the value of $\frac{1}{12} OA \cdot OB \cdot OC \cdot OD$ (where, O is the origin) is equal to

JEE Main & AIEEE Archive

(Compilation of Last 13 Years Questions)

1 If PS is the median of the triangle with vertices $P(2,2)$, $Q(6,-1)$ and $R(7,3)$, then equation of the line passing through $(1,-1)$ and parallel to PS is **(JEE Main 2014)**

a. $4x-7y-11=0$
b. $2x+9y+7=0$
c. $4x+7y+3=0$
d. $2x-9y-11=0$

2 Let a, b, c and d be non-zero numbers. If the point of intersection of the lines $4ax+2ay+c=0$ and $5bx+2by+d=0$ lies in the fourth quadrant and is equidistant from the two axes, then **(JEE Main 2014)**

a. $2bc-3ad=0$ b. $2bc+3ad=0$
c. $2ad-3bc=0$ d. $3bc+2ad=0$

3 A ray of light along $x+\sqrt{3}y=\sqrt{3}$ gets reflected upon reaching X-axis, the equation of the reflected ray is **(JEE Main 2013)**

a. $y=x+\sqrt{3}$ b. $\sqrt{3}y=x-\sqrt{3}$
c. $y=\sqrt{3}x-\sqrt{3}$ d. $\sqrt{3}y=x-1$

4 A line is drawn through the point $(1,2)$ to meet the coordinate axes at P and Q such that it forms a ΔOPQ, where O is the origin, if the area of the ΔOPQ is least, then the slope of the line PQ is **(AIEEE 2012)**

a. $-\frac{1}{4}$ b. -4
c. -2 d. $-\frac{1}{2}$

5 The lines $L_1: y-x=0$ and $L_2: 2x+y=0$ intersect the line $L_3: y+2=0$ at P and Q, respectively. The bisector of the acute angle between L_1 and L_2 intersects L_3 at R.

Statement I The ratio $PR:RQ$ equals $2\sqrt{2}:\sqrt{5}$.

Statement II In any triangle, bisector of an angle divides the triangle into two similar triangles. **(AIEEE 2011)**

a. Both Statement I and Statement II are correct and Statement II is the correct explanation of Statement I
b. Both Statement I and Statement II are correct and Statement II is not the correct explanation of Statement I
c. Statement I is correct but Statement II is incorrect
d. Statement II is correct but Statement I is incorrect

6 The curve that passes through the point (2, 3) and has the property that the segment of any tangent to it lying between the coordinate axes is bisected by the point of contact, is given by **(AIEEE 2011)**

a. $\left(\frac{x}{2}\right)^2+\left(\frac{y}{3}\right)^2=2$ b. $2y-3x=0$
c. $y=\frac{6}{x}$ d. $x^2+y^2=13$

7 The lines $x+y=|a|$ and $ax-y=1$ intersect each other in the first quadrant. Then, the set of all possible values of a in the interval **(AIEEE 2011)**

a. $(-1, 1]$ b. $(0, \infty)$
c. $[1, \infty)$ d. $(-1, \infty)$

8 The line L given by $\frac{x}{5}+\frac{y}{b}=1$ passes through the point (13, 32). The line K is parallel to L and has the equation $\frac{x}{c}+\frac{y}{3}=1$. Then, the distance between L and K is **(AIEEE 2010)**

a. $\frac{23}{\sqrt{15}}$ b. $\sqrt{17}$
c. $\frac{17}{\sqrt{15}}$ d. $\frac{23}{\sqrt{17}}$

9 The lines $p(p^2+1)x-y+q=0$ and $(p^2+1)^2x+(p^2+1)y+2q=0$ are perpendicular to a common line for **(AIEEE 2009)**

a. exactly one value of p
b. exactly two values of p
c. more than two values of p
d. no value of p

10 The perpendicular bisector of the line segment joining $P(1, 4)$ and $Q(k, 3)$ has y intercept -4. Then, a possible value of k is **(AIEEE 2008)**

a. -4 b. 1
c. 2 d. -2

11 Let $P=(-1,0)$, $Q=(0,0)$ and $R=(3,3\sqrt{3})$ be three points. The equation of the bisector of the $\angle PQR$ is **(AIEEE 2007)**

a. $\sqrt{3}x+y=0$ b. $x+\frac{\sqrt{3}}{2}y=0$
c. $\frac{\sqrt{3}}{2}x+y=0$ d. $x+\sqrt{3}y=0$

12 A straight line through the point $A(3, 4)$ is such that its intercept between the axes is bisected at A. Its equation is **(AIEEE 2006)**

a. $3x-4y+7=0$
b. $4x+3y=24$
c. $3x+4y=25$
d. $x+y=7$

13 If (a, a^2) falls inside the angle made by the lines $y=\frac{x}{2}$, $x>0$ and $y=3x$, $x>0$, then a belongs to **(AIEEE 2006)**

a. $(3, \infty)$ b. $\left(\frac{1}{2}, 3\right)$
c. $\left(-3, -\frac{1}{2}\right)$ d. $\left(0, \frac{1}{2}\right)$

14 The line parallel to the X-axis and passing through the intersection of the lines $ax + 2by + 3b = 0$ and $bx - 2ay - 3a = 0$, where $(a, b) \neq (0, 0)$ is **(AIEEE 2005)**

a. above the X-axis at a distance of (2/3) from it
b. above the X-axis at a distance of (3/2) from it
c. below the X-axis at a distance of (2/3) from it
d. below the X-axis at a distance of (3/2) from it

15 If non-zero numbers a, b and c are in HP, then the straight line $\frac{x}{a} + \frac{y}{b} + \frac{1}{c} = 0$ always passes through a fixed point. That point is **(AIEEE 2005)**

a. $\left(1, -\frac{1}{2}\right)$ b. $(1, -2)$ c. $(-1, -2)$ d. $(-1, 2)$

16 Let $A(2, -3)$ and $B(-2, 1)$ be the vertices of a ΔABC. If the centroid of this triangle moves on the line $2x + 3y = 1$, then the locus of the vertex C is the line **(AIEEE 2004)**

a. $2x + 3y = 9$ b. $2x - 3y = 7$
c. $3x + 2y = 5$ d. $3x - 2y = 3$

17 The equation of the straight line passing through the point (4, 3) and making intercepts on the coordinate axes whose sum is -1, is **(AIEEE 2004)**

a. $\frac{x}{2} + \frac{y}{3} = -1$ and $\frac{x}{-2} + \frac{y}{1} = -1$
b. $\frac{x}{2} - \frac{y}{3} = -1$ and $\frac{x}{-2} + \frac{y}{1} = -1$
c. $\frac{x}{2} + \frac{y}{3} = 1$ and $\frac{x}{-2} + \frac{y}{1} = 1$
d. $\frac{x}{2} - \frac{y}{3} = 1$ and $\frac{x}{-2} + \frac{y}{1} = 1$

18 A square of side a lies above the X-axis and has one vertex at the origin. The side passing through the origin makes an angle $\alpha \left(0 < \alpha < \frac{\pi}{4}\right)$ with the positive direction of X-axis. The equation of its diagonal not passing through the origin is **(AIEEE 2003)**

a. $y(\cos\alpha - \sin\alpha) - x(\sin\alpha - \cos\alpha) = a$
b. $y(\cos\alpha + \sin\alpha) + x(\sin\alpha - \cos\alpha) = a$
c. $y(\cos\alpha + \sin\alpha) + x(\sin\alpha + \cos\alpha) = a$
d. $y(\cos\alpha + \sin\alpha) + x(\cos\alpha - \sin\alpha) = a$

19 Three straight lines

$$2x + 11y - 5 = 0,$$
$$24x + 7y - 20 = 0$$
and $$4x - 3y - 2 = 0$$ **(AIEEE 2002)**

a. form a triangle
b. are only concurrent
c. are concurrent with one line bisecting the angle between the other two
d. None of the above

20 A straight line through the point (2, 2) intersects the lines $\sqrt{3}x + y = 0$ and $\sqrt{3}x - y = 0$ at the points A and B. The equation to the line AB so that the ΔOAB is equilateral, is **(AIEEE 2002)**

a. $x - 2 = 0$
b. $y - 2 = 0$
c. $x + y - 4 = 0$
d. None of the above

JEE Advanced & IIT JEE Archive

(Compilation of Last 10 Years Questions)

1 For $a > b > c > 0$, the distance between (1, 1) and the point of intersection of the lines $ax + by + c = 0$ and $bx + ay + c = 0$ is less than $2\sqrt{2}$. Then,

(2014 Adv., Only One Option Correct Type)

a. $a + b - c > 0$
b. $a - b + c < 0$
c. $a - b + c > 0$
d. $a + b - c < 0$

2 A straight line L through the point (3, −2) is inclined at an angle 60° to the line $\sqrt{3}x + y = 1$. If L also intersects the X-axis, then the equation of L is **(2011, Only One Option Correct Type)**

a. $y + \sqrt{3}x + 2 - 3\sqrt{3} = 0$
b. $y - \sqrt{3}x + 2 + 3\sqrt{3} = 0$
c. $\sqrt{3}y - x + 3 + 2\sqrt{3} = 0$
d. $\sqrt{3}y + x - 3 + 2\sqrt{3} = 0$

3 The locus of the orthocentre of the triangle formed by the lines

$$(1 + p)x - py + p(1 + p) = 0$$
$$(1 + q)x - qy + q(1 + q) = 0$$

and $$y = 0$$

where $p \neq q$, is **(2009, Only One Option Correct Type)**

a. a hyperbola
b. a parabola
c. an ellipse
d. a straight line

4 Consider the lines given by

$$L_1 : x + 3y - 5 = 0$$
$$L_2 : 3x - ky - 1 = 0$$

and, $$L_3 : 5x + 2y - 12 = 0$$

Match the Column I with Column II and mark the correct option from the codes given below.

	Column I		Column II
i.	L_1, L_2, L_3 are concurrent, if	p.	$k = -9$
ii.	One of L_1, L_2, L_3 is parallel to at least one of the other two, if	q.	$k = -\frac{6}{5}$
iii.	L_1, L_2, L_3 form a triangle, if	r.	$k = \frac{5}{6}$
iv.	L_1, L_2, L_3 do not form a triangle, if	s.	$k = 5$

(2008, Matching Type)

Codes

	i	ii	iii	iv
a.	s	p,q	r	p,q,s
b.	p	q	r,s	p,q
c.	s	q	p	r
d.	p,q	r,s	p	p,q

5 Lines $L_1 : y - x = 0$ and $L_2 : 2x + y = 0$ intersect the line $L_3 : y + 2 = 0$ at P and Q, respectively. The bisector of the acute angle between L_1 and L_2 intersects L_3 at R.

Statement I The ratio $PR : RQ$ equals $2\sqrt{2} : \sqrt{5}$.

Statement II In any triangle, bisector of an angle divides the triangle into two similar triangles. **(2007, Statement Type)**

a. Both Statement I and Statement II are correct and Statement II is the correct explanation of Statement I
b. Both Statement I and Statement II are correct and Statement II is not the correct explanation of Statement I
c. Statement I is correct but Statement II is incorrect
d. Statement II is correct but Statement I is incorrect

3. Pair of Straight Line

DPP

DAILY PRACTICE PROBLEMS

DPP-1 Homogeneous Pair of Straight Lines

Only One Option Correct Type

Working Space

Direction *(Q. Nos. 1-44) This section contains 44 multiple choice questions. Each question has four choices (a), (b), (c) and (d), out of which ONLY ONE option is correct.*

1 If $3x + 4y = 0$ is one of the lines represented by $6x^2 - xy + 4cy^2 = 0$, then c is equal to **[AIEEE 2004]**

a. 3
b. -1
c. 1
d. -3

2 If the sum of the slopes of the lines represented by $x^2 - 2cxy - 7y^2 = 0$ is four times to their product, then c is equal to

a. 2
b. -1
c. 1
d. -2

3 If slope of lines represented by $ax^2 + 2hxy + by^2 = 0$ are in the ratio 1 : 3, then $h^2 : ab$ equals

a. $\frac{3}{4}$
b. $\frac{1}{4}$
c. $\frac{4}{3}$
d. None of these

4 If one of the pair of lines $ax^2 + 2hxy + by^2 = 0$ bisects the angle between coordinates axes in positive quadrant, then

a. $a - b = 2|h|$
b. $a + b = -2h$
c. $a + b = 2|h|$
d. $(a - b)^2 = 4h^2$

5 If one of the lines of $my^2 + (1 - m^2)xy - mx^2 = 0$ is a bisector of the angle between the lines $xy = 0$, then m is equal to

a. 1
b. 2
c. $-\frac{1}{2}$
d. -2

6 The combined equation of the bisectors of the angle between the lines represented by $(x^2 + y^2)\sqrt{3} = 4xy$, is

a. $y^2 - x^2 = 0$
b. $xy = 0$
c. $x^2 + y^2 = 2xy$
d. None of these

7 If the acute angles between the pairs of lines $3x^2 - 7xy + 4y^2 = 0$ and $6x^2 - 5xy + y^2 = 0$ are θ_1 and θ_2 respectively, then

a. $\theta_1 = \theta_2$
b. $2\theta_1 = \theta_2$
c. $\theta_1 = 2\theta_2$
d. None of the above

Working Space

8 Equation of the pair of straight lines through origin and perpendicular to the pair of straight lines $5x^2 - 7xy - 3y^2 = 0$, is

a. $3x^2 - 7xy - 5y^2 = 0$
b. $3x^2 + 7xy + 5y^2 = 0$
c. $3x^2 - 7xy + 5y^2 = 0$
d. $3x^2 + 7xy - 5y^2 = 0$

9 If $x^2 - 2mxy - y^2 = 0$ and $x^2 - 2nxy - y^2 = 0$ represent pair of straight lines such that each pair represents bisector between the lines of the other pair, then

a. $mn + 1 = 0$
b. $mn - 1 = 0$
c. $m + n = 0$
d. $m - n = 0$

10 The equation of the lines represented by the equation $ab(x^2 - y^2) + (a^2 - b^2)xy = 0$, are

a. $ax - by = 0, bx + ay = 0$
b. $ax - by = 0, bx - ay = 0$
c. $ax + by = 0, bx + ay = 0$
d. $ax + by = 0, bx - ay = 0$

11 The equation $(x - 5)^2 + (x - 5)(y - 6) - 2(y - 6)^2 = 0$ represents

a. a circle
b. two straight lines passing through origin
c. two straight lines passing through the point (5, 6)
d. None of the above

12 The equation of lines passing through the origin and parallel to the lines $y = m_1x + c_1$ and $y = m_2x + c_2$, is

a. $m_1m_2x^2 - (m_1 + m_2)xy + y^2 = 0$
b. $m_1m_2x^2 + (m_1 + m_2)xy + y^2 = 0$
c. $m_1m_2y^2 - (m_1 + m_2)xy + x^2 = 0$
d. $m_1m_2y^2 + (m_1 + m_2)xy + x^2 = 0$

13 The lines represented by the equation $ax^2(b - c) - xy(ab - bc) + cy^2(a - b) = 0$, are

a. $a(b - c)x - c(a - b)y = 0, x + y = 0$
b. $x + y = 0, x - y = 0$
c. $a(b - c)x - c(a - b)y = 0, x - y = 0$
d. None of the above

14 The nature of straight lines represented by equation $4x^2 + 12xy + 9y^2 = 0$, is

a. real and coincident
b. real and different
c. imginary and different
d. None of the above

15 The lines $a^2x^2 + bcy^2 = a(b + c)xy$ will be coincident, if

a. $a = 0$ or $b = c$
b. $a = b$ or $a = c$
c. $c = 0$ or $a = b$
d. $a = b + c$

16 If the equation $2x^2 - 2hxy + 2y^2 = 0$ represents two coincident straight lines passing through the origin, then h is equal to

a. ± 6
b. $\sqrt{6}$
c. $-\sqrt{6}$
d. ± 2

17 If the slope of one of the lines represented by the equation $ax^2 + 2hxy + by^2 = 0$ is λ times that of the other, then

a. $4\lambda h = ab(1 + \lambda)$
b. $\lambda h = ab(1 + \lambda)^2$
c. $4\lambda h^2 = ab(1 + \lambda)^2$
d. None of these

Working Space

18 If one of the lines represented by the equation $ax^2 + 2hxy + by^2 = 0$ is coincident with one of the lines represented by $a'x^2 + 2h'xy + b'y^2 = 0$, then

a. $(ab' - a'b)^2 = 4(ah' - a'h)(hb' - h'b)$
b. $(ab' + a'b)^2 = 4(ah' - a'h)(hb' - h'b)$
c. $(ab' - a'b)^2 = (ah' - a'h)(hb' - h'b)$
d. None of the above

19 The equation of the lines represented by the equation $ax^2 + (a + b)xy + by^2 + x + y = 0$, are

a. $ax + by + 1 = 0,\ x + y = 0$
b. $ax + by - 1 = 0,\ x + y = 0$
c. $ax + by + 1 = 0,\ x - y = 0$
d. None of the above

20 The equation $y^2 - x^2 + 2x - 1 = 0$ represents

a. a pair of straight lines
b. a circle
c. a parabola
d. an ellipse

21 The equation of one of the lines represented by the equation $x^2 - 2xy \cot\theta - y^2 = 0$, is

a. $x - y\cot\theta = 0$
b. $x + y\tan\theta = 0$
c. $y\sin\theta + x(\cos\theta + 1) = 0$
d. $x\cos\theta + y(\sin\theta + 1) = 0$

22 The pair of straight lines passing through the point (1, 2) and perpendicular to the pair of straight lines $3x^2 - 8xy + 5y^2 = 0$, is

a. $(5x + 3y + 11)(x + y + 3) = 0$
b. $(5x + 3y - 11)(x + y - 3) = 0$
c. $(3x + 5y - 11)(x + y + 3) = 0$
d. $(3x - 5y + 11)(x + y - 3) = 0$

23 If in the general quadratic equation $f(x, y) = 0$, $\Delta = 0$ and $h^2 = ab$, then the equation represents

a. two parallel straight lines
b. two perpendicular straight lines
c. two coincident lines
d. None of the above

24 The gradient of one of the lines $x^2 + 2hxy + 2y^2 = 0$ is twice that of the other, then h is equal to

a. ± 3
b. $\pm 3/2$
c. ± 2
d. ± 1

25 If the sum of slopes of the pair of lines represented by $4x^2 + 2hxy - 7y^2 = 0$ is equal to the product of the slopes, then value of h is

a. -6
b. -2
c. -4
d. 4

26 The area of triangle (in sq units) formed by the lines $x^2 - 4y^2 = 0$ and $x = a$, is

a. $2a^2$
b. $\frac{a^2}{2}$
c. $\frac{\sqrt{3}a^2}{2}$
d. $\frac{2a^2}{\sqrt{3}}$

Working Space

27 If the lines represented by the equation $6x^2 + 41xy - 7y^2 = 0$ make angles α and β with X-axis, then $\tan\alpha \cdot \tan\beta$ is equal to

a. $-\frac{6}{7}$ b. $\frac{6}{7}$

c. $\frac{7}{6}$ d. $-\frac{7}{6}$

28 Difference of slopes of the lines represented by equation $x^2(\sec^2\theta - \sin^2\theta) - 2xy\tan\theta + y^2\sin^2\theta = 0$, is

a. 4 b. 3

c. 2 d. None of these

29 If two sides of a triangle are represented by $x^2 - 7xy + 6y^2 = 0$ and centroid is (1, 0), then the equation of third side is

a. $2x + 7y + 3 = 0$ b. $2x - 7y + 3 = 0$

c. $2x + 7y - 3 = 0$ d. $2x - 7y - 3 = 0$

30 If the lines represented by the equation $ax^2 - bxy - y^2 = 0$ make angles α and β with the X-axis, then $\tan(\alpha + \beta)$ is equal to

a. $\frac{b}{(1+a)}$ b. $\frac{-b}{(1+a)}$

c. $\frac{a}{(1+b)}$ d. None of these

31 Angle between the lines represented by the equation $x^2 + 2xy\sec\theta + y^2 = 0$, is

a. θ b. 2θ

c. $\frac{\theta}{2}$ d. None of these

32 If the angle between the lines represented by the equation $y^2 + kxy - x^2\tan^2 A = 0$ is $2A$, then k is equal to

a. 0 b. 1

c. 2 d. $\tan A$

33 The lines $(lx + my)^2 - 3(mx - ly)^2 = 0$ and $lx + my + n = 0$ form

a. an isosceles triangle b. a right angled triangle

c. an equilateral triangle d. None of these

34 If $(a + 3b)(3a + b) = 4h^2$, then the angle between the lines represented by $ax^2 + 2hxy + by^2 = 0$, is

a. 30° b. 45°

c. 60° d. $\tan^{-1}(1/2)$

35 If the lines represented by the equation $2x^2 - 3xy + y^2 = 0$ make angles α and β with X-axis, then $\cot^2\alpha + \cot^2\beta$ is equal to

a. 0 b. $\frac{3}{2}$

c. $\frac{7}{4}$ d. $\frac{5}{4}$

36 If the lines $(p - q)x^2 + 2(p + q)xy + (q - p)y^2 = 0$ are mutually perpendicular, then

a. $p = q$ b. $p = 0$

c. $q = 0$ d. p and q may have any value

Working Space

37 If two of the three lines represented by the equation $ax^3 + bx^2y + cxy^2 + dy^3 = 0$ are perpendicular, then

a. $a^2 + d^2 = 2ac$
b. $a^2 + d^2 = 2bd$
c. $a^2 + ac + bd + d^2 = 0$
d. $a^2 + d^2 = 4bc$

38 The line $x - 2y = 0$ will be a bisector of the angle between the lines represented by the equation $x^2 - 2hxy - 2y^2 = 0$, if h is equal to

a. $\frac{1}{2}$
b. 2
c. -2
d. $-\frac{1}{2}$

39 The equation of the bisectors of angle between the lines represented by the equation $(y - mx)^2 = (x + my)^2$, is

a. $mx^2 + (m^2 - 1)xy - my^2 = 0$
b. $mx^2 - (m^2 - 1)xy - my^2 = 0$
c. $mx^2 + (m^2 - 1)xy + my^2 = 0$
d. None of the above

40 If $y = mx$ is one of the bisectors of the angles between the lines $ax^2 - 2hxy + by^2 = 0$, then

a. $h(1+m^2) + m(a-b) = 0$
b. $h(1-m^2) + m(a+b) = 0$
c. $h(1-m^2) + m(a-b) = 0$
d. $h(1+m^2) + m(a+b) = 0$

41 If the bisectors of angles represented by $ax^2 + 2hxy + by^2 = 0$ and $a'x^2 + 2h'xy + b'y^2 = 0$ are same, then

a. $(a-b)h' = (a'-b')h$
b. $(a-b)h = (a'-b')h'$
c. $(a+b)h' = (a'-b')h$
d. $(a-b)h' = (a'+b')h$

42 If $r(1-m^2) + m(p-q) = 0$, then a bisector of the angle between the lines represented by the equation $px^2 - 2rxy + qy^2 = 0$, is

a. $y = x$
b. $y = -x$
c. $y = mx$
d. $ym = x$

43 One bisector of the angle between the lines given by $a(x-1)^2 + 2h(x-1)y + by^2 = 0$ is $2x + y - 2 = 0$, then other bisector is

a. $x - 2y + 1 = 0$
b. $2x + y - 1 = 0$
c. $x + 2y - 1 = 0$
d. $x - 2y - 1 = 0$

44 The figure formed by the lines $x^2 + 4xy + y^2 = 0$ and $x - y = 4$, is

a. a right angled triangle
b. an isosceles triangle
c. an equilateral triangle
d. None of the above

Comprehension Type

Working Space

Direction *(Q. Nos. 45-47) This section contains a paragraph, describing theory, experiments, data, etc. Three questions related to the paragraphs have been given. Each question has only one correct answer among the four given options (a), (b), (c) and (d).*

Passage

The homogeneous equation of second degree $ax^2 + 2hxy + by^2 = 0$ represents a pair of straight lines through the origin. These lines are distinct, if $h^2 > ab$ and coincident, if $h^2 = ab$ and the lines do not exist, if $h^2 < ab$.

45 The difference of the tangent of the angles which the lines $x^2(\sec^2\theta - \sin^2\theta) - 2xy\tan\theta + y^2\sin\theta = 0$ makes with X-axis, is

a. -1
b. 0
c. 1
d. 2

46 If the pair of straight lines $x^2 - 2hxy - y^2 = 0$ and $x^2 - 2gxy - y^2 = 0$ are such that each pair bisects the angle between the other pair, then the value of $(1 - gh)$, is

a. 0
b. 2
c. -2
d. -1

47 If two of the three lines given by $ax^3 + bx^2y + cxy^2 + dy^3 = 0$ (where, $a \neq 0$) make complementary angles with X-axis, then

a. $a^2 + ac + bd + d^2 = 0$
b. $a^2 - ac - bd + d^2 = 0$
c. $a^2 - ac + bd - d^2 = 0$
d. $a^2 + ac - bd - d^2 = 0$

One Integer Value Correct Type

Direction *(Q. Nos. 48-50) This section contains 3 questions. When worked out will result in an integer from 0 to 9 (both inclusive).*

48 The equation of the pair of bisectors of the angles between two straight lines, is $12x^2 - 7xy - 12y^2 = 0$. If the equation of one line is $2y - x = 0$ and the equation of the other line is $mx + ny = 0$, then $(m + n)$ is equal to

49 The greatest slope along the graph represented by the equation $4x^2 - y^2 + 2y - 1 = 0$, is

50 The area bounded by the angle bisectors of the lines $x^2 - y^2 + 2y = 1$ and $x + y = 3$ is

DPP-2 General Equation of Second Degree

Only One Option Correct Type

Working Space

Direction (*Q. Nos.* 1-20) *This section contains 20 multiple choice questions. Each question has four choices (a), (b), (c) and (d), out of which ONLY ONE option is correct.*

1 If $6x^2 + 11xy - 10y^2 + x + 31y + k = 0$ represents a pair of straight lines, then k is equal to

a. -15
b. 6
c. -10
d. -4

2 The equation of the perpendiculars drawn from the origin to the lines represented by the equation $2x^2 - 10xy + 12y^2 + 5x - 16y - 3 = 0$, is

a. $6x^2 + 5xy + y^2 = 0$
b. $6y^2 + 5xy + x^2 = 0$
c. $6x^2 - 5xy + y^2 = 0$
d. None of the above

3 The pair of lines passing through the origin and parallel to the lines represented by the equation $2x^2 - xy - 6y^2 + 7x + 21y - 15 = 0$, is

a. $2x^2 - xy - 6y^2 = 0$
b. $6x^2 - xy + 2y^2 = 0$
c. $6x^2 - xy - 2y^2 = 0$
d. $2x^2 + xy - 6y^2 = 0$

4 The equation $2x^2 + 4xy - py^2 + 4x + qy + 1 = 0$ will represent two mutually perpendicular straight lines, if

a. $p = 1$ and $q = 2$ or 6
b. $p = 2$ and $q = 0$ or 6
c. $p = 2$ and $q = 0$ or 8
d. $p = -2$ and $q = -2$ or 8

5 The equation $xy + a^2 = a(x + y)$ represents

a. a parabola
b. a pair of straight lines
c. an ellipse
d. two parallel straight lines

6 The equation $4x^2 + 12xy + 9y^2 + 2gx + 2fy + c = 0$ will represent two real parallel lines, if

a. $g = 4, f = 9, c = 0$
b. $g = 2, f = 3, c = 1$
c. $g = 2, f = 3, c \in R$
d. $g = 4, f = 9, c > 1$

7 If the equation $ax^2 + by^2 + cx + cy = 0$ represents a pair of straight lines, then

a. $a(b + c) = 0$
b. $b(c + a) = 0$
c. $c(a + b) = 0$
d. $a + b + c = 0$

Working Space

8 The equations of the lines represented by the equation $ax^2 + (a+b)xy + by^2 + x + y = 0$ are

a. $ax + by + 1 = 0, x + y = 0$
b. $ax + by - 1 = 0, x + y = 0$
c. $ax + by - 1 = 0, x - y = 0$
d. None of the above

9 If the equation $\lambda x^2 + 2y^2 - 5xy + 5x - 7y + 3 = 0$ represents two straight lines, then the value of λ will be

a. 3
b. 2
c. 8
d. -8

10 The lines represented by the equation $9x^2 + 24xy + 16y^2 + 21x + 28y + 6 = 0$, are

a. parallel
b. coincident
c. perpendicular
d. None of these

11 If the angle between the two lines represented by $2x^2 + 5xy + 3y^2 + 6x + 7y + 4 = 0$, is $\tan^{-1} m$, then m is equal to

a. $\frac{1}{5}$
b. 1
c. $\frac{7}{5}$
d. 7

12 The acute angle formed between the lines joining the origin to the points of intersection of the curves $x^2 + y^2 - 2x - 1 = 0$ and $x + y = 1$, is

a. $\tan^{-1}\left(-\frac{1}{2}\right)$
b. $\tan^{-1}(2)$
c. $\tan^{-1}\left(\frac{1}{2}\right)$
d. $60°$

13 The point of intersection of the lines represented by the equation $2x^2 + 3y^2 + 7xy + 8x + 14y + 8 = 0$, is

a. (0, 2)
b. (1, 2)
c. (– 2, 0)
d. (– 2, 1)

14 Distance between the lines represented by the equation $x^2 + 2\sqrt{3}\,xy + 3y^2 - 3x - 3\sqrt{3}y - 4 = 0$, is

a. $\frac{5}{2}$
b. $\frac{5}{4}$
c. 5
d. 0

15 The equation of second degree $x^2 + 2\sqrt{2}\,xy + 2y^2 + 4x + 4\sqrt{2}y + 1 = 0$ represented a pair of straight lines, the distance between them is

a. 4
b. $\frac{4}{3}$
c. 2
d. $2\sqrt{3}$

16 The distance between the parallel lines $9x^2 - 6xy + y^2 + 18x - 6y + 8 = 0$, is

a. $\frac{1}{\sqrt{10}}$
b. $\frac{2}{\sqrt{10}}$
c. $\frac{4}{\sqrt{10}}$
d. $\sqrt{10}$

Working Space

17 Two lines are given by $(x-2y)^2+k(x-2y)=0$. The value of k, so that the distance between them is 3, is

a. $\frac{1}{\sqrt{5}}$

b. $\pm\frac{2}{\sqrt{5}}$

c. $\pm 3\sqrt{5}$

d. None of these

18 The area (in sq units) of the quadrilateral formed by two pair of lines $l^2x^2-m^2y^2-n(lx+my)=0$ and $l^2x^2-m^2y^2+n(lx+my)=0$, is

a. $\frac{n^2}{2|lm|}$

b. $\frac{n^2}{|lm|}$

c. $\frac{n}{2|lm|}$

d. $\frac{n^2}{4|lm|}$

19 The locus of the point $P(x, y)$ satisfying the relation $\sqrt{(x-3)^2+(y-1)^2}+\sqrt{(x+3)^2+(y-1)^2}=6$, is

a. point

b. pair of coincident straight lines

c. circle

d. ellipse

20 The circumcentre of a triangle formed by the lines $xy+2x+2y+4=0$ and $x+y+2=0$, is

a. (0, 0)

b. (−2, −2)

c. (−1, −1)

d. (−1, −2)

Comprehension Type

Direction (Q. Nos. 21-25) *This section contains 2 paragraphs, each describing theory, experiments, data, etc. Five questions related to the paragraphs have been given. Each question has only one correct answer among the four given options (a), (b), (c) and (d).*

Passage I

If $3x^2+8xy-3y^2=0$ represents a pair of lines AB and BC, whereas the equation $3x^2+8xy-3y^2+2x-4y-1=0$ represents the lines CD and DA, then

21 The possible equation of CD, is (in sq units)

a. $x+y+1=0$

b. $x+3y+1=0$

c. $2x-4y-1=0$

d. $3x-y=0$

22 Area of quadrilateral $ABCD$, is

a. $\frac{1}{10}$ sq units

b. $\frac{1}{5}$ sq units

c. $\frac{1}{20}$ sq units

d. $\frac{1}{15}$ sq units

Passage II

Consider the general equation of degree 2, as

$$\lambda x^2-10xy+12y^2+5x-16y-3=0$$

23 The value of λ for which the line pair of straight lines is

a. 1

b. 2

c. $\frac{3}{2}$

d. 3

Working Space

24 For the value of λ obtained in above question, if $L_1 = 0$ and $L_2 = 0$ are the lines denoted by the given line pair, then the product of the abscissa and ordinate of their point of intersection is

a. 18 b. 28
c. 35 d. 25

25 If θ is the acute angle between $L_1 = 0$ and $L_2 = 0$, then θ lies in the interval

a. $(45°, 60°)$ b. $(30°, 45°)$
c. $(15°, 30°)$ d. $(0°, 15°)$

One Integer Value Correct Type

Direction (Q. Nos. 26-30) This section contains 5 questions. When worked out will result in an integer from 0 to 9 (both inclusive).

26 If $ax^2 + 6xy + by^2 - 10x + 10y - 6 = 0$ represents two perpendicular lines, then $|a|$ is equal to

27 If $x^2 - 3xy + \lambda y^2 + 3x - 5y + 2 = 0$ represents a pair of lines, where λ is a real number. If θ be the angle between these lines, then $\text{cosec}^2\theta$ equals

28 If the equation $x^2 + y^2 + 2gx + 2fy + 1 = 0$ represents a pair of lines, then $f^2 + g^2$ is equal to

29 Let PQR be a right angled isosceles triangle, right angled at $P(2, 1)$. If the equation of the line QR is $2x + y = 3$, then the equation representing the pair of lines PQ and PR is $3x^2 - 3y^2 + 8xy - 20x - 10y + \lambda = 0$. Thus, $(\lambda - 20)$ is equal to

30 The equation $\lambda^2x^2 + (\lambda^2 - 5\lambda + 4)xy + (3\lambda - 2)y^2 - 8x + 12y - 4 = 0$ will represent a circle, if λ is equal to

DPP-3 Equation of Line Joining the Origin to the Point of Intersection of a Curve and a Line

Only One Option Correct Type

Working Space

Direction (Q. Nos. 1-11) *This section contains 11 multiple choice questions. Each question has four choices (a), (b), (c) and (d), out of which ONLY ONE option is correct.*

1 The equation of pair of straight lines joining the point of intersection of the curve $x^2 + y^2 = 4$ and $y - x = 2$ to the origin, is

a. $x^2 + y^2 = (y - x)^2$
b. $x^2 + y^2 + (y - x)^2 = 0$
c. $x^2 + y^2 = 4(y - x)^2$
d. $x^2 + y^2 + 4(y - x)^2 = 0$

2 The equation of pair of lines joining to the points of intersection of $x^2 + y^2 = 9$ and $x + y = 3$, is

a. $(x + y)^2 = 9$
b. $x^2 + (3 - x)^2 = 9$
c. $xy = 0$
d. $(3 - x)^2 + y^2 = 9$

3 The lines joining the origin to the points of intersection of the line $3x - 2y = 1$ and the curve $3x^2 + 5xy - 3y^2 + 2x + 3y = 0$, are

a. parallel to each other
b. perpendicular to each other
c. inclined at 45° to each other
d. None of the above

4 The lines joining the points of intersection of the curve $(x - k)^2 + (y - h)^2 = c^2$ and the line $hx + ky = 2hk$ to the origin are perpendicular, then

a. $c = h \pm k$
b. $c^2 = h^2 + k^2$
c. $c^2 = (h + k)^2$
d. $4c^2 = h^2 + k^2$

5 The angle between the lines joining the points of intersection of the line $y = 3x + 2$ and the curve $x^2 + 2xy + 3y^2 + 4y + 8y - 11 = 0$ to the origin, is

a. $\tan^{-1}\left(\frac{3}{2\sqrt{2}}\right)$
b. $\tan^{-1}\left(\frac{2\sqrt{2}}{3}\right)$
c. $\tan^{-1}(\sqrt{3})$
d. $\tan^{-1}(2\sqrt{2})$

6 The lines joining the origin to the points of intersection of the line $y = mx + c$ and the circle $x^2 + y^2 = a^2$ will be mutually perpendicular, if

a. $a^2(m^2 + 1) = c^2$
b. $a^2(m^2 - 1) = c^2$
c. $a^2(m^2 + 1) = 2c^2$
d. $a^2(m^2 - 1) = 2c^2$

Working Space

7 The pair of straight lines joining the origin to the points of intersection of the line $y = 2\sqrt{2}x + c$ and the circle $x^2 + y^2 = 2$ are at right angles, if

a. $c^2 - 4 = 0$
b. $c^2 - 8 = 0$
c. $c^2 - 9 = 0$
d. $c^2 - 10 = 0$

8 The lines joining the points of intersection of line $x + y = 1$ and curve $x^2 + y^2 - 2y + \lambda = 0$ to the origin are perpendicular, then the value of λ will be

a. $\frac{1}{2}$
b. $-\frac{1}{2}$
c. $\frac{1}{\sqrt{2}}$
d. 0

9 If the lines joining the origin to the points of intersection of the line $fx - gy = \lambda$ and the curve $x^2 + hxy - y^2 + gx + fy = 0$ be mutually perpendicular, then

a. $\lambda = h$
b. $\lambda = g$
c. $\lambda = fg$
d. λ may have any value

10 The equation of the line joining origin to the points of intersection of the curve $x^2 + y^2 = a^2$ and $x^2 + y^2 - ax - ay = 0$, is

a. $x^2 - y^2 = 0$
b. $xy = 0$
c. $xy - x^2 = 0$
d. $y^2 + xy = 0$

11 If the straight lines joining the origin and the points of intersection of the curve $5x^2 + 12xy - 6y^2 + 4x - 2y + 3 = 0$ and $x + ky - 1 = 0$ are equally inclined to the X-axis, then the value of k is

a. 1
b. -1
c. 2
d. 3

One Integer Value Correct Type

Direction (Q. Nos. 12-15) *This section contains 4 questions. When worked out will result in an integer from 0 to 9 (both inclusive).*

12 If the equation $2x^2 + 2hxy + 6y^2 - 4x + 5y - 6 = 0$ represents a pair of straight lines, then the length of intercept on the X-axis cut by the lines is

13 The straight line joining the origin to the points of intersection of the line $2x + y = 1$ and curve $3x^2 + 4xy - 4x + 1 = 0$ include an angle $\frac{\pi}{k}$, then k is equal to

14 If the distance of two lines passing through origin from the point (x, y) is d, then the equation of lines is $(xy_1 - yx_1)^2 = kd^2 (x^2 + y^2)$. Thus, k is equal to

15 If the pair of straight lines given by $Ax^2 + 2Hxy + By^2 = 0$, (where, $H^2 > AB$) forms an equilateral triangle with $ax + by + c = 0$. Then, $(A + 3B)(3A + B) = kH^2$. Thus, k is equal to

SET - 1

Revisal Problems for JEE Main

Only One Option Correct Type

Working Space

Direction (Q. Nos. 1-25) *This section contains 25 multiple choice questions. Each question has four choices (a), (b), (c) and (d), out of which ONLY ONE option is correct.*

1 If the slope of one of the lines represented by $ax^2 + 2hxy + by^2 = 0$ is the square of the other, then

a. $a^2b + ab^2 - 6abh + 8h^3 = 0$
b. $a^2b + ab^2 + 6abh + 8h^3 = 0$
c. $a^2b + ab^2 - 3abh + 8h^3 = 0$
d. $a^2b + ab^2 - 6abh - 8h^3 = 0$

2 If the equation $hxy + gx + fy + c = 0$ represents a pair of straight lines, then

a. $fh = cg$
b. $fg = ch$
c. $h^2 = gf$
d. $fgh = c$

3 The equation of pair of straight lines perpendicular to the pair $ax^2 + 2hxy + by^2 = 0$, is

a. $ax^2 - 2hxy + by^2 = 0$
b. $bx^2 + 2hxy + ay^2 = 0$
c. $ay^2 - 2hxy + bx^2 = 0$
d. $ay^2 - bx^2 = 0$

4 The value of h for which the equation $3x^2 + 2hxy - 3y^2 - 40x + 30y - 75 = 0$ represents a pair of straight lines, are

a. 4, 4
b. 4, 6
c. 4, − 4
d. 0, 4

5 If $4ab = 3h^2$, then the ratio of slopes of the lines represented by the equation $ax^2 + 2hxy + by^2 = 0$, will be

a. $\sqrt{2} : 1$
b. $\sqrt{3} : 1$
c. $2 : 1$
d. $1 : 3$

6 Which of the following second degree equation represented a pair of straight lines?

a. $x^2 - xy - y^2 = 1$
b. $-x^2 + xy - y^2 = 1$
c. $4x^2 - 4xy + y^2 = 4$
d. $x^2 + y^2 = 4$

7 The equation of the lines passing through the origin and having slopes 3 and $-\frac{1}{3}$, is

a. $3y^2 + 8xy - 3x^2 = 0$
b. $3x^2 + 8xy - 3y^2 = 0$
c. $3y^2 - 8xy + 3x^2 = 0$
d. $3x^2 + 8xy + 3y^2 = 0$

8 The equation of one of the lines represented by the equation $pq(x^2 - y^2) + (p^2 - q^2)xy = 0$, is

a. $px + qy = 0$
b. $px - qy = 0$
c. $p^2x + q^2y = 0$
d. $q^2x - p^2y = 0$

9 The condition of representing the coincident lines by the general quadratic equation $f(x, y) = 0$, is

a. $\Delta = 0, h^2 = ab$
b. $\Delta = 0$ and $a + b = 0$
c. $\Delta = 0, h^2 = ab, g^2 = ac, f^2 = bc$
d. $h^2 = ab, g^2 = ac$ and $f^2 = bc$

10 If the slope of one line of the pair of lines represented by $ax^2 + 4xy + y^2 = 0$ is 3 times the slope of the other line, then a is

a. 1
b. 2
c. 3
d. 4

11 Separate equations of lines, for a pair of lines, whose equation is $x^2 + xy - 12y^2 = 0$, are

a. $x + 4y = 0$ and $x + 3y = 0$
b. $2x - 3y = 0$ and $x - 4y = 0$
c. $x - 6y = 0$ and $x - 3y = 0$
d. $x + 4y = 0$ and $x - 3y = 0$

Working Space

12 If the equation $2x^2 + 7xy + 3y^2 - 9x - 7y + k = 0$ represents a pair of lines, then k is equal to

a. 4 b. 2 c. 0 d. 3

13 If the sum of the slopes of the lines given by $x^2 - 2cxy - 7y^2 = 0$ is four times their product, then c has the value

a. −2 b. −1 c. 2 d. 1

14 If the acute angles between the pairs of lines $3x^2 - 7xy + 4y^2 = 0$ and $6x^2 - 5xy + y^2 = 0$ be θ_1 and θ_2 respectively, then

a. $\theta_1 = \theta_2$ b. $\theta_1 = 2\theta_2$ c. $2\theta_1 = \theta_2$ d. None of these

15 Angle between the line joining the origin to the points of intersection of the curves $2x^2 + 3y^2 + 10x = 0$ and $3x^2 + 5y^2 + 16x = 0$, is

a. $\tan^{-1}\left(\frac{3}{2}\right)$ b. $\tan^{-1}\left(\frac{4}{5}\right)$ c. 90° d. None of these

16 The lines represented by the equation $x^2 + 2\sqrt{3}\,xy + 3y^2 - 3x - 3\sqrt{3}\,y - 4 = 0$, are

a. perpendicular to each other b. parallel
c. inclined at 45° to each other d. None of these

17 If the angle between the pair of straight lines represented by the equation $x^2 - 3xy + \lambda y^2 + 3x - 5y + 2 = 0$ is $\tan^{-1}(1/3)$, where λ is non-negative real number, then λ is equal to

a. 2 b. 0 c. 3 d. 1

18 If the bisectors of the angles between the pairs of lines given by the equation $ax^2 + 2hxy + by^2 = 0$ and $ax^2 + 2hxy + by^2 + \lambda(x^2 + y^2) = 0$ are coincident, then λ is

a. a b. b
c. h d. any real number

19 If the equation $ax^2 + 2hxy + by^2 = 0$ has the one line as the bisector of angle between the coordinate axes, then

a. $(a - b)^2 = h^2$ b. $(a + b)^2 = h^2$
c. $(a - b)^2 = 4h^2$ d. $(a + b)^2 = 4h^2$

20 The equation $8x^2 + 8xy + 2y^2 + 26x + 13y + 15 = 0$ represents a pair of straight lines. The distance between them is (in sq units)

a. $\frac{7}{\sqrt{5}}$ units b. $\frac{7}{2\sqrt{5}}$ units c. $\frac{\sqrt{7}}{5}$ units d. None of these

21 Distance between the pair of lines represented by the equation $x^2 - 6xy + 9y^2 + 3x - 9y - 4 = 0$, is (in sq units)

a. $\frac{15}{\sqrt{10}}$ units b. $\frac{1}{2}$ units c. $\sqrt{\frac{5}{2}}$ units d. $\frac{1}{\sqrt{10}}$ units

22 The orthocentre of the triangle formed by the lines $xy = 0$ and $x + y = 1$, is

a. (0, 0) b. $\left(\frac{1}{2},\frac{1}{2}\right)$ c. $\left(\frac{1}{3},\frac{1}{3}\right)$ d. $\left(\frac{1}{4},\frac{1}{4}\right)$

23 Area of the triangle formed by the lines $y^2 - 9xy + 18x^2 = 0$ and $y = 9$, is

a. $\frac{27}{4}$ sq units b. 27 sq units
c. $\frac{27}{2}$ sq units d. None of these

24 The equations to a pair of opposite sides of a parallelogram are $x^2 - 5x + 6 = 0$ and $y^2 - 6y + 5 = 0$. The equation to its diagonals are

a. $x + 4y = 13$, $y = 4x - 7$ b. $4x + y = 13$, $4y = x - 7$
c. $4x + y = 13$, $y = 4x - 7$ d. $y - 4x = 13$, $y + 4x = 7$

25 If the lines $ax^2 + 2hxy + by^2 = 0$ represents the adjacent sides of a parallelogram, then the equation of second diagonal if one is $lx + my = 1$, will be

a. $(am + hl)x = (bl + hm)y$ b. $(am - hl)x = (bl - hm)y$
c. $(am - hl)x = (bl + hm)y$ d. None of these

SET - 2

Revisal Problems
for JEE Advanced

Only One Option Correct Type

Working Space

Direction (Q. Nos. 1-11) *This section contains 11 multiple choice questions. Each question has four choices (a), (b), (c) and (d), out of which ONLY ONE option is correct.*

1 The three lines whose combined equation is $y^3 - 4x^2y = 0$ form a triangle which is

a. isosceles b. equilateral
c. right angled d. None of these

2 If the distance of a point (x_1, y_1) from each of two straight lines, which pass through the origin of coordinates, is δ. Then, the two lines are given by

a. $(x_1 y - xy_1)^2 = \delta^2(x + y)^2$ b. $(x_1 y + xy_1)^2 = \delta^2(x^2 + y^2)$
c. $(x_1 y - xy_1)^2 = \delta^2(x^2 - y^2)$ d. None of these

3 The equation of straight line through the point (x_1, y_1) and parallel to the lines given by $ax^2 + 2hxy + by^2 = 0$, is

a. $a(y - y_1)^2 + 2h(x - x_1)(y - y_1) + b(x - x_1)^2 = 0$
b. $a(y - y_1)^2 - 2h(x - x_1)(y - y_1) + b(x - x_1)^2 = 0$
c. $b(y - y_1)^2 + 2h(x - x_1)(y - y_1) + a(x - x_1)^2 = 0$
d. None of the above

4 The combined equation of the pair of lines through the point (1, 0) and perpendicular to the lines represented by $2x^2 - xy - y^2 = 0$, is

a. $2x^2 - xy - y^2 - x + y - 1 = 0$ b. $2x^2 - xy - y^2 - x + y - 1 = 0$
c. $2y^2 + xy - x^2 + 2x - y - 1 = 0$ d. None of these

5 The equation $\lambda(x^3 - 3xy^2) + y^3 - 3x^2y = 0$ represents three straight lines passing through the origin such that

a. they are equally inclined to one another
b. two of which are at the right angles
c. two of which are coincident
d. None of the above

6 The equation $x^3 + x^2y - xy^2 - y^3 = 0$ represents three straight lines passing through the origin such that

a. two of them are coincident and two of them are perpendicular
b. two of them are coincident but no two are perpendicular
c. two of them are perpendicular but no two are coincident
d. None of the above

7 The product of perpendiculars drawn from the origin to the lines represented by the equation $ax^2 + 2hxy + by^2 + 2gx + 2fy + c = 0$, will be

a. $\dfrac{ab}{\sqrt{a^2 - b^2 + 4h^2}}$ b. $\dfrac{bc}{\sqrt{b^2 - c^2 + 4h^2}}$
c. $\dfrac{c}{\sqrt{(a - b)^2 + 4h^2}}$ d. None of these

8 The square of distance of the point of intersection of the lines represented by the equation $ax^2 + 2hxy + by^2 + 2gx + 2fy + c = 0$ and origin, is

a. $\dfrac{c(a + b) - f^2 - g^2}{ab - h^2}$ b. $\dfrac{c(a - b) + f^2 + g^2}{\sqrt{ab - h^2}}$
c. $\dfrac{c(a + b) - f^2 - g^2}{ab + h^2}$ d. None of these

Working Space

9 Two of the lines represented by the equation $ay^4 + bxy^3 + cx^2y^2 + dx^3y + ex^4 = 0$ will be perpendicular, then

a. $(b+d)(ad+be)+(e-a)^2(a+c+e)=0$

b. $(b+d)(ad+be)+(e+a)^2(a+c+e)=0$

c. $(b-d)(ad-be)+(e-a)^2(a+c+e)=0$

d. $(b-d)(ad-be)+(a+e)^2(a+c+e)=0$

10 The lines represented by the equation $ax^2 + 2hxy + by^2 + 2gx + 2fy + c = 0$ will be equidistant from origin, if

a. $f^2 + g^2 = c(b-a)$

b. $f^4 + g^4 = c(bf^2 + ag^2)$

c. $f^4 - g^4 = c(bf^2 - ag^2)$

d. $f^2 + g^2 = af^2 + bg^2$

11 Consider a parallelogram whose sides are represented by the lines $2x + 3y = 0$, $2x + 3y - 5 = 0$, $3x - 4y = 0$ and $3x - 4y = 3$. The equation of the diagonal not passing through the origin, is

a. $21x - 11y + 15 = 0$

b. $9x - 11y + 15 = 0$

c. $21x - 29y - 15 = 0$

d. $21x - 11y - 15 = 0$

Matching List Type

Direction (Q. No. 12) Choices for the correct combination of elements from Column I and Column II are given as options (a), (b), (c) and (d), out of which one is correct.

12 Match the Column I with Column II and select the correct option from the given codes.

	Column I		Column II
i.	Four lines $3x - 4y - 11 = 0$, $3x - 4y - 9 = 0$, $4x + 3y + 3 = 0$ and $4x + 3y - 17 = 0$ enclose a figure which is	p.	a quadrilateral which is neither a parallelogram nor a trapezium nor a kite.
ii.	The lines $2x + y = 1$, $x + 2y = 1$, $2x + y = 3$ and $x + 2y = 3$ form a figure which is	q.	a parallelogram which is neither a rectangle nor a rhombus.
iii.	If O is the origin, P is the intersection of the lines $2x^2 - 7xy + 3y^2 + 5x + 10y - 25 = 0$, A and B are the points in which these lines are cut by the line $x + 2y = 5$, then the points O, A, P, B (in order) are the vertices of	r.	a rhombus which is not a square.
		s.	a square

Codes

	i	ii	iii			i	ii	iii
a.	s	r	q		b.	p	r	s
c.	p	q	r		d.	s	p	q

Comprehension Type

Direction (Q. Nos. 13-18) This section contains 2 paragraphs, each describing theory, experiments, data, etc. Six questions related to the paragraphs have been given. Each question has only one correct answer among the four given options (a), (b), (c) and (d).

Passage I

A straight line passing through $O(0, 0)$ cuts the lines $x = \alpha$, $y = \beta$ and $x + y = 8$ at A, B and C respectively such that $OA \cdot OB \cdot OC = 48\sqrt{2}$ and $f(\alpha, \beta) \le 0$, where

$$f(x, y) = \left| \frac{y}{x} - \frac{3}{2} \right| + (3x - 2y)^6 + \sqrt{ex + 2y - 2e - 6}$$

13 The point of intersection of lines $x = \alpha$ and $y = \beta$, is

a. (3, 2)

b. (2, 3)

c. (3, 4)

d. (5, 3)

Working Space

14 Equation of line OA, is

a. $x - y = 0$
b. $x - \sqrt{2}y = 0$
c. $\sqrt{2}x - y = 0$
d. $2x - y = 0$

15 $OA + OB + OC$ equals

a. $2\sqrt{2}$
b. $4\sqrt{2}$
c. $7\sqrt{2}$
d. $9\sqrt{2}$

Passage II

Consider a line pair $ax^2 + 3xy - 2y^2 - 5x + 5y + c = 0$ representing perpendicular lines intersecting each other at C and forming a ΔABC with the X-axis.

16 If x_1 and x_2 are intercepts on the X-axis and y_1 and y_2 are the intercepts on the Y-axis, then the sum $(x_1 + x_2 + y_1 + y_2)$ is equal to

a. 6
b. 5
c. 4
d. 3

17 Distance between the orthocentre and circumcentre of the ΔABC, is

a. 2
b. 3
c. $\frac{7}{4}$
d. $\frac{9}{4}$

18 If the circle $x^2 + y^2 - 4y + k = 0$ is orthogonal with the circumcircle of the ΔABC, then k equals

a. $\frac{1}{2}$
b. 1
c. 2
d. $\frac{3}{2}$

One Integer Value Correct Type

Direction (Q. Nos. 19-23) *This section contains 5 questions. When worked out will result in an integer from 0 to 9 (both inclusive).*

19 The combined equation of three sides of a triangle is $(x^2 - y^2)(2x + 3y - 6) = 0$. If $(-2, a)$ is the interior point of the triangle, then the integer value of a is

20 If $3x^2 + 2\lambda xy + 3y^2 + (6 - \lambda)x + (2\lambda - 6)y - 21 = 0$ is the equation of the circle, then its radius is

21 The value of λ for which the lines joining the point of intersection of curves C_1 and C_2 to the origin are equally inclined to the axis of X. $C_1 : \lambda x^2 + 3y^2 - 2\lambda xy + 9x = 0$ and $C_2 : 3x^2 - 4y^2 + 8xy - 3x = 0$, then λ is equal to

22 Two conics $a_1x^2 + 2h_1xy + b_1y^2 = c_1$ and $a_2x^2 + 2h_2xy + b_2y^2 = c_2$ intersect in 4 concylic points, where $(a_1 - b_1)h_2 = \lambda(a_2 - b_2)h_1$, then λ is equal to

23 The equation of the locus of foot of perpendiculars drawn from the origin to the line passing through a fixed point (a, b) is $x^2 + y^2 - ax - by = \lambda$, then λ is equal to

JEE Archive

(Compilation of Last 10 Years Questions of JEE Main, AIEEE, JEE Advanced & IIT JEE)

1 If one of the line of $my^2 + (1-m^2)xy - mx^2 = 0$ is a bisector of the angle between the lines $xy = 0$, then m is **(AIEEE 2007)**

a. $-\frac{1}{2}$ b. -2

c. ± 1 d. 2

2 If the pair of lines $ax^2 + 2(a+b)xy + by^2 = 0$ lie along diameters of a circle and divide the circle into four sectors such that the area of one of the sectors is thrice the area of another sector, then **(AIEEE 2005)**

a. $3a^2 + 2ab + 3b^2 = 0$

b. $3a^2 + 10ab + 3b^2 = 0$

c. $3a^2 - 2ab + 3b^2 = 0$

d. $3a^2 - 10ab + 3b^2 = 0$

3 If the sum of the slopes of the lines given by $x^2 - 2cxy - 7y^2 = 0$ is four times their product, then c has the value **(AIEEE 2004)**

a. 1 b. –1

c. 2 d. –2

4 If one of the lines given by $6x^2 - xy + 4cy^2 = 0$ is $3x + 4y = 0$, then c is equal to **(AIEEE 2004)**

a. 1 b. –1 c. 3 d. –3

5 If the pair of straight lines $x^2 - 2pxy - y^2 = 0$ and $x^2 - 2qxy - y^2 = 0$ be such that each pair bisects the angle between the other pair, then **(AIEEE 2003)**

a. $p = q$ b. $p = -q$

c. $pq = 1$ d. $pq = -1$

6 Let a and b be non-zero and real numbers. Then, the equation $(ax^2 + by^2 + c)(x^2 - 5xy + 6y^2) = 0$ represents **(IIT JEE 2008, Only one Option Correct Type)**

a. four straight lines, when $c = 0$ and a, b are of the same sign

b. two straight lines and a circle, when $a = b$ and c is of sign opposite to that of a

c. two straight lines and a hyperbola, when a and b are of the same sign and c is of sign opposite to that of a

d. a circle and an ellipse, when a and b are of the same sign and c is of sign opposite to that of a

4. Circle

DPP-1 Equation of Circle

Only One Option Correct Type

Working Space

Direction (Q. Nos. 1-27) *This section contains 27 multiple choice questions. Each question has four choices (a), (b), (c) and (d), out of which ONLY ONE option is correct.*

1 If the equation $px^2 + (2-q)xy + 3y^2 - 6qx + 30y + 6q = 0$ represent a circle, then the values of p and q are

a. 2, 2   b. 3, 1
c. 3, 2   d. 3, 4

2 The circumcircle of the quadrilateral formed by the lines $x = a$, $x = 2a$, $y = a$ and $y = -a$, is

a. $x^2 + y^2 - 3ax - a^2 = 0$ b. $x^2 - y^2 + 3ax + a^2 = 0$
c. $x^2 + y^2 - 3ax + a^2 = 0$ d. $x^2 + y^2 + 3ax + a^2 = 0$

3 The x-coordinates of two points A and B are roots of equation $x^2 + 2x - a^2 = 0$ and y-coordinate are roots of equation $y^2 + 4y - b^2 = 0$. Then, equation of the circle which has diameter AB, is

a. $(x-1)^2 + (y-2)^2 = 5 + a^2 + b^2$
b. $(x+1)^2 + (y+2)^2 = \sqrt{5 + a^2 + b^2}$
c. $(x+1)^2 + (y+2)^2 = a^2 + b^2$
d. $(x+1)^2 + (y+2)^2 = 5 + a^2 + b^2$

4 The equation of the circle with centre (α, β) and passing through the origin is

a. $x^2 + y^2 - \alpha x - \beta y = 0$ b. $x^2 + y^2 + 2\alpha x + 2\beta y = 0$
c. $x^2 + y^2 - 2\alpha x - 2\beta y = 0$ d. $x^2 + y^2 + \alpha x + \beta y = 0$

5 Centre of circle passing through the points (4, 5), (3, 4) and (5, 2), is

a. $\left(\frac{9}{2}, \frac{7}{2}\right)$ b. $\left(\frac{7}{2}, \frac{9}{2}\right)$
c. $\left(\frac{7}{2}, \frac{7}{2}\right)$ d. $\left(\frac{9}{2}, \frac{9}{2}\right)$

6 The radius (in units) of the circle $x^2 + y^2 + 2x\cos\theta + 2y\sin\theta = 8$, is

a. 1   b. 3   c. $2\sqrt{3}$ d. $\sqrt{10}$

7 The equation of the circle whose centre is at origin and which passes through the vertices of an equilateral triangle the length of whose median is $3a$, is

a. $x^2 + y^2 = a^2$ b. $x^2 + y^2 = 16a^2$
c. $x^2 + y^2 = 4a^2$ d. $x^2 + y^2 = 9a^2$

8 The equation of the circumcircle of the triangle by the lines $x = 0$, $y = 0$, $\frac{x}{a} + \frac{y}{b} = 1$ is

a. $x^2 + y^2 + ax - by = 0$ b. $x^2 + y^2 - ax + by = 0$
c. $x^2 + y^2 - ax - by = 0$ d. $x^2 + y^2 + ax + by = 0$

Working Space

9 The equation of the circle passing through the origin with $3x + y = 14$ and $2x + 5y = 18$ as its two diameters will be

a. $x^2 + y^2 + 8x + 4y = 0$
b. $x^2 + y^2 + 8x - 4y = 0$
c. $x^2 + y^2 - 8x - 4y = 0$
d. $x^2 + y^2 - 8x + 4y = 0$

10 The equation of the circle passing through the vertices of the quadrilateral whose sides are $x = 1$, $x = 2$, $y = 1$ and $y = 2$, is

a. $x^2 + y^2 - 3x - 3y + 2 = 0$
b. $x^2 + y^2 + 3x + 3y + 2 = 0$
c. $x^2 + y^2 - 3x - 3y + 4 = 0$
d. None of these

11 The equation of the circle which passes through (1, 0) and (0, 1) and has its radius as small as possible, is

a. $2x^2 + 2y^2 - 3x - 3y + 1 = 0$
b. $x^2 + y^2 - x - y = 0$
c. $x^2 + y^2 - 2x - 2y + 1 = 0$
d. $x^2 + y^2 - 3x - 3y + 2 = 0$

12 Centre of circles of radius 5 touching the line $3x + 4y = 7$ at (1, 1), is

a. (4, 5) and (−2, − 3)
b. (−4,− 5) and (2, 3)
c. (−4, − 5) and (6, 7)
d. None of these

13 The lines $2x - 3y = 5$ and $3x - 4y = 7$ are diameters of a circle having area as 154 sq units. Then, the equation of the circle, is **[AIEEE 2003]**

a. $x^2 + y^2 - 2x + 2y = 62$
b. $x^2 + y^2 + 2x - 2y = 62$
c. $x^2 + y^2 + 2x - 2y = 47$
d. $x^2 + y^2 - 2x + 2y = 47$

14 A variable circle passes through the fixed point $A(p, q)$ and touches X-axis. The locus of the other end of the diameter through A, is **[AIEEE 2004]**

a. $(y - p)^2 = 4qx$
b. $(x - q)^2 = 4py$
c. $(x - p)^2 = 4qy$
d. $(y - q)^2 = 4px$

15 If the lines $2x + 3y + 1 = 0$ and $3x - y - 4 = 0$ lie along diameters of a circle of circumference 10π, then the equation of the circle is **[AIEEE 2004]**

a. $x^2 + y^2 + 2x + 2y - 23 = 0$
b. $x^2 + y^2 - 2x - 2y - 23 = 0$
c. $x^2 + y^2 - 2x + 2y - 23 = 0$
d. $x^2 + y^2 + 2x - 2y - 23 = 0$

16 The intercept on the line $y = x$, the circle $x^2 + y^2 - 2x = 0$ is AB. Equation of the circle AB as a diameter, is

a. $x^2 + y^2 + x + y = 0$
b. $x^2 + y^2 - x + y = 0$
c. $x^2 + y^2 - x - y = 0$
d. $x^2 + y^2 + x - y = 0$

17 If an equilateral triangle is inscribed in the circle $x^2 + y^2 = a^2$, the length of its such side is (in units)

a. $\sqrt{2}a$
b. $\frac{\sqrt{3}}{2}a$
c. $\sqrt{3}a$
d. None of these

18 The image of the centre of the circle $x^2 + y^2 = a^2$ with respect to the mirror $x + y = 1$, is

a. $\left(\frac{1}{\sqrt{2}}, \sqrt{2}\right)$
b. $(\sqrt{2}, \sqrt{2})$
c. $(\sqrt{2}, 2\sqrt{2})$
d. None of these

Working Space

19 A circle is inscribed in an equilateral triangle of side a. The area of any square inscribed in the circle is (in sq units)

a. $\frac{a^2}{3}$ b. $\frac{2a^2}{3}$

c. $\frac{a^2}{6}$ d. $\frac{a^2}{12}$

20 The ΔPQR is inscribed in the circle $x^2 + y^2 = 25$. If Q and R have coordinates $(3, 4)$ and $(-4, 3)$ respectively, then $\angle QPR$ is **[IIT JEE 2000]**

a. $\frac{\pi}{2}$ b. $\frac{\pi}{3}$

c. $\frac{\pi}{4}$ d. $\frac{\pi}{6}$

21 The centre of the circle inscribed in the square formed by lines $x^2 - 8x + 12 = 0$ and $y^2 - 14y + 45 = 0$, is **[IIT JEE 2003]**

a. $(4, 7)$ b. $(7, 4)$

c. $(9, 4)$ d. $(4, 9)$

22 A square is inscribed in the circle $x^2 + y^2 - 2x + 4y - 93 = 0$ with its sides parallel to the coordinate of its vertices are

a. $(-6, 9), (-6, -5), (8, -9), (8, 5)$ b. $(-6, -9), (-6, 5), (8, -9), (8, -5)$

c. $(-6, -9), (-6, 5), (8, 9), (8, 5)$ d. $(-6, -9), (-6, 5), (8, 5), (8, -9)$

23 Three equal circles each of radius r touches one another. The radius of the circle touching all the three given circles internally is

a. $(2+\sqrt{3})r$ b. $\frac{(2+\sqrt{3})r}{\sqrt{3}}$

c. $\frac{(2-\sqrt{3})}{\sqrt{3}}r$ d. $(2-\sqrt{3})r$

24 If a line segment $AM = a$, moves in the plane XOY remaining parallel to OX, so that the left end point A slides along the circle $x^2 + y^2 = a^2$, the locus of M is

a. $x^2 + y^2 = 4a^2$ b. $x^2 + y^2 = 2ax$

c. $x^2 + y^2 = 2ay$ d. $x^2 + y^2 - 2ax - 2ay = 0$

25 If the lines $3x - 4y - 7 = 0$ and $2x - 3y - 5 = 0$ are two diameters of a circle of area 49π sq units. The equation of the circle is **[AIEEE 2006]**

a. $x^2 + y^2 + 2x - 2y - 62 = 0$ b. $x^2 + y^2 - 2x + 2y - 62 = 0$

c. $x^2 + y^2 - 2x + 2y - 47 = 0$ d. $x^2 + y^2 + 2x - 2y - 47 = 0$

26 The point diametrically opposite to the point $P(1, 0)$ on the circle $x^2 + y^2 + 2x + 4y - 3 = 0$, is **[AIEEE 2008]**

a. $(3, -4)$ b. $(-3, 4)$

c. $(-3, -4)$ d. $(3, 4)$

27 The circle passing through the point $(-1, 0)$ and touching the Y-axis at $(0, 2)$ also passes through the point

a. $\left(\frac{-3}{2}, 0\right)$ b. $\left(\frac{-5}{2}, 0\right)$

c. $\left(\frac{-5}{2}, \frac{3}{2}\right)$ d. $(-4, 0)$

Working Space

One or More than One Options Correct Type

Direction (Q. Nos. 28-31) This section contains 4 multiple choice questions. Each question has four choices (a), (b), (c) and (d), out of which ONE or MORE THAN ONE are correct.

28 Let L_1 be a line passing through the origin and L_2 be the line $x + y = 1$. If the intercepts made by the circle $x^2 + y^2 - x + 3y = 0$ on L_1 and L_2 are equal, then the equation of L_1 can be

a. $x + y = 0$
b. $x - y = 0$
c. $x + 7y = 0$
d. $x - 7y = 0$

29 $\frac{x - x_1}{\cos\theta} = \frac{y - y_1}{\sin\theta} = r$, represents

a. equation of straight line, if θ is constant and r is variable
b. equation of circle, if r is constant and θ is variable
c. a straight line passing through a fixed point and having a known slope
d. a circle with a known centre and a given radius

30 Which of the following lines have the intercepts of equal length on the circle, $x^2 + y^2 - 2x + 4y = 0$?

a. $3x - y = 0$
b. $x + 3y = 0$
c. $x + 3y + 10 = 0$
d. $3x - y - 10 = 0$

31 Locus of intersection of the two straight lines passing through (1, 0) and (−1, 0) respectively and including an angle of 45° can be a circle with

a. centre (1, 0) and radius $\sqrt{2}$ units
b. centre (1, 0) and radius 2 units
c. centre (0, 1) and radius $\sqrt{2}$ units
d. centre (0, − 1) and radius $\sqrt{2}$ units

Comprehension Type

Direction (Q. Nos. 32-37) This section contains 2 paragraphs, each describing theory, experiments, data, etc. Six questions related to the paragraphs have been given. Each question has only one correct answer among the four given options (a), (b), (c) and (d).

Passage I

The centres of the circles touching both the coordinate axes lies on the straight line $x - y = 0$ and $x + y = 0$ according to the quadrant in which the circle lies.

32 The number of circles whose centres are in the first quadrant, touching the axes and passing through a fixed point (h, k), is

a. always be 1
b. always be 2
c. 1 or 2
d. more than 2

33 If the difference between the radii of the circles passing through (a, b) and touching the axes is $c > 0$ and $ab > 0$, then the least value of $(a + b)$ is

a. c
b. $\sqrt{2}c$
c. $\frac{c}{\sqrt{2}}$
d. $2c$

34 If one of the points of intersection of the two circles touching the axes is (p, q), then the length of the common chord is

a. $|p \pm q|$
b. $\sqrt{2}\,|p \pm q|$
c. $2|p \pm q|$
d. $\sqrt{2}\,|pq|$

Working Space

Passage II

A circle C of radius 1 is inscribed in an equilateral ΔPQR. The points of contact of C with the sides PQ, QR, RP are D, E, F respectively. The line PQ is given by the equation $\sqrt{3}x + y = 6$ and the point D is $\left(\frac{3\sqrt{3}}{2}, \frac{3}{2}\right)$. Further, it is given that the origin and the centre of C are on the same side of the line PQ. **[IIT JEE 2008]**

35 The equation of circle C, is

a. $(x - 2\sqrt{3})^2 + (y - 1)^2 = 1$

b. $(x - 2\sqrt{3})^2 + \left(y + \frac{1}{2}\right)^2 = 1$

c. $(x - \sqrt{3})^2 + (y + 1)^2 = 1$

d. $(x - \sqrt{3})^2 + (y - 1)^2 = 1$

36 Points E and F are given by

a. $\left(\frac{\sqrt{3}}{2}, \frac{3}{2}\right), (\sqrt{3}, 0)$

b. $\left(\frac{\sqrt{3}}{2}, \frac{1}{2}\right), (\sqrt{3}, 0)$

c. $\left(\frac{\sqrt{3}}{2}, \frac{3}{2}\right), \left(\frac{\sqrt{3}}{2}, \frac{1}{2}\right)$

d. $\left(\frac{3}{2}, \frac{\sqrt{3}}{2}\right), \left(\frac{\sqrt{3}}{2}, \frac{1}{2}\right)$

37 Equation of sides RP, RQ are

a. $y = \frac{2}{\sqrt{3}}x + 1, y = -\frac{2}{\sqrt{3}}x - 1$

b. $y = \frac{1}{\sqrt{3}}x, y = 0$

c. $y = \frac{\sqrt{3}}{2}x + 1, y = -\frac{\sqrt{3}}{2}x - 1$

d. $y = \sqrt{3}x, y = 0$

One Integer Value Correct Type

Direction (Q. Nos. 38-40) This section contains 3 questions. When worked out will result in an integer from 0 to 9 (both inclusive).

38 The intercept made by the circle $x^2 + y^2 - 14x - 10y + 23 = 0$ on the line $7x + y - 4 = 0$, is

39 If $\left(m_i, \frac{1}{m_i}\right)$, $m_i > 0, i = 1, 2, 3, 4$ are four distinct points on a circle, then the value of $\prod_{i=1}^{4} m_i$ must be

40 If the line $3x - 4y - 8 = 0$ divides the circumference of the circle with centre $(2, -3)$ in the ratio 1 : 2. Then, the radius of the circle is

DPP-2 Equation of Tangent, Equation of Normal and Length of Tangent

Only One Option Correct Type

Working Space

Direction *(Q. Nos. 1-23) This section contains 23 multiple choice questions. Each question has four choices (a), (b), (c) and (d), out of which ONLY ONE option is correct.*

1 The tangent lines to the circle $x^2 + y^2 - 6x + 4y = 12$, which are parallel to the line $4x + 3y + 5 = 0$ are given by

a. $4x + 3y - 7 = 0, 4x + 3y + 15 = 0$
b. $4x + 3y - 31 = 0, 4x + 3y + 19 = 0$
c. $4x + 3y - 17 = 0, 4x + 3y + 13 = 0$
d. None of the above

2 The gradient of the tangent line at the point $(a\cos\alpha, a\sin\alpha)$ to the circle $x^2 + y^2 = a^2$, is

a. $-\tan x$ b. $\tan x$
c. $\cot\alpha$ d. $-\cot\alpha$

3 The equation of the normal at the point $(4, -1)$ of the circle $x^2 + y^2 - 40x + 10y = 153$, is

a. $x + 4y = 0$ b. $x - 4y = 0$
c. $4x + y = 0$ d. $4x - y = 0$

4 The equation of the normal to the circle $x^2 + y^2 - 8x - 2y + 12 = 0$ at the points whose ordinate is -1, will be

a. $2x - y - 7 = 0, 2x + y - 9 = 0$
b. $2x + y - 7 = 0, 2x + y + 9 = 0$
c. $2x + y + 7 = 0, 2x + y + 9 = 0$
d. $2x - y + 7 = 0, 2x - y + 9 = 0$

5 The value of c for which $x + y = 2$ touches circle $x^2 + y^2 - 2cx = 0$ and $(1, 0)$ lies inside the circle, is

a. 2 b. $\sqrt{2} - 1$
c. $2(\sqrt{2} - 1)$ d. $2(\sqrt{2} + 1)$

6 If the line $3x + 4y = m$ touches the circle $x^2 + y^2 = 10x$, then m is equal to

a. $-40, 10$ b. $40, -10$
c. $40, 10$ d. $-40, -10$

7 The equation of tangent to the circle $x^2 + y^2 = 25$, which is inclined at 60° angle with X-axis, will be

a. $y = \sqrt{3}x \pm 10$ b. $\sqrt{3}y = x \pm 10$
c. $y = \sqrt{3}x \pm 2$ d. None of these

Working Space

8 Tangents drawn from the point $P(1, 8)$ to the circle $x^2 + y^2 - 6x - 4y - 11 = 0$ touch the circle at the points A and B. The equation of the circumcircle of the ΔPAB, is **[IIT JEE 2008]**

a. $x^2 + y^2 + 4x - 6y + 19 = 0$
b. $x^2 + y^2 - 4x - 10y + 19 = 0$
c. $x^2 + y^2 - 2x + 6y - 29 = 0$
d. $x^2 + y^2 - 6x - 4y + 19 = 0$

9 Equation of tangent to the circle $x^2 + y^2 = a^2$. Which is parallel to $y = mx + c$?

a. $ym = x \pm a\sqrt{1+m^2}$
b. $y = mx \pm \sqrt{1+m^2}$
c. $y = mx \pm a\sqrt{1+m^2}$
d. None of these

10 If $x + 2y = 5$ is tangent to the circle $x^2 + y^2 = 5$, then equation of normal at their point of contact is

a. $2x + y = 5$
b. $2x - y = 0$
c. $x + 2y = 0$
d. None of these

11 The line $y = x + c$ will intersect the circle $x^2 + y^2 = 1$ in two coincident points, if

a. $c = -\sqrt{2}$
b. $c = \sqrt{2}$
c. $c = \pm\sqrt{2}$
d. None of these

12 If the circle $x^2 + y^2 = a^2$ cuts off a chord of length $2b$ from the line $y = mx + c$, then

a. $(1-m^2)(a^2-b^2)=c^2$
b. $(1+m^2)(a^2-b^2)=c^2$
c. $(1-m^2)(a^2+b^2)=c^2$
d. None of these

13 The equation of tangent at the point on the circle $x^2 + y^2 - 4x - 2y - 20 = 0$, nearest from the point $P(10, 7)$, is

a. $3x - 4y = 5$
b. $4x + 3y - 36 = 0$
c. $4x + 3y + 14 = 0$
d. None of these

14 A line is drawn through a fixed point $P(\alpha, \beta)$ to cut the $x^2 + y^2 = r^2$ at A and B. Then, $PA \cdot PB$ is equal to

a. $(\alpha + \beta)^2 - r^2$
b. $\alpha^2 + \beta^2 - r^2$
c. $(\alpha - \beta)^2 + r^2$
d. None of these

15 The length of the tangent drawn from any point on the circle $x^2 + y^2 + 2gx + 2fy + \alpha = 0$ to the circle $x^2 + y^2 + 2gx + 2fy + \beta = 0$, is

a. $\sqrt{\beta - \alpha}$
b. $\sqrt{\alpha\beta}$
c. $\sqrt{\alpha - \beta}$
d. $\sqrt{\frac{\alpha}{\beta}}$

16 Area of circle in which a chord of length $2\sqrt{3}$ units, subtends angle 120° at its centre is

a. π sq units
b. 2π sq units
c. 4π sq units
d. None of these

17 Let C be the circle with centre $(0, 0)$ and radius 3 units. The equation of the locus of the mid-points of the chords of the circle C that subtends an angle of $\frac{2\pi}{3}$ at its centre, is **[AIEEE 2006]**

a. $x^2 + y^2 = 1$
b. $x^2 + y^2 = \frac{27}{4}$
c. $x^2 + y^2 = \frac{9}{4}$
d. $x^2 + y^2 = \frac{3}{2}$

Working Space

18 If the tangent at the point P on the circle $x^2 + y^2 + 6x + 6y = 2$ meets the straight line $5x - 2y + 6 = 0$ at a point Q on the Y-axis, then the length of PQ is (in units) **[IIT JEE 2002]**

a. 4 b. $2\sqrt{5}$

c. 5 d. 3

19 If $a > 2b > 0$, then the positive value of m for which $y = mx - b\sqrt{1+m^2}$ is a common tangent to $x^2 + y^2 = b^2$ and $(x-a)^2 + y^2 = b^2$, is **[IIT JEE 2002]**

a. $\frac{2b}{\sqrt{a^2 - 4b^2}}$ b. $\frac{\sqrt{a^2 - 4b^2}}{2b}$

c. $\frac{2b}{a - 2b}$ d. $\frac{b}{a - 2b}$

20 Tangent to the curve $y = x^2 + 6$ at a point $P(1, 7)$ touches the circle $x^2 + y^2 + 16x + 12y + c = 0$ at a point Q. Then, the coordinates of Q are **[IIT JEE 2005]**

a. $(-6, -11)$ b. $(-9, -13)$

c. $(-10, -15)$ d. $(-6, -7)$

21 Two parallel chords of a circle of radius 2 are at a distance $(\sqrt{3} + 1)$ apart. If the chords subtends at the centre, angles of $\frac{\pi}{k}$ and $\frac{2\pi}{k}$, where $k > 0$, then the value of $[k]$, is (where [·] denotes Greatest Integer Function) **[IIT JEE 2012]**

a. 2 b. 3

c. 1 d. 0

22 The circle passing through $(1, -2)$ and touching the axis of X at $(3, 0)$, also passes through the point **[JEE Main 2013]**

a. $(-5, 2)$ b. $(2, -5)$

c. $(5, -2)$ d. $(-2, -5)$

23 The line $2x - y + 1 = 0$ is tangent to the circle at the point $(2, 5)$ and the centre of the circle lies on $x - 2y = 4$. The radius of the circle is (in units)

a. $3\sqrt{5}$ b. $5\sqrt{3}$

c. $2\sqrt{5}$ d. $5\sqrt{2}$

Statement Type

Direction (Q. Nos. 24-29) *This section is based on Statement I and Statement II. Select the correct answer from the codes given below:*

a. Both Statement I and Statement II are correct and Statement II is the correct explanation of Statement I
b. Both Statement I and Statement II are correct and Statement II is not the correct explanation of Statement I
c. Statement I is correct but Statement II is incorrect
d. Statement II is correct but Statement I is incorrect

24 Tangents are drawn from the point $(17, 7)$ to the circle $x^2 + y^2 = 169$.

Statement I The tangents are mutually perpendicular.

Statement II The locus of the points from which mutually perpendicular tangents can be drawn to the given circle is $x^2 + y^2 = 338$. **[IIT JEE 2007]**

Working Space

25 Consider

$$L_1 : 2x + 3y + (p - 3) = 0$$
$$L_2 : 2x + 3y + (p + 3) = 0$$

where, p is real number and $C : x^2y^2 + 6x - 10y + 30 = 0$.

Statement I If lines L_1 is a chord of circle C, then line L_2 is not always a diameter of circle C.

Statement II If line L_1 is a diameter of circle C, then line L_2 is not a chord of circle C.

[IIT JEE 2008]

26 Consider the lines, $L = (k + 7)x - (k - 1)y - 4(k - 5) = 0$, where k is a parameter and the circle $C : x^2 + y^2 + 4x + 12y - 60 = 0$.

Statement I Every member of L intersects the circle C at an angle of 90°.

Statement II Every member of L is tangent to the circle C.

27 **Statement I** Angle between the tangents drawn from the point $P(13, 6)$ to the circle $S : x^2 + y^2 - 6x + 8y - 75 = 0$ is 90°.

Statement II Point P lies on the director's circle of S.

28 **Statement I** Only one normal can be drawn through the point $P(2, -3)$ to the circle $x^2 + y^2 - 4x + 8y - 16 = 0$.

Statement II Passing through any point lying inside a given circle only one normal can be drawn.

29 Consider the circle $C : x^2 + y^2 - 2x - 2y - 23 = 0$ and a point $P(3, 4)$.

Statement I No normal can be drawn to the circle C, passing through $P(3, 4)$.

because

Statement II Point P lies inside the given circle C.

Comprehension Type

Direction (Q. Nos. 30-35) *This section contains 2 paragraphs, each describing theory, experiments, data, etc. Six questions related to the paragraphs have been given. Each question has only one correct answer among the four given options (a), (b), (c) and (d).*

Passage I

Let A, B and C be the sets of real numbers (x, y) defined as

$$A : \{(x, y) : y \ge 1\}$$
$$B : \{(x, y) : x^2 + y^2 - 4x - 2y - 4 = 0\}$$
$$C : \{(x, y) : x + y = \sqrt{2}\}$$

30 Number of elements in the $A \cap B \cap C$, is

a. 0 b. 1
c. 2 d. infinite

31 $(x + 1)^2 + (y - 1)^2 + (x - 5)^2 + (y - 1)^2$ has the value equal to

a. 16 b. 25
c. 36 d. 49

32 If the locus of the point of intersection of the pair of perpendicular tangents to the circle B is the curve S, then the area enclosed between B and S, is

a. 6π b. 8π
c. 9π d. 18π

Working Space

Passage II

To the circle $x^2 + y^2 = 4$ two tangents are drawn from $P(-4, 0)$ which touch the circle at Q and R and a rhombus $PQP'R$ is completed. Now, answer the following questions :

33 Circumcentre of ΔPQR is at

a. $(-2, 0)$ b. $(2, 0)$

c. $\left(\frac{\sqrt{3}}{2}, 0\right)$ d. $(1, 0)$

34 Ratio of area of $\Delta PQP'$ and $\Delta P'QR$, is

a. $2:1$ b. $1:2$

c. $\sqrt{3}:2$ d. $1:1$

35 The area of rhombus $PQP'R$, is (in sq units)

a. $6\sqrt{3}$ b. $2\sqrt{2}$

c. $3\sqrt{3}$ d. $\frac{\sqrt{3}}{2}$

One Integer Value Correct Type

Direction (Q. Nos. 36-40) This section contains 5 questions. When worked out will result in an integer from 0 to 9 (both inclusive).

36 If pair of tangents to a circle in first quadrant has equation $6x^2 - 5xy + y^2 = 0$ and if one point of contact is $(1, 2)$, where the radius of that circle is $\sqrt{\alpha} - \sqrt{\beta}$, then $\alpha - \beta$ equals

37 Circles are drawn through $(1, 1)$ touching the circle $x^2 + y^2 - 6x + 2y + 1 = 0$. If r_1 and r_2 are the radii of smallest and largest circles. Then, the value of $(r_1 + r_2)^2 - (r_2 - r_1)^2$ equals

38 Suppose that the equation of the circle having $(-3, 5)$ and $(5, -1)$ as end points of a diameter is $(x - a)^2 + (y - b)^2 = r^2$. Then, $a + b + r$ is $(r > 0)$

39 Number of points in which the graphs of $|y| = x + 1$ and $(x - 1)^2 + y^2 = 4$ intersect, is

40 Number of value(s) of A for which the system of equations $x^2 = y^2$ and $(x - A)^2 + y^2 = 1$ has exactly 3 solutions, is

DPP-3 Angle between Circles, Radical Axis and Length of Tangent

Only One Option Correct Type

Working Space

Direction *(Q. Nos. 1-21) This section contains 21 multiple choice questions. Each question has four choices (a), (b), (c) and (d), out of which ONLY ONE option is correct.*

1 The locus of a point which moves such that the tangents from it to the two circles $x^2 + y^2 - 5x - 3 = 0$ and $3x^2 + 3y^2 + 2x + 4y - 6 = 0$ are equal, is given by

a. $2x^2 + 2y^2 + 7x - 4y - 3 = 0$
b. $17x + 4y + 3 = 0$
c. $4x^2 + 4y^2 - 3x + 4y - 9 = 0$
d. $13x - 4y + 15 = 0$

2 A point circle has its centre at $(2, -2)$. The radical axis of this circle with the circle $3(x^2 + y^2) - 6x + 9y + 1 = 0$, is

a. $6x - 3y + 23 = 0$
b. $6x + 3y - 23 = 0$
c. $6x + 3y + 23 = 0$
d. $6x - 3y - 23 = 0$

3 The radical centre of circles $(x - p)^2 + y^2 = p^2$, $x^2 + (y - q)^2 = q^2$ and $x^2 + y^2 = p^2 + q^2$, is

a. $\left(\frac{p^2+q^2}{2q}, \frac{p^2+q^2}{2p}\right)$
b. $\left(\frac{p^2+q^2}{p}, \frac{p^2+q^2}{q}\right)$
c. $\left(\frac{p^2+q^2}{2p}, \frac{p^2+q^2}{2q}\right)$
d. None of these

4 The equation of common tangent of circles $x^2 + y^2 = 4$ and $x^2 + y^2 - 8x + 12 = 0$ at the point, where they touch is

a. $x = -2$
b. $x = 2$
c. $y = 2$
d. $y = -2$

5 The radical axis of two circles and the lines joining their centres are

a. parallel
b. perpendicular
c. neither parallel nor perpendicular
d. None of these

6 Given the circles $x^2 + y^2 - 4x - 5 = 0$ and $x^2 + y^2 + 6x - 2y + 6 = 0$. Let P be a point (α, β) such that the tangents from P to both the circles are equal. Then,

a. $2\alpha + 10\beta + 11 = 0$
b. $2\alpha - 10\beta + 11 = 0$
c. $10\alpha - 2\beta + 11 = 0$
d. $10\alpha + 2\beta + 11 = 0$

7 Two equal circles of radius r, intersect such that each passes through the centre of the other. The length of their common chord is

a. $\sqrt{2}r$
b. $2r$
c. $\sqrt{3}r$
d. $\frac{\sqrt{3}r}{2}$

Working Space

8 The radical centre of the three circles $x^2 + y^2 = a^2$, $(x - c)^2 + y^2 = a^2$ and $x^2 + (y - b)^2 = a^2$, is

a. $\left(\frac{a}{2}, \frac{b}{2}\right)$
b. $\left(\frac{b}{2}, \frac{c}{2}\right)$
c. $\left(\frac{c}{2}, \frac{b}{2}\right)$
d. None of these

9 The equation of radical axis of the circle $x^2 + y^2 + x - y + 2 = 0$ and $3x^2 + 3y^2 - 4x - 12 = 0$ will be

a. $2x^2 + 2y^2 - 5x + y - 14 = 0$
b. $7x - 3y + 18 = 0$
c. $5x - y + 14 = 0$
d. None of these

10 The point from which the tangents to the circles $x^2 + y^2 - 8x + 40 = 0$, $5x^2 + 5y^2 - 25x + 80 = 0$ and $x^2 + y^2 - 8x + 16y + 160 = 0$ are equal length is

a. $\left(8, \frac{15}{2}\right)$
b. $\left(-8, \frac{15}{2}\right)$
c. $\left(8, \frac{-15}{2}\right)$
d. None of these

11 If the circles $x^2 + y^2 + 2ax + cy + a = 0$ and $x^2 + y^2 - 3ax + dy - 1 = 0$ intersecting at two distinct points P and Q, then the line $5x + by - a = 0$ passes through P and Q **[AIEEE 2005]**

a. exactly two values of a
b. infinitely many values of a
c. no value of a
d. exactly one value of a

12 The distance from the centre of the circle $x^2 + y^2 = 2x$ to the straight line passing through the points of intersection of two circles $x^2 + y^2 + 5x - 8y + 1 = 0$ and $x^2 + y^2 - 3x + 7y - 25 = 0$, is (in units) **[IIT JEE 1995]**

a. 1
b. 3
c. 2
d. $\frac{1}{3}$

13 If the circles $x^2 + y^2 + 2x + 2ky + 6 = 0$ and $x^2 + y^2 + 2ky + k = 0$ intersect orthogonally, then k is **[IIT JEE 2000]**

a. 2 or $\frac{-3}{2}$
b. -2 or $\frac{3}{2}$
c. 2 or $\frac{3}{2}$
d. -2 or $\frac{-3}{2}$

14 If the circles of same radius r and centres at (3, 4) and (7, 8) cut orthogonally, then r is (in units)

a. 2
b. 3
c. 4
d. 5

15 A circle touches the X-axis and also touches the circle with centre at (0, 3) and radius 2. The locus of the centre of the circle is **[AIEEE 2005]**

a. a parabola
b. a hyperbola
c. a circle
d. None of these

16 If a circle passes through the point (1, 2) and cuts the circle $x^2 + y^2 = 4$ orthogonally, then the equation of the locus of its centre is

a. $x^2 + y^2 - 2x - by - 7 = 0$
b. $x^2 + y^2 - 3x - 8y + 1 = 0$
c. $2x + 4y - 9 = 0$
d. $2x + 4y - 1 = 0$

Working Space

17 If a circle passes through the point (a, b) and cuts the circle $x^2 + y^2 = p^2$ orthogonally, then the equation of the locus of its centre is **[AIEEE 2005]**

a. $2ax + 2by - (a^2 + b^2 + p^2) = 0$

b. $x^2 + y^2 - 2ax - 3by + (a^2 - b^2 - p^2) = 0$

c. $2ax + 2by - (a^2 - b^2 + p^2) = 0$

d. $x^2 + y^2 - 3ax - 4by + (a^2 + b^2 - p^2) = 0$

18 The line $2x - y + 1 = 0$ is tangent to the circle at the point (2, 5) and the centre of the circles lies on $x - 2y = 4$. The radius of the circle is (in units)

a. $3\sqrt{5}$ b. $5\sqrt{3}$

c. $2\sqrt{5}$ d. $5\sqrt{2}$

19 The locus of the centres of the circles which cut the circles $x^2 + y^2 + 4x - 6y + 9 = 0$ and $x^2 + y^2 - 5x + 4y - 2 = 0$ orthogonally is

a. $9x + 10y - 7 = 0$ b. $x - y + 2 = 0$

c. $9x - 10y + 11 = 0$ d. None of these

20 If the angle between the tangents drawn from P to the circle $x^2 + y^2 + 4x - 6y + 9\sin^2\alpha + 13\cos^2\alpha = 0$ is 2α, then the locus of P, is

a. $x^2 + y^2 + 4x - 6y + 14 = 0$ b. $x^2 + y^2 + 4x - 6y - 9 = 0$

c. $x^2 + y^2 + 4x - 6y - 4 = 0$ d. $x^2 + y^2 + 4x - 6y + 9 = 0$

21 The angle at which the circles $(x - 1)^2 + y^2 = 10$ and $x^2 + (y - 2)^2 = 5$ intersects is

a. $\frac{\pi}{6}$ b. $\frac{\pi}{4}$ c. $\frac{\pi}{3}$ d. $\frac{\pi}{2}$

Statement Type

Direction (Q. Nos. 22-24) This section is based on Statement I and Statement II. Select the correct answer from the codes given below :

a. Both Statement I and Statement II are correct and Statement II is the correct explanation of Statement I

b. Both Statement I and Statement II are correct and Statement II is not the correct explanation of Statement I

c. Statement I is correct but Statement II is incorrect

d. Statement II is correct but Statement I is incorrect

22 Consider the circles, $S_1 : x^2 + y^2 + 2x - 4 = 0$ and $S_2 : x^2 + y^2 - y + 1 = 0$.

Statement I Tangents from the point $P(0, 5)$ on S_1 and S_2 are equal.

Statement II Point $P(0, 5)$ lies on the radical axis of the two circles.

23 **Statement I** The circle $C_1 : x^2 + y^2 - 6x - 4y + 9 = 0$ bisects the circumference of the circle $C_2 : x^2 + y^2 - 8x - 6y + 23 = 0$.

Statement II Centre of the circle C_1 lies on the circumference of C_2.

24 Let C_1 denotes a family of circles with centre on X-axis and touching the Y-axis at the origin and C_2 denotes a family of circles with centre of Y-axis and touching the X-axis at the origin.

Statement I Every member of C_1 intersects any member of C_2 at right angles at the point other than origin.

Statement II If two circles intersect at 90° at one point of their intersection, then they most intersect at 90° on the other point of intersection also.

Comprehension Type

Working Space

Direction (Q. Nos. 25-27) *This section contains a paragraph, describing theory, experiments, data, etc. Three questions related to the paragraphs have been given. Each question has only one correct answer among the four given options (a), (b), (c) and (d).*

Passage

Consider 3 circles $S_1 : x^2 + y^2 + 2x - 3 = 0$

$S_2 : x^2 + y^2 - 1 = 0$

$S_3 : x^2 + y^2 + 2y - 3 = 0$

25 The radius of the circle which bisects the circumferences of the circles $S_1 = 0$, $S_2 = 0, S_3 = 0$, is (in units)

a. 2 b. $2\sqrt{2}$
c. 3 d. $\sqrt{10}$

26 If the circle $S = 0$ is orthogonal to $S_1 = 0, S_2 = 0$ and $S_3 = 0$ and has its centre at (a, b) and radius equals to r, then the value of $(a + b + r)$ equals

a. 0 b. 1
c. 2 d. 3

27 The radius of the circle touching $S_1 = 0$ and $S_2 = 0$ at (1, 0) and passing through (3, 2), is (in units)

a. 1 b. $\sqrt{12}$
c. 2 d. $2\sqrt{2}$

One or More than One Options Correct Type

Direction (Q. Nos. 28 and 29) *This section contains 2 multiple choice questions. Each question has four choices (a), (b), (c) and (d), out of which ONE or MORE THAN ONE are correct.*

28 If $(a, 0)$ is a point on a diameter of the circle $x^2 + y^2 = 4$. Then, $x^2 - 4x - a^2 = 0$ has

a. exactly one real root in $(-1, 0]$
b. exactly one real root in $[2, 5]$
c. distinct roots greater than 5
d. distinct roots greater than -1 and less than 5

29 $C_1 : x^2 + y^2 = 25, C_2 : x^2 + y^2 - 2x - 4y - 7 = 0$ be two circles intersecting at the points A and B.

a. Equation of common chord must be $x + 2y - 9 = 0$
b. Equation of common chord must be $x + 2y + 7 = 0$
c. Tangents at A and B to the circle C_1 intersect at $\left(\frac{25}{9}, \frac{50}{9}\right)$
d. Tangents at A and B to the circle C_1 intersect at (1, 2)

DPP-4 Location of Two Circles, Number of Circles, Tangents and Family of Circles

Only One Option Correct Type

Direction (Q. Nos. 1-19) *This section contains 19 multiple choice questions. Each question has four choices (a), (b), (c) and (d), out of which ONLY ONE option is correct.*

Working Space

1 The number of common tangents to the circles $x^2 + y^2 = 4$ and $x^2 + y^2 - 8x + 12 = 0$, is

a. 1
b. 2
c. 3
d. 4

2 The coordinates of the point at which the circle $x^2 + y^2 - 4x - 2y - 4 = 0$ and $x^2 + y^2 - 12x - 8y - 36 = 0$ touch each other, are **[IIT JEE 1993]**

a. (3, – 2)
b. (–2, 3)
c. (3, 2)
d. None of these

3 Given that two circles $x^2 + y^2 = r^2$ and $x^2 + y^2 - 10x + 16 = 0$, the value of r such that they intersect in real and distinct point is given by **[IIT JEE 1994]**

a. $2 < r < 8$
b. $r = 2$ or $r = 8$
c. $r < 2$ or $r > 8$
d. None of these

4 The number of common tangents to the circles $x^2 + y^2 = 4$ and $x^2 + y^2 - 6x - 8y = 24$, is **[IIT JEE 1998]**

a. 0
b. 1
c. 3
d. 4

5 Two circles $x^2 + y^2 + 2hx = r_1^2 - h^2$ and $x^2 + y^2 + 2kx = r_2^2 - k^2$ touch externally, if

a. $h^2 + k^2 = r_1^2 - r_2^2$
b. $h^2 + k^2 = r_1^2 + r_2^2$
c. $h^2 + k^2 = r_1^2 + r_2^2 - 2r_1r_2$
d. $h^2 + k^2 = r_1^2 + r_2^2 + 2r_1r_2$

6 Circles $x^2 + y^2 + 4x + d = 0$, $x^2 + y^2 + 4fy + d = 0$ touch each other, if

a. $f = \pm 2\sqrt{4 + d}$
b. $d = \pm \dfrac{2}{\sqrt{4 - d}}$
c. $f = \pm \sqrt{\dfrac{d}{4 + d}}$
d. $f = \pm \sqrt{\dfrac{d}{4 - d}}$

7 The number of common tangents to the circles $x^2 + y^2 - 2x - 4y + 1 = 0$ and $x^2 + y^2 - 12x - 16y + 91 = 0$, is

a. 1
b. 2
c. 3
d. 4

Working Space

8 The two circles $x^2 + y^2 - 2x - 4y = 0$ and $x^2 + y^2 - 8y - 4 = 0$

a. intersect each other
b. touch each other internally
c. touch each other externally
d. None of the above

9 If the circles $x^2 + y^2 + 2x - 8y + 8 = 0$ and $x^2 + y^2 + 10x - 2y + 22 = 0$ touch each other, then

a. $\left(\frac{-17}{5}, \frac{11}{5}\right)$ b. $\left(\frac{11}{3}, 2\right)$

c. $\left(\frac{17}{5}, \frac{11}{5}\right)$ d. $\left(\frac{-11}{3}, 2\right)$

10 If a circle S_1 bisects the circumference of another circle S_2, then their radical axis

a. passes through the centre of S_1
b. passes through the centre of S_2
c. bisects the line joining their centres
d. None of the above

11 The locus of a point such that tangents drawn from it to the circle $x^2 + y^2 = a^2$ are at 120°, is

a. $4x^2 + 4y^2 = 3a^2$ b. $3x^2 + 3y^2 = 4a^2$

c. $x^2 + y^2 = \frac{2a^2}{3}$ d. None of these

12 Let A be the centre of the circle $x^2 + y^2 - 2x - 4y - 20 = 0$ and $B(1, 7)$ and $D(4, -2)$ are points on the circle, then if tangents are drawn at B and D, which meets at C, then area of quadrilateral $ABCD$, is (in sq units)

a. 150 b. 75

c. $\frac{75}{2}$ d. None of these

13 The equation of the circle whose diameter is the common chord of the circle, $x^2 + y^2 + 3x + 2y + 1 = 0$ and $x^2 + y^2 + 3x + 4y + 2 = 0$, is

a. $x^2 + y^2 + 3x + y + 5 = 0$
b. $x^2 + y^2 + x + 3y + 7 = 0$
c. $x^2 + y^2 + 2x + 3y + 1 = 0$
d. $2(x^2 + y^2) + 6x + 2y + 1 = 0$

14 The circle $x^2 + y^2 + 2gx + 2fy + c = 0$ bisects the circumference of the circle $x^2 + y^2 + 2ax + 2by + d = 0$, then

a. $2a(g - a) + 2b(f - b) = (c - d)$
b. $2a(g + a) + 2b(f + b) = (c + d)$
c. $2g(g - a) + 2f(f - b) = (d - c)$
d. $2g(g + a) + 2f(f + b) = (c + d)$

15 If the circle $C_1 : x^2 + y^2 = 16$ intersects another circle of C_2 of radius 5 in such a manner that the common chord is of maximum length and has a slope equal to $\frac{3}{4}$, then the coordinates of the centre of C_2, are

a. $\left(\frac{-9}{5}, \frac{-12}{5}\right)$ b. $\left(\frac{9}{5}, \frac{12}{5}\right)$

c. $\left(\mp\frac{9}{5}, \pm\frac{12}{5}\right)$ d. None of these

Working Space

16 The centre of the circle passing through (0, 0) and (1, 0) and touching the circle $x^2 + y^2 = 9$, is **[AIEEE 2002]**

a. $\left(\frac{1}{2}, \frac{1}{2}\right)$ b. $\left(\frac{1}{2}, -\sqrt{2}\right)$

c. $\left(\frac{3}{2}, \frac{1}{2}\right)$ d. $\left(\frac{1}{2}, \frac{3}{2}\right)$

17 The locus of the centre of a circle which touches externally. The circle $x^2 + y^2 - 6x - 6y + 14 = 0$ and also touches the Y-axis, is given by the equation

a. $x^2 - 6x - 10y + 14 = 0$ b. $x^2 - 10x - 6y + 14 = 0$

c. $y^2 - 6x - 10y + 14 = 0$ d. $y^2 - 10x - 6y + 14 = 0$

18 A circle is given by $x^2 + (y-1)^2 = 1$, another circle C touches it externally and also the X-axis, then the locus of its centre is **[IIT JEE 2005]**

a. $\{(x, y): x^2 = 4y\} \cup \{(x, y): y \le 0\}$ b. $\{(x, y): x^2 + (y-1)^2 = 4\} \cup \{(x, y): y \le 0\}$

c. $\{(x, y): x^2 = y\} \cup \{(0, y): y \le 0\}$ d. $\{(x, y): x^2 = 4y\} \cup \{(0, y): y \le 0\}$

19 Let $ABCD$ be a quadrilateral with area 18, with side AB parallel to the side CD and $AB = 2CD$. Let AD be the perpendicular to AB and CD. If a circle is drawn inside the quadrilateral $ABCD$ touching all the sides, then its radius is (in units) **[IIT JEE 2007]**

a. 3 b. 2 c. $\frac{3}{2}$ d. 1

Comprehension Type

Direction (Q. Nos. 20-30) *This section contains 4 paragraphs, each describing theory, experiments, data, etc. Eleven questions related to the paragraphs have been given. Each question has only one correct answer among the four given options (a), (b), (c) and (d).*

Passage I

Consider the family of circles $x^2 + y^2 - 2x - 2\lambda y - 8 = 0$, passing through two fixed points A and B. Also, $S = 0$ is a circle of this family, the tangent to which at A and B intersect on the line $x + 2y + 5 = 0$.

20 The distance between the points A and B is (in units)

a. 4 b. $4\sqrt{2}$ c. 6 d. 8

21 The area of an equilateral triangle inscribed in S is (in sq units)

a. $27\frac{\sqrt{3}}{4}$ b. $9\frac{\sqrt{3}}{2}$ c. $27\frac{\sqrt{3}}{2}$ d. $9\sqrt{3}$

22 If the circle $x^2 + y^2 - 10x + 2y + C = 0$ is orthogonal to $S = 0$, then the value of C equals

a. 8 b. 9 c. 10 d. 12

Passage II

Let $ABCD$ be a square of side length 2 units. C_2 is the circle through vertices A, B, C, D and C_1 is the circle touching all sides of the square $ABCD$. L is the line through A. **[IIT JEE 2006]**

23 If P is the point on C_1 and Q on C_2, then $\frac{PA^2 + PB^2 + PC^2 + PD^2}{QA^2 + QB^2 + QC^2 + QD^2}$, is equal to

a. 0.75 b. 1.25 c. 1 d. 0.5

Working Space

24 A circle touches the line L and the circle C_1 externally such that both the circles are on the same side of the line, then the locus of the centre of the circle is

a. ellipse
b. hyperbola
c. parabola
d. parts of straight line

25 A line M through A is drawn parallel to BD. Point S moves such that its distance from the line BD and the vertex A are equal. If locus of parabola cuts M at T_2 and T_3 and AC at T_1, then ar $\Delta T_1 T_2 T_3$ is

a. $\frac{1}{2}$ sq unit
b. $\frac{2}{3}$ sq unit
c. 1 sq unit
d. 2 sq units

Passage III

A circle C of radius 1 is inscribed in an equilateral ΔPQR. The points of contact of C with the sides PQ, QR, RP are D, E, F respectively. The line PQ is given by the equation $\sqrt{3}x + y - 6 = 0$ and the point D is $\left(\frac{3\sqrt{3}}{2}, \frac{3}{2}\right)$. Further, it is given that the origin and the centre of C are on the same side of the line PQ. **[IIT JEE 2008]**

26 The equation of circle C, is

a. $(x - 2\sqrt{3})^2 + (y - 1)^2 = 1$
b. $(x - 2\sqrt{3})^2 + \left(y + \frac{1}{2}\right)^2 = 1$
c. $(x - \sqrt{3})^2 + (y + 1)^2 = 1$
d. $(x - \sqrt{3})^2 + (y - 1)^2 = 1$

27 Points E and F are given by

a. $\left(\frac{\sqrt{3}}{2}, \frac{3}{2}\right), (\sqrt{3}, 0)$
b. $\left(\frac{\sqrt{3}}{2}, \frac{1}{2}\right), (\sqrt{3}, 0)$
c. $\left(\frac{\sqrt{3}}{2}, \frac{3}{2}\right), \left(\frac{\sqrt{3}}{2}, \frac{1}{2}\right)$
d. $\left(\frac{3}{2}, \frac{\sqrt{3}}{2}\right), \left(\frac{\sqrt{3}}{2}, \frac{1}{2}\right)$

28 Equations of the sides QR, RP are

a. $y = \frac{2}{\sqrt{3}}x + 1, y = \frac{2}{3}x - 1$
b. $y = \frac{1}{3}x, y = 0$
c. $y = \frac{\sqrt{3}}{2}x + 1, y = -\frac{\sqrt{3}}{2}x - 1$
d. $y = \sqrt{3}x, y = 0$

Passage IV

A tangent PT is drawn to the circle $x^2 + y^2 = 4$ at the point $P(\sqrt{3}, 1)$. A straight line L, perpendicular to PT is a tangent to the circle $(x - 3)^2 + y^2 = 1$.

29 A possible equation of L, is

a. $x - \sqrt{3}y = 1$
b. $x + \sqrt{3}y = 1$
c. $x - \sqrt{3}y = -1$
d. $x + \sqrt{3}y = 5$

30 A common tangent to the two circles is

a. $x = 4$
b. $y = 2$
c. $x + \sqrt{3}y = 4$
d. $x + 2\sqrt{2}y = 6$

Matching List Type

Direction (Q. No. 31) *Choices for the correct combination of elements from Column I and Column II are given as options (a), (b), (c) and (d), out of which one is correct.*

31 Match the Column I with Column II and mark the correct option from the codes given below.

	Column I		Column II
i.	Two intersecting circles	p.	have a common tangent.
ii.	Two mutually external circles	q.	have a common normal.
iii.	Two circles, one strictly inside the other	r.	do not have a common tangent.
iv.	Two branches of a hyperbola	s.	do not have a common normal.

	i	ii	iii	iv
a.	p,q	p,q	q,r	q,r
b.	p	q,r	s,r	q,p
c.	p,q	p,q	s	r,s
d.	q,r	p,q	r,s	s,p

One Integer Value Correct Type

Direction (Q. Nos. 32-35) *This section contains 4 questions. When worked out will result in an integer from 0 to 9 (both inclusive).*

32 The centres of two circles C_1 and C_2 each of unit radius are at a distance of 6 units from each other. Let P be the mid-point of the line segment joining the centres of C_1 and C_2 and C be a circle touching circles C_1 and C_2 externally. If a common tangents to C_1 and C passing through P is also a common tangent to C_2 and C_1, then the radius of the circle is (in units) **[IIT JEE 2009]**

33 Let C_1 and C_2 are circles defined by $x^2 + y^2 - 20x + 64 = 0$ and $x^2 + y^2 + 30x + 144 = 0$. The length of the shortest line segment PQ that is tangent to C_1 at P and C_2 at Q is k, then $\frac{k}{4}$ is

34 If the straight line $y = mx$ is outside the circle $x^2 + y^2 - 20y + 90 = 0$, then sum of all integral values of m, is

35 The centre of the smallest circle touching the circles $x^2 + y^2 - 2y - 3 = 0$ and $x^2 + y^2 - 8x - 18y + 93 = 0$ is (h, k), then $|h - k|$ is

Working Space

DPP-5 $S' = T,\ S \cdot S' = T^2,\ T = 0$ and Pole, Polar

Only One Option Correct Type

Working Space

Direction (Q. Nos. 1-22) *This section contains 22 multiple choice questions. Each question has four choices (a), (b), (c) and (d), out of which ONLY ONE option is correct.*

1 Equation of pair of tangents from (0, 1) on circle $x^2 + y^2 = \frac{1}{4}$, is

a. $\left(x^2 + y^2 - \frac{1}{4}\right) \cdot \left(\frac{3}{4}\right) = \left(y - \frac{1}{4}\right)^2$

b. $\left(x^2 - y^2 - \frac{1}{4}\right) \cdot \left(\frac{3}{4}\right) = \left(y - \frac{1}{4}\right)^2$

c. $\left(x^2 + y^2 - \frac{1}{4}\right) \cdot \left(\frac{3}{4}\right) = \left(y + \frac{1}{4}\right)^2$

d. None of these

2 The pole of the line $9x + y - 28 = 0$ w.r.t. circle $2x^2 + 2y^2 - 3x + 5y - 7 = 0$, is

a. (3, 1)
b. (1, 3)
c. (3, – 1)
d. (–3, 1)

3 If line $x + 2y = 1$ is the polar of circle $x^2 + y^2 = 5$, then pole is

a. (5, 10)
b. (5, 5)
c. (–5, 10)
d. (–5, – 10)

4 If two distinct chords from the point (p, q) on the circle $x^2 + y^2 = px + qy$ (where, $pq \neq 0$) are bisected by the X-axis, then

a. $p^2 = q^2$
b. $p^2 = 8q^2$
c. $p^2 < 8q^2$
d. $p^2 > 8q^2$

5 A diameter of the circle $x^2 + y^2 = 2x + 6y - 6$ is a chord of the another circle with centre (2, 1). The radius of this circle *C*, is (in units) **[IIT JEE 2004]**

a. 1
b. 3
c. 2
d. $\sqrt{3}$

6 Tangents are drawn from the point on the circle $x^2 + y^2 = a^2$ to the circle $x^2 + y^2 = b^2$ and the chord of contact touch the circle $x^2 + y^2 = c^2$, then *a, b, c* will be in

a. AP
b. GP
c. HP
d. None of these

7 The locus of the mid-point of the chord of tangents drawn from points lying on the straight line $4x - 5y = 20$ to the circle $x^2 + y^2 = 9$, is **[IIT JEE 2012]**

a. $20(x^2 + y^2) - 36x + 45y = 0$

b. $20(x^2 + y^2) + 36x - 45y = 0$

c. $36(x^2 + y^2) - 20x + 45y = 0$

d. $36(x^2 + y^2) + 20x - 45y = 0$

Working Space

8 If the pole of the polar w.r.t. the circle $x^2 + y^2 = c^2$ is on the circle $x^2 + y^2 = 9c^2$. Then, this polar will be the tangent of the circle.

a. $x^2 + y^2 = \frac{9}{4}c^2$ b. $x^2 + y^2 = \frac{c^2}{9}$

c. $x^2 + y^2 = 4c^2$ d. None of these

9 The locus of the mid-point of the chord of length $2l$ of the circle $x^2 + y^2 = a^2$, will be

a. $x^2 + y^2 = l^2 + a^2$ b. $2x^2 + 2y^2 = a^2 + l^2$

c. $2x^2 + 2y^2 = l^2 - a^2$ d. $x^2 + y^2 = a^2 - l^2$

10 The equation of the pair of tangents drawn to the circle $x^2 + y^2 - 2x + 4y + 3 = 0$ from $(6, -5)$ is

a. $7x^2 + 23y^2 + 30xy + 66x + 50y - 73 = 0$

b. $7x^2 + 23y^2 - 30xy - 66x - 50y + 73 = 0$

c. $7x^2 + 23y^2 + 30xy - 66x - 50y - 73 = 0$

d. None of the above

11 If OA and OB are the tangents to the circle $x^2 + y^2 - 6x - 8y + 21 = 0$ drawn from the origin O, then AB equals

a. $\sqrt{\frac{17}{3}}$ b. $\frac{4}{5}\sqrt{21}$

c. 11 d. None of these

12 Two lines $l_1x + m_1y + n_1 = 0$ and $l_2x + m_2y + n_2 = 0$ are conjugate lines with respect to the circle $x^2 + y^2 = a^2$, if

a. $l_1l_2 + m_1m_2 + n_1n_2 = 0$ b. $l_1l_2 + m_1m_2 = n_1n_2$

c. $a^2(l_1l_2 + m_1m_2) = n_1n_2$ d. $l_1l_2 + m_1m_2 = a^2n_1n_2$

13 The equation of the diameter of the circle $x^2 + y^2 + 2x - 4y - 11 = 0$ which bisects the chords intercepted on the line $2x - y + 3 = 0$, is

a. $x + y - 7 = 0$ b. $2x - y - 5 = 0$

c. $x + 2y - 3 = 0$ d. None of these

14 The equation of chord of circle $x^2 + y^2 - 8x + 6y + 9 = 0$ from the point $(8, 1)$, is

a. $x + y + 5 = 0$ b. $x + y = 5$

c. $x - y + 5 = 0$ d. $x + y = 7$

15 The pole of the line $lx + my + n = 0$ with respect to the circle $x^2 + y^2 = a^2$, is

a. $\left(\frac{-la^2}{n}, \frac{ma^2}{n}\right)$ b. $\left(\frac{la^2}{n}, \frac{ma^2}{n}\right)$

c. $\left(\frac{-la^2}{n}, \frac{-ma^2}{n}\right)$ d. $\left(\frac{la^2}{n}, \frac{-ma^2}{n}\right)$

16 From the origin, chords are drawn to the circle $(x - 1)^2 + y^2 = 1$, then the equation of locus of middle points of these chords, is

a. $x^2 + y^2 = 1$ b. $x^2 + y^2 = x$ c. $x^2 + y^2 = y$ d. None of these

17 The lengths of the tangents from the points A and B to a circle are l_1 and l_2, respectively. If points are conjugate with respect to the circle, then AB^2 is equal to

a. $l_1 + l_2$ b. $l_1^2 + l_2^2$ c. $|l_1^2 - l_2^2|$ d. None of these

Working Space

18 The angle between the tangents from (α, β) to the circle $x^2 + y^2 = a^2$, is

a. $\tan^{-1}\left(\frac{a}{\sqrt{S_1}}\right)$

b. $2\tan^{-1}\left(\frac{a}{\sqrt{S_1}}\right)$

c. $2\tan^{-1}\left(\frac{\sqrt{S_1}}{a}\right)$

d. None of these

19 The equation of the locus of the middle point of a chord of the circle $x^2 + y^2 - 2x - 2y = 0$ such that the pair of lines joining the origin to the point of intersection of the chord and the circle are equally inclined to the X-axis is

a. $x + y = 2$

b. $x - y = 2$

c. $2x - y = 1$

d. None of these

20 Three concentric circles of which the biggest is $x^2 + y^2 = 1$, have their radii in an AP. If the line $y = x + 1$ cuts all the circles in real and distinct points. The interval in which the common difference of the AP will lie is

a. $\left(0, \frac{1}{4}\right)$

b. $\left(0, \frac{1}{2\sqrt{2}}\right)$

c. $\left(0, \frac{2-\sqrt{2}}{4}\right)$

d. None of these

21 The locus of poles whose polar with respect to $x^2 + y^2 = a^2$ always passes through $(k, 0)$, is

a. $kx = a^2$

b. $kx + a^2 = 0$

c. $ky + a^2 = 0$

d. $ky = a^2$

22 The locus of the middle points of the system of chords of the circle $x^2 + y^2 = 16$ which are parallel to the line $2y = 4x + 5$, is

a. $x = 2y$

b. $x + 2y = 0$

c. $y + 2x = 0$

d. $y = 2x$

Comprehension Type

Direction *(Q. Nos. 23-28) This section contains 2 paragraphs, each describing theory, experiments, data, etc. Six questions related to the paragraphs have been given. Each question has only one correct answer among the four given options (a), (b), (c) and (d).*

Passage I

Consider a circle $x^2 + y^2 = 4$ and a point $P(4, 2)$. θ denotes the angle enclosed by the tangents from P on the circle and A, B are the points of contact of the tangents from P on the circle.

23 The value of θ lies in the interval.

a. $(0, 15°)$ b. $(15°, 30°)$ c. $(30°, 45°)$ d. $(45°, 60°)$

24 The intercept made by a tangent on the X-axis, is

a. $\frac{9}{4}$ b. $\frac{10}{4}$ c. $\frac{11}{4}$ d. $\frac{12}{4}$

25 Locus of the middle points of the portion of the tangents to the circle terminated by the coordinate axes, is

a. $\frac{1}{x^2} + \frac{1}{y^2} = 1$

b. $\frac{1}{x^2} + \frac{1}{y^2} = \frac{1}{4}$

c. $\frac{1}{x^2} + \frac{1}{y^2} = \frac{1}{9}$

d. $\frac{1}{x^2} - \frac{1}{y^2} = \frac{1}{16}$

Working Space

Passage II

Consider the two quadratic polynomials

$$C_a : y = \frac{x^2}{4} - ax + (a^2 + a - 2) \text{ and } C : y = 2 - \frac{x^2}{4}$$

26 If the origin lies between the zeroes of the polynomial C_a, then the number of integral value(s) of a is

a. 1
b. 2
c. 3
d. more than 3

27 If a varies, then the equation of the locus of the vertex of C_a, is

a. $x - 2y - 4 = 0$
b. $2x - y - 4 = 0$
c. $x - 2y + 4 = 0$
d. None of these

28 For $a = 3$, if the lines $y = m_1 x + c_1$ and $y = m_2 x + c_2$ are common tangents to the graph of C_a and C, then the value of $(m_1 + m_2)$ is equal to

a. −6
b. −3
c. $\frac{1}{2}$
d. None of these

One Integer Value Correct Type

Direction (Q. Nos. 29 and 30) *This section contains 2 questions. When worked out will result in an integer from 0 to 9 (both inclusive).*

29 A variable circle C has the equation $x^2 + y^2 - 2(t^2 - 3t + 1)x - 2(t^2 + 2t)y + t = 0$, where t is a parameter. If the power of point $P(a, b)$ w.r.t. the circle C is constant, then the ordered pair $(a + b)$, is

30 Locus of middle points of a system of parallel chords with slope 2 of the circle $x^2 + y^2 - 4x - 2y - 4 = 0$, has the equation $ax + by + c = 0$, then $|a + b + c|$ is

SET - 1

Revisal Problems for JEE Main

Only One Option Correct Type

Working Space

Direction *(Q. Nos. 1-45) This section contains 45 multiple choice questions. Each question has four choices (a), (b), (c) and (d), out of which ONLY ONE option is correct.*

1 The equations of the circumcircle of the triangle formed by the lines $y + \sqrt{3}x = 6$, $y - \sqrt{3}x = 6$ and $y = 0$, is

a. $x^2 + y^2 + 4x = 0$ b. $x^2 + y^2 - 4y = 0$
c. $x^2 + y^2 - 4y = 12$ d. None of these

2 The radius of a circle with centre (a, b) and passing through the centre of the circle $x^2 + y^2 - 2gx + f^2 = 0$, is

a. $\sqrt{(a-g)^2 + b^2}$ b. $\sqrt{a^2 + (b+g)^2}$
c. $\sqrt{a^2 + (b-g)^2}$ d. $\sqrt{(a+g)^2 + b^2}$

3 The equation of the circle of radius 5 in the first quadrant which touches the X-axis and the line $4y = 3x$, is

a. $x^2 + y^2 - 30x - 10y + 225 = 0$ b. $x^2 + y^2 + 30x + 10y - 225 = 0$
c. $x^2 + y^2 - 30x - 10y - 225 = 0$ d. $x^2 + y^2 + 30x + 10y + 225 = 0$

4 A rectangle $ABCD$ is inscribed in a circle with a diameter lying along the line $3y = x + 10$. If A and B are the points $(-6, 7)$ and $(4, 7)$, respectively. Then, the area of the rectangle is (in sq units)

a. 40 b. 80
c. 20 d. 160

5 The parametric coordinates of a point on the circle $x^2 + y^2 - 2x + 2y - 2 = 0$, are

a. $(1 + 2\cos\alpha, 1 + 2\sin\alpha)$ b. $(1 - 2\cos\alpha, 1 - 2\sin\alpha)$
c. $(1 + 2\cos\alpha, -1 + 2\sin\alpha)$ d. $(-1 + 2\cos\alpha, 1 + 2\sin\alpha)$

6 If $y = 2x$ is a chord of the circle $x^2 + y^2 = 10x$, then equation of the circle whose diameter is this chord is

a. $x^2 + y^2 + 2x + 4y = 0$ b. $x^2 + y^2 + 2x - 4y = 0$
c. $x^2 + y^2 - 2x - 4y = 0$ d. None of these

7 Locus of the point of intersection of lines $x\cos\alpha + y\sin\alpha = a$ and $x\sin\alpha - y\cos\alpha = a$, $(\alpha \in R)$, is

a. $x^2 + y^2 = a^2$ b. $x^2 + y^2 = 2a^2$
c. $x^2 + y^2 + 2x + 2y = a^2$ d. None of these

8 If the straight line $y = mx$ is outside the circle $x^2 + y^2 - 20y + 90 = 0$, then

a. $m > 3$ b. $m < 3$
c. $|m| > 3$ d. $|m| < 3$

9 The point (1, 2) lies inside the circle $x^2 + y^2 - 4x - 2y + c = 0$, then

a. $c < 3$ b. $c > 3$
c. $2 < c < 3$ d. None of these

10 The equation of the circle which passes through the origin has its centre on the line $x + y = 4$ and cuts the circle $x^2 + y^2 - 4x + 2y + 4 = 0$ orthogonally, is

a. $x^2 + y^2 - 2x - 6y = 0$ b. $x^2 + y^2 - 6x - 3y = 0$
c. $x^2 + y^2 - 4x - 4y = 0$ d. None of these

Working Space

11 The distance from the centre of the circle $x^2 + y^2 = 2x$ to the straight line passing through the points of intersection of the two circles $x^2 + y^2 + 5x - 8y + 1 = 0$, $x^2 + y^2 - 3x + 7y - 25 = 0$, is (in units)

a. 1 b. 2
c. 3 d. None of these

12 The length of common chord of the circle $x^2 + y^2 + 4x + 6y + 4 = 0$ and $x^2 + y^2 + 6x + 4y + 4 = 0$, is (in units)

a. $\sqrt{10}$ b. $\sqrt{22}$
c. $\sqrt{34}$ d. $\sqrt{38}$

13 The equation of circle passing through the points of intersection of circle $x^2 + y^2 = 6$ and $x^2 + y^2 - 6x + 8 = 0$ and the point (1, 1), is

a. $x^2 + y^2 - 3x + 1 = 0$ b. $x^2 + y^2 - 6x + 4 = 0$
c. $x^2 + y^2 - 4y + 2 = 0$ d. None of these

14 If the circle $x^2 + y^2 + 4x + 22y + c = 0$ bisects the circumference of the circle $x^2 + y^2 - 2x + 8y - d = 0$, then $(c + d)$ is equal to

a. 60 b. 50
c. 40 d. 56

15 If a circle centred at (4, 3) touches a line $x + y = 1$, then its point of contact is

a. (1, 2) b. (1, 0)
c. (0, 1) d. (0, 2)

16 The abscissa of two points A and B are the roots of the equation $x^2 + 2ax - b^2 = 0$ and their ordinates are the roots of the equation $y^2 + 2py - q^2 = 0$. The radius of the circle with AB as a diameter will be

a. $\sqrt{a^2 + b^2 + p^2 + q^2}$ b. $\sqrt{b^2 + q^2}$
c. $\sqrt{a^2 + b^2 - p^2 - q^2}$ d. $\sqrt{a^2 + p^2}$

17 Two vertices of an equilateral triangle are (–1, 0) and (1, 0) and its third vertex lies above the X-axis, then its circumcircle, is

a. $x^2 + \left(y - \frac{1}{\sqrt{3}}\right)^2 = \frac{4}{3}$ b. $x^2 + \left(y + \frac{1}{\sqrt{3}}\right)^2 = \frac{4}{3}$
c. $x^2 + \left(y - \frac{1}{3}\right)^2 = \frac{4}{3}$ d. None of these

18 Equation $ax^2 + 2hxy + by^2 + 2gx + 2fy + c = 0$ represents a circle, if

a. $a = b \neq 0$ and $h = 0$ b. $a + b = 0$
c. $a = b$ and $c = 0$ d. $f = g = 0$

19 Parametric coordinates of a curve is $x = 3 + 2(\cos^2 t - \sin^2 t)$ and $y = 4 + 4\sin t \cos t$, then curve is

a. straight line b. circle
c. parabola d. ellipse

20 If the chord $y = mx + 1$ of the circle $x^2 + y^2 = 1$ subtends an angle of measure 45° at the major segment of the circle, then value of m, is **[AIEEE 2002]**

a. 2 b. –2
c. 1 d. None of these

21 The centre of a set of circles, each of radius 3, lies on the circle $x^2 + y^2 = 25$. The locus of any point in the set is **[AIEEE 2002]**

a. $4 \leq x^2 + y^2 \leq 64$ b. $x^2 + y^2 \leq 25$
c. $x^2 + y^2 \geq 25$ d. $3 \leq x^2 + y^2 \leq 9$

22 Circles are drawn through the point (2, 0) to cut intercept of length 5 units on the X-axis. If their centres lie in the first quadrant, then their equation is **[IIT JEE 1992]**

a. $x^2 + y^2 - 9x + 2fy + 14 = 0$ b. $3x^2 + 3y^2 + 27x - 2fy + 42 = 0$
c. $x^2 + y^2 - 2fx - 9y + 14 = 0$ d. None of these

Working Space

23 Let L_1 be a straight line passing through the origin and L_2 be the straight line $x + y = 1$. If the intercepts made by the circle $x^2 + y^2 - x + 3y = 0$ on L_1 and L_2 are equal, then which of the following equation can represent L_1?

a. $x + y = 0$ b. $x - y = 0$
c. $x - 7y = 0$ d. None of these

24 Let PQ and RS be tangent at the extremities of the diameter PR of a circle of radius r. If PS and RQ intersect at a point X on the circumference of the circle, then $(2r)$ equals **[IIT JEE 2001]**

a. $\sqrt{PQ \cdot RS}$ b. $\frac{PQ + RS}{2}$
c. $\frac{2PQ \cdot RS}{PQ + RS}$ d. $\sqrt{\frac{PQ^2 + RS^2}{2}}$

25 In an equilateral triangle, 3 coins of radii 1 unit each are kept so that they touch each other and also the sides of the triangle. Area of the triangle, is (in sq units) **[IIT JEE 2005]**

a. $4 + 2\sqrt{3}$ b. $6 + 4\sqrt{3}$
c. $12 + \frac{7\sqrt{3}}{4}$ d. $3 + \frac{7\sqrt{3}}{4}$

26 Consider a family of circles which are passing through the point $(-1, 1)$ and are tangent to the X-axis. If (h, k) are the coordinates of the centre of the circles, then the set of values of k is given by the interval **[AIEEE 2007]**

a. $\frac{-1}{2} \le k \le \frac{1}{2}$ b. $k \le \frac{1}{2}$
c. $0 < k < \frac{1}{2}$ d. $k \ge \frac{1}{2}$

27 Let a and b be non-zero real numbers. Then, the equation $(ax^2 + by^2 + c)(x^2 - 5xy + 6y^2) = 0$ represents **[IIT JEE 2008]**

a. four straight lines, when $c = 0$ and a, b are the same sign
b. two straight lines and a circle, when $a = b$ and c is of sign opposite of that of a
c. two straight lines and a hyperbola, when a and b are of the same sign and c is of sign opposite to that of a
d. a circle and an ellipse, when a and b are of the same sign opposite to that of a

28 The tangents to $x^2 + y^2 = a^2$ having inclinations α and β intersect at P. If $\cot\alpha + \cot\beta = 0$, then locus of P is

a. $x - y = 0$ b. $x + y = 0$
c. $xy = 0$ d. None of these

29 If from any point P on the circle $x^2 + y^2 + 2gx + 2fy + c = 0$, tangents are drawn to the circle $x^2 + y^2 + 2gx + 2fy + c\sin^2\alpha + (g^2 + f^2)\cos^2\alpha = 0$, then the angle between the tangents is

a. α b. 2α
c. $\pi/2$ d. None of these

30 The possible value of p for which $x\cos\alpha + y\sin\alpha = p$ is tangent to the circle $x^2 + y^2 - 2qx\cos\alpha - 2qy\sin\alpha = 0$, is/are

a. q and $2q$ b. 0 and q
c. 0 and $2q$ d. None of these

31 A line meets the coordinate axes at A and B. A circle is circumscribed about the ΔOAB. If m and n are the distance of the tangents to the circle at the origin from the points A and B respectively, then diameter of the circle is

a. $n(m + n)$ b. $(m + n)$
c. $m(m + n)$ d. $\frac{1}{2}(m + n)$

32 If the area of the triangle formed by the tangent and normal of circle $x^2 + y^2 = 4$ at its point $(1, \sqrt{3})$ together with the positive direction of X-axis, is (in sq units)

a. $2\sqrt{3}$ b. $\sqrt{3}$
c. $4\sqrt{3}$ d. $\frac{\sqrt{3}}{2}$

Working Space

33 A circle is given by $x^2 + (y-1)^2 = 1$, the another circle C touches it externally and also touches the X-axis, then the locus of its centre is **[IIT JEE 2005]**

a. $\{(x, y) : x^2 = 4y\} \cup \{(x, y) : y \le 0\}$
b. $\{(x, y) : x^2 + (y-1)^2 = 4\} \cup \{(x, y) : y \le 0\}$
c. $\{(x, y) : x^2 = y\} \cup \{(0, y) : y \le 0\}$
d. $\{(x, y) : x^2 = 4y\} \cup \{(0, y) : y \le 0\}$

34 The locus of centres of all the circles passing through (1, 2) and cutting $x^2 + y^2 = 16$ orthogonally, is

a. $4x - 2y + 21 = 0$
b. $x + 2y + 11 = 0$
c. $2x + 4y - 21 = 0$
d. $2x - y + 11 = 0$

35 If the squares of the lengths of the tangents from a point P to the circles $x^2 + y^2 = a^2$, $x^2 + y^2 = b^2$ and $x^2 + y^2 = c^2$ are in an AP, then

a. a, b, c are in GP
b. a, b, c are in an AP
c. a^2, b^2, c^2 are in an AP
d. a^2, b^2, c^2 are in GP

36 Radical axis of the circles $3x^2 + 3y^2 - 7x + 8y + 11 = 0$ and $x^2 + y^2 - 3x - 4y + 5 = 0$, is

a. $x + 10y = 2$
b. $x + 10y + 2 = 0$
c. $x + 10y + 8 = 0$
d. $x + 10y + 8 = 0$

37 Equation of radical axis of two circles $x^2 + y^2 = 16$ and $(x-4)^2 + (y-5)^2 = 25$, is

a. $3x + 5y = 8$
b. $4x + 5y = 16$
c. $5x - y = 5$
d. None of these

38 If a circle passes through the point (a, b) and cuts the circle $x^2 + y^2 = 4$ orthogonally, then the locus of its centre is **[AIEEE 2004]**

a. $2ax - 2by + (a^2 + b^2 + 4) = 0$
b. $2ax + 2by - (a^2 + b^2 + 4) = 0$
c. $2ax + 2by + (a^2 + b^2 + 4) = 0$
d. $2ax - 2by - (a^2 + b^2 + 4) = 0$

39 If a line $L : x \cos 30^\circ + y \sin 30^\circ + 1 = 0$, cuts the circle $S : x^2 + y^2 - a^2 = 0$ at the points A and B and equation of circle with AB as diameter is $S + \lambda L = 0$, then equals

a. 2
b. 1
c. −2
d. −1

40 The equation of the circle on the common chord of circle $(x-a)^2 + y^2 = a^2$ and $x^2 + (y+b)^2 = b^2$ as diameter is

a. $x^2 + y^2 = bx + ay$
b. $x^2 + y^2 = 2ab\,(bx + ay)$
c. $(a^2 + b^2)(x^2 + y^2) = 2ab(bx - ay)$
d. None of these

41 The circle on the chord $x \cos \alpha + y \sin \alpha = p$ of the circle $x^2 + y^2 = a^2$ as diameter has the equation

a. $x^2 + y^2 - a^2 - 2p\,(x\cos\alpha + y\sin\alpha - p) = 0$
b. $x^2 + y^2 + a^2 + 2p\,(x\cos\alpha + y\sin\alpha + p) = 0$
c. $x^2 + y^2 - a^2 + 2p\,(x\cos\alpha + y\sin\alpha - p) = 0$
d. $x^2 + y^2 - a^2 - 2p\,(x\cos\alpha - y\sin\alpha + p) = 0$

42 The equation of the circle which passes through $(2a, 0)$ and whose radical axis with respect to the circle $x^2 + y^2 = a^2$ is $x = \frac{a}{2}$, will be

a. $x^2 + y^2 - 2ax = 0$
b. $x^2 + y^2 - 2ay = 0$
c. $x^2 + y^2 + 2ax = 0$
d. None of these

43 The equation of the circle whose diameter is the common chord of the circle $x^2 + y^2 + 2ax + c = 0$ and $x^2 + y^2 + 2by + c = 0$, will be

a. $x^2 + y^2 + \dfrac{2ab^2}{a^2 + b^2}x + \dfrac{2a^2b}{a^2 + b^2}y + c = 0$

b. $x^2 + y^2 - \dfrac{2ab^2}{a^2 + b^2}x + \dfrac{2a^2b}{a^2 + b^2}y + c = 0$

c. $x^2 + y^2 - \dfrac{2ab^2}{a^2 + b^2}x - \dfrac{2a^2b}{a^2 + b^2}y + c = 0$

d. None of the above

Working Space

44 If the circle $x^2 + y^2 = 4$ bisects the circumference of the circle $x^2 + y^2 - 2x + 6y + a = 0$, then a equals

a. 4 b. −4
c. 16 d. −16

45 A variable chord is drawn through the origin to the circle $x^2 + y^2 - 2ax = 0$. The locus of the centre of the circle drawn on this chord as diameter is

a. $x^2 + y^2 + ax = 0$ b. $x^2 + y^2 - ax = 0$
c. $x^2 + y^2 + ay = 0$ d. $x^2 + y^2 - ay = 0$

Statement Type

Direction *(Q. Nos. 46-50) This section is based on Statement I and Statement II. Select the correct answer from the codes given below :*

a. Both Statement I and Statement II are correct and Statement II is the correct explanation of Statement I
b. Both Statement I and Statement II are correct and Statement II is not the correct explanation of Statement I
c. Statement I is correct but Statement II is incorrect
d. Statement II is correct but Statement I is incorrect

46 **Statement I** From the point (1, 5) as its centre, only one circle can be drawn touching the circle $x^2 + y^2 - 2x = 7$ because

Statement II Point (1, 5) lie outside the circle $x^2 + y^2 - 2x = 7$.

47 **Statement I** If the circles $x^2 + y^2 - 2x + 4y + 4 = 0$ and $x^2 + y^2 + 4x - 2y + c = 0$ intersect such that the common chord is tangent, then $c = -14$ because

Statement II If two circles intersect, then the common chord is longest, if it is a diameter of the smaller circle.

48 Passing through a point $A(6, 8)$ a variable secant line L is drawn to the circle $S : x^2 + y^2 - 6x - 8y + 5 = 0$. From the point of intersection of L with S, a pair of tangent lines are drawn which intersect at P.

Statement I Locus of the point P has the equation $3x + 4y - 4 = 0$ because

Statement II Point A lies outside the circle.

49 Given the lines $y + 2x = 3$ and $y + 2x = 5$ cut the axes at A, B and C, D respectively.

Statement I $ABDC$ forms quadrilateral and point (2, 3) lies inside the quadrilateral because

Statement II Point lies on the same side of the lines.

50 Let C be a circle with centre O and HK is the chord of contact of pair of the tangents from point A. OA intersects the circle C at P, Q and B is the mid-point of HK, then

Statement I AB is the harmonic mean of AP and AQ because

Statement II AK is the GM of AB and AO and OA is AM of AP and AQ.

SET - 2

Revisal Problems
for JEE Advanced

Only One Option Correct Type

Working Space

Direction (Q. Nos. 1-35) *This section contains 35 multiple choice questions. Each question has four choices (a), (b), (c) and (d), out of which ONLY ONE option is correct.*

1 Consider three non-collinear points A, B, C with coordinates (0, 6), (5, 5) and (−1, 1), respectively. Equation of a line tangent to the circle circumscribing the ΔABC and passing through the origin is

a. $2x - 3y = 0$ b. $3x + 2y = 0$
c. $3x - 2y = 0$ d. $2x + 3y = 0$

2 Let C be a circle $x^2 + y^2 = 1$. The line l intersects C at the point (−1, 0) and the point P. Suppose that the slope of the line l is a rational number m. Number of choices for m for which both the coordinates of P are rational, is

a. 3 b. 4
c. 5 d. infinitely many

3 The circle with equation $x^2 + y^2 = 1$ intersects the line $y = 7x + 5$ at two distinct points A and B. Let C the be point at which the positive X-axis intersects the circle. The $\angle ACB$, is

a. $\tan^{-1}\left(\frac{4}{3}\right)$ b. $\tan^{-1}\left(\frac{3}{4}\right)$
c. $\tan^{-1}(1)$ d. $\tan^{-1}\left(\frac{3}{2}\right)$

4 In the XY-plane, the length of the shortest path from (0, 0) to (12, 16) that does not go inside the circle $(x - 6)^2 + (y - 8)^2 = 25$, is

a. $10\sqrt{3}$ b. $10\sqrt{5}$
c. $10\sqrt{3} + \frac{5\pi}{6}$ d. $10\sqrt{3} + \frac{5\pi}{3}$

5 ΔABC is right angled at A. The circle with centre A and radius AB cuts BC and AC internally at D and E, respectively. If $BD = 20$ and $DC = 16$, then the length AC equals (in units)

a. $6\sqrt{21}$ b. $6\sqrt{26}$
c. 30 d. 32

6 Combined equation to the pair of tangents drawn from the origin to the circle $x^2 + y^2 + 4x + 6y + 9 = 0$, is

a. $3(x^2 + y^2) = (x + 2y)^2$
b. $2(x^2 + y^2) = (3x + y)^2$
c. $9(x^2 + y^2) = (2x + 3y)^2$
d. $x^2 + y^2 = (2x + 3y)^2$

7 The equation of a line inclined at an angle $\frac{\pi}{4}$ to the X-axis, such that the two circles $x^2 + y^2 = 4$, $x^2 + y^2 - 10x - 14y + 65 = 0$ intercept equal lengths on it, is

a. $2x - 2y - 3 = 0$ b. $2x - 2y + 3 = 0$
c. $x - y + 6 = 0$ d. $x - y - 6 = 0$

8 Suppose that two circles C_1 and C_2 in a plane have no points in common. Then,

a. there is no line tangent to both C_1 and C_2
b. there are exactly four lines tangent to both C_1 and C_2
c. there are no lines tangent to both C_1 and C_2 or there are exactly two lines tangent to both C_1 and C_2
d. there are no lines tangent to both C_1 and C_2 or there are exactly four lines tangent to both C_1 and C_2

Working Space

9 A variable line moves in such a way that the product of the perpendiculars from $(a, 0)$ and $(0, 0)$ is equal to k^2. The locus of the feet of the perpendicular from $(0, 0)$ upon the variable line is a circle, the square of whose radius is (given, $|a| < 2|k|$)

a. $\frac{a^4}{4} + k^2$ b. $\frac{a^2 + k^2}{4}$ c. $a^2 + \frac{k^2}{4}$ d. $\frac{a^2 + k^2}{2}$

10 The shortest distance from the line $3x + 4y = 25$ to the circle $x^2 + y^2 = 6x - 8y$ is equal to (in units)

a. $\frac{7}{5}$ b. $\frac{9}{5}$ c. $\frac{11}{5}$ d. $\frac{32}{5}$

11 Two lines $p_1x + q_1y + r_1 = 0$ and $p_2x + q_2y + r_2 = 0$ are conjugate lines w.r.t. the circle $x^2 + y^2 = a^2$, if

a. $p_1p_2 + q_1q_2 = r_1r_2$ b. $p_1p_2 + q_1q_2 + r_1r_2 = 0$
c. $a^2(p_1p_2 + q_1q_2) = r_1r_2$ d. $p_1p_2 + q_1q_2 = a^2r_1r_2$

12 Let C be the circle of radius unity centred at the origin. If two positive numbers x_1 and x_2 are such that the line passing through $(x_1, -1)$ and $(x_2, 1)$ is tangent to C, then

a. $x_1x_2 = 1$ b. $x_1x_2 = -1$
c. $x_1 + x_2 = 1$ d. $x_1x_2 = \frac{1}{4}$

13 Tangents are drawn from any point on the circle $x^2 + y^2 = R^2$ to the circle $x^2 + y^2 = r^2$. If the line joining the points of intersection of these tangents with the first circle also touch the second, then R equals

a. $\sqrt{2}r$ b. $2r$
c. $\frac{2r}{2 - \sqrt{3}}$ d. $\frac{4r}{3 - \sqrt{5}}$

14 The distance between the chords of contact of the tangents to the circle $x^2 + y^2 + 2gx + 2fy + c = 0$ from the origin and the point (g, f), is

a. $\sqrt{g^2 + f^2}$ b. $\frac{\sqrt{g^2 + f^2 - c}}{2}$
c. $\frac{g^2 + f^2 - c}{2\sqrt{g^2 + f^2}}$ d. $\frac{\sqrt{g^2 + f^2 + c}}{2\sqrt{g^2 + f^2}}$

15 The locus of the center of the circles such that the point $(2, 3)$ is the mid-point of the chord $5x + 2y = 16$, is

a. $2x - 5y + 11 = 0$ b. $2x + 5y - 11 = 0$
c. $2x + 5y + 11 = 0$ d. None of these

16 The points $A(a, 0)$, $B(0, b)$, $C(c, 0)$ and $D(0, d)$ are such that $ac = bd$ and a, b, c, d are all non-zero. Then, the points

a. form a parallelogram b. do not lie on a circle
c. form a trapezium d. are concyclic

17 The locus of the mid-points of the chords of the circle $x^2 + y^2 + 4x - 6y - 12 = 0$ which subtend an angle of $\frac{\pi}{3}$ radians at its circumference is

a. $(x - 2)^2 + (y + 3)^2 = 6.25$ b. $(x + 2)^2 + (y - 3)^2 = 6.25$
c. $(x + 2)^2 + (y - 3)^2 = 18.75$ d. $(x + 2)^2 + (y + 3)^2 = 18.75$

18 If two distinct chords, drawn from the point (p, q) on the circle $x^2 + y^2 = px + qy$, where $pq \neq 0$ are bisected by the X-axis, then

a. $p^2 = q^2$ b. $p^2 = 8q^2$
c. $p^2 < 9q^2$ d. $p^2 > 8q^2$

19 A circle of radius unity is centred at origin. Two particles start moving at the same time from the point $(1, 0)$ and move around the circle in opposite direction. One of the particles moves counter clockwise with constant speed v and the other moves clock-wise with constant speed of $3v$. After leaving $(1, 0)$, the two particles meets first at a point P and continue untill they meet next at point Q. The coordinates of the point Q are

a. $(1, 0)$ b. $(0, 1)$ c. $(0, -1)$ d. $(-1, 0)$

20 The value of c for which the set,
$\{(x, y)|x^2 + y^2 + 2x \leq 1|\} \cap \{(x, y)|x - y + c \geq 0|\}$ contains only one point in common is

a. $(-\infty, -1] \cup [3, \infty)$
b. $\{-1, 3\}$
c. $\{-3\}$
d. $\{-1\}$

Working Space

21 If $\left(a, \frac{1}{a}\right), \left(b, \frac{1}{b}\right), \left(c, \frac{1}{c}\right)$ and $\left(d, \frac{1}{d}\right)$ are four distinct points on a circle of radius 4 units, then $abcd$ is equal to

a. 4
b. $\frac{1}{4}$
c. 1
d. 16

22 A square is inscribed in the circle $x^2 + y^2 - 2x + 4x + 3 = 0$. Its sides are parallel to the coordinate axes, then one vertex of the square is

a. $(1 + \sqrt{2}, -2)$
b. $(1 - \sqrt{2}, -2)$
c. $(1, -2 + \sqrt{2})$
d. None of these

23 Two circles are drawn through the point $(1, 0)$ and $(2, -1)$ to touch the axis of Y. They intersect at an angle

a. $\cot^{-1}\left(\frac{3}{4}\right)$
b. $\cos^{-1}\left(\frac{4}{5}\right)$
c. $\frac{\pi}{2}$
d. $\tan^{-1}(1)$

24 A foot of normal from the point $(4, 3)$ to a circle is $(2, 1)$ and a diameter of the circle has the equation $2x - y - 2 = 0$. Then, the equation of the circle is

a. $x^2 + y^2 - 4y + 2 = 0$
b. $x^2 + y^2 - 4y + 1 = 0$
c. $x^2 + y^2 - 2x - 1 = 0$
d. None of these

25 A rhombus is inscribed in the region common to the two circles $x^2 + y^2 - 4x - 12 = 0$ and $x^2 + y^2 + 4x - 12 = 0$ with two of its vertices on the line joining the centres of the circles. The area of the rhombus is

a. $8\sqrt{3}$ sq units
b. $4\sqrt{3}$ sq units
c. $16\sqrt{3}$ sq units
d. None of these

26 AB is diameter of a circle. CD is a chord parallel to AB and $2CD = AB$. The tangent at B meets the line AC produced at E, then AE is equal to

a. AB
b. $\sqrt{2}\, AB$
c. $2\sqrt{2}AB$
d. $2AB$

27 A circle of constant radius a passes through origin O and cuts the axes of coordinates at points P and Q, then the equation of the locus of the foot of perpendicular from O to PQ, is

a. $(x^2 + y^2) \cdot \left(\frac{1}{x^2} + \frac{1}{y^2}\right) = 4a^2$
b. $(x^2 + y^2)^2 \cdot \left(\frac{1}{x^2} + \frac{1}{y^2}\right) = a^2$
c. $(x^2 + y^2)^2 \cdot \left(\frac{1}{x^2} + \frac{1}{y^2}\right) = 4a^2$
d. $(x^2 + y^2) \cdot \left(\frac{1}{x^2} + \frac{1}{y^2}\right) = a^2$

28 If a circle of constant radius $3k$ passes through the origin O and meets coordinate axes at A and B, then the locus of the centroid of the ΔOAB, is

a. $x^2 + y^2 = 4k^2$
b. $x^2 + y^2 = 9k^2$
c. $x^2 + y^2 = 16k^2$
d. None of these

29 The locus of the centre of circle which touches externally the circle $x^2 + y^2 - 6x - 6y + 14 = 0$ and also touches the Y-axis, is

a. $x^2 - 6x - 10y + 14 = 0$
b. $x^2 - 10x - 6y + 14 = 0$
c. $y^2 - 6x - 10y + 14 = 0$
d. $y^2 - 10x - 6y + 14 = 0$

30 A circle is drawn touching the X-axis and centre at the point which is the reflection of (a, b) in the line $y = x$. The equation of the circle, is

a. $x^2 + y^2 - 2bx - 2ay + a^2 = 0$
b. $x^2 + y^2 - 2ax - 2by + b^2 = 0$
c. $x^2 + y^2 - 2bx - 2ay + b^2 = 0$
d. $x^2 + y^2 - 2ax - 2by + a^2 = 0$

Working Space

31 $A(1, 0)$ and $B(0, 1)$ are two fixed points on the circle $x^2 + y^2 = 1$. C is a variable point on this circle. As C moves, the locus of the orthocentre of the ΔABC, is

a. $x^2 + y^2 - 2x - 2y + 1 = 0$
b. $x^2 + y^2 - x - y = 0$
c. $x^2 + y^2 = 4$
d. $x^2 + y^2 + 2x - 2y + 1 = 0$

32 The locus of the mid-points of the chords of the circle $x^2 + y^2 - 2x - 4y - 11 = 0$ which subtend $60°$ at the centre is

a. $x^2 + y^2 - 4x - 2y - 7 = 0$
b. $x^2 + y^2 + 4x + 2y - 7 = 0$
c. $x^2 + y^2 - 2x - 4y - 7 = 0$
d. $x^2 + y^2 + 2x + 4y + 7 = 0$

33 In a right ΔABC, right angled at A, on the leg AC as diameter, a semi-circle is described. The chord joining A with the point of intersection D of the hypotenuse and the semi-circle, then the length AC equals

a. $\dfrac{AB \cdot AD}{\sqrt{AB^2 + AD^2}}$
b. $\dfrac{AB \cdot AD}{AB + AD}$
c. $\sqrt{AB \cdot CD}$
d. $\dfrac{AB \cdot AD}{\sqrt{AB^2 - AD^2}}$

34 A straight line l_1 with equation $x - 2y + 10 = 0$ meets the circle with the equation $x^2 + y^2 = 100$ at B in the first quadrant. A line through B, perpendicular to l_1 cuts the Y-axis at $P(0, t)$. The value of y, is

a. 12
b. 15
c. 20
d. 25

35 B and C are fixed points having coordinates (3, 0) and (–3, 0), respectively. If the vertical $\angle BAC$ is 90°, then the locus of the centroid of ΔABC has the equation

a. $x^2 + y^2 = 1$
b. $x^2 + y^2 = 2$
c. $x^2 + y^2 = \dfrac{1}{9}$
d. $x^2 + y^2 = \dfrac{1}{9}$

One or More than One Options Correct Type

Direction (Q. Nos. 36-42) ***This section contains 7 multiple choice questions. Each question has four choices (a), (b), (c) and (d), out of which ONE or MORE THAN ONE are correct.***

36 A family of linear function is given by $f(x) = 1 + c(x + 3)$, where $c \in R$. If a member of this family meets a unit circle centred at origin at two coincident points, then c can be equal to

a. $\dfrac{-3}{4}$
b. 0
c. $\dfrac{3}{4}$
d. 1

37 A circle passes through the points (–1, 1), (0, 6) and (5, 5). The points on this circle, the tangent(s) at which is/are parallel to the straight line joining the origin to its centre is/are

a. (1, – 5)
b. (5, 1)
c. (–5, – 1)
d. (–1, 5)

38 Consider the circles $S_1 : x^2 + y^2 = 4$ and $S_2 : x^2 + y^2 - 2x - 4y + 4 = 0$. Which of the following statements are correct?

a. Number of common tangents to these circles is 2
b. If the power of a variable point P w.r.t. these two circles is same, then P moves on the line $x + 2y - 4 = 0$
c. Sum of the y-intercepts of both the circles is 6
d. The circles S_1 and S_2 are orthogonal

39 If $al^2 - bm^2 + 2dl + 1 = 0$, where a, b, d are fixed real numbers such that $a + b = d^2$. Then, the line $lx + my + 1 = 0$ touches a fixed circle

a. which cuts the X-axis orthogonally
b. with radius equal to b
c. on which the length of the tangent from the origin is $\sqrt{d^2 - b}$
d. None of the above

40 Point M moved along the circle $(x - 4)^2 + (y - 8)^2 = 20$. Then it broke away from it and moving along a tangent to the circle, cuts the X-axis at the point (–2, 0). The coordinates of the point on the circle at which the moving point broke away can be

a. $\left(\dfrac{-3}{5}, \dfrac{46}{5}\right)$
b. $\left(\dfrac{-2}{5}, \dfrac{44}{5}\right)$
c. (6, 4)
d. (3, 5)

Working Space

41 Three distinct lines are drawn in a plane. Suppose there exists exactly n circles in the plane tangent to all the three lines, then the possible values of n is/are

a. 0 b. 1
c. 2 d. 4

42 Consider the circles,

$$S_1 : x^2 + y^2 + 2x + 4y + 1 = 0$$
$$S_2 : x^2 + y^2 - 4x + 3 = 0$$
$$S_3 : x^2 + y^2 + 6y + 5 = 0$$

Which of this following statements are correct?

a. Radical centre of S_1, S_2 and S_3 lies in I quadrant
b. Radical centre of S_1, S_2 and S_3 lies in IV quadrant
c. Radius of the circle intersecting S_1, S_2 and S_3 orthogonally is 1
d. Circle orthogonal to S_1, S_2 and S_3 has its x and y intercept equal to zero

Matching List Type

Direction *(Q. Nos. 43-45) Choices for the correct combination of elements from Column I and Column II are given as options (a), (b), (c) and (d), out of which one is correct.*

43 Match the Column I with Column II and mark the correct option of codes given below.

	Column I		Column II
i.	If the straight line $y = kx$, for all $k \in I$ touches or passes outside the circle $x^2 + y^2 - 20y + 90 = 0$, then $\|k\|$ can have the value	p.	1
ii.	Two circles $x^2 + y^2 + px + py - 7 = 0$ and $x^2 + y^2 - 10x + 2py + 1 = 0$ intersect each other orthogonally, then the value of p is	q.	2
iii.	If the equation $x^2 + y^2 + 2\lambda x + 4 = 0$ and $x^2 + y^2 - 4\lambda y + 8 = 0$ represent real circles, then the value of λ can be	r.	3
iv.	Each side of a square is of length 4. The centre of the square is (3, 7) one diagonal of the square is parallel to $y = x$. The possible abscissae of the vertices of the square can be	s.	5

	i	ii	iii	iv
a.	p,q,r	q,r	q,r,s	p,s
b.	p,q	q,r	q,s	s
c.	p,q,r	q,r	p,s	q,r,s
d.	p,q	q,r	q	s,r

44 Match the Column I with Column II and mark the correct option of codes given below.

	Column I		Column II
i.	Two intersecting circles	p.	have a common tangent
ii.	Two circles touching each other	q.	have a common normal
iii.	Two non-concentric circles, one strictly inside the other	r.	do not have a common normal
iv.	Two concentric circles of different radii	s.	do not have a radical axis

	i	ii	iii	iv
a.	p	q	r	s
b.	p,q	p,q	q	q,s
c.	p,q	r,s	s	q,s
d.	q,r	p,q	s	r,s

45 Match the Column I with Column II and mark the correct option of codes given below.

Working Space

	Column I		Column II
i.	The circle $x^2 + y^2 - 6x - 10y + C = 0$ does not touch or intersect the coordinate axes and the point (1, 4) is inside the circle. The number of integral values of C, is	p.	12
ii.	The chords of contact of pair of tangents to the circle $x^2 + y^2 = 1$ drawn from any point on $2x + y = 4$ pass through the point (α, β), then the value of $(3\alpha + 2\beta)$, is	q.	6
iii.	If M and m are the maximum and minimum value of y/x for pair of real number (x, y) which satisfy the equation $(x-3)^2 + (y-3)^2 = 6$, then the value of $(M + m)$, is equal to	r.	4
iv.	Sum of the radii of the circles touching the coordinate axes and the line $3x + 4y = 12$, is	s.	3
v.	Sum of the abscissa and ordinate of the centre of the circle touching the line $3x + y + 2 = 0$ at the point (–1, 1) and passing through the point (3, 5), is	t.	2

	i	ii	iii	iv	v
a.	s	t	q	p	r
b.	p	q	r	s	t
c.	s	q	p	r	t
d.	t	q	p	r	s

Comprehension Type

Direction (Q. Nos. 46-53) *This section contains 3 paragraphs, each describing theory, experiments, data, etc. Eight questions related to the paragraphs have been given. Each question has only one correct answer among the four given options (a), (b), (c) and (d).*

Passage I

Let $C_1 : x^2 + y^2 - 2x + 2y + 1 = 0$ and $C_2 : x^2 + y^2 - 2x + 2y = 0$ are two given circles. From a moving point P on C_2, tangents are drawn to C_1 at A and B.

46 The locus of orthocentre of ΔPAB, is

a. $x^2 + y^2 - 2x + 2y + 1 = 0$
b. $x^2 + y^2 - 2x + 2y = 0$
c. $x^2 + y^2 + 2x - 2y = 0$
d. $x^2 + y^2 - 2x - 2y + 1 = 0$

47 The locus of circumcentre of ΔPAB, is

a. $2x^2 + 2y^2 - 3x + 3y = 0$
b. $x^2 + y^2 - 2x + 2y + 3 = 0$
c. $x^2 + y^2 - 3x + 3y + 1 = 0$
d. $2x^2 + 2y^2 - 4x + 4y + 3 = 0$

Passage II

Consider a line pair $ax^2 + 3xy - 2y^2 - 5x + 5y + c = 0$ representing perpendicular lines intersecting each other at C and forming a ΔABC with the X-axis.

48 If x_1 and x_2 are intercepts on the X-axis and y_1 and y_2 are the intercepts on the Y-axis, then the sum $(x_1 + x_2 + y_1 + y_2)$ is equal to

a. 6
b. 5
c. 4
d. 3

49 Distance between the orthocentre and circumcentre of ΔABC, is (in units)

a. 2
b. 3
c. $\frac{7}{4}$
d. $\frac{9}{4}$

50 If the circle $x^2 + y^2 = 4y - k$ is orthogonal with the circumcircle of the ΔABC, then k equals

a. $\frac{1}{2}$
b. 1
c. 2
d. $\frac{3}{2}$

Working Space

Passage III

Let C be a circle of radius r with centre O. Let P be a point outside C and D be a point on C. A line through P intersects C at Q and R, S is the mid-point of QR.

51 For different choices of line through P, the curve on which S lies, is

a. a straight line
b. an arc of the circle with P as centre
c. an arc of the circle with PS as diameter
d. an arc of circle with OP as diameter

52 Let P is situated at a distance d from centre O, then which of the following does not equal to the product $(PQ) \cdot (PR)$?

a. $d^2 - r^2$
b. PT^2, where T is a point on C and PT is tangent to C
c. $PS^2 - (QS)(RS)$
d. $(PS)^2$

53 Let XYZ be an equilateral triangle inscribed in C. If α, β, γ denote the distance of D from the vertices X, Y, Z respectively, the value of $(\beta + \gamma - \alpha)(\alpha + \gamma - \beta)(\alpha + \beta - \gamma)$, is

a. 0
b. $\dfrac{\alpha\beta r}{8}$
c. $\dfrac{\alpha^3 + \beta^3 + \gamma^3 - 3\alpha b\gamma}{6}$
d. None of these

One Integer Value Correct Type

Direction *(Q. Nos. 54-60) This section contains 7 questions. When worked out will result in an integer from 0 to 9 (both inclusive).*

54 The centre of variable circle $x^2 + y^2 + 2gx + 2fy + c = 0$ lies on the line $2x - 2y + 9 = 0$ and the variable circle cuts the circle $x^2 + y^2 = 4$ orthogonally. If the variable circle passes through two fixed points (a, b) and (c, d), where $(b < d)$, then the value of $(2b + d)$ is equal to

55 The graph of $x^2 + y^2 = 4 + 12x + 6y$ and $x^2 + y^2 = kx + 4x + 12y$, where k satisfies $a \le k \le b$ and for no other values of k. The value of $|b + 3a|$ is

56 Let C_1 and C_2 are circles defined by $x^2 + y^2 - 20x + 64 = 0$ and $x^2 + y^2 + 30x + 144 = 0$. The length of the shortest line segment PQ that is tangent of C_1 at P and C_2 at Q is l, then the value of $(l - 15)$, is

57 Let C be a circle that intersects each of the circles $(x + 2)^2 + y^2 = 4$, $(x - 4)^2 + (y - 2)^2 = 4$ and $(x - 4)^2 + (y + 2)^2 = 4$ in exactly one point does not contain any of these circles inside it. If the radius r of C is of the form of where p and q are prime, then $(p + q)$ is

58 Two perpendicular tangents are drawn to the circle $x^2 + y^2 = 1$, then the maximum value of square of ordinate of the point from where tangents are drawn is

59 Six points $(x_i, y_i)\, i = 1, 2, 3, \cdots; 6$ are taken on circle $x^2 + y^2 = 4$ such that $\sum_{i=1}^{6} x_i = 8$ and $\sum_{i=1}^{6} x_i = 4$. The line segment joining orthocentre of triangle made by any three points and the centroid of the triangle made by other three points passes through a fixed points (h, k), then $(h + k)$ is

60 Let $f(x, y) = 0$ be the equation of circle. If $f(0, \lambda) = 0$ has equal roots $\lambda = 1, 1$ and $f(\lambda, 0) = 0$ has roots $\lambda = \frac{1}{5}, 5$, then $(5r - 4)$ is (where, r is radius of circle)

JEE Main & AIEEE Archive

(Compilation of Last 13 Years Questions)

1 Let C be the circle with centre at (1, 1) and radius 1. If T is the circle centred at $(0, k)$ passing through origin and touching the circle C externally, then the radius of T is equal to **(JEE Main 2014)**

a. $\frac{\sqrt{3}}{\sqrt{2}}$ b. $\frac{\sqrt{3}}{2}$

c. $\frac{1}{2}$ d. $\frac{1}{4}$

2 The circle passing through $(1, -2)$ and touching the axis of X at $(3, 0)$ also passes through the point **(JEE Main 2013)**

a. $(-5, 2)$ b. $(2, -5)$

c. $(5, -2)$ d. $(-2, 5)$

3 The length of the diameter of the circle which touches the X-axis at the point (1, 0) and passes through the point (2, 3) is **(AIEEE 2012)**

a. $\frac{10}{3}$ b. $\frac{3}{5}$

c. $\frac{6}{5}$ d. $\frac{5}{3}$

4 The two circles $x^2 + y^2 = ax$ and $x^2 + y^2 = c^2$, $(c > 0)$ touch each other, if **(AIEEE 2011)**

a. $|a| = c$

b. $a = 2c$

c. $|a| = 2c$

d. $2|a| = c$

5 The equation of the circle passing through the points (1, 0) and (0, 1) and having the smallest radius is **(AIEEE 2011)**

a. $x^2 + y^2 + x + y - 2 = 0$

b. $x^2 + y^2 - 2x - 2y + 1 = 0$

c. $x^2 + y^2 - x - y = 0$

d. $x^2 + y^2 + 2x + 2y - 7 = 0$

6 The circle $x^2 + y^2 = 4x + 8y + 5$ intersects the line $3x - 4y = m$ at two distinct points, if **(AIEEE 2010)**

a. $-85 < m < -35$

b. $-35 < m < 15$

c. $15 < m < 65$

d. $35 < m < 85$

7 The point diametrically opposite to the point $P(1, 0)$ on the circle $x^2 + y^2 + 2x + 4y - 3 = 0$ is **(AIEEE 2008)**

a. $(3, 4)$

b. $(3, -4)$

c. $(-3, 4)$

d. $(-3, -4)$

8 Consider a family of circles which are passing through the point $(-1, 1)$ and are tangent to X-axis. If (h, k) are the coordinates of the centre of the circles, then the set of values of k is given by the interval **(AIEEE 2007)**

a. $0 < k < \frac{1}{2}$ b. $k \geq \frac{1}{2}$

c. $-\frac{1}{2} \leq k \leq \frac{1}{2}$ d. $k \leq \frac{1}{2}$

9 If the lines $3x - 4y - 7 = 0$ and $2x - 3y - 5 = 0$ are two diameters of a circle of area 49π sq units, then the equation of the circle is **(AIEEE 2006)**

a. $x^2 + y^2 + 2x - 2y - 62 = 0$

b. $x^2 + y^2 - 2x + 2y - 62 = 0$

c. $x^2 + y^2 - 2x + 2y - 47 = 0$

d. $x^2 + y^2 + 2x - 2y - 47 = 0$

10 Let C be the circle with centre (0, 0) and radius 3 units. The equation of the locus of the mid-points of the chords of the circle C that subtend an angle of $\frac{2\pi}{3}$ at its centre, is **(AIEEE 2006)**

a. $x^2 + y^2 = 1$ b. $x^2 + y^2 = \frac{27}{4}$

c. $x^2 + y^2 = \frac{9}{4}$ d. $x^2 + y^2 = \frac{3}{2}$

11 If the circles $x^2 + y^2 + 2ax + cy + a = 0$ and $x^2 + y^2 - 3ax + dy - 1 = 0$ intersect at two distinct points P and Q, then the line $5x + by - a = 0$ passes through P and Q for **(AIEEE 2005)**

a. exactly two values of a

b. infinitely many values of a

c. no value of a

d. exactly one value of a

12 If a circle passes through the point (a, b) and cuts the circle $x^2 + y^2 = p^2$ orthogonally, then the equation of the locus of its centre is **(AIEEE 2005)**

a. $2ax + 2by - (a^2 + b^2 + p^2) = 0$

b. $x^2 + y^2 - 2ax - 3by + (a^2 - b^2 - p^2) = 0$

c. $2ax + 2by - (a^2 - b^2 + p^2) = 0$

d. $x^2 + y^2 - 3ax - 4by + (a^2 + b^2 - p^2) = 0$

13 If a circle passes through the point (a, b) and cuts the circle $x^2 + y^2 = 4$ orthogonally, then the locus of its centre is **(AIEEE 2004)**

a. $2ax + 2by + (a^2 + b^2 + 4) = 0$

b. $2ax + 2by - (a^2 + b^2 + 4) = 0$

c. $2ax - 2by + (a^2 + b^2 + 4) = 0$

d. $2ax - 2by - (a^2 + b^2 + 4) = 0$

14 A variable circle passes through the fixed point $A(p, q)$ and touches X-axis. The locus of the other end of the diameter through A is **(AIEEE 2004)**

a. $(x-p)^2 = 4qy$ b. $(x-q)^2 = 4py$
c. $(y-p)^2 = 4qx$ d. $(y-q)^2 = 4px$

15 If the lines $2x + 3y + 1 = 0$ and $3x - y - 4 = 0$ lie along diameters of a circle of circumference 10π, then the equation of the circle is **(AIEEE 2004)**

a. $x^2 + y^2 - 2x + 2y - 23 = 0$
b. $x^2 + y^2 - 2x - 2y - 23 = 0$
c. $x^2 + y^2 + 2x + 2y - 23 = 0$
d. $x^2 + y^2 + 2x - 2y - 23 = 0$

16 The intercept on the line $y = x$ by the circle $x^2 + y^2 - 2x = 0$ is AB. Equation of the circle on AB as a diameter is **(AIEEE 2004)**

a. $x^2 + y^2 - x - y = 0$
b. $x^2 + y^2 - x + y = 0$
c. $x^2 + y^2 + x + y = 0$
d. $x^2 + y^2 + x - y = 0$

17 If the two circles $(x-1)^2 + (y-3)^2 = r^2$ and $x^2 + y^2 - 8x + 2y + 8 = 0$ intersect at two distinct points, then **(AIEEE 2003)**

a. $2 < r < 8$ b. $r < 2$
c. $r = 2$ d. $r > 2$

18 The lines $2x - 3y = 5$ and $3x - 4y = 7$ are diameters of a circle having area as 154 sq units. Then, the equation of the circle is **(AIEEE 2003)**

a. $x^2 + y^2 + 2x - 2y = 62$
b. $x^2 + y^2 + 2x - 2y = 47$
c. $x^2 + y^2 - 2x + 2y = 47$
d. $x^2 + y^2 - 2x + 2y = 62$

19 The greatest distance of the point $P(10, 7)$ from the circle $x^2 + y^2 - 4x - 2y - 20 = 0$ is **(AIEEE 2002)**

a. 10 units
b. 15 units
c. 5 units
d. None of the above

JEE Advanced & IIT JEE Archive

(Compilation of Last 10 Years Questions)

1 A circle S passes through the point $(0, 1)$ and is orthogonal to the circles $(x-1)^2 + y^2 = 16$ and $x^2 + y^2 = 1$. Then,

(2014, One or More than One Options Correct Type)

a. radius of S is 8   b. radius of S is 7

c. centre of S is $(-7, 1)$ d. centre of S is $(-8, 1)$

2 Circle(s) touching X-axis at a distance 3 from the origin and having an intercept of length $2\sqrt{7}$ on Y-axis is (are)

(2013, One or More than One Options Correct Type)

a. $x^2 + y^2 - 6x + 8y + 9 = 0$

b. $x^2 + y^2 - 6x + 7y + 9 = 0$

c. $x^2 + y^2 - 6x - 8y + 9 = 0$

d. $x^2 + y^2 - 6x - 7y + 9 = 0$

Passage for Q. Nos. (3-4)

A tangent PT is drawn to the circle $x^2 + y^2 = 4$ at the point $P(\sqrt{3}, 1)$. A straight line L, perpendicular to PT is a tangent to the circle $(x-3)^2 + y^2 = 1$.

(2012, Comprehension Type)

3 A common tangent of the two circles is

a. $x = 4$ b. $y = 2$

c. $x + \sqrt{3}y = 4$ d. $x + 2\sqrt{2}\,y = 6$

4 A possible equation of L is

a. $x - \sqrt{3}y = 1$ b. $x + \sqrt{3}y = 1$

c. $x - \sqrt{3}y = -1$ d. $x + \sqrt{3}y = 5$

5 The locus of the mid-point of the chord of contact of tangents drawn from points lying on the straight line $4x - 5y = 20$ to the circle $x^2 + y^2 = 9$ is **(2012, Only One Option Correct Type)**

a. $20(x^2 + y^2) - 36x + 45y = 0$

b. $20(x^2 + y^2) + 36x - 45y = 0$

c. $36(x^2 + y^2) - 20y + 45y = 0$

d. $36(x^2 + y^2) + 20x - 45y = 0$

6 The circle passing through the point $(-1, 0)$ and touching the Y-axis at $(0, 2)$ also passes through the point **(2011, Only One Option Correct Type)**

a. $\left(-\frac{3}{2}, 0\right)$ b. $\left(-\frac{5}{2}, 2\right)$

c. $\left(-\frac{3}{2}, \frac{5}{2}\right)$ d. $(-4, 0)$

7 Two parallel chords of a circle of radius 2 are at a distance $\sqrt{3} + 1$ apart. If the chords subtend at the centre, angles of π/k and $\frac{2\pi}{k}$, where $k > 0$, then the value of $[k]$ is...... **(2010, Integer Type)**

8 The centres of two circles C_1 and C_2 each of unit radius are at a distance of 6 unit from each other. Let P be the mid-point of the line segment joining the centres of C_1 and C_2 and C be a circle touching circles C_1 and C_2 externally. If a common tangents to C_1 and C passing through P is also a common tangent to C_2 and C, then the radius of the circle C is... **(2009, Integer Type)**

9 Tangents drawn from the point $P(1, 8)$ to the circle $x^2 + y^2 - 6x - 4y - 11 = 0$ touch the circle at the points A and B. The equation of the circumcircle of the ΔPAB is

(2009, Only One Option Correct Type)

a. $x^2 + y^2 + 4x - 6y + 19 = 0$

b. $x^2 + y^2 - 4x - 10y + 19 = 0$

c. $x^2 + y^2 - 2x + 6y - 29 = 0$

d. $x^2 + y^2 - 6x - 4y + 19 = 0$

10 Consider the two curves

$C_1 : y^2 = 4x$

$C_2 : x^2 + y^2 - 6x + 1 = 0$, then

(2008, Only One Option Correct Type)

a. C_1 and C_2 touch each other only at one point

b. C_1 and C_2 touch each other exactly at two points

c. C_1 and C_2 intersect (but do not touch) at exactly two points

d. C_1 and C_2 neither intersect nor touch each other

Passage for Q. Nos. (11-13)

A circle C of radius 1 is inscribed in an equilateral ΔPQR. The points of contact of C with the sides PQ, QR, RP are D, E, F respectively. The line PQ is given by the equation $\sqrt{3}\,x + y - 6 = 0$ and the point D is $\left(\frac{3\sqrt{3}}{2}, \frac{3}{2}\right)$.

Further, it is given that the origin and the centre of C are on the same side of the line PQ.

(2008, Comprehension Type)

11 The equation of circle C is

a. $(x - 2\sqrt{3})^2 + (y-1)^2 = 1$

b. $(x - 2\sqrt{3})^2 + \left(y + \frac{1}{2}\right)^2 = 1$

c. $(x - \sqrt{3})^2 + (y+1)^2 = 1$

d. $(x - \sqrt{3})^2 + (y-1)^2 = 1$

12 Points E and F are given by

a. $\left(\frac{\sqrt{3}}{2}, \frac{3}{2}\right), (\sqrt{3}, 0)$ b. $\left(\frac{\sqrt{3}}{2}, \frac{1}{2}\right), (\sqrt{3}, 0)$

c. $\left(\frac{\sqrt{3}}{2}, \frac{3}{2}\right), \left(\frac{\sqrt{3}}{2}, \frac{1}{2}\right)$ d. $\left(\frac{3}{2}, \frac{\sqrt{3}}{2}\right), \left(\frac{\sqrt{3}}{2}, \frac{1}{2}\right)$

13 Equations of the sides QR, RP are

a. $y = \frac{2}{\sqrt{3}}x + 1$, $y = -\frac{2}{\sqrt{3}}x - 1$

b. $y = \frac{1}{\sqrt{3}}x$, $y = 0$

c. $y = \frac{\sqrt{3}}{2}x + 1$, $y = -\frac{\sqrt{3}}{2}x - 1$

d. $y = \sqrt{3}x$, $y = 0$

14 Consider $L_1 : 2x + 3y + p - 3 = 0$

$L_2 : 2x + 3y + p + 3 = 0$

where p is a real number and

$C : x^2 + y^2 - 6x + 10y + 30 = 0$

Statement I If line L_1 is a chord of circle C, then line L_2 is not always a diameter of circle C.

Statement II If line L_1 is a diameter of circle C, then line L_2 is not a chord of circle C.

(2008, Statement Type)

a. Both Statement I and Statement II are correct and Statement II is the correct explanation of Statement I
b. Both Statement I and Statement II are correct and Statement II is not the correct explanation of Statement I
c. Statement I is correct but Statement II is incorrect
d. Statement II is correct but Statement I is incorrect

15 Let $ABCD$ be a quadrilateral with area 18, with side AB parallel to the side CD and $AB = 2CD$. Let AD be perpendicular to AB and CD. If a circle is drawn inside the quadrilateral $ABCD$ touching all the sides, then its radius is **(2007, Only One Option Correct Type)**

a. 3 b. 2 c. $\frac{3}{2}$ d. 1

16 Tangents are drawn from the point (17, 7) to the circle $x^2 + y^2 = 169$.

Statement I The tangents are mutually perpendicular.

Statement II The locus of the points from which a mutually perpendicular tangents can be drawn to the given circle is $x^2 + y^2 = 338$.

(2007, Statement Type)

a. Both Statement I and Statement II are correct and Statement II is the correct explanation of Statement I
b. Both Statement I and Statement II are correct and Statement II is not the correct explanation of Statement I
c. Statement I is correct but Statement II is incorrect
d. Statement II is correct but Statement I is incorrect

17 Match the conditions/expressions in Column I with statement in Column II. **(2007, Matching Type)**

	Column I		Column II
i	Two intersecting circles	p.	have a common tangent
ii	Two mutually external circles	q.	have a common normal
iii	Two circles, one strictly inside the other	r.	do not have a common tangent
iv	Two branches of a hyperbola	s.	do not have a common normal

Codes

	i	ii	iii	iv
a.	p,r	p	s	r,s
b.	p,q	p,q	q,r	q,r
c.	p	q	r	p,q
d.	p,s	r,s	p,q	q,r

Passage for Q. Nos. (18-20)

Let $ABCD$ be a square of side length 2 unit. C_2 is the circle through vertices A, B, C, D and C_1 is the circle touching all the sides of square $ABCD$. L is the line through A. **(2006, Comprehension Type)**

18 If P is a point of C_1 and Q is a point on C_2, then $\frac{PA^2 + PB^2 + PC^2 + PD^2}{QA^2 + QB^2 + QC^2 + QD^2}$ is equal to

a. 0.75 b. 1.25 c. 1 d. 0.5

19 A circle touches the line L and the circle C_1 externally such that both the circles are on the same side of the line, then the locus of centre of the circle is

a. ellipse b. hyperbola
c. parabola d. parts of straight line

20 A line M through A is drawn parallel to BD. Point S moves such that its distances from the line BD and the vertex A are equal. If locus of S cuts M at T_2 and T_3 and AC at T_1, then area of $\Delta T_1T_2T_3$ is

a. $\frac{1}{2}$ sq unit b. $\frac{2}{3}$ sq unit
c. 1 sq unit d. 2 sq unit

21 The locus of the centre of circle which touches $(y-1)^2 + x^2 = 1$ externally and also touches X-axis, is **(2005, Only One Option Correct Type)**

a. $\{x^2 = 4y, y \geq 0\} \cup \{(0, y), y<0\}$
b. $x^2 = y$
c. $y = 4x^2$
d. $y^2 = 4x \cup (0, y), y \in R$

Answers with Explanations

Answers

1. Point

DPP-1 Distance Formula

1. (i) $\sqrt{a^2+b^2}$,

(ii) $\sqrt{a^2+2b^2+c^2-2ab-2bc}$,

(iii) $\left|2a\sin\frac{\alpha-\beta}{2}\right|$,

(iv) $|a(m_2-m_1)|\sqrt{(m_2+m_1)^2+4}$

2. *a* 3. *d* 4. *b* 5. *c* 6. *c* 7. *a* 8. *c* 9. *c* 10. *b*
11. *c* 12. *b* 13. *c* 14. *c* 15. *c*

DPP-2 Section Formula

1. *c* 2. *c* 3. *a* 4. *a* 5. *d* 6. *c* 7. *b* 8. *a* 9. *d* 10. *b*
11. *a* 12. *a* 13. *c* 14. *c* 15. *a*

DPP-3 Area

1. (i) 1 sq unit,

(ii) $2ac$ sq units,

(iii) $\left|-2ab\sin\left(\frac{\phi_1-\phi_2}{2}\right)\sin\left(\frac{\phi_2-\phi_3}{2}\right)\sin\left(\frac{\phi_3-\phi_1}{2}\right)\right|$

(iv) $a^2|(m_1-m_2)(m_2-m_3)(m_3-m_1)|$ sq units,

(v) $\frac{1}{2}a^2\left|\frac{(m_1-m_2)(m_2-m_3)(m_3-m_1)}{m_1m_2m_3}\right|$ sq units

2. (i) $\frac{41}{2}$ sq units,

(ii) 96 sq units

3. 36.5 sq units 4. *c* 5. *b* 6. *b* 7. *b* 8. *a* 9. *c* 10. *b* 11. *c*
12. *a* 13. *a* 14. *b* 15. *a* 16. *c* 17. *b* 18. *a* 19. *d* 20. *b* 21. *b*
22. *b* 23. *c* 24. *b* 25. *b* 26. *a* 27. 7 28. 3 29. 0 30. 1

DPP-4 Centroid, Circumcentre, Orthocentre and Incentre of Triangle

1. *a* 2. *a* 3. *b* 4. *a* 5. *c* 6. *a* 7. *c* 8. *a* 9. *a* 10. *b*
11. *d* 12. *c* 13. *b* 14. *a* 15. *d* 16. *a* 17. *b* 18. *b* 19. *d* 20. *c*
21. *b* 22. 4 23. 0 24. 6 25. 5

DPP-5 Shifting of Origin, Rotation of Axes and Locus

1. *b* 2. *a* 3. *b* 4. *a* 5. *b* 6. *a* 7. *d* 8. *d* 9. *c* 10. *d*
11. *a* 12. *c* 13. *b* 14. *b* 15. *c* 16. *c* 17. *c* 18. *b* 19. *b* 20. *b*

Revisal Problems

1. *a* 2. *b* 3. *a* 4. *c* 5. *b* 6. *b* 7. *b* 8. *b* 9. *d* 10. *c*
11. *b* 12. *a* 13. *b* 14. *c* 15. *a* 16. *a* 17. *b* 18. 1 19. 0 20. 7

JEE Main & AIEEE Archive

1. *b* 2. *c* 3. *c* 4. *a* 5. *c* 6. *b* 7. *a* 8. *b* 9. *a* 10. *d*

2. Straight Line

DPP-1 Slope and Types of Straight Line

1. *d* 2. *b* 3. *c* 4. *a* 5. *c* 6. *a* 7. *b* 8. *a* 9. *d* 10. *d*
11. *c* 12. *b* 13. *c* 14. *a* 15. *c* 16. *b* 17. *d* 18. *a* 19. *a* 20. *a*
21. *d* 22. *a* 23. *a* 24. *b* 25. *d* 26. *c* 27. *c* 28. *a* 29. *a* 30. *a*
31. *a,b* 32. *a,b* 33. *a,c* 34. *a,b* 35. *b,d* 36. *c* 37. *d* 38. *b* 39. 0 40. 1

DPP-2 Angle between Straight Lines, Image and Foot of Perpendicular

1. (i) $\frac{\pi}{2}$, (ii) $\tan^{-1}\left(\frac{23}{10}\right)$, (iii) $\frac{\pi}{3}$, (iv) $\tan^{-1}\left(\frac{4}{3}\right)$, (v) $\tan^{-1}\left|\frac{4m^2n^2}{m^4-n^4}\right|$

2. (i) –4 (ii) $3x-4y+11=0$, $3x-4y-9=0$, $4x+3y+3=0$, $4x+3y-17=0$

3. $121y-88x=371$, $33y-24x=1043$ 4. *c* 5. *a* 6. *d* 7. *d* 8. *a* 9. *d*
10. *d* 11. *d* 12. *a* 13. *a* 14. *d* 15. *c* 16. *d* 17. *c* 18. *d* 19. *d*
20. *b* 21. *d* 22. *a* 23. *d* 24. *b* 25. *c* 26. *d* 27. *b* 28. *b* 29. *b*
30. *c* 31. *a* 32. *b* 33. *c* 34. *a* 35. *b*

DPP-3 Equation of Straight Line Making an Angle 'α' with Given Straight Line

1. *a* 2. *c* 3. *b* 4. *a* 5. *b* 6. *a* 7. *d* 8. *d* 9. *c* 10. *d*
11. *a* 12. *d* 13. *a* 14. *d* 15. *a,b* 16. *b,c* 17. *a,b,c,d* 18. *a,d* 19. 0
20. 5

DPP-4 Angle Bisector and Family of Concurrent Lines

1. *b* 2. *a* 3. *a* 4. *a* 5. *b* 6. *c* 7. *c* 8. *b* 9. *b* 10. *a*
11. *b* 12. *b* 13. *a* 14. *c* 15. *b* 16. *a* 17. *b* 18. *b* 19. *d* 20. *a*
21. *b* 22. *a* 23. *b* 24. *a* 25. *b* 26. *a,c,d* 27. *a,b* 28. *a,d* 29. *a,c,d* 30. *b,d*
31. *a,b* 32. *a,c* 33. *a,c* 34. *d* 35. *d* 36. *c* 37. *b* 38. *b* 39. *c* 40. *c*

DPP-5 Locus

1. *c* 2. *c* 3. *b* 4. *d* 5. *a* 6. *d* 7. *a* 8. *c* 9. *c* 10. *a*
11. *a* 12. *a* 13. *a* 14. *b* 15. *d* 16. *d* 17. *b* 18. *d* 19. *d* 20. *a*
21. *a* 22. *a* 23. *a* 24. *c* 25. *d* 26. *a* 27. *d* 28. *a* 29. *b* 30. *c*

Revisal Problems (JEE Main)

1. *a* 2. *a* 3. *a* 4. *b* 5. *d* 6. *a* 7. *a* 8. *a* 9. *c* 10. *b*
11. *d* 12. *a* 13. *b* 14. *b* 15. *a* 16. *d* 17. *d* 18. *d* 19. *b* 20. *a*
21. *b* 22. *c* 23. *c* 24. *d* 25. *c* 26. *b* 27. *a* 28. *a* 29. *b* 30. *b*
31. *b* 32. *b* 33. *a* 34. *d* 35. *c* 36. *c* 37. *c* 38. *a* 39. *a* 40. *a*
41. *b* 42. *a* 43. *b* 44. *d* 45. *c*

Revisal Problems (JEE Advanced)

1. *d* 2. *d* 3. *c* 4. *b* 5. *a* 6. *a* 7. *b* 8. *a* 9. *d* 10. *b*
11. *b* 12. *c* 13. *d* 14. *a* 15. *a* 16. *d* 17. *c* 18. *c* 19. *d* 20. *a*
21. *d* 22. *a* 23. *c* 24. *a* 25. *d* 26. *d* 27. *b* 28. *a* 29. *b* 30. *a*
31. *d* 32. *d* 33. *c* 34. *d* 35. *a* 36. *c* 37. *d* 38. *c* 39. *b* 40. *d*
41. *c* 42. *c* 43. *d* 44. *d* 45. *c* 46. *a* 47. *b* 48. *a* 49. *d* 50. *c*
51. *b* 52. *c* 53. *b* 54. *c* 55. 6 56. 6 57. 4 58. 1 59. 9 60. 8

JEE Main & AIEEE Archive

1. *b* 2. *c* 3. *b* 4. *c* 5. *c* 6. *c* 7. *c* 8. *d* 9. *a* 10. *a*
11. *a* 12. *b* 13. *b* 14. *d* 15. *b* 16. *a* 17. *d* 18. *d* 19. *c* 20. *b*

JEE Advanced & IIT JEE Archive

1. *a* 2. *b* 3. *d* 4. *a* 5. *c*

3. Pair of Straight Line

DPP-1 Homogeneous Pair of Straight Lines

1. *d*	2. *d*	3. *c*	4. *b*	5. *a*	6. *a*	7. *a*	8. *d*	9. *b*	10. *a*
11. *c*	12. *a*	13. *c*	14. *a*	15. *a*	16. *d*	17. *c*	18. *a*	19. *a*	20. *a*
21. *c*	22. *b*	23. *a*	24. *b*	25. *b*	26. *b*	27. *a*	28. *c*	29. *d*	30. *b*
31. *a*	32. *a*	33. *c*	34. *c*	35. *d*	36. *d*	37. *c*	38. *c*	39. *a*	40. *c*
41. *a*	42. *c*	43. *d*	44. *c*	45. *d*	46. *b*	47. *c*	48. 3	49. 2	50. 2

DPP-2 General Equation of Second Degree

1. *a*	2. *a*	3. *a*	4. *c*	5. *d*	6. *c*	7. *c*	8. *a*	9. *b*	10. *a*
11. *a*	12. *b*	13. *c*	14. *a*	15. *c*	16. *b*	17. *c*	18. *a*	19. *b*	20. *c*
21. *b*	22. *a*	23. *b*	24. *c*	25. *d*	26. 4	27. 10	28. 1	29. 5	30. 1

DPP-3 Equation of Line Joining the Origin to the Point of Intersection of a Curve and a Line

1. *a*	2. *c*	3. *b*	4. *b*	5. *b*	6. *c*	7. *c*	8. *d*	9. *d*	10. *b*
11. *b*	12. 4	13. 2	14. 1	15. 4					

Revisal Problems (JEE Main)

1. *a*	2. *b*	3. *c*	4. *a*	5. *d*	6. *c*	7. *b*	8. *b*	9. *c*	10. *c*
11. *d*	12. *a*	13. *c*	14. *a*	15. *c*	16. *b*	17. *a*	18. *d*	19. *d*	20. *b*
21. *c*	22. *a*	23. *a*	24. *c*	25. *b*					

Revisal Problems(JEE Advanced)

1. *d*	2. *d*	3. *c*	4. *c*	5. *a*	6. *a*	7. *c*	8. *a*	9. *a*	10. *c*
11. *d*	12. *a*	13. *b*	14. *a*	15. *d*	16. *b*	17. *c*	18. *d*	19. 3	20. 3
21. 12	22. 1	23. 0							

JEE Archive

1. *c*	2. *a*	3. *c*	4. *d*	5. *d*	6. *b*

4. Circle

DPP-1 Equation of Circle

1. *c*	2. *c*	3. *d*	4. *c*	5. *a*	6. *b*	7. *c*	8. *c*	9. *c*	10. *c*
11. *b*	12. *a*	13. *d*	14. *c*	15. *c*	16. *c*	17. *c*	18. *d*	19. *c*	20. *c*
21. *a*	22. *d*	23. *b*	24. *b*	25. *c*	26. *c*	27. *d*	28. *b,c*	29. *a,b,c,d*	
30. *a, b, c, d*		31. *c, d*	32. *c*	33. *c*	34. *b*	35. *d*	36. *a*	37. *d*	38. 2
39. 1	40. 4								

DPP-2 Equation of Tangent, Equation of Normal and Length of Tangent

1. *b*	2. *d*	3. *a*	4. *a*	5. *c*	6. *b*	7. *a*	8. *b*	9. *c*	10. *b*
11. *c*	12. *b*	13. *b*	14. *d*	15. *a*	16. *c*	17. *c*	18. *c*	19. *a*	20. *d*
21. *b*	22. *c*	23. *a*	24. *a*	25. *c*	26. *c*	27. *a*	28. *c*	29. *d*	30. *b*
31. *c*	32. *c*	33. *a*	34. *d*	35. *a*	36. 5	37. 1	38. 8	39. 3	40. 2

DPP-3 Angle between Circles, Radical Axis and Length of Tangent

1. *b*	2. *d*	3. *c*	4. *b*	5. *b*	6. *c*	7. *c*	8. *c*	9. *b*	10. *c*
11. *c*	12. *c*	13. *a*	14. *c*	15. *a*	16. *c*	17. *a*	18. *a*	19. *d*	20. *d*
21. *b*	22. *a*	23. *b*	24. *a*	25. *c*	26. *d*	27. *c*	28. *a,b,d*	29. *a,c*	

DPP-4 Location of Two Circles, Number of Circles, Tangents and Family of Circles

1. *c* 2. *d* 3. *a* 4. *b* 5. *d* 6. *d* 7. *d* 8. *b* 9. *a* 10. *a*
11. *b* 12. *b* 13. *d* 14. *a* 15. *c* 16. *b* 17. *d* 18. *d* 19. *b* 20. *c*
21. *c* 22. *d* 23. *a* 24. *c* 25. *c* 26. *d* 27. *a* 28. *d* 29. *a* 30. *d*
31. *a* 32. 8 33. 5 34. 0 35. 3

DPP-5 $S' = T, S \cdot S' = T^2, T = 0$ and Pole, Polar

1. *a* 2. *c* 3. *a* 4. *d* 5. *b* 6. *b* 7. *a* 8. *b* 9. *d* 10. *a*
11. *b* 12. *c* 13. *c* 14. *b* 15. *c* 16. *b* 17. *b* 18. *b* 19. *a* 20. *c*
21. *a* 22. *b* 23. *d* 24. *b* 25. *a* 26. *b* 27. *a* 28. *b* 29. 0 30. 1

Revisal Problems (JEE Main)

1. *c* 2. *a* 3. *a* 4. *b* 5. *c* 6. *c* 7. *b* 8. *d* 9. *a* 10. *c*
11. *b* 12. *c* 13. *a* 14. *b* 15. *b* 16. *a* 17. *a* 18. *a* 19. *b* 20. *c*
21. *a* 22. *a* 23. *b* 24. *a* 25. *b* 26. *d* 27. *b* 28. *c* 29. *b* 30. *c*
31. *b* 32. *a* 33. *d* 34. *c* 35. *c* 36. *a* 37. *b* 38. *b* 39. *a* 40. *c*
41. *a* 42. *a* 43. *a* 44. *c* 45. *b* 46. *d* 47. *a* 48. *d* 49. *d* 50. *a*

Revisal Problems (JEE Advanced)

1. *d* 2. *d* 3. *c* 4. *d* 5. *b* 6. *c* 7. *a* 8. *d* 9. *a* 10. *a*
11. *c* 12. *a* 13. *b* 14. *c* 15. *a* 16. *d* 17. *b* 18. *d* 19. *d* 20. *d*
21. *c* 22. *d* 23. *a* 24. *c* 25. *a* 26. *d* 27. *c* 28. *a* 29. *d* 30. *c*
31. *a* 32. *c* 33. *d* 34. *c* 35. *a* 36. *a,b* 37. *b, d* 38. *a,b,d* 39. *a,c* 40. *b,c*
41. *a,c,d* 42. *b,c,d* 43. *a* 44. *b* 45. *b* 46. *b* 47. *d* 48. *b* 49. *c* 50. *d*
51. *d* 52. *d* 53. *a* 54. 5 55. 4 56. 5 57. 7 58. 2 59. 3 60. 9

JEE Main & AIEEE Archive

1. *d* 2. *c* 3. *a* 4. *a* 5. *c* 6. *b* 7. *d* 8. *b* 9. *c* 10. *c*
11. *c* 12. *a* 13. *b* 14. *a* 15. *a* 16. *a* 17. *a* 18. *c* 19. *b*

JEE Advanced & IIT JEE Archive

1. *b,c* 2. *a,c* 3. *d* 4. *a* 5. *a* 6. *d* 7. 3 8. 8 9. *b* 10. *b*
11. *d* 12. *a* 13. *d* 14. *c* 15. *b* 16. *a* 17. *b* 18. *a* 19. *c* 20. *c*
21. *a*

1. Point

DPP-1 Distance Formula

1. We know that, the distance between two points (x_1, y_1) and (x_2, y_2) is

$$d = \sqrt{(x_2 - x_1)^2 + (y_2 - y_1)^2}$$

(i) $(a, 0)$ and $(0, b)$

$$d = \sqrt{(0 - a)^2 + (b - 0)^2} = \sqrt{a^2 + b^2}$$

(ii) $(b + c, c + a)$ and $(c + a, a + b)$

$$d = \sqrt{(c + a - b - c)^2 + (a + b - c - a)^2}$$
$$= \sqrt{(a - b)^2 + (b - c)^2}$$
$$= \sqrt{a^2 + b^2 - 2ab + b^2 + c^2 - 2bc}$$
$$[\because (a - b)^2 = a^2 + b^2 - 2ab]$$
$$= \sqrt{a^2 + 2b^2 + c^2 - 2ab - 2bc}$$

(iii) $(a\cos\alpha, a\sin\alpha)$ and $(a\cos\beta, a\sin\beta)$

$$d = \sqrt{(a\cos\beta - a\cos\alpha)^2 + (a\sin\beta - a\sin\alpha)^2}$$
$$= \sqrt{\begin{matrix} a^2\cos^2\beta + a^2\cos^2\alpha - 2a^2\cos\beta\cos\alpha \\ + a^2\sin^2\beta + a^2\sin^2\alpha - 2a^2\sin\alpha\sin\beta \end{matrix}}$$
$$= \sqrt{\begin{matrix} a^2(\sin^2\alpha + \cos^2\alpha) + a^2(\sin^2\beta + \cos^2\beta) \\ -2a^2(\cos\beta\cos\alpha + \sin\alpha\sin\beta) \end{matrix}}$$
$$= \sqrt{a^2 + a^2 - 2a^2\cos(\alpha - \beta)}$$
$$[\because \cos A\cos B + \sin A\sin B = \cos(A - B)]$$
$$= \sqrt{2a^2 - 2a^2\cos(\alpha - \beta)} = \sqrt{2a^2[1 - \cos(\alpha - \beta)]}$$
$$= \sqrt{2a^2 \cdot 2\sin^2\frac{(\alpha - \beta)}{2}} \qquad \left[\because 1 - \cos\theta = 2\sin^2\frac{\theta}{2}\right]$$
$$= \sqrt{4a^2\sin^2\frac{(\alpha - \beta)}{2}} = \left|2a\sin\frac{(\alpha - \beta)}{2}\right|$$

(iv) $(am_1^2, 2am_1)$ and $(am_2^2, 2am_2)$

$$d = \sqrt{(am_2^2 - am_1^2)^2 + (2am_2 - 2am_1)^2}$$
$$= \sqrt{a^2(m_2^2 - m_1^2)^2 + 4a^2(m_2 - m_1)^2}$$
$$= \sqrt{a^2(m_2 + m_1)^2(m_2 - m_1)^2 + 4a^2(m_2 - m_1)^2}$$
$$= \sqrt{a^2(m_2 - m_1)^2[(m_2 + m_1)^2 + 4]}$$
$$= |a(m_2 - m_1)|\sqrt{\{(m_2 + m_1)^2 + 4\}}$$

2. (a) Given, distance between the points $(x_i, 2)$ and $(3, 4)$ is 8.

$$\Rightarrow \quad \sqrt{(3 - x_i)^2 + (4 - 2)^2} = 8$$
$$\left[\because d = \sqrt{(x_2 - x_1)^2 + (y_2 - y_1)^2}\right]$$

On squaring both sides, we get

$$(3 - x_i)^2 + 4 = 64$$
$$\Rightarrow \quad 9 + x_i^2 - 6x_i + 4 = 64$$
$$\Rightarrow \quad x_i^2 - 6x_i + 13 = 64$$
$$\Rightarrow \quad x_i^2 - 6x_i - 51 = 0$$
$$\Rightarrow \quad x_i = \frac{6 \pm \sqrt{(-6)^2 - 4 \times 1(-51)}}{2}$$
$$= \frac{6 \pm \sqrt{36 + 204}}{2} = \frac{6 \pm \sqrt{240}}{2}$$
$$= \frac{2(3 \pm \sqrt{60})}{2} = 3 \pm \sqrt{60} = 3 \pm 2\sqrt{15}$$

3. (d) Let y be the ordinate of the other extremity.

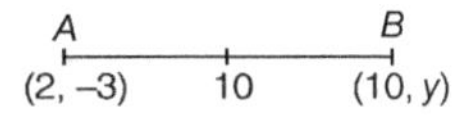

The distance between the points $(2, -3)$ and $(10, y)$ is 10 units.

$$\Rightarrow \quad (10 - 2)^2 + (y + 3)^2 = (10)^2$$
$$\Rightarrow \quad (8)^2 + (y + 3)^2 = 100$$
$$\Rightarrow \quad 64 + y^2 + 9 + 6y = 100$$
$$\Rightarrow \quad y^2 + 6y - 27 = 0$$
$$\Rightarrow \quad y^2 + 9y - 3y - 27 = 0$$
$$\Rightarrow \quad y(y + 9) - 3(y + 9) = 0$$
$$\Rightarrow \quad (y + 9)(y - 3) = 0$$
$$\Rightarrow \quad y = 3, -9$$

Sum of all possible values of $y = 3 - 9 = -6$

4. (b) The abscissa of the point P will be the sum of infinite series

$$1 + \frac{1}{4} + \frac{1}{16} + \cdots$$
$$= 1 + \frac{1}{4} + \frac{1}{4^2} + \cdots$$
$$= \frac{1}{1 - \frac{1}{4}} = \frac{1}{\frac{3}{4}} = \frac{4}{3}$$

and the ordinate of the point P will be the sum of infinite series

$$\frac{1}{2} - \frac{1}{8} + \frac{1}{32} - \cdots$$
$$\frac{1}{2} - \frac{1}{2^3} + \frac{1}{2^5} + \cdots$$
$$= \frac{\frac{1}{2}}{1 + \frac{1}{4}} = \frac{1}{2} \times \frac{4}{5} = \frac{2}{5}$$

Hence, the coordinates of the point P is $\left(\frac{4}{3}, \frac{2}{5}\right)$.

5. (c) Let $A(1, 5)$, $B(-3, 1)$ and $C(3, -5)$ be the vertices of ΔABC.

Then, $AB = \sqrt{(-3 - 1)^2 + (1 - 5)^2}\left[\because d = \sqrt{(x_2 - x_1)^2 + (y_2 - y_1)^2}\right]$

$$= \sqrt{(-4)^2 + (-4)^2]} = \sqrt{16 + 16}$$
$$= \sqrt{32}$$
$$BC = \sqrt{(3 + 3)^2 + (-5 - 1)^2}$$
$$= \sqrt{(6)^2 + (-6)^2}$$
$$= \sqrt{36 + 36}$$
$$= \sqrt{72}$$
$$CA = \sqrt{(1 - 3)^2 + (-5 - 5)^2}$$
$$= \sqrt{(-2)^2 + (-10)^2}$$
$$= \sqrt{4 + 100}$$
$$= \sqrt{104}$$

Now, $AB^2 + BC^2 = (\sqrt{32})^2 + (\sqrt{72})^2$

$$= 32 + 72$$
$$= 104 = CA^2$$

Hence, ΔABC is a right angled triangle.

6. (c) Let $A(4, -4)$, $B(-4, 4)$ and $C(x, y)$ be vertices of an equilateral triangle, then using distance formula, we have

$$AB = \sqrt{(-4-4)^2 + (4+4)^2}$$
$$= \sqrt{(-8)^2 + (8)^2} = \sqrt{64 + 64}$$
$$= \sqrt{128}$$
$$BC = \sqrt{(x+4)^2 + (y-4)^2}$$
$$= \sqrt{x^2 + 16 + 8x + y^2 + 16 - 8y}$$
$$= \sqrt{x^2 + y^2 + 8x - 8y + 32}$$
$$AC = \sqrt{(x-4)^2 + (y+4)^2}$$
$$= \sqrt{x^2 + 16 - 8x + y^2 + 16 + 8y}$$
$$= \sqrt{x^2 + y^2 - 8x + 8y + 32}$$

Since, ΔABC is equilateral.

$\therefore \quad AB = BC = AC$

On taking first two equalities, we have

$$AB = BC$$
$$\Rightarrow \quad AB^2 = BC^2$$
$$\Rightarrow \quad 128 = x^2 + y^2 + 8x - 8y + 32 \quad \ldots\text{(i)}$$

On taking last two equalities, we have

$$BC = AC$$
$$\Rightarrow \quad BC^2 = AC^2$$
$$\Rightarrow \quad x^2 + y^2 + 8x - 8y + 32 = x^2 + y^2 - 8x + 8y + 32$$
$$\Rightarrow \quad 8x - 8y = -8x + 8y$$
$$\Rightarrow \quad 16x = 16y$$
$$\Rightarrow \quad x = y \quad \ldots\text{(ii)}$$

On using Eq. (ii) in Eq. (i), we have

$$128 = 2x^2 + 32$$
$$\Rightarrow \quad 2x^2 = 96$$
$$\Rightarrow \quad x^2 = 48$$
$$\Rightarrow \quad x = \pm 4\sqrt{3}$$

Hence $\quad x = y = 4\sqrt{3}$

or $\quad x = y = -4\sqrt{3}$

7. (a) From the figure, M and N are the foot of perpendicular of point $(-3, 4)$ on Y and X-axes, respectively.

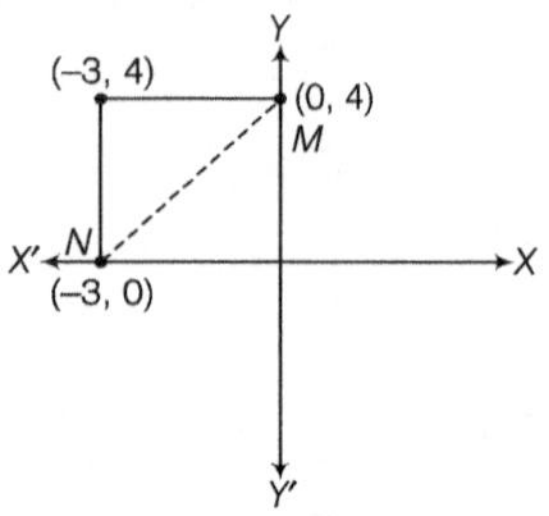

Now, distance $\quad MN = \sqrt{(0+3)^2 + (4-0)^2}$

$$= \sqrt{(3)^2 + (4)^2} = \sqrt{9 + 16}$$
$$= \sqrt{25} = 5 \text{ units}$$

8. (c) Given, $A(1, 1)$, $B(-1, -1)$ and $C(\sqrt{-3}, k)$ are the vertices of an equilateral triangle, then

$$AB = \sqrt{(-1-1)^2 + (-1-1)^2}$$
$$= \sqrt{(-2)^2 + (-2)^2}$$
$$= \sqrt{4 + 4} = \sqrt{8}$$
$$BC = \sqrt{(-\sqrt{3}+1)^2 + (k+1)^2} = \sqrt{(1-\sqrt{3})^2 + (k+1)^2}$$
$$= \sqrt{1 + 3 - 2\sqrt{3} + k^2 + 1 + 2k}$$
$$= \sqrt{k^2 + 2k + 5 - 2\sqrt{3}}$$
$$AC = \sqrt{(-\sqrt{3}-1)^2 + (k-1)^2}$$
$$= \sqrt{3 + 1 + 2\sqrt{3} + k^2 + 1 - 2k}$$
$$= \sqrt{k^2 - 2k + 5 + 2\sqrt{3}}$$

Since, ΔABC is equilateral.

$\therefore \quad AB = BC = AC$

On taking last two equalities, we have

$$BC = AC$$
$$\sqrt{k^2 + 2k + 5 - 2\sqrt{3}} = \sqrt{k^2 - 2k + 5 + 2\sqrt{3}}$$
$$\Rightarrow \quad k^2 + 2k + 5 - 2\sqrt{3} = k^2 - 2k + 5 + 2\sqrt{3}$$
$$\Rightarrow \quad 4k = 4\sqrt{3}$$
$$\therefore \quad k = \sqrt{3}$$

9. (c) Let $O(0, 0)$, $A(2, 2\sqrt{3})$ and $B(a, b)$ be the vertices of an equilateral triangle, then

$$OA = \sqrt{(2)^2 + (2\sqrt{3})^2} = \sqrt{4 + 12} = \sqrt{16} = 4$$
$$AB = \sqrt{(a-2)^2 + (b - 2\sqrt{3})^2}$$
$$= \sqrt{a^2 + 4 - 4a + b^2 + 12 - 4\sqrt{3}b}$$
$$= \sqrt{a^2 + b^2 - 4a - 4\sqrt{3}b + 16}$$

and $\quad OB = \sqrt{a^2 + b^2}$

Since, ΔABC is equilateral.

$\therefore \quad OA = OB = AB$

On taking first two equalities, we have

$$OA = OB$$
$$4 = \sqrt{a^2 + b^2}$$
$$\Rightarrow \quad a^2 + b^2 = 16 \quad \ldots\text{(i)}$$

On taking last two equalities, we have

$$OB = AB$$
$$\Rightarrow \quad \sqrt{a^2 + b^2} = \sqrt{a^2 + b^2 - 4a - 4\sqrt{3}b + 16}$$
$$\Rightarrow \quad a^2 + b^2 = a^2 + b^2 - 4a - 4\sqrt{3}b + 16$$
$$\Rightarrow \quad 4a + 4\sqrt{3}b - 16 = 0$$
$$\Rightarrow \quad a + \sqrt{3}b - 4 = 0 \quad \ldots\text{(ii)}$$

On using Eq. (ii) in Eq. (i), we have

$$(4 - \sqrt{3}b)^2 + b^2 = 16$$
$$\Rightarrow \quad 16 + 3b^2 - 8\sqrt{3}b + b^2 = 16$$
$$\Rightarrow \quad 4b^2 - 8\sqrt{3}b = 0$$
$$\Rightarrow \quad 4b(b - 2\sqrt{3}) = 0$$
$$\Rightarrow \quad b = 0 \text{ or } b = 2\sqrt{3}$$

Now, when $b = 0 \quad \Rightarrow \quad a = 4$

and when $b = 2\sqrt{3} \quad \Rightarrow \quad a = -2$

$\therefore$ Coordinates may be $(4, 0)$ or $(-2, 2\sqrt{3})$.

10. (b) Any point on Y-axis have coordinates $(0, y)$.

Since, point $(0, y)$ is equidistant from $(3, 2)$ and $(-1, 3)$.

$\therefore$ Distance between $(0, y)$ and $(3, 2)$,

$$d_1\text{(say)} = \sqrt{(3)^2 + (2-y)^2} = \sqrt{9 + 4 + y^2 - 4y}$$
$$= \sqrt{y^2 - 4y + 13}$$

and distance between $(0, y)$ and $(-1, 3)$,

$$d_2\text{ (say)} = \sqrt{(-1)^2 + (3-y)^2} = \sqrt{1 + 9 + y^2 - 6y}$$
$$= \sqrt{y^2 - 6y + 10}$$

According to the question,

$$d_1 = d_2$$

$$\Rightarrow \quad \sqrt{y^2 - 4y + 13} = \sqrt{y^2 - 6y + 10}$$

$$\Rightarrow \quad y^2 - 4y + 13 = y^2 - 6y + 10$$

$$\Rightarrow \quad -4y + 13 = -6y + 10$$

$$\Rightarrow \quad 2y = -3$$

$$\Rightarrow \quad y = -\frac{3}{2}$$

$\therefore$ Required coordinates of the point lying on Y-axis is $\left(0, -\frac{3}{2}\right)$.

11. (c) Let coordinates of vertex A be (x, y).

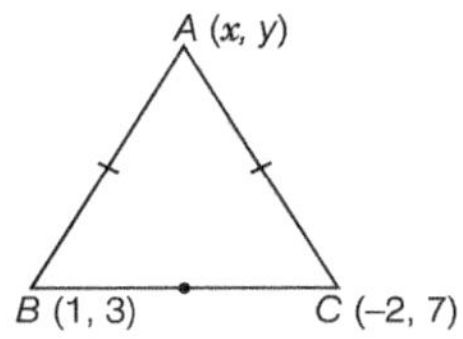

$$\therefore \quad AB = \sqrt{(x-1)^2 + (y-3)^2}$$

$$AC = \sqrt{(x+2)^2 + (y-7)^2}$$

Since, ΔABC is isosceles, therefore

$$AB^2 = AC^2$$

$$\Rightarrow \quad (x-1)^2 + (y-3)^2 = (x+2)^2 + (y-7)^2$$

$$\Rightarrow \quad x^2 + 1 - 2x + y^2 + 9 - 6y = x^2 + 4 + 4x + y^2 + 49 - 14y$$

$$\Rightarrow \quad 6x - 8y + 43 = 0 \qquad \ldots(i)$$

Only option (c) satisfy Eq. (i),

as $6 \times \frac{5}{6} - 8 \times 6 + 43 = 5 - 48 + 43 = 0$

Hence, option (c) is correct.

12. (b) Let AB subtends an angle θ at origin.

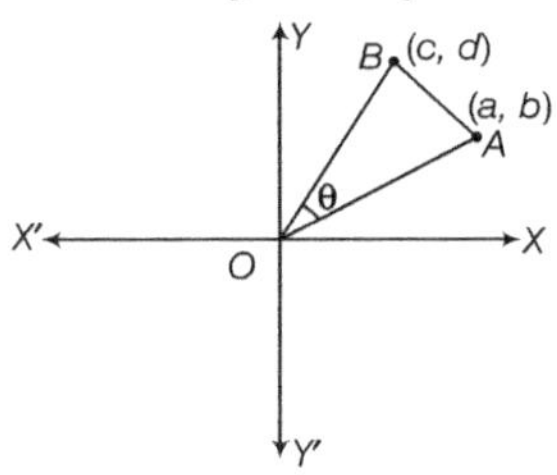

$$\therefore \quad \cos\theta = \frac{OA^2 + OB^2 - AB^2}{2 \cdot OA \cdot OB} \qquad \ldots(i)$$

Now, $\quad OA = \sqrt{a^2 + b^2}$

$$OB = \sqrt{c^2 + d^2}$$

and $\quad AB = \sqrt{(a-c)^2 + (b-d)^2}$

$$= \sqrt{a^2 + c^2 - 2ac + b^2 + d^2 - 2bd}$$

On putting these values in Eq. (i), we have

$$\cos\theta = \frac{a^2 + b^2 + c^2 + d^2 - a^2 - c^2 + 2ac - b^2 - d^2 + 2bd}{2\sqrt{a^2 + b^2}\,\sqrt{c^2 + d^2}}$$

$$= \frac{2ac + 2bd}{2\sqrt{(a^2 + b^2)(c^2 + d^2)}}$$

$$= \frac{ac + bd}{\sqrt{(a^2 + b^2)(c^2 + d^2)}}$$

13. (c) Let $A(a, a)$, $B(-a, -a)$ and $C(-a\sqrt{3}, a\sqrt{3})$.

Now, $\quad AB = \sqrt{(-a - a)^2 + (-a - a)^2}$

$$= \sqrt{(-2a)^2 + (-2a)^2} = \sqrt{4a^2 + 4a^2} = \sqrt{8a^2}$$

$$BC = \sqrt{(-a\sqrt{3} + a)^2 + (a\sqrt{3} + a)^2}$$

$$= \sqrt{3a^2 + a^2 - 2a^2\sqrt{3} + a^2 + 3a^2 + 2\sqrt{3}a^2} = \sqrt{8a^2}$$

$$CA = \sqrt{(-a\sqrt{3} - a)^2 + (a\sqrt{3} - a)^2}$$

$$= \sqrt{3a^2 + a^2 + 2\sqrt{3}a^2 + 3a^2 + a^2 - 2\sqrt{3}a^2}$$

$$= \sqrt{8a^2}$$

$\because \quad AB = BC = CA$

Hence, $\Delta\, ABC$ is an equilateral triangle.

14. (c) Given, $\sqrt{x^2 + y^2 + 8y + 16} - \sqrt{x^2 + y^2 - 6x + 9} = 5$

$$\Rightarrow \quad \sqrt{x^2 + (y+4)^2} - \sqrt{(x-3)^2 + y^2} = 5$$

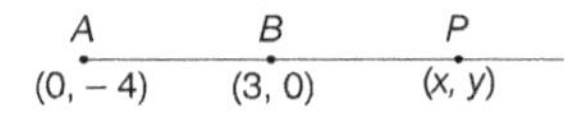

Since, $PA - PB = AB$ $\qquad [\because AB = \sqrt{9 + 16} = 5]$

Thus, locus of P is an infinite ray.

15. (c) According to the question,

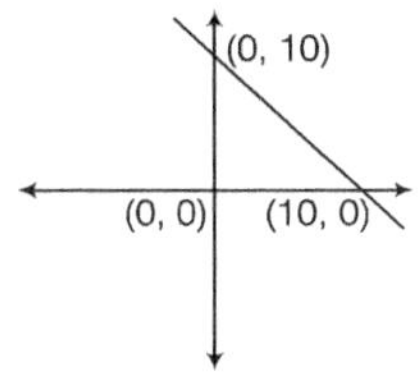

$$x + y < 10$$

$$\Rightarrow \quad x + y + k = 10$$

where, $0 < x < 10$

$$0 < y < 10$$

$$0 < k < 10$$

To find the possible combinations of x and y, we have

Coefficient of x^{10} in $(x + x^2 + x^3 + \cdots + x^9)^3$

$$= \text{Coeff. of } x^7 \text{ in } (1 + x + x^2 + \cdots + x^8)^3$$

$$= \text{Coeff. of } x^7 \text{ in } \left(\frac{1 - x^9}{1 - x}\right)^3$$

$$= \text{Coeff. of } x^7 \text{ in } (1 - x^9)^3 (1 - x)^{-3}$$

$$= \text{Coeff. of } x^7 \text{ in } (1 - x)^{-3}$$

$$= {}^9C_2 = \frac{9!}{2!\,7!} = \frac{9 \times 8}{2} = 36$$

Alternate Method

As, $\quad x + y < 10$

	Number of integer points
When $x = 1 \Rightarrow y < 9$, i.e. $y = \{1, 2, 3, \ldots, 8\}$	8
When $x = 2 \Rightarrow y < 8$, i.e. $y = \{1, 2, 3, \ldots, 7\}$	7
When $x = 3 \Rightarrow y < 7$, i.e. $y = \{1, 2, 3, \ldots, 6\}$	6
:	:
:	:
When $x = 8 \Rightarrow y < 2$, i.e. $y = \{1\}$	1

$\therefore \quad$ Total, $= 1 + 2 + 3 + \cdots + 8$

$$= \frac{8}{2}(1 + 8) = 36$$

DPP-2 Section Formula

1. (c) Let C and D be the middle points of PR and QS respectively, then coordinates of point C is $\left(\frac{1+5}{2}, \frac{2+7}{2}\right)$ i.e. $\left(3, \frac{9}{2}\right)$ and coordinates of point D is $\left(\frac{a+4}{2}, \frac{b+6}{2}\right)$.

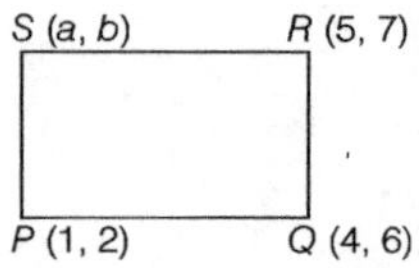

Since, $PQRS$ is a parallelogram, therefore PR and QS will bisect each other.

Hence, C and D are same point,

i.e. $\frac{a+4}{2} = 3$ and $\frac{b+6}{2} = \frac{9}{2}$

$\Rightarrow$ $a + 4 = 6$ and $b + 6 = 9$

$\Rightarrow$ $a = 2$ and $b = 3$

Hence, coordinates of fourth vertex S is (2, 3).

2. (c) Let coordinates of the point C be (x, y).

Given, $AC = 2BC$

$\Rightarrow$ $AC : BC = 2 : 1$

It means point C divides the line segment AB in the ratio 2 : 1.

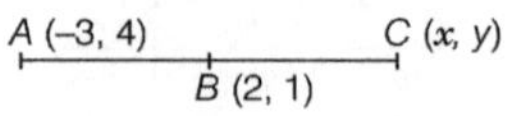

By section formula, we have

[If point (x, y) divides the line segment joining (x_1, y_1) and (x_2, y_2) in the ratio $m : n$ externally, then $x = \frac{mx_2 - nx_1}{m-n}$, $y = \frac{my_2 - ny_1}{m-n}$]

$$x = \frac{2 \times 2 - 1 \times (-3)}{2-1} = \frac{4+3}{1} = 7$$

$$y = \frac{2 \times 1 - 1 \times 4}{2-1} = \frac{2-4}{1} = -2$$

Hence, coordinates of point C is $(7, -2)$.

3. (a) Let $P(x, y)$ be a point which trisected the join of $A(1, -2)$ and $B(-3, 4)$, then the ratio in which point P trisect AB is 2 : 1 or 1 : 2.

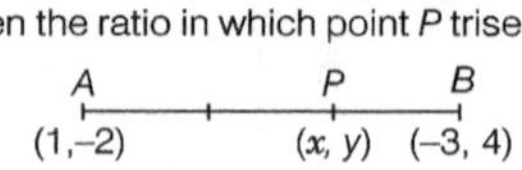

Case I When point P trisect AB in 2 : 1.

The coordinates of point P is

$$x = \frac{2 \times (-3) + 1 \times 1}{2+1} = \frac{-6+1}{3} = \frac{-5}{3}$$

$$y = \frac{2 \times 4 + 1 \times (-2)}{2+1} = \frac{8-2}{3} = 2$$

Hence, coordinates of $P\left(-\frac{5}{3}, 2\right)$.

Case II When point P trisect the AB in 1 : 2, then coordinates of point P is

$$x = \frac{1 \times (-3) + 2 \times 1}{1+2} = \frac{-3+2}{3} = -\frac{1}{3}$$

$$y = \frac{1 \times 4 + 2 \times (-2)}{1+2} = \frac{4-4}{3} = 0$$

Hence, coordinates of P is $\left(-\frac{1}{3}, 0\right)$.

4. (a) Since, $BC \parallel MN$

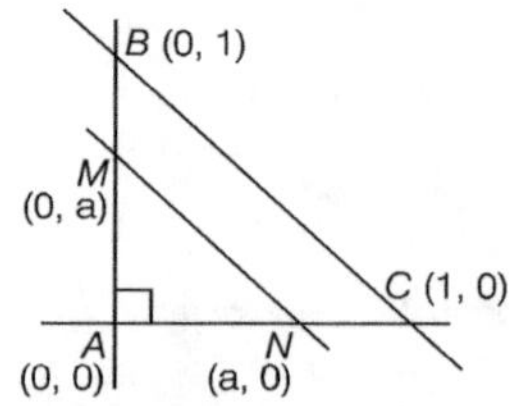

$\therefore$ ΔAMN is similar to ΔABC.

$$\Rightarrow \quad \frac{ar(\Delta AMN)}{ar(\Delta ABC)} = \frac{AM^2}{AB^2} \quad \ldots(i)$$

Now, $ar(\Delta ABC) = \frac{1}{2} \times 1 \times 1 = \frac{1}{2}$ sq unit ...(ii)

and $ar(\Delta AMN) = \frac{1}{2}\, ar(\Delta ABC)$

$= \frac{1}{2} \times \frac{1}{2} = \frac{1}{4}$ sq unit ...(iii)

From Eqs. (i), (ii) and (iii), we have

$$\frac{\frac{1}{4}}{\frac{1}{2}} = \frac{AM^2}{AB^2} \quad \Rightarrow \quad \frac{1}{4} \times \frac{2}{1} = \frac{AM^2}{AB^2}$$

$$\Rightarrow \quad \frac{AM^2}{AB^2} = \frac{1}{2} \quad \Rightarrow \quad \frac{AM}{AB} = \frac{1}{\sqrt{2}}$$

$$\Rightarrow \quad \frac{a}{1} = \frac{1}{\sqrt{2}}$$

$$\therefore \quad a = \frac{1}{\sqrt{2}}$$

Hence, the coordinates of M are $\left(0, \frac{1}{\sqrt{2}}\right)$.

5. (d) Let the line $2x + 3y + 7 = 0$ divides the line segment AB in the ratio $m : n$ at point $P(x, y)$.

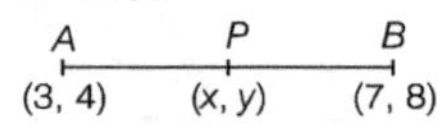

$\therefore$ The coordinates of point P will be

$$x = \frac{m \times 7 + n \times 3}{m+n}$$

$$\Rightarrow \quad x = \frac{7m + 3n}{m+n}$$

$$y = \frac{m \times 8 + n \times 4}{m+n}$$

$$\Rightarrow \quad y = \frac{8m + 4n}{m+n}$$

Since, point P lies on the line $2x + 3y + 7 = 0$

$$\therefore \quad 2\left(\frac{7m+3n}{m+n}\right) + 3\left(\frac{8m+4n}{m+n}\right) + 7 = 0$$

$\Rightarrow$ $2(7m + 3n) + 3(8m + 4n) + 7(m + n) = 0$

$\Rightarrow$ $14m + 6n + 24m + 12n + 7m + 7n = 0$

$\Rightarrow$ $45m + 25n = 0$

$\Rightarrow$ $45m = -25n$

$\Rightarrow$ $\frac{m}{n} = -\frac{25}{45} = -\frac{5}{9}$

Hence, line $2x + 3y + 7 = 0$ divides AB in the ratio 5 : 9 externally.

Alternate Method

Let the line segment joining (3, 4) and (7, 8) is divided by the line $2x + 3y + 7 = 0$ in the ratio of $\lambda : 1$.

$$\Rightarrow \quad \lambda = -\frac{(2\times 3 + 3\times 4 + 7)}{(2\times 7 + 3\times 8 + 7)}$$

$$= \frac{-25}{45}$$

$$= -\frac{5}{9}$$

i.e. 5 : 9 externally.

6. (c) Distance between points P (7, 7) and Q (3, 4)

$$PQ = \sqrt{(3-7)^2 + (4-7)^2}$$

$$= \sqrt{(-4)^2 + (-3)^2}$$

$$= \sqrt{16 + 9}$$

$$= \sqrt{25} = 5$$

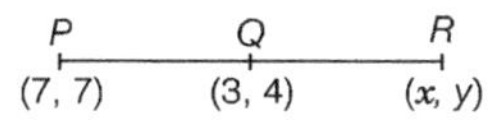

$\because \quad PR = 10$

$\Rightarrow \quad PQ + QR = 10$

$\Rightarrow \quad 5 + QR = 10$

$\Rightarrow \quad QR = 5$

$\therefore \quad \frac{PQ}{QR} = \frac{5}{5} = \frac{1}{1}$

$PQ : QR = 1 : 1$

Hence, point Q divides PR in the ratio 1 : 1 internally.

$\therefore$ By section formula, we have

$$3 = \frac{1\times x + 1\times 7}{1+1}$$

$\Rightarrow \quad 3 = \frac{x + 7}{2}$

$\Rightarrow \quad x = -1$

and $\quad 4 = \frac{1\times y + 1\times 7}{2}$

$\Rightarrow \quad y + 7 = 8$

$\Rightarrow \quad y = 1$

Hence, coordinates of point R is (– 1, 1).

7. (b) Let X-axis divides the join of (2, – 3) and (5, 6) in the ratio $m : n$.

$\therefore$ Coordinates of any point on the X-axis is $(x, 0)$.

i.e. $\quad 0 = \frac{m\times 6 + n(-3)}{m+n}$

$\Rightarrow \quad 6m - 3n = 0$

$\Rightarrow \quad 6m = 3n$

$\Rightarrow \quad \frac{m}{n} = \frac{3}{6} = \frac{1}{2}$

$\therefore \quad m : n = 1 : 2$

8. (a) Let P and Q be the mid-points of AC and BD, respectively.

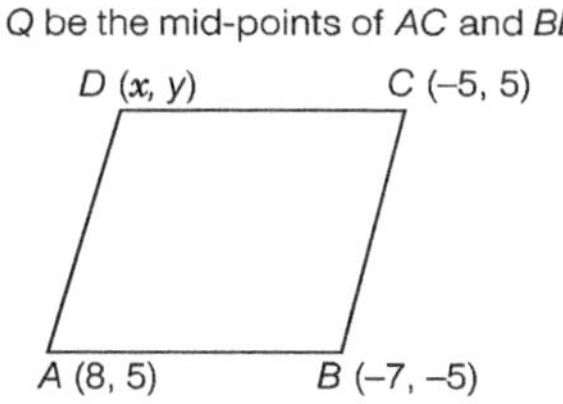

$\therefore$ Coordinates of point P is $\left(\frac{8-5}{2}, \frac{5+5}{2}\right)$ i.e. $\left(\frac{3}{2}, 5\right)$ and coordinates of point Q is $\left(\frac{x-7}{2}, \frac{y-5}{2}\right)$.

$\because$ $ABCD$ is a parallelogram, therefore its diagonal bisect each other.

Hence, point P and Q are same,

i.e. $\quad \frac{x-7}{2} = \frac{3}{2}$ and $\frac{y-5}{2} = 5$

$\Rightarrow \quad x - 7 = 3$ and $y - 5 = 10$

$\Rightarrow \quad x = 10$ and $y = 15$

Hence, coordinates of fourth vertex D is (10, 15).

9. (d) We know that, the points (x_1, y_1) and (x_2, y_2) lie on the same side of $ax + by + c = 0$.

If $ax_1 + by_1 + c = 0$ and $ax_2 + by_2 + c = 0$ have same sign.

Given line, $\quad 3x - 8y - 7 = 0 \quad$ ...(i)

Consider points (– 1, – 1) and (3, 7)

Now, $\quad 3(-1) - 8(-1) - 7 = -3 + 8 - 7 = -2 < 0$

and $\quad 3(3) - 8(7) - 7 = 9 - 56 - 7 = -54 < 0$

Hence, points (– 1, – 1) and (3, 7) lies on same side of $3x - 8y - 7 = 0$.

10. (b) Since, (0, 0) and (1, 1) lies on the same side of $a^2x + aby + 1 = 0$ for all $a \in R$, $b > 0$.

$\therefore \quad a^2(0) + ab(0) + 1 = 1 > 0$

and $\quad a^2(1) + ab(1) + 1 = a^2 + ab + 1 > 0$

$\Rightarrow \quad A = 1 > 0$ and $D < 0$

$[\because$ If $f(x) = ax^2 + bx + c > 0 \Rightarrow a > 0, D < 0]$

$\Rightarrow \quad b^2 - 4 < 0$

$\Rightarrow \quad b^2 < 4$

$\Rightarrow \quad b < 2 \quad [\because b > 0]$

$\Rightarrow \quad b \in (0, 2)$

11. (a) Let S (α, β) be the harmonic conjugates of the points R (5, 1).

Suppose R divides PQ in the ratio $\lambda : 1$ internally, then S divides PQ in the ratio $\lambda : 1$ externally, then

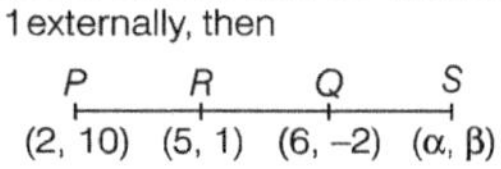

$$5 = \frac{6\lambda + 2}{\lambda + 1}$$

$\Rightarrow \quad 5\lambda + 5 = 6\lambda + 2$

$\Rightarrow \quad \lambda = 3$

Also, $\quad 1 = \frac{-2\lambda + 10}{\lambda + 1}$

$\Rightarrow \quad \lambda + 1 = -2\lambda + 10$

$\Rightarrow \quad 3\lambda = 9$

$\therefore \quad \lambda = 3$

Now, $\quad \alpha = \frac{3\times 6 - 1\times 2}{3-1} = 8$

$$\beta = \frac{3\times(-2) - 1\times 10}{3-1} = -8$$

Hence, harmonic conjugates of R (5, 1) is (8, – 8).

12. (a) Since, coordinates of variable point M $(\beta, \beta + 1)$.

It satisfy the equation $y = x + 1$.

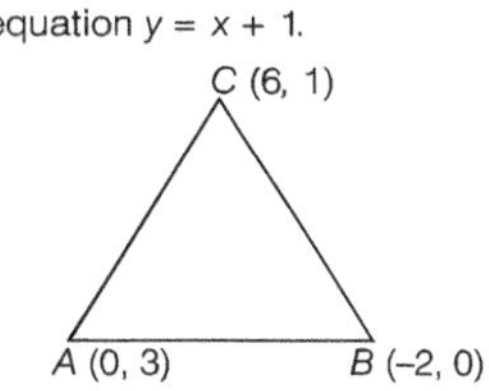

as when $x = \beta$ and $y = \beta + 1$

13. (c) Equation of BC is

$$y - 0 = \frac{1-0}{6+2}(x+2)$$

[Equation of line joining (x_1, y_1) and (x_2, y_2) is

$$y - y_1 = \frac{y_2 - y_1}{x_2 - x_1}(x - x_1)]$$

$$\Rightarrow \quad y = \frac{1}{8}(x+2)$$

$$\Rightarrow \quad 8y = x + 2$$

$$\Rightarrow \quad x - 8y + 2 = 0$$

$\because$ Points A and M lie on the same side of BC.

$\therefore \quad 0 - 8 \times 3 + 2 = -24 + 2 = -22 < 0$, at $A(0, 3)$

$\Rightarrow \quad \beta - 8(\beta + 1) + 2 < 0$, at $M(\beta, \beta + 1)$

[$\because$ if (x_1, y_1) and (x_2, y_2) lies on the same side of $ax + by + c = 0$. Then, $ax_1 + by_1 + c = 0$ and $ax_2 + by_2 + c = 0$ have same sign]

$$\Rightarrow \quad \beta - 8\beta - 8 + 2 < 0$$

$$\Rightarrow \quad -7\beta - 6 < 0$$

$$-7\beta < 6$$

$$\therefore \quad \beta > -\frac{6}{7}$$

14. (c) Since, point M lies within the ΔABC.

$\therefore$ M and A lies on the same side of BC.

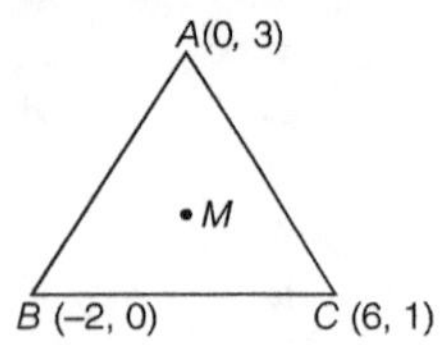

Similarly, M and B lies on the same side of AC.

Equation of AC, $(y - 3) = \frac{1-3}{6-0}(x - 0)$

$$y - 3 = -\frac{2}{6}(x)$$

$$\Rightarrow \quad y - 3 = -\frac{1}{3}x$$

$$\Rightarrow \quad 3y - 9 = -x$$

$$\Rightarrow \quad x + 3y - 9 = 0$$

Now, $\quad (-2) + 3(0) - 9 = -11 < 0$

$\Rightarrow \quad \beta + 3(\beta + 1) - 9 < 0$, at $B(-2, 0)$

$\Rightarrow \quad \beta + 3\beta + 3 - 9 < 0$, at $M(\beta, \beta + 1)$

$$\Rightarrow \quad 4\beta - 6 < 0$$

$$\Rightarrow \quad \beta < \frac{6}{4} = \frac{3}{2}$$

Hence, $-\frac{6}{7} < \beta < \frac{3}{2}$

15. (a)

(i) If point $\left(\frac{10}{7}, \frac{33}{7}\right)$ divides the join of (1, 3) and (2, 7) in the ratio $p : q$, then by section formula, we have

$$\frac{10}{7} = \frac{2 \times p + 1 \times q}{p + q}$$

$$\Rightarrow \quad \frac{10}{7} = \frac{2p + q}{p + q}$$

$$\Rightarrow \quad 10p + 10q = 14p + 7q$$

$$\Rightarrow \quad -4p = -3q$$

$$\Rightarrow \quad \frac{p}{q} = \frac{3}{4}$$

$$\Rightarrow \quad p : q = 3 : 4$$

$$\therefore \quad p + q = 3 + 4 = 7$$

Hence **(i) → (s)**

(ii)

(1, 3) —————— (2, 7)

If point (– 2, – 9) divides the line segment joining the points (1, 3) and (2, 7) is $p : q$, then

$$-2 = \frac{2 \times p + 1 \times q}{p + q}$$

$$\Rightarrow \quad -2p - 2q = 2p + q$$

$$\Rightarrow \quad -4p = q + 2q$$

$$\Rightarrow \quad -4p = 3q$$

$$\Rightarrow \quad \frac{p}{q} = -\frac{3}{4}$$

Hence, point (– 2, – 9) divides the join of (1, 3) and (2, 7) in the ratio 3 : 4 externally.

$$\therefore \quad p + q = 3 + 4 = 7$$

Hence, **(ii) → (s)**

(iii) If (p, q) divides the join of (– 1, 2) and (4, – 5) in the ratio 2 : 3, then

2 : 3

(–1, 2) —— (p, q) —— (4, –5)

$$p = \frac{4 \times 2 + (-1)3}{2 + 3}$$

$$\Rightarrow \quad p = \frac{8 - 3}{5} = \frac{5}{5} = 1$$

and $\quad q = \frac{(-5) \times 2 + 2 \times 3}{2 + 3}$

$$\Rightarrow \quad q = \frac{-10 + 6}{5} = -\frac{4}{5}$$

$$\therefore \quad p + q = 1 - \frac{4}{5} = \frac{1}{5}$$

Hence, **(iii) → (p)**

(iv) Since, (p, q) divides externally the line joining (– 1, 2) and (4, – 5) in the ratio 2 : 3, then

$$p = \frac{2 \times 4 - 3 \times (-1)}{2 - 3}$$

$$\Rightarrow \quad p = \frac{8 + 3}{-1}$$

$$\Rightarrow \quad p = -11$$

$$q = \frac{2 \times (-5) - 3(2)}{2 - 3}$$

$$\Rightarrow \quad q = \frac{-10 - 6}{-1}$$

$$\Rightarrow \quad q = \frac{-16}{-1} = 16$$

$$\therefore \quad p + q = -11 + 16 = 5$$

Hence, **(iv) → (r)**

DPP-3 **Area**

1. We know that, the area of triangle having vertices (x_1, y_1) (x_2, y_2) and (x_3, y_3) is

$$\Delta = \frac{1}{2}|x_1(y_2 - y_3) + x_2(y_3 - y_1) + x_3(y_1 - y_2)|$$

$$\Rightarrow \quad \Delta = \frac{1}{2}\begin{vmatrix} x_1 & y_1 & 1 \\ x_2 & y_2 & 1 \\ x_3 & y_3 & 1 \end{vmatrix}$$

(i) Given, $(x_1, y_1) \equiv (0, 4)$, $(x_2, y_2) \equiv (3, 6)$ and $(x_3, y_3) \equiv (-8, -2)$

$$\text{Area} = \frac{1}{2}|0(6 + 2) + 3(-2 - 4) - 8(4 - 6)|$$

$$= \frac{1}{2}|3(-6) - 8(-2)| = \frac{1}{2}|-18 + 16|$$

$$= \frac{1}{2}|-2| = 1 \text{ sq unit}$$

(ii) $(x_1, y_1) \equiv (a, b + c)$, $(x_2, y_2) \equiv (a, b - c)$
$(x_3, y_3) \equiv (-a, c)$

$$\text{Area} = \frac{1}{2}|a(b - c - c) + a(c - b - c) - a(b + c - b + c)|$$

$$= \frac{1}{2}|a(b - 2c) + a(-b) - a(2c)|$$

$$= \frac{1}{2}|ab - 2ac - ab - 2ac|$$

$$= \frac{1}{2}|-4ac| = 2ac \text{ sq units}$$

(iii) The vertices of triangle are
$(a\cos\phi_1, b\sin\phi_1)$, $(a\cos\phi_2, b\sin\phi_2)$
and $(a\cos\phi_3, b\sin\phi_3)$

$$\text{Area} = \frac{1}{2}\begin{vmatrix} a\cos\phi_1 & b\sin\phi_1 \\ a\cos\phi_2 & b\sin\phi_2 \\ a\cos\phi_3 & b\sin\phi_3 \\ a\cos\phi_1 & b\sin\phi_1 \end{vmatrix}, \text{ using stair-method.}$$

$$= \frac{1}{2}|(ab\cos\phi_1\sin\phi_2 + ab\cos\phi_2\sin\phi_3 + ab\cos\phi_3\sin\phi_1) - (ab\sin\phi_1\cos\phi_2 + ab\sin\phi_2\cos\phi_3 + ab\sin\phi_3\cos\phi_1|$$

$$= \frac{1}{2}|ab\{(\cos\phi_1\sin\phi_2 - \sin\phi_1\cos\phi_2) + (\cos\phi_2\sin\phi_3 - \sin\phi_2\cos\phi_3) + (\cos\phi_3\sin\phi_1 - \sin\phi_3\cos\phi_1)\}|$$

$$= \frac{1}{2}|ab\{-\sin(\phi_1 - \phi_2) - \sin(\phi_2 - \phi_3) - \sin(\phi_3 - \phi_1)\}|$$

$[\because \sin A\cos B - \cos A\sin B = \sin(A - B)]$

$$= \frac{1}{2}|-ab\{\sin(\phi_1 - \phi_2) + \sin(\phi_2 - \phi_3) + \sin(\phi_3 - \phi_1)\}|$$

$$= \frac{1}{2}\left|-ab\left\{2\sin\left(\frac{\phi_1 - \phi_3}{2}\right)\cos\left(\frac{\phi_1 - 2\phi_2 + \phi_3}{2}\right) + \sin(\phi_3 - \phi_1)\right\}\right|$$

$$= \frac{1}{2}\left|-ab\left[2\sin\left(\frac{\phi_1 - \phi_3}{2}\right)\left\{\cos\left(\frac{\phi_1 - 2\phi_2 + \phi_3}{2}\right) - \cos\left(\frac{\phi_3 - \phi_1}{2}\right)\right\}\right]\right|$$

$$\left[\because \sin C + \sin D = 2\sin\frac{C + D}{2}\cos\frac{C - D}{2} \text{ and } \cos C - \cos D = 2\sin\frac{C + D}{2}\sin\frac{D - C}{2}\right]$$

$$= \frac{1}{2}\left|-ab\left[2\sin\left(\frac{\phi_1 - \phi_3}{2}\right)\left\{2\sin\left(\frac{\phi_3 - \phi_2}{2}\right)\sin\left(\frac{\phi_2 - \phi_1}{2}\right)\right\}\right]\right|$$

$$= \frac{1}{2}\left|-ab\left[4\sin\left(\frac{\phi_1 - \phi_3}{2}\right)\sin\left(\frac{\phi_3 - \phi_2}{2}\right)\sin\left(\frac{\phi_2 - \phi_1}{2}\right)\right]\right|$$

$$= \left|-2ab\sin\frac{(\phi_1 - \phi_2)}{2}\sin\left(\frac{\phi_2 - \phi_3}{2}\right)\sin\left(\frac{\phi_3 - \phi_1}{2}\right)\right| \text{ sq units}$$

(iv) Given, vertices of triangles are
$(am_1^2, 2am_1)$, $(am_2^2, 2am_2)$ and $(am_3^2, 2am_3)$

$$\therefore \text{Area} = \frac{1}{2}|(am_1^2 \times 2am_2 + am_2^2 \times 2am_3 + am_3^2 \times 2am_1) - (am_2^2 \times 2am_1 + am_3^2 \times 2am_2 + am_1^2 \times 2am_3)|$$

$$= \frac{1}{2}|(2a^2m_1^2m_2 + 2a^2m_2^2m_3 + 2a^2m_3^2m_1) - (2a^2m_1m_2^2 + 2a^2m_2m_3^2 + 2a^2m_1^2m_3)|$$

$$= \frac{1}{2}|2a^2\{m_1^2m_2 + m_2^2m_3 + m_3^2m_1) - (m_1m_2^2 + m_2m_3^2 + m_3m_1^2)|$$

$$= a^2|(m_1^2m_2 - m_1^2m_3) + (m_2^2m_3 - m_2m_3^2) + (m_3^2m_1 - m_1m_2^2)|$$

$$= a^2|m_1^2(m_2 - m_3) + m_2m_3(m_2 - m_3) + m_1(m_3^2 - m_2^2)|$$

$$= a^2|(m_2 - m_3)\{m_1^2 + m_2m_3 - m_1(m_2 + m_3)\}|$$

$$= a^2|(m_2 - m_3)\{m_1^2 + m_2m_3 - m_1m_2 - m_1m_3\}|$$

$$= a^2|(m_2 - m_3)\{(m_1^2 - m_1m_2) - (m_1m_3 - m_2m_3)\}|$$

$$= a^2|(m_2 - m_3)\{m_1(m_1 - m_2) - m_3(m_1 - m_2)\}|$$

$$= a^2|(m_2 - m_3)\{(m_1 - m_2)(m_1 - m_3)\}|$$

$$= a^2|(m_1 - m_2)(m_2 - m_3)(m_3 - m_1)| \text{ sq units}$$

(v) $(x_1, y_1) = \left(am_1, \frac{a}{m_1}\right)$

$(x_2, y_2) = \left(am_2, \frac{a}{m_2}\right)$

$(x_3, y_3) = \left(am_3, \frac{a}{m_3}\right)$

$$\text{Area} = \frac{1}{2}\left|am_1\left(\frac{a}{m_2} - \frac{a}{m_3}\right) + am_2\left(\frac{a}{m_3} - \frac{a}{m_1}\right) + am_3\left(\frac{a}{m_1} - \frac{a}{m_2}\right)\right|$$

$$= \frac{1}{2}\left|a^2m_1\left(\frac{1}{m_2} - \frac{1}{m_3}\right) + a^2m_2\left(\frac{1}{m_3} - \frac{1}{m_1}\right) + a^2m_3\left(\frac{1}{m_1} - \frac{1}{m_2}\right)\right|$$

$$= \frac{1}{2}\left|a^2\left\{m_1\left(\frac{m_3 - m_2}{m_2m_3}\right) + m_2\left(\frac{m_1 - m_3}{m_1m_3}\right) + m_3\left(\frac{m_2 - m_1}{m_1m_2}\right)\right\}\right|$$

$$= \frac{1}{2}\left|a^2\left\{\frac{m_1^2(m_3 - m_2) + m_2^2(m_1 - m_3) + m_3^2(m_2 - m_3)}{m_1m_2m_3}\right\}\right|$$

$$= \frac{1}{2}a^2\left|\frac{(m_1 - m_2)(m_2 - m_3)(m_3 - m_1)}{m_1m_2m_3}\right| \text{ sq units}$$

2. (i) Let $A(1, 1)$, $B(3, 4)$, $C(5, -2)$ and $D(4, -7)$ be the vertices of quadrilateral.

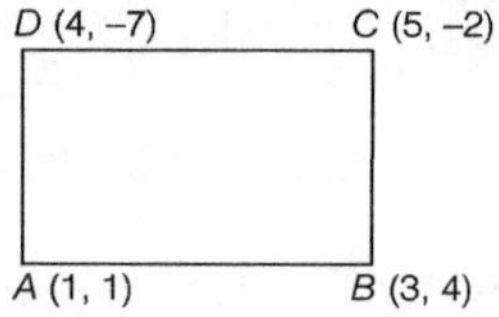

We know that,

$\therefore$ Area of quadrilateral with vertices $(x_1, y_1), (x_2, y_2), (x_3, y_3)$ and (x_4, y_4) is

$$\frac{1}{2}\left[\left|\begin{vmatrix} x_1 & y_1 \\ x_2 & y_2 \end{vmatrix} + \begin{vmatrix} x_2 & y_2 \\ x_3 & y_3 \end{vmatrix} + \begin{vmatrix} x_3 & y_3 \\ x_4 & y_4 \end{vmatrix} + \begin{vmatrix} x_4 & y_4 \\ x_1 & y_1 \end{vmatrix}\right|\right]$$

$$\text{Area} = \frac{1}{2}\left|\begin{vmatrix} 1 & 1 \\ 3 & 4 \end{vmatrix} + \begin{vmatrix} 3 & 4 \\ 5 & -2 \end{vmatrix} + \begin{vmatrix} 5 & -2 \\ 4 & -7 \end{vmatrix} + \begin{vmatrix} 4 & -7 \\ 1 & 1 \end{vmatrix}\right|$$

$$= \frac{1}{2}|(4 - 3) + (-6 - 20) + (-35 + 8) + (4 + 7)|$$

$$= \frac{1}{2}|1 - 26 - 27 + 11| = \frac{1}{2}|-41| = \frac{41}{2} \text{ sq units}$$

(ii) $$\text{Area} = \frac{1}{2}\left|\begin{vmatrix} -1 & 6 \\ -3 & -9 \end{vmatrix} + \begin{vmatrix} -3 & -9 \\ 5 & -8 \end{vmatrix} + \begin{vmatrix} 5 & -8 \\ 3 & 9 \end{vmatrix} + \begin{vmatrix} 3 & 9 \\ -1 & 6 \end{vmatrix}\right|$$

$$= \frac{1}{2}|(9 + 18) + (24 + 45) + (45 + 24) + (18 + 9)|$$

$$= \frac{1}{2}|27 + 69 + 69 + 27| = \frac{1}{2} \times 192 = 96 \text{ sq units}$$

3. We know that, the area of pentagon having vertices $(x_1, y_1)(x_2, y_2)(x_3, y_3)(x_4, y_4)$ and (x_5, y_5) is

$$= \frac{1}{2}\left|\begin{vmatrix} x_1 & y_1 \\ x_2 & y_2 \end{vmatrix} + \begin{vmatrix} x_2 & y_2 \\ x_3 & y_3 \end{vmatrix} + \begin{vmatrix} x_3 & y_3 \\ x_4 & y_4 \end{vmatrix} + \begin{vmatrix} x_4 & y_4 \\ x_5 & y_5 \end{vmatrix} + \begin{vmatrix} x_5 & y_5 \\ x_1 & y_1 \end{vmatrix}\right|$$

Here, $(x_1, y_1) \equiv (0, 4)$

$(x_2, y_2) \equiv (3, 0)$

$(x_3, y_3) \equiv (6, 1)$

$(x_4, y_4) \equiv (7, 5)$

$(x_5, y_5) \equiv (4, 9)$

$$\text{Area} = \frac{1}{2}\left|\begin{vmatrix} 0 & 4 \\ 3 & 0 \end{vmatrix} + \begin{vmatrix} 3 & 0 \\ 6 & 1 \end{vmatrix} + \begin{vmatrix} 6 & 1 \\ 7 & 5 \end{vmatrix} + \begin{vmatrix} 7 & 5 \\ 4 & 9 \end{vmatrix} + \begin{vmatrix} 4 & 9 \\ 0 & 4 \end{vmatrix}\right|$$

$$= \frac{1}{2}|(0 - 12) + (3 - 0) + (30 - 7) + (63 - 20) + 16 - 0)|$$

$$= \frac{1}{2}|-12 + 3 + 23 + 43 + 16| = \frac{1}{2}|73| = 36.5 \text{ sq units}$$

4. (c) Given, points $(k, 2 - 2k)$, $(1 - k, 2k)$ and $(-4 - k, 6 - 2k)$ are collinear.

We know that, three points $A(x_1, y_1)$, $B(x_2, y_2)$ and $C(x_3, y_3)$ are collinear, iff area of $\Delta ABC = 0$

i.e. $$\frac{1}{2}\begin{vmatrix} x_1 & y_1 & 1 \\ x_2 & y_2 & 1 \\ x_3 & y_3 & 1 \end{vmatrix} = 0 \text{ or } \begin{vmatrix} x_1 & y_1 & 1 \\ x_2 & y_2 & 1 \\ x_3 & y_3 & 1 \end{vmatrix} = 0$$

$$\therefore \quad \begin{vmatrix} k & 2 - 2k & 1 \\ 1 - k & 2k & 1 \\ -4 - k & 6 - 2k & 1 \end{vmatrix} = 0$$

$\Rightarrow$ $k(2k - 6 + 2k) - (2 - 2k)\{(1 - k) + (4 + k)\}$ $+ \{(1 - k)(6 - 2k) + 2k(4 + k)\} = 0$

$\Rightarrow$ $k(4k - 6) - (2 - 2k)(5) + \{6(1 - k)$ $- 2k(1 - k) + 8k + 2k^2\} = 0$

$\Rightarrow$ $k(4k - 6) - 10(1 - k) + \{6 - 6k - 2k + 2k^2 + 8k + 2k^2\} = 0$

$\Rightarrow$ $4k^2 - 6k - 10 + 10k + 4k^2 + 6 = 0$

$\Rightarrow$ $8k^2 + 4k - 4 = 0$

$\Rightarrow$ $2k^2 + k - 1 = 0$

$\Rightarrow$ $(2k - 1)(k + 1) = 0$

$\therefore$ $k = -1$ or $\frac{1}{2}$

5. (b) Given, points $(a, 0)$, $(0, b)$ and $(1, 1)$ are collinear.

$$\therefore \quad \begin{vmatrix} a & 0 & 1 \\ 0 & b & 1 \\ 1 & 1 & 1 \end{vmatrix} = 0$$

$\Rightarrow$ $a(b - 1) + 1(0 - b) = 0$

$\Rightarrow$ $a(b - 1) - b = 0$

$\Rightarrow$ $ab - a - b = 0 \quad \Rightarrow \quad a + b = ab$

$\Rightarrow$ $\frac{1}{b} + \frac{1}{a} = 1 \quad \Rightarrow \quad \frac{1}{a} + \frac{1}{b} = 1$

6. (b) Given, D, E and F are the mid-points of sides AB, BC and AC.

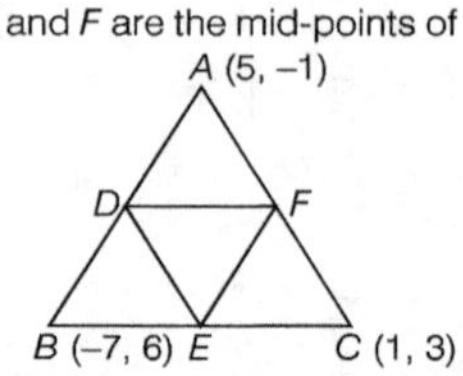

$\therefore$ Coordinates of point D is $\left(\frac{5 - 7}{2}, \frac{-1 + 6}{2}\right)$ i.e. $\left(-1, \frac{5}{2}\right)$

Coordinates of point E is $\left(\frac{-7 + 1}{2}, \frac{6 + 3}{2}\right)$ i.e. $\left(-3, \frac{9}{2}\right)$

Coordinates of point F is $\left(\frac{1 + 5}{2}, \frac{3 - 1}{2}\right)$ i.e. $(3, 1)$

$$\therefore \quad \text{Area of } \Delta DEF = \frac{1}{2}\begin{vmatrix} -1 & \frac{5}{2} & 1 \\ -3 & \frac{9}{2} & 1 \\ 3 & 1 & 1 \end{vmatrix}$$

$$= \frac{1}{2}\left|-1\left(\frac{9}{2} - 1\right) - \frac{5}{2}(-3 - 3) + 1\left(-3 - \frac{27}{2}\right)\right|$$

$$= \frac{1}{2}\left|-\frac{7}{2} + 15 - \frac{33}{2}\right| = \frac{5}{2} \text{ sq units}$$

7. (b) Given, $(x_1, y_1) \equiv (1, 5)$, $(x_2, y_2) \equiv (4, a)$, $(x_3, y_3) \equiv (2, 3)$ and area of triangle = 3 sq units

$$\Rightarrow \quad \frac{1}{2}\begin{vmatrix} x_1 & y_1 & 1 \\ x_2 & y_2 & 1 \\ x_3 & y_3 & 1 \end{vmatrix} = 3$$

$$\Rightarrow \quad \begin{vmatrix} x_1 & y_1 & 1 \\ x_2 & y_2 & 1 \\ x_3 & y_3 & 1 \end{vmatrix} = \pm 6$$

$$\Rightarrow \quad \begin{vmatrix} 1 & 5 & 1 \\ 4 & a & 1 \\ 2 & 3 & 1 \end{vmatrix} = \pm 6$$

$\Rightarrow$ $1(a - 3) - 5(4 - 2) + 1(12 - 2a) = \pm 6$

$\Rightarrow$ $a - 3 - 10 + 12 - 2a = \pm 6$

$\Rightarrow$ $-a - 1 = \pm 6$

On taking positive sign, we have

$-a - 1 = 6$

$\Rightarrow$ $-a = 7 \Rightarrow a = -7$

On taking negative sign, we have

$-a - 1 = -6$

$\Rightarrow$ $-a = -5$

$\therefore$ $a = 5$

8. (a) Since, point P divides the line segment AB in the ratio $\lambda : 1$.

$\therefore$ Coordinates of point $P\left(\frac{3\lambda-5}{\lambda+1}, \frac{5\lambda+1}{\lambda+1}\right)$

Now, area of $\Delta PQR = 2$

$$\Rightarrow \frac{1}{2}\begin{vmatrix} \frac{3\lambda-5}{\lambda+1} & \frac{5\lambda+1}{\lambda+1} & 1 \\ 1 & 5 & 1 \\ 7 & 2 & 1 \end{vmatrix} = 2$$

$$\Rightarrow \begin{vmatrix} \frac{3\lambda-5}{\lambda+1} & \frac{5\lambda+1}{\lambda+1} & 1 \\ 1 & 5 & 1 \\ 7 & 2 & 1 \end{vmatrix} = \pm 4$$

$$\Rightarrow \frac{3\lambda-5}{\lambda+1}(5-2) - \frac{(5\lambda+1)}{\lambda+1}(1-7) + (2-35) = \pm 4$$

$$\Rightarrow \frac{3(3\lambda-5)}{\lambda+1} + \frac{6(5\lambda+1)}{\lambda+1} - 33 = \pm 4$$

$$\Rightarrow 3(3\lambda-5) + 6(5\lambda+1) - 33(\lambda+1) = \pm 4$$

$$\Rightarrow 9\lambda - 15 + 30\lambda + 6 - 33\lambda - 33 = \pm 4$$

$$\Rightarrow 6\lambda - 42 = \pm 4$$

On taking the positive sign, we have

$$6\lambda - 42 = 4$$

$$\Rightarrow 6\lambda = 46$$

$$\Rightarrow \lambda = \frac{46}{6} = \frac{23}{3}$$

$$\Rightarrow \lambda : 1 = 23 : 3$$

$$\therefore \lambda = \frac{23}{3}$$

On taking negative sign, we have

$$6\lambda - 42 = -4$$

$$\Rightarrow 6\lambda = 38$$

$$\Rightarrow \lambda = \frac{38}{6} = \frac{19}{3}$$

$$\Rightarrow \lambda : 1 = 19 : 3$$

$$\therefore \lambda = \frac{19}{3}$$

9. (c) Let $O(0, 0)$, $A(2q, 0)$, $B(2q, 2p)$ and $C(0, 2p)$ be the vertices of a rectangle as shown in the following figure.
(Here, we consider $p, q > 0$)

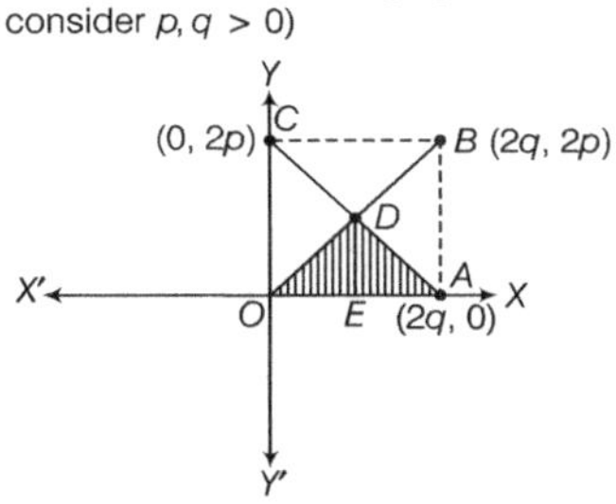

From the figure, it is clear that the common area of ΔOAC and OAB is ΔOAD.

Since, diagonals of rectangle bisect each other.

$\therefore$ D will be the mid-point of OB and AC.

$\therefore$ Coordinates of point $D\left(\frac{0+2q}{2}, \frac{0+2p}{2}\right)$, i.e. (q, p)

Now, area of $\Delta ODA = \frac{1}{2} \times OA \times DE$

$$= \frac{1}{2} \times 2q \times p = pq \text{ sq units}$$

10. (b) Since, $(1, 1)$, $(0, \sec^2\theta)$, $(\text{cosec}^2\theta, 0)$ are collinear.

Thus,

$$\frac{1}{2}\begin{vmatrix} 1 & 1 & 1 \\ 0 & \sec^2\theta & 1 \\ \text{cosec}^2\theta & 0 & 1 \end{vmatrix} = 0$$

$$\Rightarrow \sec^2\theta + \text{cosec}^2\theta - \sec^2\theta\cos^2\theta = 0$$

$$\Rightarrow \frac{1}{\cos^2\theta} + \frac{1}{\sin^2\theta} - \frac{1}{\sin^2\theta\cos^2\theta} = 0$$

$$\Rightarrow \frac{\sin^2\theta + \cos^2\theta - 1}{\sin^2\theta\cos^2\theta} = 0$$

$\Rightarrow$ $0 = 0$, but $\sin^2\theta \neq 0$ and $\cos^2\theta \neq 0$

$\therefore$ $\theta \neq n\pi$ and $\theta \neq (2n+1)\frac{\pi}{2}$

On combining, $\theta \neq \frac{n\pi}{2}$.

11. (c) $|x| + |y| = 1$

$$\Rightarrow \begin{cases} -x + y = 1, & x < 0, y > 0 \\ -x - y = 1, & x < 0, y < 0 \\ x - y = 1, & x > 0, y < 0 \\ x + y = 1, & x > 0, y > 0 \end{cases}$$

The figure represented by these four lines is given below :

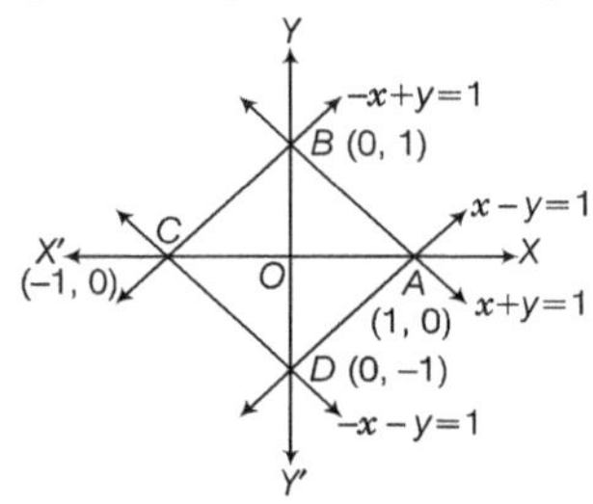

Area of quadrilateral $ABCD$ = Area of ΔABC + Area of ΔADC

$$= \frac{1}{2} \times AC \times OB + \frac{1}{2} \times AC \times OD$$

$$= \frac{1}{2} \times 2 \times 1 + \frac{1}{2} \times 2 \times 1 = 2 \text{ sq units}$$

12. (a) Given, points $A(6, 3)$, $B(-3, 5)$, $C(4, -2)$ and $D(x, 3x)$.

$$\frac{\text{ar}(\Delta DBC)}{\text{ar}(\Delta ABC)} = \frac{1}{2} \quad \ldots\text{(i)}$$

Now, area of $\Delta DBC = \frac{1}{2}\begin{vmatrix} x & 3x & 1 \\ -3 & 5 & 1 \\ 4 & -2 & 1 \end{vmatrix}$

$$= \frac{1}{2}|x(5+2) - 3x(-3-4) + 1(6-20)|$$

$$= \frac{1}{2}|7x + 21x - 14|$$

$$= \frac{1}{2}|28x - 14| \quad \ldots\text{(ii)}$$

and $\text{ar}(\Delta ABC) = \frac{1}{2}\begin{vmatrix} 6 & 3 & 1 \\ -3 & 5 & 1 \\ 4 & -2 & 1 \end{vmatrix}$

$$= \frac{1}{2}|6(5+2) - 3(-3-4) + 1(6-20)|$$

$$= \frac{1}{2}|42 + 21 - 14|$$

$$\text{ar}(\Delta ABC) = \frac{1}{2}|49| = \frac{49}{2} \quad \ldots\text{(iii)}$$

Using Eqs. (ii) and (iii) in Eq. (i), we get

$$\frac{\frac{1}{2}|28x-14|}{\frac{1}{2}|49|}=\frac{1}{2}$$

$\Rightarrow \quad |28x-14|=\frac{49}{2}$

$\Rightarrow \quad 28x-14=\pm\frac{49}{2}$

On taking positive sign, we have

$$28x=\frac{49}{2}+14$$

$\Rightarrow \quad 28x=\frac{77}{2}$

$\therefore \quad x=\frac{77}{2\times 28}=\frac{11}{8}$

On taking negative sign, we have

$$28x=-\frac{49}{2}+14$$

$\therefore \quad x=\frac{-21}{2\times 28}=\frac{-3}{8}$

13. (a) Given points $A(6, 3)$, $B(-3, 5)$, $C(4, -2)$ and $P(x, y)$.

$\therefore$ Area of $\Delta PBC=\frac{1}{2}|x(5+2)+(-3)(-2-y)+4(y-5)|$

$=\frac{1}{2}|7x+6+3y+4y-20|$

$=\frac{1}{2}|7x+7y-14|$

and area of $\Delta ABC=\frac{1}{2}|6(5+2)+(-3)(-2-3)+4(3-5)|$

$=\frac{1}{2}|42+15-8|=\frac{1}{2}|49|$

Now, $\frac{\text{ar}(\Delta PBC)}{\text{ar}(\Delta ABC)}=\frac{\frac{1}{2}|7x+7y-14|}{\frac{1}{2}|49|}$

$=\frac{7|x+y-2|}{49}$

$=\left|\frac{x+y-2}{7}\right|$

14. (b) Let $A(x_1, y_1)$, $B(x_2, y_2)$ and $C(x_3, y_3)$ be the vertices of a ΔABC, where x_1, x_2, x_3 and y_1, y_2, y_3 all are integers, then the area of ΔABC is given by

$\Delta=\frac{1}{2}[x_1(y_2-y_3)+x_2(y_3-y_1)+x_3(y_1-y_2)]=$ an integer

[$\because$ all $x_i, y_i, i=1, 2, 3$, are integers]

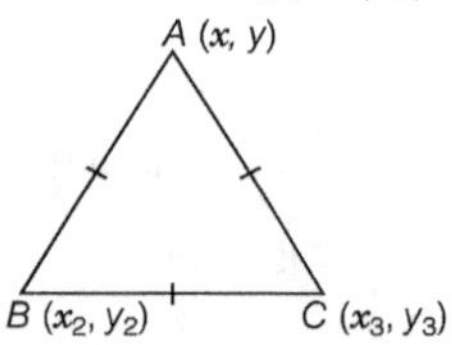

If possible, let the ΔABC be an equilateral triangle, then its area is given by

$\Delta=\frac{\sqrt{3}}{4}(\text{Side})^2=\frac{\sqrt{3}}{4}(AB)^2$ [$\because AB=BC=CA$]

$=\frac{\sqrt{3}}{4}$ (an integer) [$\because$ vertices are integer, $\therefore AB^2$ is an integer]

= An irrational number

This contradict the fact that the area is an integer number. Hence, the triangle cannot be equilateral.

15. (a) Given, points $(a-2, a-4)$, $(a, a+1)$ and $(a+4, 16)$ are collinear, then

$$\frac{1}{2}\begin{vmatrix} a-2 & a-4 & 1 \\ a & a+1 & 1 \\ a+4 & 16 & 1 \end{vmatrix}=0$$

$$\Rightarrow \begin{vmatrix} a-2 & a-4 & 1 \\ a & a+1 & 1 \\ a+4 & 16 & 1 \end{vmatrix}=0$$

$\Rightarrow (a-2)\{(a+1)-16\}-(a-4)\{a-(a+4)\}$
$+1\{16a-(a+1)(a+4)\}=0$

$\Rightarrow (a-2)(a-15)-(a-4)(-4)$
$+1\{16a-(a^2+4a+a+4)\}=0$

$\Rightarrow (a-2)(a-15)-(a-4)(-4)+\{16a-a^2-5a-4\}=0$

$\Rightarrow \quad (a^2-17a+30)+4a-16+11a-a^2-4=0$

$\Rightarrow \quad -2a+10=0$

$\therefore \quad a=5$

16. (c) Let third vertex of triangle be (h, k).

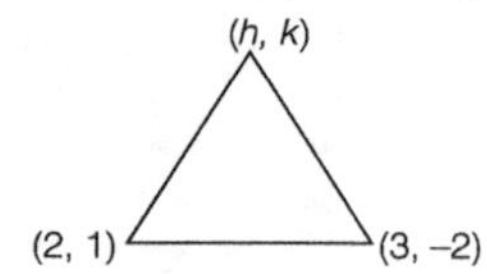

$\because$ Area of triangle = 5 sq units [given]

$$\Rightarrow \frac{1}{2}\begin{vmatrix} h & k & 1 \\ 2 & 1 & 1 \\ 3 & -2 & 1 \end{vmatrix}=5$$

$$\Rightarrow \begin{vmatrix} h & k & 1 \\ 2 & 1 & 1 \\ 3 & -2 & 1 \end{vmatrix}=\pm 10$$

$\Rightarrow \quad h(1+2)-k(2-3)+1(-4-3)=\pm 10$

$\Rightarrow \quad 3h+k-7=\pm 10$

On taking positive sign, we have

$3h+k=17$ ...(i)

On taking negative sign, we have

$3h+k=-3$ ...(ii)

Also, vertex (h, k) lies on the line $y=x+3$

$\Rightarrow \quad k=h+3$

$\Rightarrow \quad h-k=-3$ ...(iii)

On solving Eqs. (i) and (iii), we get

$$h=\frac{7}{2}, k=\frac{13}{2}$$

Hence, third vertex is $\left(\frac{7}{2}, \frac{13}{2}\right)$.

On solving Eqs. (ii) and (iii), we get

$$h=\frac{-3}{2} \text{ and } k=\frac{3}{2}$$

Hence, third vertex is $\left(\frac{-3}{2}, \frac{3}{2}\right)$.

17. (b) Given, area of ΔABC = 20 sq units

Let the third vertex C be (h, k).

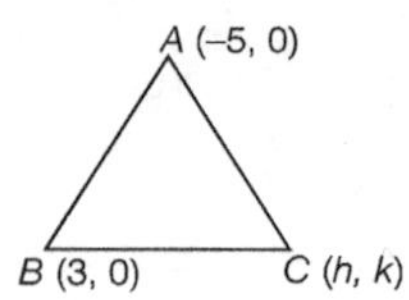

$$\therefore \frac{1}{2}\begin{vmatrix} -5 & 0 & 1 \\ 3 & 0 & 1 \\ h & k & 1 \end{vmatrix}=20$$

$$\Rightarrow \begin{vmatrix} -5 & 0 & 1 \\ 3 & 0 & 1 \\ h & k & 1 \end{vmatrix}=\pm 40$$

$\Rightarrow \quad -5(0-k)+1(3k-0)=+40$

$\Rightarrow \quad 5k+3k=\pm 40$

On taking positive sign, we have

$$8k=40$$

$$k=5 \quad \dots(i)$$

On taking negative sign, we have

$$k=-5 \quad \dots(ii)$$

Now, point (h, k) lies on the line $x-y=2$.

$$\therefore \quad h-k=2 \quad \dots(iii)$$

On using Eq. (i) in Eq. (iii), we have

$$h=7$$

Hence, point C will be (7, 5).

On using Eq. (ii) in Eq. (iii), we have

$$h=-3$$

Hence, point C will be $(-3, -5)$.

18. (a) Let r be the common ratio of two GP's.

Then, $\quad x_1, x_2=x_1r, x_3=x_1r^2$

and $\quad y_1, y_2=y_1r, y_3=y_1r^2$

Now, $$\frac{1}{2}\begin{vmatrix} x_1 & y_1 & 1 \\ x_2 & y_2 & 1 \\ x_3 & y_3 & 1 \end{vmatrix}=\frac{1}{2}\begin{vmatrix} x_1 & y_1 & 1 \\ x_1r & y_1r & 1 \\ x_1r^2 & y_1r^2 & 1 \end{vmatrix}$$

$$=\frac{1}{2}x_1y_1\begin{vmatrix} 1 & 1 & 1 \\ r & r & 1 \\ r^2 & r^2 & 1 \end{vmatrix}$$

[taking common x_1 from c_1 and y_1 from c_2]

$$=\frac{1}{2}x_1y_1\times 0 \quad [\because c_1 \text{ and } c_2 \text{ are identical}]$$

Therefore, (x_1, y_1) (x_2, y_2) and (x_3, y_3) are collinear.

19. (d) Given lines are

$$y=mx \quad \dots(i)$$

$$y=mx+1 \quad \dots(ii)$$

$$y=nx \quad \dots(iii)$$

$$y=nx+1 \quad \dots(iv)$$

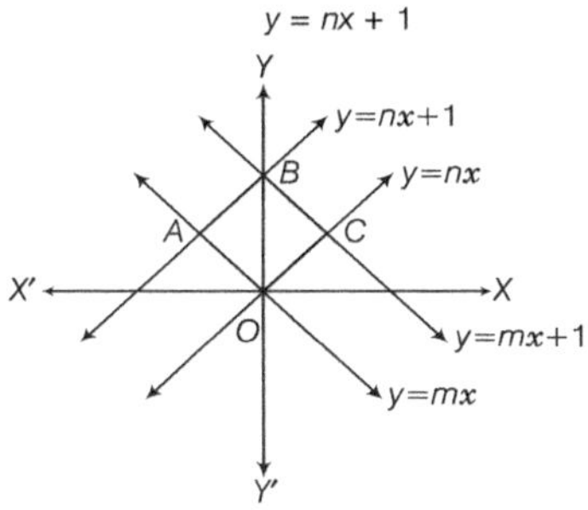

$y=mx$ is a line through origin, $y=mx+1$ is a line parallel to above line having y-intercept 1. Similarly, $y=nx$ is a line through origin and $y=nx+1$ is a line parallel to $y=nx$ having y-intercept 1.

The vertices are $O(0, 0)$, $A\left(\frac{1}{m-n}, \frac{m}{m-n}\right)$, $B(0, 1)$.

Now, area of parallelogram $= 2 \times$ Area ΔOAB

$$=2\times\frac{1}{2}\begin{vmatrix} 0 & 0 & 1 \\ 0 & 1 & 1 \\ \frac{1}{m-n} & \frac{m}{m-n} & 1 \end{vmatrix}$$

$$=\frac{1}{|m-n|}$$

20. (b) Given lines are

$$-3x+4y=1 \quad \dots(i)$$

$$-3x+4y=3 \quad \dots(ii)$$

$$-4x+3y=-1 \quad \dots(iii)$$

$$-4x+3y=-2 \quad \dots(iv)$$

We know that, area of parallelogram formed by the lines

$$a_1x+b_1y+c_1=0, a_1x+b_1y+d_1=0,$$

$$a_2x+b_2y+c_2=0 \text{ and } a_2x+b_2y+d_2=0$$

is $$\left|\frac{(d_1-c_1)(d_2-c_2)}{a_1b_2-a_2b_1}\right|$$

Here, $\quad c_1=-1, d_1=-3, c_2=1, d_2=2$

and $\quad a_1=-3, b_1=4, a_2=-4, b_2=3$

$$\therefore \text{ Area of parallelogram} = \left|\frac{(-3+1)(2-1)}{(-3)(3)-(-4)(4)}\right|=\left|\frac{-2}{-9+16}\right|$$

$$=\frac{2}{7} \text{ sq units}$$

21. (b) Given inequality $\quad 2|x|+3|y|\le 6$

The given inequality is equivalent to the following system of inequaties

$$2x+3y\le 6 \text{ when } x\ge 0, y\ge 0$$

$$2x-3y\le 6 \text{ when } x\ge 0, y\le 0$$

$$-2x+3y\le 6 \text{ when } x\le 0, y\ge 0$$

$$-2x-3y\le 6 \text{ when } x\le 0, y\le 0$$

which represents a rhombus with sides

$$2x\pm 3y=6 \text{ and } 2x\pm 3y=-6$$

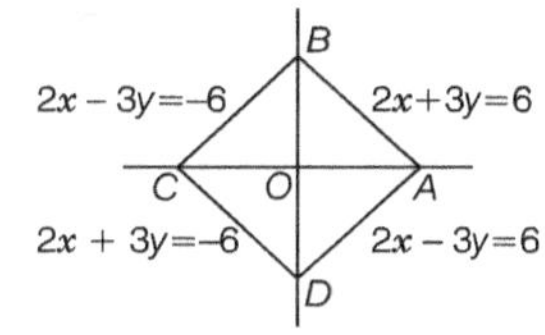

Length of diagonal is 6 and 4 units along X and Y-axes. Therefore, the required area $=\frac{1}{2}\times 6\times 4=12$ sq units.

$$\left[\because \text{ area of rhombus} = \frac{1}{2}d_1\cdot d_2\right]$$

22. (b) Given system of equation

$$a|x|+b|y|+c=0$$

The given equation is equivalent to the following system of equations

$$-ax+by=-c \text{ when } x<0, y>0$$

$$-ax-by=-c \text{ when } x<0, y<0$$

$$ax-by=-c \text{ when } x>0, y<0$$

$$ax+by=-c \text{ when } x>0, y>0$$

which represents a rhombus with sides

$$ax\pm by=c \text{ and } ax\pm by=-c$$

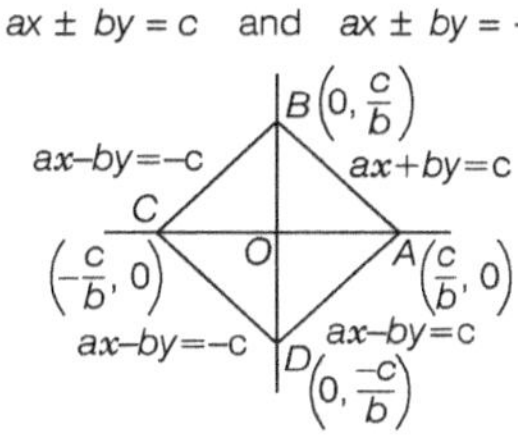

Length of diagonal is $\frac{2c}{a}$ and $\frac{2c}{b}$ units along X and Y-axes.

$$\therefore \text{ Required area} = \frac{1}{2}\times\frac{2c}{a}\times\frac{2c}{b}=\frac{2c^2}{ab} \text{ sq units} \quad (\text{if } ab>0)$$

$$\left[\because \text{ area of rhombus} = \frac{1}{2}\cdot d_1\cdot d_2\right]$$

23. (c) Given, points $(t_1, 2at_1 + at_1^3), (t_2, 2at_2 + at_2^3)$ and $(t_3, 2at_3 + at_3^3)$ are collinear.

$$\therefore \quad \begin{vmatrix} t_1 & 2at_1 + at_1^3 & 1 \\ t_2 & 2at_2 + at_2^3 & 1 \\ t_3 & 2at_3 + at_3^3 & 1 \end{vmatrix} = 0$$

$$\Rightarrow \quad \begin{vmatrix} t_1 & 2at_1 & 1 \\ t_2 & 2at_2 & 1 \\ t_3 & 2at_3 & 1 \end{vmatrix} + \begin{vmatrix} t_1 & at_1^3 & 1 \\ t_2 & at_2^3 & 1 \\ t_3 & at_3^3 & 1 \end{vmatrix} = 0$$

$$\Rightarrow \quad 2a\begin{vmatrix} t_1 & t_1 & 1 \\ t_2 & t_2 & 1 \\ t_3 & t_3 & 1 \end{vmatrix} + a\begin{vmatrix} t_1 & t_1^3 & 1 \\ t_2 & t_2^3 & 1 \\ t_3 & t_3^3 & 1 \end{vmatrix} = 0$$

$$\Rightarrow \quad 2a \times 0 + a\begin{vmatrix} t_1 & t_1^3 & 1 \\ t_2 & t_2^3 & 1 \\ t_3 & t_3^3 & 1 \end{vmatrix} = 0$$

[$\because c_1$ and c_2 of first determinant is identical]

$$\Rightarrow \quad a\begin{vmatrix} t_1 & t_1^3 & 1 \\ t_2 & t_2^3 & 1 \\ t_3 & t_3^3 & 1 \end{vmatrix} = 0$$

Applying $R_1 \to R_1 - R_3$ and $R_2 \to R_2 - R_3$, we have

$$\Rightarrow \quad a\begin{vmatrix} t_1 - t_3 & t_1^3 - t_3^3 & 0 \\ t_2 - t_3 & t_2^3 - t_3^3 & 0 \\ t_3 & t_3^3 & 1 \end{vmatrix} = 0$$

$$\Rightarrow \quad a[(t_1 - t_3)(t_2^3 - t_3)^3 - (t_2 - t_3)(t_1^3 - t_3^3)] = 0$$

$$\Rightarrow a[(t_1 - t_3)(t_2 - t_3)(t_1^2 + t_2^2 + t_1t_2)$$
$$-(t_2 - t_3)(t_1 - t_3)(t_1^2 + t_3^2 + t_1t_3)] = 0$$

$$\Rightarrow a[(t_1 - t_3)(t_2 - t_3)\{t_1^2 + t_2^2 + t_1t_2 - t_1^2 - t_3^2 - t_1t_3\}] = 0$$

$$\Rightarrow \quad a[(t_1 - t_3)(t_2 - t_3)\{t_2^2 - t_3^2 + t_1(t_2 - t_3)\}] = 0$$

$$\Rightarrow \quad a[(t_1 - t_3)(t_2 - t_3)(t_2 - t_3)\{t_2 + t_3 + t_1\}] = 0$$

$$\Rightarrow \quad t_1 + t_2 + t_3 = 0 \quad \left[\because a \neq 0, t_1 \neq t_2 \neq t_3\right]$$

24. (b) Given, $O(0, 0)$, $A(a\cos\theta, b\sin\theta)$, $B(a\cos\theta, -b\sin\theta)$

$\therefore$ Area of ΔOAB

$$\Delta = \frac{1}{2}\begin{vmatrix} 0 & 0 & 1 \\ a\cos\theta & b\sin\theta & 1 \\ a\cos\theta & -b\sin\theta & 1 \end{vmatrix}$$

$$\Delta = \frac{1}{2}[|-ab\sin\theta\cos\theta - ab\sin\theta\cos\theta|]$$

$$\Delta = \frac{1}{2}2ab\sin\theta\cos\theta = \frac{1}{2}ab\sin2\theta \quad \ldots\text{(i)}$$

For area to be maximum, $\sin 2\theta = 1 \Rightarrow \theta = \frac{\pi}{4}$

$$\therefore \quad \Delta_{max} = \frac{ab}{2}, \quad \theta = \frac{\pi}{4}$$

Hence, area of triangle is maximum when $\theta = \frac{\pi}{4}$

and $\quad \Delta_{max} = \frac{1}{2}ab\sin\frac{2\pi}{4} = \frac{ab}{2}$

25. (b) Given,

vertices of triangle are $A(1, 2)$, $B(3, 4)$ and $C(x, y)$.

$\therefore$ Area of triangle = 1 sq unit

$$\frac{1}{2}\begin{vmatrix} 1 & 2 & 1 \\ 3 & 4 & 1 \\ x & y & 1 \end{vmatrix} = 1$$

$$\Rightarrow \quad \frac{1}{2}|1(4 - y) - 2(3 - x) + 1(3y - 4x)| = 1$$

$$\Rightarrow \quad \frac{1}{2}|4 - y - 6 + 2x + 3y - 4x| = 1$$

$$\Rightarrow \quad \frac{1}{2}|-2x + 2y - 2| = 1$$

$$\Rightarrow \quad -2x + 2y - 2 = \pm 2$$

$$\Rightarrow \quad x - y = -2 \quad \text{or} \quad x - y = 0$$

***Case* I** When $x = (y - 2)$, then

$$(x - 1)(x - 3) + (y - 2)(y - 4) = 0$$

$$\Rightarrow \quad (y - 2 - 1)(y - 2 - 3) + (y - 2)(y - 4) = 0$$

$$\Rightarrow \quad (y - 3)(y - 5) + (y - 2)(y - 4) = 0$$

$$\Rightarrow \quad y^2 - 8y + 15 + y^2 - 6y + 8 = 0$$

$$\Rightarrow \quad 2y^2 - 14y + 23 = 0$$

$$\therefore \quad y = \frac{7 \pm \sqrt{3}}{2}$$

Hence, coordinates of point C are

$$\left(\frac{3 + \sqrt{3}}{2}, \frac{7 + \sqrt{3}}{2}\right) \text{or} \left(\frac{7 - \sqrt{3}}{2}, \frac{3 - \sqrt{3}}{2}\right).$$

***Case* II** When $x - y = 0$, then

$$(x - 1)(x - 3) + (y - 2)(y - 4) = 0$$

$$\Rightarrow \quad (x - 1)(x - 3) + (x - 2)(x - 4) = 0$$

$$\Rightarrow \quad x^2 - 4x + 3 + x^2 - 6x + 8 = 0$$

$$\Rightarrow \quad 2x^2 - 10x + 11 = 0$$

$$\therefore \quad x = \frac{5 \pm \sqrt{3}}{2}$$

Hence, coordinates of point C is

$$\left(\frac{5 + \sqrt{3}}{2}, \frac{5 + \sqrt{3}}{2}\right) \text{ or } \left(\frac{5 - \sqrt{3}}{2}, \frac{5 - \sqrt{3}}{2}\right).$$

Thus, there are four positions of point C.

26. (a) Given system of lines are

$$2x - 3y + a = 0 \quad \ldots\text{(i)}$$
$$2x - 3y + 3a = 0 \quad \ldots\text{(ii)}$$
$$3x - 2y - a = 0 \quad \ldots\text{(iii)}$$
$$3x - 2y - 2a = 0 \quad \ldots\text{(iv)}$$

We know that, area of parallelogram formed by the lines $a_1x + b_1y + c_1 = 0$, $a_1x + b_1y + d_1 = 0$ and $a_2x + b_2y + c_2 = 0$, $a_2x + b_2y + d_2 = 0$ is

$$\left|\frac{(d_1 - c_1)(d_2 - c_2)}{a_1b_2 - a_2b_1}\right|$$

Here, $\quad a_1 = 2, b_1 = -3, c_1 = a, d_1 = 3a$

and $\quad a_2 = 3, b_2 = -2, c_2 = -a, d_2 = -2a$

$$\therefore \quad \left|\frac{(3a - a)(-2a + a)}{-4 + 9}\right| = 10$$

$$\Rightarrow \quad \left|\frac{2a(-a)}{5}\right| = 10$$

$$\Rightarrow \quad \left|\frac{-2a^2}{5}\right| = 10$$

$$\Rightarrow \quad \frac{2a^2}{5} = 10$$

$$\therefore \quad a = \pm 5$$

27. (7) Given lines are

$$7x - 2y + 10 = 0$$

$$\Rightarrow \quad y = \frac{7}{2}x + 5$$

$$7x + 2y - 10 = 0$$

$$\Rightarrow \quad y = -\frac{7}{2}x + 5$$

$$y + 2 = 0 \quad \Rightarrow \quad y = -2$$

We know that, the area of a triangle formed by the lines $y = m_1x + c_1$, $y = m_2x + c_2$ and $y = m_3x + c_3$ is

$$= \frac{1}{2}\left|\sum \frac{(c_1 - c_2)^2}{(m_1 - m_2)}\right|$$

$$\therefore \text{ Required area} = \frac{1}{2}\left|\frac{(5-5)^2}{\frac{7}{2}+\frac{7}{2}} + \frac{(5+2)^2}{-\frac{7}{2}+0} + \frac{(-2-5)^2}{\left(0-\frac{7}{2}\right)}\right|$$

$$= \frac{1}{2}\left|\frac{49}{\left(\frac{-7}{2}\right)} + \frac{49}{\left(\frac{-7}{2}\right)}\right|$$

$$= \frac{1}{2}\left|-2 \times 49 \times \frac{2}{7}\right|$$

$$= \frac{1}{2}|-28| = 14$$

But it is given that area of triangle is $= 2k$

$\Rightarrow \quad 2k = 14$

$\therefore \quad k = 7$

28. (3) Given, point $P(3, 2)$ gets reflected through X-axis.

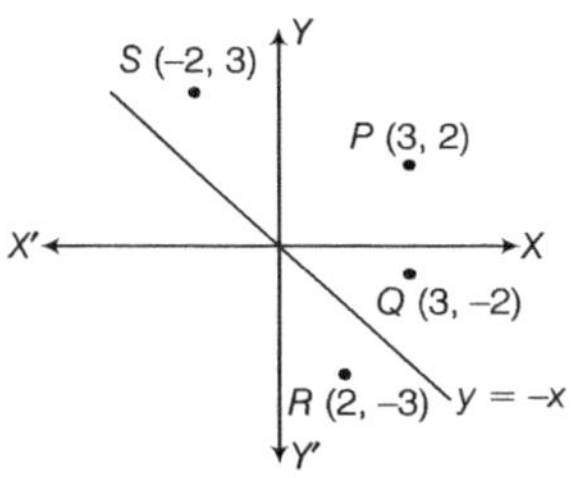

Reflection of point $P(3, 2)$ is $Q\ (3, -2)$

i.e. $\quad P(3,2) \to Q\ (3, -2)$.

When point $Q(3, -2)$ gets reflected about the line $y = -x$, its coordinate will be

$R\ (2, -3)$

i.e. $\quad Q(3, -2) \to R(2, -3)$

When point R gets reflected through origin.

Then it will be $S\ (-2, 3)$

i.e. $\quad R(2, -3) \to S\ (-2, 3)$

$\therefore$ Area of $PQRS$

$$= \frac{1}{2}\left|\begin{vmatrix}3 & 2\\3 & -2\end{vmatrix} + \begin{vmatrix}3 & -2\\2 & -3\end{vmatrix} + \begin{vmatrix}2 & -3\\-2 & 3\end{vmatrix} + \begin{vmatrix}-2 & 3\\3 & 2\end{vmatrix}\right|$$

$$= \frac{1}{2}|-6-6+(-9)+4+6-6+(-4)-9|$$

$$= \frac{1}{2}|-12-18|$$

$$= \frac{1}{2}\times|-30| = 15$$

But it is given that, area of $PQRS = 5k = 15 \Rightarrow \quad k = 3$

29. (0) Given lines are

$$L_1 : ax + ky - al - a^3m = 0$$

$$L_2 : bx + ky - bl - b^3m = 0$$

$$L_3 : cx + ky - lc - c^3m = 0$$

$\because$ These lines are concurrent.

$$\therefore \quad \begin{vmatrix} a & k & -(al + a^3m)\\ b & k & -(bl + b^3m)\\ c & k & -(cl + c^3m)\end{vmatrix} = 0$$

$$\Rightarrow \quad \begin{vmatrix} a & k & -al\\ b & k & -bl\\ c & k & -cl\end{vmatrix} + \begin{vmatrix} a & k & -a^3m\\ b & k & -b^3m\\ c & k & -c^3m\end{vmatrix} = 0$$

$$\Rightarrow \quad -kl\begin{vmatrix} a & 1 & a\\ b & 1 & b\\ c & 1 & c\end{vmatrix} - km\begin{vmatrix} a & 1 & a^3\\ b & 1 & b^3\\ c & 1 & c^3\end{vmatrix} = 0$$

$$\Rightarrow \quad -kl \times 0 - km\begin{vmatrix} a & 1 & a^3\\ b & 1 & b^3\\ c & 1 & c^3\end{vmatrix} = 0$$

[since, $C_1 = C_3$ first determinant is zero]

$$\Rightarrow \quad km\begin{vmatrix} 1 & a & a^3\\ 1 & b & b^3\\ 1 & c & c^3\end{vmatrix} = 0$$

Applying $R_1 \to R_1 - R_3$ and $R_2 \to R_2 - R_3$, we get

$$\Rightarrow \quad km\begin{vmatrix} 0 & a-c & a^3-c^3\\ 0 & b-c & b^3-c^3\\ 1 & c & c^3\end{vmatrix} = 0$$

$$\Rightarrow \quad km\,[1(a-c)(b^3-c^3) - (b-c)(a^3-c^3)] = 0$$

$$\Rightarrow \quad km[(a-c)(b-c)\{b^2+c^2+bc-a^2-c^2-ac\}] = 0$$

$$\Rightarrow \quad km\,[(a-c)(b-c)(b-a)\{b+a+c\}] = 0$$

$$\Rightarrow \quad km\,(a-c)(b-c)(b-a)(a+b+c) = 0$$

$$\Rightarrow \quad a+b+c = 0 \quad [\because k, m \neq 0, a \neq b \neq c]$$

30. (1) Let ABC be an equilateral triangle with side a, i.e. $AB = BC = AC = a$

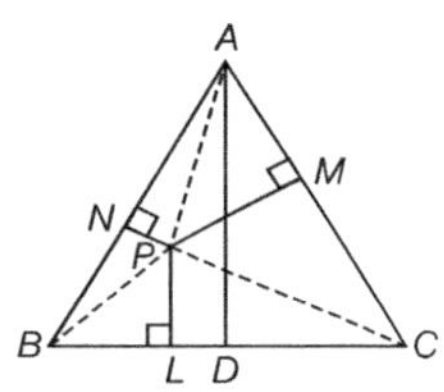

Given, $PL + PM + PN = 2015$

Now, area of $\Delta ABC = ar(\Delta PBC) + ar(\Delta CPA) + ar(\Delta APB)$

$$= \frac{1}{2}\times a \times AD = \frac{1}{2}\times BC \times PL + \frac{1}{2}\times PM \times AC + \frac{1}{2}\times AB \times PN$$

$$\Rightarrow \quad \frac{1}{2}\times a \times AD = \frac{1}{2}a\cdot PL + \frac{1}{2}a\cdot AC + \frac{1}{2}a\cdot PN$$

$$\Rightarrow \quad \frac{1}{2}\times a \times AD = \frac{1}{2}a\,(PL + AC + PN)$$

$$\Rightarrow \quad AD = PL + AC + PN$$

$$\Rightarrow \quad AD = 2015$$

$$\therefore \quad \frac{\text{Length of altitude}}{2015} = \frac{AD}{2015}$$

$$= \frac{2015}{2015} = 1$$

DPP-4 Centroid, Circumcentre, Orthocentre and Incentre of a Triangle

1. (a) Given, $A(-4, 6)$, $B(2, 3)$ and $C(-2, -5)$ are the vertices of a triangle.

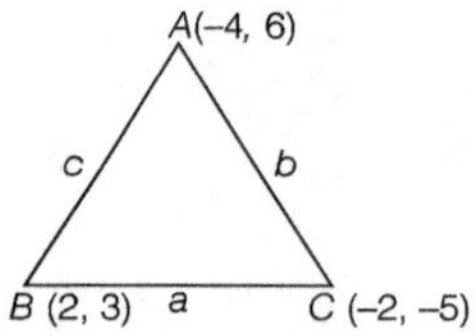

We know that, the coordinates of incentre of a triangle having vertices (x_1, y_1), (x_2, y_2) and (x_3, y_3), is

$$\frac{ax_1 + bx_2 + cx_3}{a + b + c}, \frac{ay_1 + by_2 + cy_3}{a + b + c}$$

Now, $a = BC = \sqrt{(-4)^2 + (-8)^2} = \sqrt{16 + 64} = \sqrt{80} = 4\sqrt{5}$

$b = AC = \sqrt{(2)^2 + (-11)^2} = \sqrt{4 + 121} = \sqrt{125} = 5\sqrt{5}$

$c = AB = \sqrt{(6)^2 + (-3)^2} = \sqrt{36 + 9} = \sqrt{45} = 3\sqrt{5}$

$$\therefore \quad \text{Incentre} = \left(\frac{4\sqrt{5} \times (-4) + 5\sqrt{5}\,(2) + 3\sqrt{5}\,(-2)}{12\sqrt{5}}, \frac{4\sqrt{5}\,(6) + 5\sqrt{5}\,(3) + 3\sqrt{5}\,(-5)}{12\sqrt{5}}\right)$$

$= (-1, 2)$

2. (a) Let $H(x, y)$ be the orthocentre of ΔABC.

$\because \quad AM \perp BC$

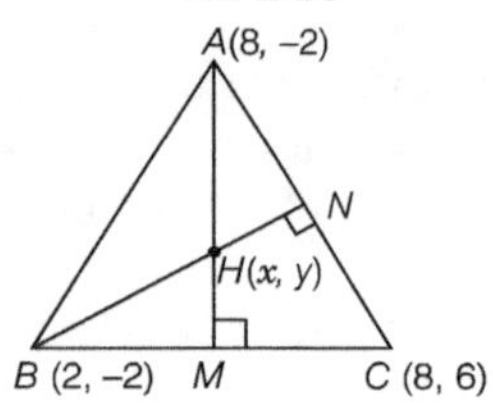

$$\text{Slope of } AM = -\frac{1}{\text{Slope of } BC} = -\frac{1}{\frac{6+2}{6}} = -\frac{6}{8} = -\frac{3}{4}$$

[since, product of slope of two perpendicular lines $= -1$ and slope of a line joining point (x_1, y_1), (x_2, y_2) is $\frac{y_2 - y_1}{x_2 - x_1}$]

Equation of AM $\quad y + 2 = -\frac{3}{4}(x - 8)$

$\Rightarrow \quad 4y + 8 = -3x + 24$

$\Rightarrow \quad 3x + 4y = 16 \quad \ldots$(i)

Again, since $BN \perp AC$

$$\text{Slope of } BN = -\frac{1}{\text{Slope of } AC} = -\frac{1}{\frac{8}{0}} = 0$$

Equation of BN

$y + 2 = 0 \quad \ldots$(ii)

On solving Eqs. (i) and (ii), we get

$3x - 8 = 16$

$3x = 24$

$x = 8$

Hence, coordinates of orthocentre is $(8, -2)$.

Alternate Method

$$m_{AB} = 0, m_{AC} = \frac{1}{0}$$

$\therefore \quad AB \perp AC$

$\angle A$ is right angled.

Since, $\angle BAC$ is right angled at A and the vertex containing 90° is orthocentre.

Hence, $A(8, -2)$ is orthocentre.

3. (b) Given, vertices of triangle are

$(a, b - c)$, $(b, c - a)$ and $(c, a - b)$

Let (x, y) be the coordinates of centroid.

$$\therefore \quad x = \frac{a + b + c}{3}, \quad y = \frac{b - c + c - a + a - b}{3}$$

$y = 0$

$\therefore$ Coordinates of centroid is $\left(\frac{a + b + c}{3}, 0\right)$.

Hence, centroid lies on X-axis.

4. (a) Let $H(x, y)$ be the orthocentre of ΔABC

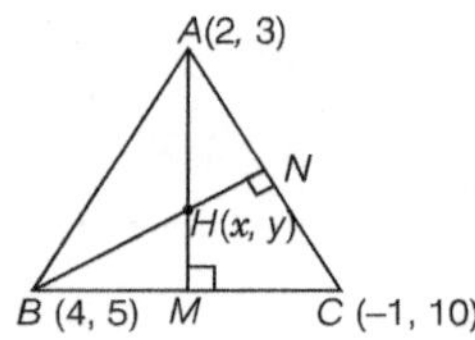

$\because \quad AM \perp BC$

$$\text{Slope of } AM = -\frac{1}{\text{Slope } BC} = -\frac{1}{\frac{5}{-5}} = 1$$

Equation of AM, $\quad y - 3 = 1(x - 2) \Rightarrow y - 3 = x - 2$

$\Rightarrow \quad x - y + 1 = 0 \quad \ldots$(i)

Again, $BN \perp AC$

$$\text{Slope of } BN = -\frac{1}{\text{Slope of } AC} = \frac{-1}{-\frac{7}{3}} = \frac{3}{7}$$

Equation of BN, $\quad y - 5 = \frac{3}{7}(x - 4)$

$7y - 35 = 3x - 12$

$3x - 7y = -23 \quad \ldots$(ii)

Since, orthocentre is the point of intersection of altitudes of a triangle. Therefore, common solution of Eqs. (i) and (ii) will give required result.

On solving Eqs. (i) and (ii), we get

$x = 4$ and $y = 5$

Hence, coordinates of orthocentre is (4, 5).

Now, distance between (4, 5) and (2, 3) is

$$= \sqrt{(-2)^2 + (-2)^2} = 2\sqrt{2}$$

Alternate Method

$$m_{AB} = \frac{5-3}{4-2} = 1, \quad m_{BC} = \frac{10-5}{-1-4} = \frac{5}{-5} = -1$$

Since, $m_{AB} \cdot m_{BC} = -1$

$\therefore$ Right angled at B

$\Rightarrow$ Orthocentre is $B(4, 5)$.

Thus, distance between $B(4, 5)$ and $P(2, 3)$ is

$$\sqrt{(4-2)^2 + (5-3)^2} = 2\sqrt{2}$$

5. (c) Let $G(x, y)$ be the centroid of triangle.

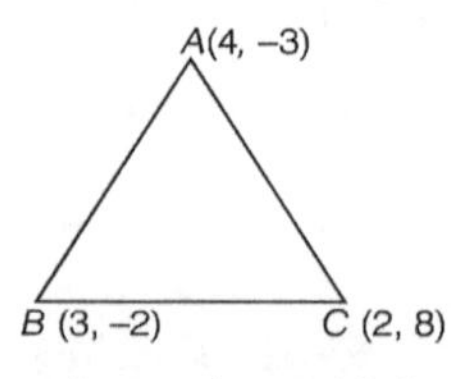

We know that, coordinates of centroid of a triangle having vertices (x_1, y_1), (x_2, y_2) and (x_3, y_3) is $\left(\frac{x_1 + x_2 + x_3}{3}, \frac{y_1 + y_2 + y_3}{3}\right)$.

$\therefore$ Coordinates of G are $\left(\frac{4 + 3 + 2}{3}, \frac{-3 - 2 + 8}{3}\right)$ i.e. (3, 1)

$\therefore$ Distance of centroid from Y-axis = Value of abscissa of centroid $= 3$

6. (a) In equilateral triangle, orthocentre coinside with centroid. And also, in equilateral triangle, median and altitudes are same.

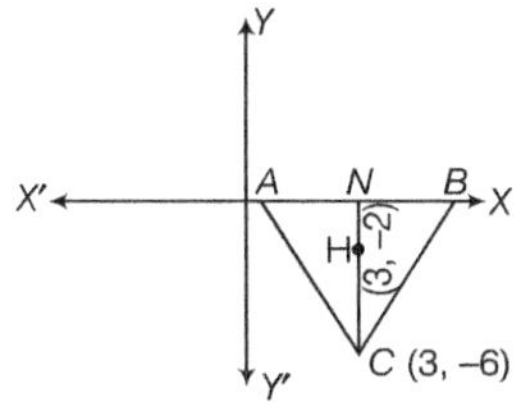

∵ Centroid divides the median in the ratio 2 : 1.

$$\therefore \quad CH : HN := 2 : 1 \Rightarrow \frac{1}{3} CN = HN$$

$$\Rightarrow \quad \frac{1}{3} CN = 2 \quad \Rightarrow \quad CN = 6$$

Hence, coordinate of vertex C is $(3, -6)$.

7. (c) Let S $(-2, 5)$ be the centroid of ΔABC.

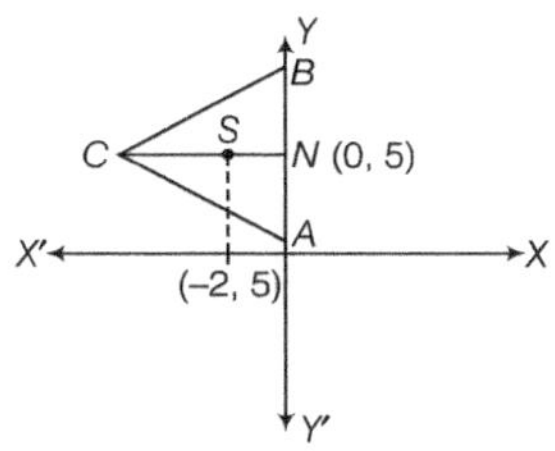

Since, triangle is equilateral, its centroid and circumcentre coincide.

∴ Point S divides CN in the ratio 2 : 1.

i.e. $\quad CS : SN = 2 : 1$

$$\Rightarrow \quad \frac{1}{3} CN = SN$$

$$\Rightarrow \quad CN = 3\,SN$$

$$\Rightarrow \quad CN = 3 \times 2 = 6 \qquad [\because SN = 2]$$

Let a be the sides of ΔABC.

$$\Rightarrow \quad CN^2 = BC^2 - BN^2$$

$$\Rightarrow \quad (6)^2 = a^2 - \left(\frac{a}{2}\right)^2$$

$$\Rightarrow \quad \frac{3}{4} a^2 = 36 \quad \Rightarrow \quad a^2 = 36 \times \frac{4}{3}$$

$$\therefore \quad a = 4\sqrt{3}$$

8. (a) Let $A(a \cos \alpha, a \sin \alpha)$ $B(a \cos \beta, a \sin \beta)$ and $C(a\cos \gamma, a\sin \gamma)$ be the vertices of triangle.

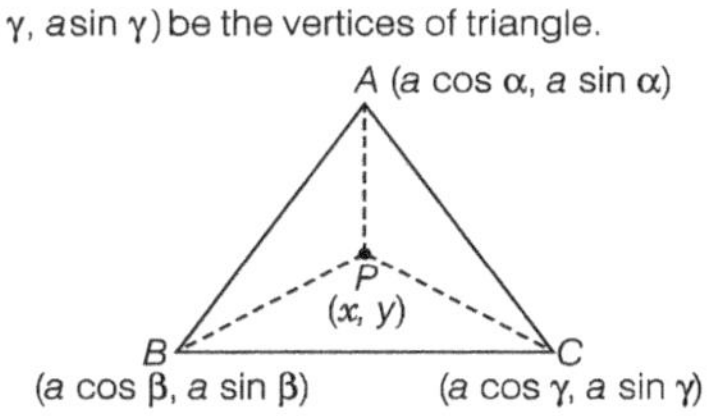

Let $P(x, y)$ be the circumcentre of ΔABC.

$$\therefore \quad PA^2 = PB^2$$

$$\Rightarrow (x - a \cos \alpha)^2 + (y - a \sin \alpha)^2 = (x - a \cos \beta)^2 + (y - a \sin \beta)^2$$

$$\Rightarrow x^2 + a^2 \cos^2 \alpha - 2ax \cos \alpha + y^2 + a^2 \sin^2 \alpha - 2ay \sin \alpha$$
$$= x^2 + a^2 \cos^2 \beta - 2ax \cos \beta + y^2 + a^2 \sin^2 \beta - 2ay \sin \beta$$

$$\Rightarrow a^2(\cos^2 \alpha + \sin^2 \alpha) - 2a(x \cos \alpha + y \sin \alpha)$$
$$= a^2 (\cos^2 \beta + \sin^2 \beta) - 2a(x \cos \beta + y \sin \beta)$$

$$\Rightarrow \quad x \cos \alpha + y \sin \alpha = x \cos \beta + y \sin \beta$$

$$\Rightarrow \quad x(\cos \alpha - \cos \beta) + y(\sin \alpha - \sin \beta) = 0 \qquad \ldots\text{(i)}$$

Now, $\quad PA^2 = PC^2$

$$\Rightarrow (x - a \cos \alpha)^2 + (y - a \sin \gamma)^2 = (x - a \cos \gamma)^2 + (y - a \sin \gamma)^2$$

$$\Rightarrow x^2 + y^2 + a^2(\cos^2 \alpha + \sin^2 \alpha) - 2a(x \cos \alpha + y \sin \alpha)$$
$$= x^2 + y^2 + a^2(\cos^2 \gamma + \sin^2 \gamma) - 2a(x \cos \gamma + y \sin \gamma)$$

$$\Rightarrow \quad x \cos \alpha + y \sin \alpha = x \cos \gamma + y \sin \gamma$$

$$\Rightarrow \quad x(\cos \alpha - \cos \gamma) + y(\sin \alpha - \sin \gamma) = 0 \qquad \ldots\text{(ii)}$$

On solving Eqs. (i) and (ii), we get

$$x = 0, y = 0$$

Hence, coordinates of circumcenter is (0, 0).

Alternate Method

Here, $A(a\cos \alpha, a\sin \alpha)$, $B(a\cos \beta, a\sin \beta)$ and $C(a\cos \gamma, a\sin \gamma)$ are equidistant from $O(0, 0)$.

i.e. $OA = \sqrt{a^2 \cos^2 \alpha + n^2 \sin^2 \alpha} = a = OB = OC$

Since, A, B and C are equidistant from origin.

$\Rightarrow$ $O(0, 0)$ is circumcentre.

9. (a) Let $A(x_1, y_1)$, $B(x_2, y_2)$ and $C(x_3, y_3)$ be the vertices of ΔABC and $D(5, 0)$ and $E(5, 12)$ and $F(0, 12)$ are the mid-points of AB, BC and AC, respectively.

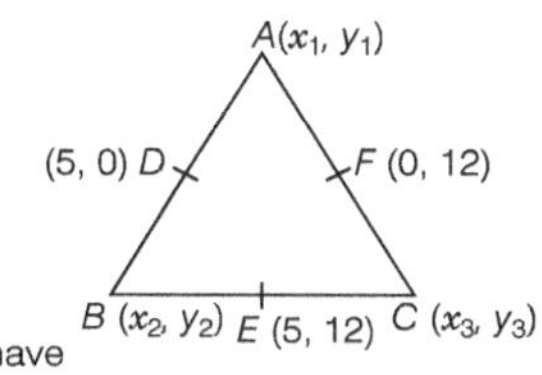

$$\therefore \quad x_1 + x_2 = 10$$
$$y_1 + y_2 = 0$$
$$x_2 + x_3 = 10$$
$$y_2 + y_3 = 24$$
$$x_1 + x_3 = 0$$
$$y_1 + y_3 = 24$$

On solving these equations, we have

$$y_1 = y_2 = 0, y_3 = 24$$
$$x_1 = x_3 = 0, x_2 = 10$$

∴ Coordinates of vertices A, B and C are respectively (0, 0), (10, 0) and (0, 24).

Since, ΔABC is right angled at $A(0, 0)$.

Hence, orthocentre is (0, 0).

10. (b) Let $H(x, y)$ be the orthocentre of ΔABC.

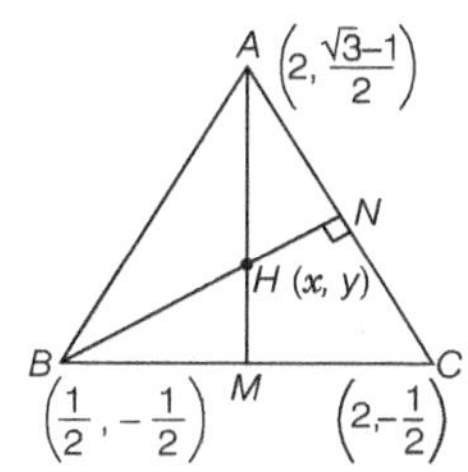

$$\because \quad AM \perp BC$$

$$\therefore \text{ Slope of } AM = \frac{-1}{\text{Slope of } BC} = \frac{-1}{\frac{0}{3/2}} = \infty$$

$$\text{Equation of } AM, \quad \frac{\left(y - \frac{\sqrt{3}+1}{2}\right)}{(x-2)} = \frac{1}{0}$$

$$\Rightarrow \quad x - 2 = 0 \qquad \ldots\text{(i)}$$

Again, $\quad BN \perp AC$

$$\text{Slope of } \quad BN = \frac{-1}{\text{Slope of } AC} = \frac{-1}{\infty} = 0$$

$$\text{Equation of } BN, \quad y + \frac{1}{2} = 0 \qquad \ldots\text{(ii)}$$

From Eqs. (i) and (ii), we get

Coordinates of orthocentre is $\left(2, -\frac{1}{2}\right)$.

Alternate Method

$$m_{CA} = \frac{\frac{\sqrt{3}-1}{2} + \frac{1}{2}}{2-2} = \frac{1}{0} \text{ and } m_{BC} = 0$$

Since, $AC \perp BC$

Hence, C is right angled.

Orthocentre is $C\left(2, \frac{-1}{2}\right)$, as in right angled triangle verterx, containing 90° is orthocentre.

11. (d) Let $A(1, \sqrt{3})$, $O(0, 0)$ and $B(2, 0)$ be the vertices of a triangle.

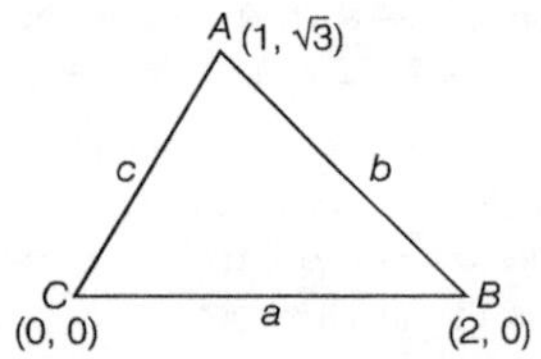

Now, $a = OB = \sqrt{(2)^2 + 0^2} = 2$

$b = AB = \sqrt{(2-1)^2 + (-\sqrt{3})^2} = 2$

$c = OA = \sqrt{(1)^2 + (\sqrt{3})^2} = 2$

$\therefore$ Coordinates of incentre are $\left(\dfrac{ax_1 + bx_2 + cx_3}{a+b+c}, \dfrac{ay_1 + by_2 + cy_3}{a+b+c}\right)$

$$= \left(\frac{2 \times 1 + 2 \times 0 + 2 \times 2}{6}, \frac{2 \times \sqrt{3} + 2 \times 0 + 2 \times 0}{6}\right)$$

$$= \left(1, \frac{1}{\sqrt{3}}\right)$$

12. (c) Let $H(x, y)$ be the orthocentre of ΔABC.

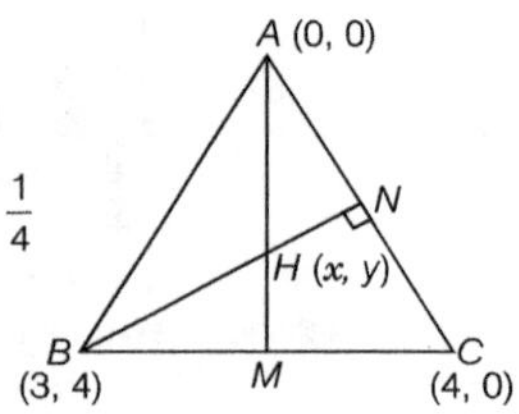

$\because$ $AM \perp BC$

Slope of $AM = \dfrac{-1}{-4} = \dfrac{1}{4}$

Equation of AM,

$$y = \frac{1}{4}x$$

$\Rightarrow$ $x - 4y = 0$ ...(i)

Slope of $BN = \dfrac{-1}{0} = \infty$

Equation of BN, $\dfrac{y-4}{x-3} = \dfrac{1}{0}$

$\Rightarrow$ $x - 3 = 0$ ...(ii)

On solving Eqs. (i) and (ii), we get

$$x = 3,\ y = \frac{3}{4}$$

Hence, coordinates of orthocentre are $\left(3, \dfrac{3}{4}\right)$.

13. (b) Let D and E be the mid-points of AB and AC.

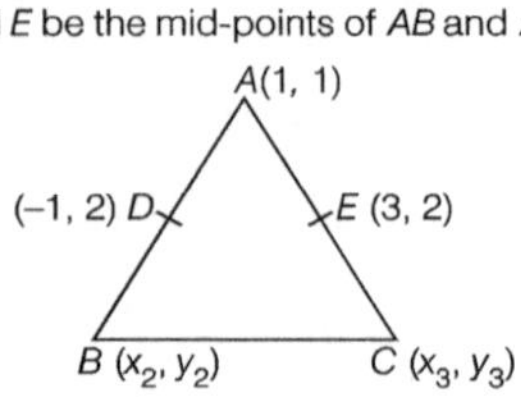

The coordinates of B are

$$-1 = \frac{1 + x_2}{2} \text{ and } 2 = \frac{1 + y_2}{2}$$

i.e. $x_2 = -3$ and $y_2 = 3$

and coordinates of C are

$$3 = \frac{1 + x_3}{2} \text{ and } 2 = \frac{1 + y_3}{2}$$

i.e. $x_3 = 5$ and $y_3 = 3$

$\therefore$ Centroid of triangle $= \left(\dfrac{x_1 + x_2 + x_3}{3}, \dfrac{y_1 + y_2 + y_3}{3}\right)$

$$= \left(\frac{1 - 3 + 5}{3}, \frac{1 + 3 + 3}{3}\right)$$

$$= \left(1, \frac{7}{3}\right)$$

14. (a) Given,

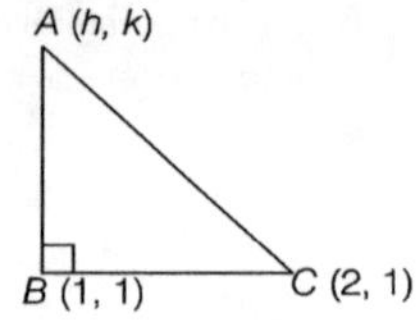

area of ΔABC = 1 sq unit

$$= \frac{1}{2}\begin{vmatrix} h & k & 1 \\ 1 & 1 & 1 \\ 2 & 1 & 1 \end{vmatrix} = 1$$

$\Rightarrow$ $\dfrac{1}{2}|h(1-1) - k(1-2) + 1(1-2)| = 1$

$\Rightarrow$ $\dfrac{1}{2}|k - 1| = 1$

$\Rightarrow$ $k - 1 = \pm 2$

On taking positive sign, we have

$$k = 3$$

On taking negative sign, we have

$$k - 1 = -2$$
$$k = -1$$

So, set of values of k is $\{-1, 3\}$.

15. (d) Let $A(12, 8)$, $B(-2, 6)$ and $C(6, 0)$ be the vertices of triangle.

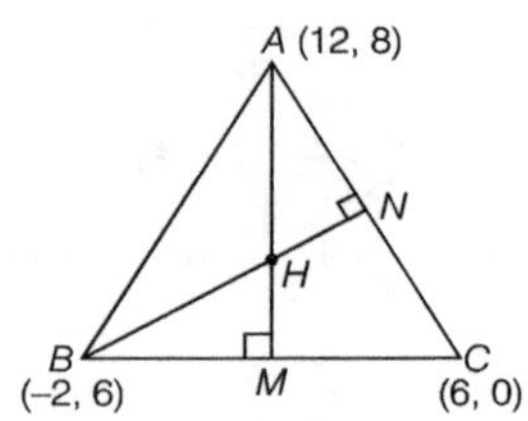

$\therefore$ Coordinates of centroid

$$\left(\frac{12 - 2 + 6}{3}, \frac{8 + 6 + 0}{3}\right), \text{ i.e. } \left(\frac{16}{3}, \frac{14}{3}\right)$$

Let $AM \perp BC$ and $BN \perp AC$.

$\therefore$ Point of intersection of AM and BN will be orthocentre of ΔABC.

Now, equation of AM

$$(y - 8) = \frac{-1}{\dfrac{0-6}{6+2}}(x - 12)$$

$\Rightarrow$ $y - 8 = \dfrac{8}{6}(x - 12)$

$\Rightarrow$ $3(y - 8) = 4(x - 12)$

$\Rightarrow$ $4x - 3y - 24 = 0$ ...(i)

and equation of BN

$$(y - 6) = \frac{-1}{\dfrac{0-8}{6-12}}(x + 2)$$

$$y - 6 = -\frac{6}{8}(x + 2)$$

$\Rightarrow$ $y - 6 = -\dfrac{3}{4}(x + 2)$

$\Rightarrow$ $4(y - 6) = -3(x + 2)$

$\Rightarrow$ $3x + 4y + 8 = 0$ ...(ii)

On solving Eqs. (i) and (ii), we get

$$x = 6,\ y = 0$$

Now, we know that, circumcentre is the point of intersection of perpendicular bisector of sides of triangle and is equidistant from the vertices of triangle.

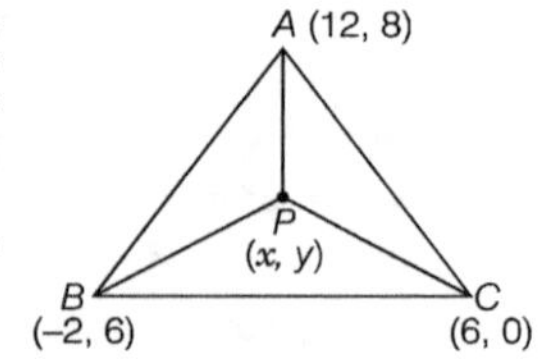

Let $P(x, y)$ be the circumcentre of the ΔABC.

$\therefore$ $PA^2 = PB^2$

$$(x - 12)^2 + (y - 8)^2 = (x + 2)^2 + (y - 6)^2$$

$\Rightarrow x^2 + 144 - 24x + y^2 + 64 - 16y$

$= x^2 + 4 + 4x + y^2 + 36 - 12y$

$\Rightarrow \quad -24x - 4x - 16y + 12y = 40 - 208$

$\Rightarrow \quad -28x - 4y = -168$

$\Rightarrow \quad 28x + 4y = 168$

$\Rightarrow \quad 7x + y = 42 \quad \ldots(iii)$

and $\quad PA^2 = PC^2$

$(x-12)^2 + (y-8)^2 = (x-6)^2 + y^2$

$\Rightarrow x^2 + 144 - 24x + y^2 + 64 - 16y = x^2 + 36 - 12x + y^2$

$\Rightarrow \quad -24x + 12x - 16y = 36 - 208$

$\Rightarrow \quad -12x - 16y = -172$

$\Rightarrow \quad 12x + 16y = 172$

$\Rightarrow \quad 3x + 4y = 43 \quad \ldots(iv)$

On solving Eqs. (iii) and (iv), we get

$x = 5, y = 7$

Product of abscissae of the centroid, orthocentre and circumcentre

$= \frac{16}{3} \times 6 \times 5 = 160$

16. (a) Since, orthocentre coincide with the vertex B of ΔABC.

$\therefore$ ΔABC is right angled triangle.

And circumcentre will be the mid-point of the hypotenuse of ΔABC.

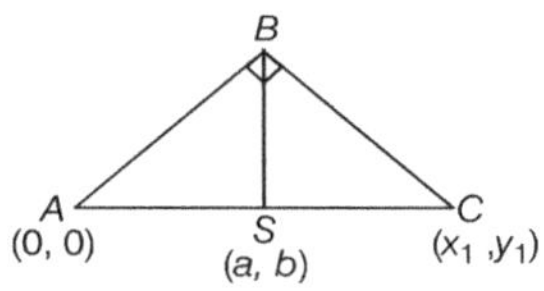

Let coordinates of point C are (x_1, y_1), then

$a = \frac{x_1 + 0}{2} \Rightarrow x_1 = 2a$

$b = \frac{y_1 + 0}{2} \Rightarrow y_1 = 2b$

Hence, coordinates of C are $(2a, 2b)$.

17. (b) Given, mid-points of a triangle are (0, 1), (1, 1) and (1, 0). Plot these points on a graph paper and make a triangle.

So, the sides of a triangle will be 2, 2 and $\sqrt{2^2 + 2^2}$, i.e. $2\sqrt{2}$.

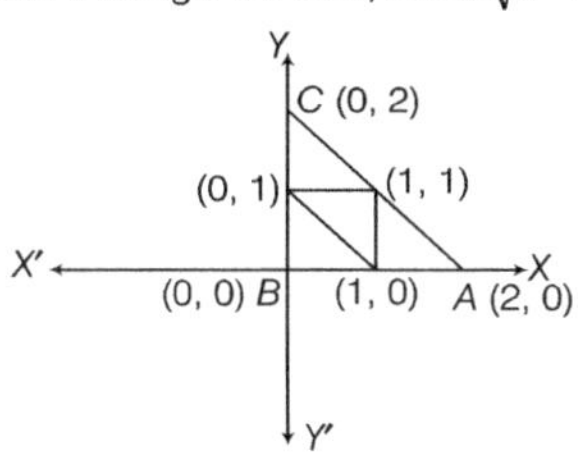

$\therefore$ x-coordinate of incentre $= \frac{2 \times 0 + 2\sqrt{2} \times 0 + 2 \cdot 2}{2 + 2 + 2\sqrt{2}}$

$= \frac{2}{2+\sqrt{2}} \times \frac{2-\sqrt{2}}{2-\sqrt{2}}$

$= 2 - \sqrt{2}$

18. (b) Since, point of intersection of medians is centroid.

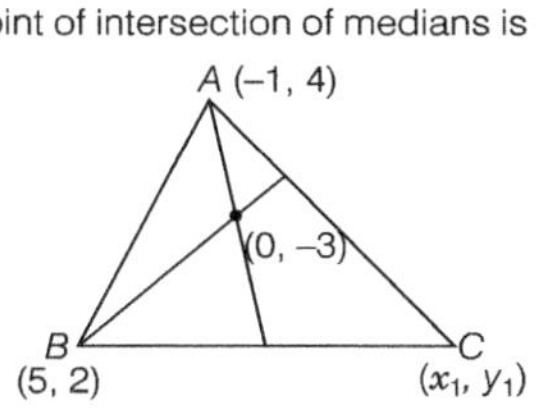

$\therefore$ Coordinates of centroid of ΔABC are $(0, -3)$.

Now, x-coordinate of centroid $= \frac{x_1 + x_2 + x_3}{3}$

$0 = \frac{x_1 - 1 + 5}{3}$

$\Rightarrow \quad 0 = \frac{x_1 + 4}{3}$

$\therefore \quad x_1 = -4$

y-coordinate of centroid $= \frac{y_1 + 4 + 2}{3}$

$-3 = \frac{y_1 + 6}{3}$

$\Rightarrow \quad y_1 + 6 = -9$

$\Rightarrow \quad y_1 = -15$

$\therefore$ Third vertex is $(-4, -15)$.

19. (d) Let $P(2, -4)$, $Q(4, -2)$ and $R(7, 1)$.

Now, $\quad PQ = \sqrt{(2)^2 + (2)^2} = 2\sqrt{2}$

$QR = \sqrt{(3)^2 + (3)^2} = 3\sqrt{2}$

$PR = \sqrt{(5)^2 + (5)^2} = 5\sqrt{2}$

$PQ^2 + QR^2 = (2\sqrt{2})^2 + (3\sqrt{2})^2$

$= 8 + 18 = 26$

and $\quad PR^2 = (5\sqrt{2})^2 = 50$

$PQ^2 + QR^2 \neq PR^2$

So, ΔPQR is not a right angled triangle, neither isosceles or equilateral.

Now, area of $\Delta PQR = \frac{1}{2}\begin{vmatrix} 2 & -4 & 1 \\ 4 & -2 & 1 \\ 7 & 1 & 1 \end{vmatrix}$

$= \frac{1}{2}|2(-2-1) + 4(4-7) + 1(4+14)|$

$= \frac{1}{2}|-6 - 12 + 18| = 0$

$\therefore$ Points are collinear.

20. (c) In ΔASB,

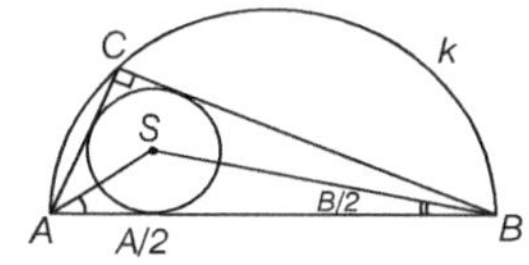

$\frac{A}{2} + \angle ASB + \frac{B}{2} = 180°$

$\Rightarrow \quad \angle ASB + \left(\frac{A+B}{2}\right) = 180°$

$\Rightarrow \quad \angle ASB + \frac{90°}{2} = 180°$

$\therefore \quad \angle ASB = 135°$

21. (b)

(i) Let $A(2, -2)$, $B(-2, 1)$ and $C(5, 2)$

$\Rightarrow \quad AB = \sqrt{(-4)^2 + (3)^2} = \sqrt{25} = 5$

$BC = \sqrt{(7)^2 + (1)^2} = \sqrt{50} = 5\sqrt{2}$

$AC = \sqrt{(3)^2 + (4)^2} = \sqrt{25} = 5$

$AB = AC$

$\Rightarrow$ ΔABC is isosceles

and $\quad AB^2 + AC^2 = (5)^2 + (5)^2 = 50 = BC^2$

$\Rightarrow$ ΔABC is right angle isosceles triangle.

Hence, $\quad$ **(i) → (q)**

(ii) Let $A(1, -2)$, $B(-3, 0)$ and $C(5, 6)$.

$AB = \sqrt{(-4)^2 + (2)^2} = \sqrt{20} = 2\sqrt{5}$

$BC = \sqrt{(8)^2 + (6)^2} = \sqrt{100} = 10$

$AC = \sqrt{(4)^2 + (8)^2} = \sqrt{80} = 4\sqrt{5}$

As, $AB^2 + AC^2 = (2\sqrt{5})^2 + (4\sqrt{5})^2 = 20 + 80 = 100 = BC^2$

$\therefore$ ΔABC is right angled triangle.

Hence, $\quad$ **(ii) → (p)**

(iii) Let $A(3, 7)$, $B(6, 5)$ $C(15, -1)$

$$AB = \sqrt{(3)^2 + (-2)^2} = \sqrt{13}$$
$$BC = \sqrt{(9)^2 + (-6)^2}$$
$$= \sqrt{117} = 3\sqrt{13}$$
$$AC = \sqrt{(12)^2 + (-8)^2} = \sqrt{208} = 4\sqrt{13}$$

As, $AB + BC = \sqrt{13} + 3\sqrt{13}$
$= 4\sqrt{13} = AC$
$AB + BC = AC$

$\therefore$ ΔABC is not possible

[$\because$ sum of any two sides of a triangle must be greater than third side]

Hence, **(iii) $\rightarrow$ (s)**

(iv) Let $A(2, 2)$, $B(-2, -2)$, $C(-2\sqrt{3}, 2\sqrt{3})$

$$AB = \sqrt{(-4)^2 + (-4)^2} = 4\sqrt{2}$$
$$BC = \sqrt{(-2\sqrt{3} + 2)^2 + (2\sqrt{3} + 2)^2}$$
$$= \sqrt{16 - 8\sqrt{3} + 16 + 8\sqrt{3}} = \sqrt{32} = 4\sqrt{2}$$
$$AC = \sqrt{(-2\sqrt{3} - 2)^2 + (2\sqrt{3} - 2)^2}$$
$$= \sqrt{16 + 8\sqrt{3} + 16 - 8\sqrt{3}} = 4\sqrt{2}$$

$\because$ $AB = BC = AC = 4\sqrt{2}$

$\therefore$ ABC is an equilateral triangle.

Hence, **(iv) $\rightarrow$ (r)**

22. (4) Let $A(7, 2)$, $B(1, 6)$, $C(a, b)$ and centroid of ΔABC is $(4, 6)$

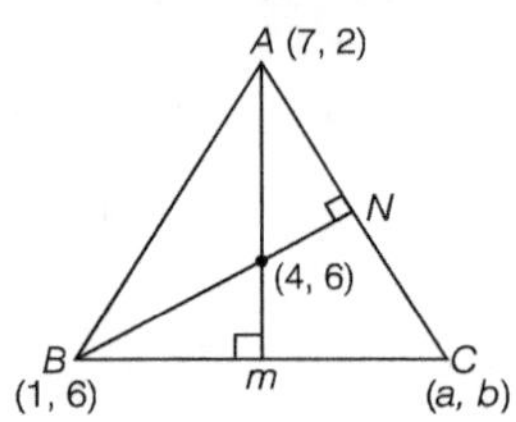

$\therefore$ $4 = \dfrac{7 + 1 + a}{3}$

$\Rightarrow$ $a = 12 - 8 = 4$

and $6 = \dfrac{2 + 6 + b}{3}$

$\Rightarrow$ $b = 18 - 8 = 10$

$\therefore$ $a + b - 10 = 4 + 10 - 10 = 4$

23. (0) Let $A(a, 1)$, $B(1, b)$ and $C(0, 0)$ be the vertices of ΔABC.

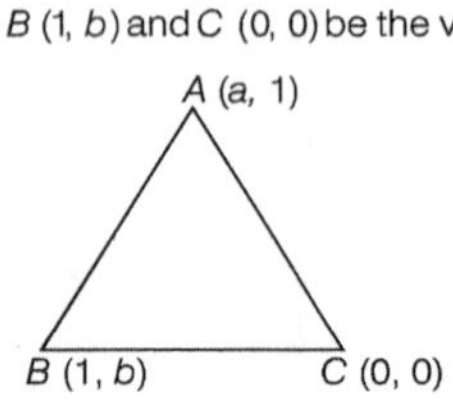

$\because$ ΔABC is equilateral.

$\therefore$ $AB = BC = AC$

$$\sqrt{(1 - a)^2 + (b - 1)^2} = \sqrt{(0 - 1)^2 + (0 - b)^2}$$
$$= \sqrt{(0 - a)^2 + (0 - 1)^2}$$

$\Rightarrow$ $\sqrt{a^2 + b^2 - 2a - 2b + 2} = \sqrt{1 + b^2} = \sqrt{a^2 + 1}$

On taking first two equalities, we have

$$a^2 + b^2 - 2a - 2b + 2 = 1 + b^2$$

$\Rightarrow$ $a^2 - 2a - 2b + 1 = 0$ ...(i)

On taking last equalities, we have

$$1 + b^2 = a^2 + 1$$

$\Rightarrow$ $b^2 = a^2$

$\Rightarrow$ $b = \pm a$

$\Rightarrow$ $b = a$ [$\because 0 < a, b < 1$] ...(ii)

From Eqs. (i) and (ii), we get

$$a^2 - 4a + 1 = 0$$
$$a = \frac{4 \pm \sqrt{16 - 4}}{2}$$
$$= \frac{4 \pm \sqrt{12}}{2} = 2 \pm \sqrt{3}$$

$a = 2 - \sqrt{3}$ [$\because 0 < a < 1$]

$\therefore$ $a - b = 2 - \sqrt{3} - (2 - \sqrt{3})$ [$\because a = b$]

$= 2 - \sqrt{3} - 2 + \sqrt{3} = 0$

24. (6) Let G be the centroid of ΔABC.

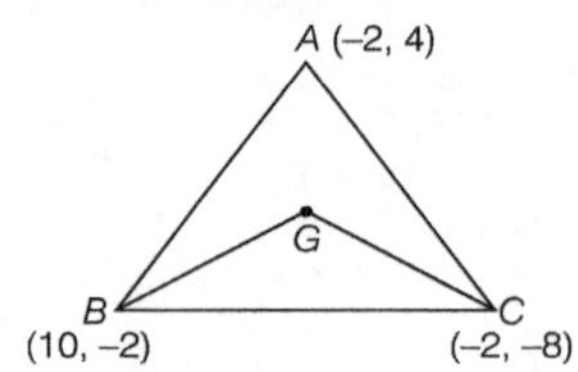

$\therefore$ Coordinates of $G\left(\dfrac{-2 + 10 - 2}{3}, \dfrac{4 - 2 - 8}{3}\right)$

i.e. $(2, -2)$

Now, area of $\Delta BGC = \dfrac{1}{2}\begin{vmatrix} 2 & -2 & 1 \\ 10 & -2 & 1 \\ -2 & -8 & 1 \end{vmatrix}$

$$\Delta_1 = \frac{1}{2}|2(-2 + 8) + 2(10 + 2) + 1(-80 - 4)|$$
$$\Delta_1 = \frac{1}{2}|12 + 24 - 84|$$
$$\Delta_1 = \frac{1}{2}|-48|$$
$$\Delta_1 = \frac{1}{2} \times 48 = 24$$

Now, $\dfrac{\Delta_1}{4} = \dfrac{24}{4} = 6$

25. (5) Since, ar (ΔPAB) = ar (ΔPBC) = ar (ΔPCA)

$\therefore$ P must be centroid of ΔABC

Here, A, B and C are $(-3, 1)$, $(2, -4)$ and $(1, 3)$, respectively.

$\therefore$ $P(a, b) = \left(\dfrac{-3 + 2 + 1}{3}, \dfrac{1 - 4 + 3}{3}\right) = (0, 0)$

$\Rightarrow$ $a = b = 0$ ...(i)

Also, δ = ar $(\Delta ABC) = \dfrac{1}{2}\begin{vmatrix} -3 & 1 & 1 \\ 2 & -4 & 1 \\ 1 & 3 & 1 \end{vmatrix} = 15$ ...(ii)

Thus, $a + b + \delta - 10 = 0 + 0 + 15 - 10 = 5$

DPP-5 Shifting of Origin, Rotation of Axes and Locus

1. (b) Given, axes are rotated through an angle of 30°.

To find Position of point $(4, -2\sqrt{3})$ with respect to new axes.

We know that, if the position of a point with respect to X and Y-axes is (x, y) and if the axes are rotated through an angle θ, then the new coordinates of the point (x, y) is (x', y'), then

$$x' = x\cos\theta + y\sin\theta$$
$$y' = -x\sin\theta + y\cos\theta$$

Here, $\theta = 30°, x = 4$, and $y = -2\sqrt{3}$

$$\therefore \quad x' = 4\cos 30° + (-2\sqrt{3})\sin 30°$$
$$= \frac{4\sqrt{3}}{2} - 2\sqrt{3} \times \frac{1}{2}$$
$$= 2\sqrt{3} - \sqrt{3} = \sqrt{3}$$
$$y' = -4\sin 30° - 2\sqrt{3}\cos 30°$$
$$= -2 - 3 = -5$$

Hence, the new coordinates of point $(4, -2\sqrt{3})$ is $(\sqrt{3}, -5)$.

2. (a) If point $(2, -1)$ gets reflected about Y-axis, then coordinates of new point will be $(-2, -1)$. Now, the position of point $(-2, -1)$ when axes are rotated through an angle 45° in negative direction (i.e. in clockwise direction).

If X and Y are the new coordinates of the point $(-2, -1)$, then

$$X = x\cos\theta - y\sin\theta$$
$$Y = x\sin\theta + y\cos\theta$$
$$\therefore \quad X = (-2)\cos 45° - (-1)\sin 45°$$
$$= -2 \times \frac{1}{\sqrt{2}} + \frac{1}{\sqrt{2}} = -\frac{1}{\sqrt{2}}$$
$$Y = (-2)\sin 45° + (-1)\cos 45°$$
$$= \frac{-2}{\sqrt{2}} - \frac{1}{\sqrt{2}} = -\frac{3}{\sqrt{2}}$$

Hence, the new coordinates of the point is $\left(-\frac{1}{\sqrt{2}}, -\frac{3}{\sqrt{2}}\right)$.

3. (b) Let origin be shifted at point (h, k) without changing the direction of coordinate axes.

Then, we replace x by $(x + h)$ and y by $(y + k)$ in the equation of given curve, then the transformed equation is

$$(x + h)^2 + (y + k)^2 - 4(x + h) + 6(y + k) - 7 = 0$$
$$\Rightarrow \quad x^2 + h^2 + 2hx + y^2 + k^2 + 2ky - 4x - 4h + 6y + 6k - 7 = 0$$
$$\Rightarrow \quad x^2 + y^2 + x(2h - 4) + y(2k + 6) + h^2 + k^2 - 4h + 6k - 7 = 0$$

Since, this equation is required to be free from the terms of first degree.

$$\therefore \quad 2h - 4 = 0 \text{ and } 2k + 6 = 0$$
$$\Rightarrow \quad h = 2 \text{ and } k = -3$$

Hence, the point to which the origin be shifted is $(2, -3)$.

4. (a) Since, origin is shifted to the point $(1, -2)$ without changing the direction of coordinate axes.

$\therefore$ Replace x by $(x + 1)$ and y by $(y - 2)$ in the given equation of curve.

$$\therefore \quad (x + 1)^2 + (y - 2)^2 = 2$$
$$\Rightarrow \quad x^2 + y^2 + 2x - 4y + 5 = 2$$
$$\Rightarrow \quad x^2 + y^2 + 2x - 4y + 3 = 0$$

5. (b) If coordinate axes are rotated through an angle θ, then replace x by $(x\cos\theta - y\sin\theta)$ and y by $(x\sin\theta + y\cos\theta)$ in the given equation of curve.

$$\therefore 5(x\cos\theta - y\sin\theta)^2 + 8(x\cos\theta - y\sin\theta)(x\sin\theta + y\cos\theta) + 5(x\sin\theta + y\cos\theta)^2 + 3(x\cos\theta - y\sin\theta) + 2(x\sin\theta + y\cos\theta) + 5 = 0$$

$$\Rightarrow x^2(8\sin\theta\cos\theta + 5\sin^2\theta + 5\cos^2\theta) + y^2(5\sin^2\theta + 5\cos^2\theta - 8\sin\theta\cos\theta) + xy(8\cos^2\theta - 8\sin^2\theta) + x(8\cos\theta + 2\sin\theta) + y(2\cos\theta - 8\sin\theta) + 5 = 0$$

$$\Rightarrow \quad x^2(5 + 4\sin 2\theta) + y^2(5 - 4\sin 2\theta) + 8(\cos 2\theta)\cdot xy + x(8\cos\theta + 2\sin\theta) + y(2\cos\theta - 8\sin\theta) + 5 = 0$$

To remove xy term, put coefficient of xy equal to zero, i.e.

$$\Rightarrow \quad 8\cos 2\theta = 0$$
$$\Rightarrow \quad \cos 2\theta = 0$$
$$\Rightarrow \quad 2\theta = \frac{\pi}{2}$$
$$\therefore \quad \theta = \frac{\pi}{4}$$

6. (a) Let the coordinates of the point be (h, k).

According to the question,

$$\sqrt{h^2 + k^2} = 3\cdot k$$
$$\Rightarrow \quad h^2 + k^2 = 9k^2$$
$$\Rightarrow \quad h^2 - 8k^2 = 0$$

Replacing h by x and k by y, we have

$$x^2 - 8y^2 = 0$$

which is the required locus.

7. (d) Let the coordinates of the point be (h, k).

According to the question,

$$k^2 + h^2 = 4$$

Replacing h by x and k by y, we have

$$x^2 + y^2 = 4$$

which is required locus.

8. (d) Let (h, k) be the coordinates of variable point P.

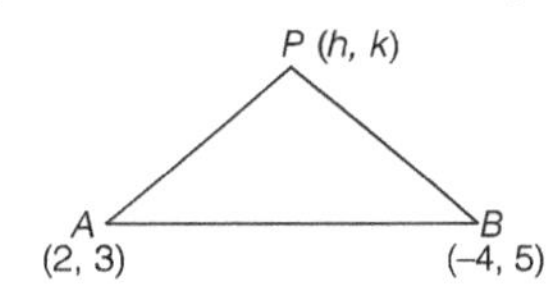

$$\therefore \text{Area of } \Delta PAB = \frac{1}{2}\begin{vmatrix} h & k & 1 \\ 2 & 3 & 1 \\ -4 & 5 & 1 \end{vmatrix} = \frac{1}{2}\begin{vmatrix} h & k & 1 \\ 2 & 3 & 1 \\ -4 & 5 & 1 \end{vmatrix} = 12$$

$$\Rightarrow \frac{1}{2}|\, h(3 - 5) - k(2 + 4) + 1(10 + 12)\,| = 12$$
$$\Rightarrow \quad \frac{1}{2}|\, -2h - 6k + 22\,| = 12$$
$$\Rightarrow \quad -2h - 6k + 22 = \pm 24$$

On taking positive sign, we have

$$-2h - 6k + 22 = 24$$
$$\Rightarrow \quad 2h + 6k + 2 = 0$$
$$\Rightarrow \quad 2x + 6y + 2 = 0$$
$$\Rightarrow \quad x + 3y + 1 = 0 \quad \ldots\text{(i)}$$

On taking negative sign, we have

$$-2h - 6k + 46 = 0$$
$$\Rightarrow \quad -2x - 6y + 46 = 0$$
$$\Rightarrow \quad x + 3y - 23 = 0 \quad \ldots\text{(ii)}$$

On multiplying Eqs. (i) and (ii), we have

$$(x + 3y + 1)(x + 3y - 23) = 0$$
$$\Rightarrow \quad x^2 + 9y^2 + 6xy - 22x - 66y - 23 = 0$$

which is required locus.

9. (c) Let (h, k) be the coordinates of moving point and $A(0, 0)$, $B(4 - 3)$.

According to the question,

$\therefore$ $$2PA = 3PB$$

$$2\sqrt{h^2 + k^2} = 3\sqrt{(h-4)^2 + (k+3)^2}$$

$$4(h^2 + k^2) = 9\{h^2 + 16 - 8h + k^2 + 9 + 6k\}$$

$$4(h^2 + k^2) = 9(h^2 + k^2) - 72h + 54k + 225$$

$\Rightarrow$ $$5(h^2 + k^2) - 72h + 54k + 225 = 0$$

$\Rightarrow$ $$5h^2 + 5k^2 - 72h + 54k + 225 = 0$$

Replacing h by x and y by k, we have

$\Rightarrow$ $$5x^2 + 5y^2 - 72x + 54y + 225 = 0$$

10. (d) Let the coordinates of the variable point C be (h, k).

Draw $CM \perp AB$.

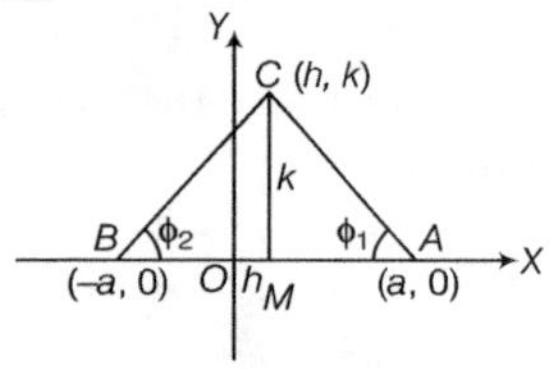

$\therefore$ $$CM = k, OM = h$$

Let $$\angle A = \phi_1 \text{ and } \angle B = \phi_2$$

$\therefore$ $$\angle A - \angle B = \phi_1 - \phi_2 = \theta$$

Now, in ΔAMC, $$\tan \phi_1 = \frac{k}{a-h}$$

In ΔBMC, $$\tan \phi_2 = \frac{k}{a+h}$$

Now, $$\tan \theta = \tan(\phi_1 - \phi_2)$$

$$\tan \theta = \frac{\tan \phi_1 - \tan \phi_2}{1 + \tan \phi_1 \cdot \tan \phi_2}$$

$$\tan \theta = \frac{\dfrac{k}{a-h} - \dfrac{k}{a+h}}{1 + \dfrac{k^2}{a^2 - h^2}}$$

$$= \frac{2hk}{a^2 - h^2 + k^2}$$

$\Rightarrow$ $$a^2 - h^2 + k^2 = 2hk \cot \theta$$

$\Rightarrow$ $$a^2 - x^2 + y^2 = 2xy \cot \theta$$

$\Rightarrow$ $$x^2 - y^2 + 2xy \cot \theta = a^2$$

which is required locus.

11. (a) Since, ΔABC is a right angled triangle.

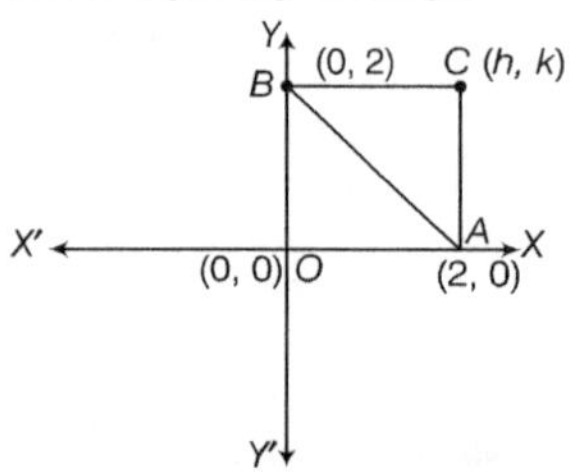

If (h, k) be the third vertex of Δ ABC

$\therefore$ $$AC^2 + BC^2 = AB^2 \quad \text{[by Pythagoras theorem]}$$

$\Rightarrow$ $$(h-2)^2 + k^2 + h^2 + (k-2)^2 = 8$$

$\Rightarrow$ $$h^2 + 4 - 4h + k^2 + h^2 + k^2 + 4 - 4k = 8$$

$\Rightarrow$ $$2h^2 + 2k^2 - 4h - 4k + 8 = 8$$

$\Rightarrow$ $$2h^2 + 2k^2 - 4h - 4k = 0$$

$\Rightarrow$ $$h^2 + k^2 - 2h - 2k = 0$$

$\therefore$ $$x^2 + y^2 - 2x - 2y = 0$$

which is required locus.

12. (c) Let (h, k) be the coordinates of variable point.

Now, distance of (h, k) from origin, i.e. $(0,0) = \sqrt{h^2 + k^2}$

and distance of the point (h, k) from the line $x - 2 = 0$ is $|h - 2|$.

Now, according to the question,

$$\sqrt{h^2 + k^2} + |h - 2| = 4 \quad \text{...(i)}$$

On taking positive sign, we have

$\Rightarrow$ $$\sqrt{h^2 + k^2} = (6 - h)$$

$\Rightarrow$ $$h^2 + k^2 = (6 - h)^2$$

$\Rightarrow$ $$h^2 + k^2 - h^2 - 36 - 12h = 0$$

$\Rightarrow$ $$k^2 - 36 = 12h$$

$\Rightarrow$ $$k^2 = 12h + 36$$

$\Rightarrow$ $$k^2 = 12(h + 3)$$

$\therefore$ $$y^2 = 12(x + 3)$$

which is a parabola.

On taking negative sign of Eq. (i), we have

$$\sqrt{h^2 + k^2} = 4 + h - 2$$

$\Rightarrow$ $$\sqrt{h^2 + k^2} = h + 2$$

$\Rightarrow$ $$h^2 + k^2 = (h + 2)^2$$

$\Rightarrow$ $$h^2 + k^2 = h^2 + 4 + 4h$$

$\Rightarrow$ $$k^2 = 4(h + 1)$$

$\therefore$ $$y^2 = 4(x + 1)$$

which is also a parabola.

Hence, locus of the point is parabola.

13. (b) Let coordinates of point P be (h, k).

Given, $\angle APB = 90°$

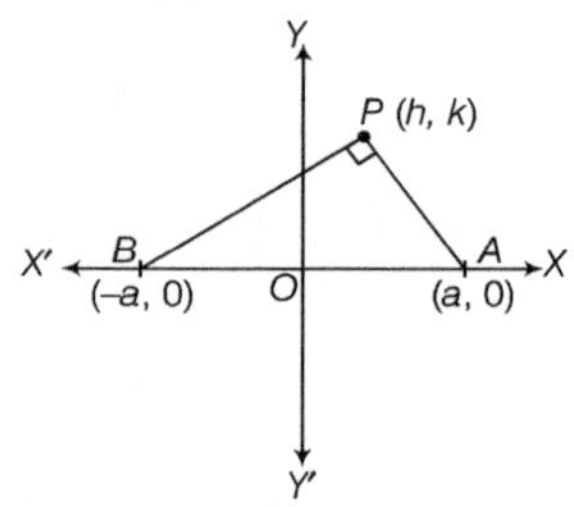

By Pythagoras theorem, we have

$$AB^2 = PA^2 + PB^2$$

$\Rightarrow$ $$(2a)^2 = (h - a)^2 + k^2 + (h + a)^2 + k^2$$

$\Rightarrow$ $$4a^2 = h^2 + a^2 - 2ah + h^2 + a^2 + 2ah + 2k^2$$

$\Rightarrow$ $$2a^2 = 2h^2 + 2k^2$$

$\Rightarrow$ $$h^2 + k^2 = a^2$$

$\therefore$ $$x^2 + y^2 = a^2$$

which is required locus.

14. (b) Let $S(h, k)$ be the variable point and $P(1, 0)$, $Q(-1, 0)$ and $R(2, 0)$ are fixed points, then

$$SQ^2 + SR^2 = 2SP^2$$

$$\Rightarrow \quad (-1-h)^2 + k^2 + (2-h)^2 + k^2 = 2[(1-h)^2 + k^2]$$

$$\Rightarrow \quad h^2 + 1 + 2h + 4 + h^2 - 4h + 2k^2 = 2(h^2 + 1 - 2h + k^2)$$

$$\Rightarrow \quad 2h^2 + 2k^2 - 2h + 5 = 2h^2 + 2k^2 - 4h + 2$$

$$\Rightarrow \quad -2h + 5 = -4h + 2$$

$$\Rightarrow \quad 2h + 3 = 0$$

$$\Rightarrow \quad 2x + 3 = 0$$

$$\therefore \quad x = -3/2$$

which is a straight line parallel to Y-axis.

15. (c) Given, a triangle whose vertices are $(a \cos t, a \sin t)$, $(b \sin t, -b \cos t)$ and $(1, 0)$.

$\therefore$ The coordinates of centroid be

$$x = \frac{a \cos t + b \sin t + 1}{3}$$

$$\Rightarrow \quad 3x - 1 = a \cos t + b \sin t \qquad \ldots(\text{i})$$

and $$y = \frac{a \sin t - b \cos t + 0}{3}$$

$$\Rightarrow \quad 3y = a \sin t - b \cos t \qquad \ldots(\text{ii})$$

On squaring and adding Eqs. (i) and (ii), we get

$$(3x-1)^2 + (3y)^2 = a^2(\cos^2 t + \sin^2 t) + b^2(\sin^2 t + \cos^2 t)$$

$$\Rightarrow (3x-1)^2 + (3y)^2 = a^2 + b^2$$

16. (c) Let $P(h, k)$ be the point which is equidistant from $A(a_1, b_1)$ and $B(a_2, b_2)$.

$$PA = PB \Rightarrow PA^2 = PB^2$$

$$(h - a_1)^2 + (k - b_1)^2 = (h - a_2)^2 + (k - b_2)^2$$

$$\Rightarrow h^2 + a_1^2 - 2a_1h + k^2 + b_1^2 - 2kb_1$$
$$= h^2 + a_2^2 - 2a_2h + k^2 + b_2^2 - 2kb_2$$

$$\Rightarrow 2(a_2 - a_1)h + 2(b_2 - b_1)k + (a_1^2 + b_1^2 - a_2^2 - b_2^2) = 0$$

Thus, the equation of locus is

$$(a_2 - a_1)x + (b_2 - b_1)y + \frac{1}{2}(a_1^2 + b_1^2 - a_2^2 - b_2^2) = 0$$

But the given equation is

$$(a_2 - a_1)x + (b_2 - b_1)y - c = 0$$

$$\therefore \quad c = -\frac{1}{2}(a_1^2 + b_1^2 - a_2^2 - b_2^2) = \frac{1}{2}(a_2^2 + b_2^2 - a_1^2 - b_1^2)$$

17. (c) Let (α, β) be the coordinates of point C.

$\therefore$ Centroid is $\left(\frac{2 - 2 + \alpha}{3}, \frac{-3 + 1 + \beta}{3}\right)$, i.e. $\left(\frac{\alpha}{3}, \frac{\beta - 2}{3}\right)$

This lies on $2x + 3y = 1$.

$$\therefore \quad 2\left(\frac{\alpha}{3}\right) + 3\left(\frac{\beta - 2}{3}\right) = 1$$

$$\Rightarrow \quad 2\alpha + 3\beta - 6 = 3$$

$$\Rightarrow \quad 2\alpha + 3\beta = 9$$

Hence, the locus of point C be $2x + 3y = 9$.

18. (b) According to the question,

$$x + y = \sqrt{(x-1)^2 + (y-1)^2}$$

$$\Rightarrow \quad x + y = \sqrt{x^2 + 1 - 2x + y^2 + 1 - 2y}$$

$$\Rightarrow \quad x + y = \sqrt{x^2 + y^2 - 2x - 2y + 2}$$

$$\Rightarrow \quad (x + y)^2 = x^2 + y^2 - 2x - 2y + 2$$

$$\Rightarrow \quad x^2 + y^2 + 2xy = x^2 + y^2 - 2x - 2y + 2$$

$$\Rightarrow \quad 2x + 2y + 2xy - 2 = 0$$

$$\Rightarrow \quad x + y + xy - 1 = 0$$

$$\Rightarrow \quad x + y + xy + 1 - 2 = 0$$

$$\Rightarrow \quad y + xy + (1 + x) - 2 = 0$$

$$\Rightarrow \quad y(1 + x) + (1 + x) - 2 = 0$$

$$\Rightarrow \quad (1 + x)(1 + y) = 2$$

19. (b) Let the cross-section of the floor and wall be taken as the coordinate axes and AB be one of the position of the stick.

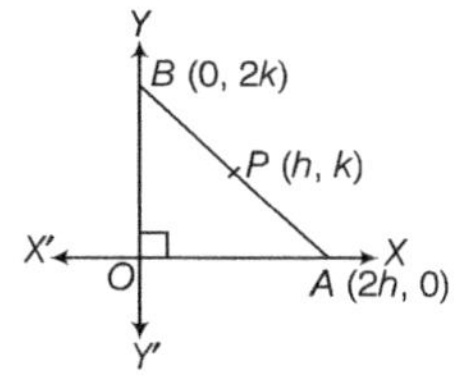

If the mid-point of AB is $P(h, k)$, then coordinates of A and B are $(2h, 0)$ and $(0, 2k)$, respectively.

But, given $\quad |AB| = 10 \Rightarrow AB^2 = 100$

$$\Rightarrow \quad (2h - 0)^2 + (0 - 2k)^2 = 100$$

$$\Rightarrow \quad 4h^2 + 4k^2 = 100$$

Changing (h, k) by (x, y), then

$$4x^2 + 4y^2 = 100 \Rightarrow x^2 + y^2 = 25$$

which is the required locus of P.

20. (b)

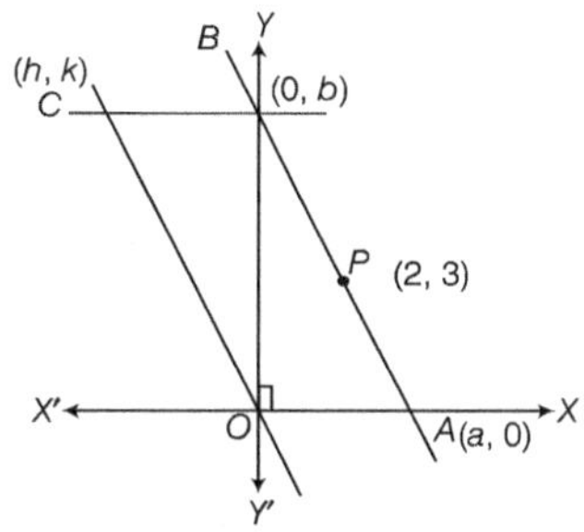

Equation of AB is $\frac{x}{a} + \frac{y}{b} = 1$

which passes through $(2, 3)$.

$$\Rightarrow \quad \frac{2}{a} + \frac{3}{b} = 1 \qquad \ldots(\text{i})$$

Since, diagonals of parallelogram bisect each other.

$$\Rightarrow \quad \left(0, \frac{b}{2}\right) = \left(\frac{a + h}{2}, \frac{k}{2}\right)$$

$$\Rightarrow \quad a + h = 0, \quad b = k \qquad \ldots(\text{ii})$$

From Eqs. (i) and (ii),

$$\frac{2}{-h} + \frac{3}{k} = 1$$

$\therefore$ Locus of C is $\frac{2}{-x} + \frac{3}{y} = 1$ or $3x - 2y = xy$

$$\Rightarrow \quad 3x - 2y - xy + 6 = 6$$

$$\Rightarrow \quad (x + 3)(3 - y) = 6$$

As, $(x, y) \in$ Integer.

$$\Rightarrow \quad \left.\begin{matrix} x + 3 = \pm 6 \\ 3 - y = \pm 1 \end{matrix}\right\} \text{or} \left.\begin{matrix} x + 3 = \pm 3 \\ 3 - y = \pm 2 \end{matrix}\right\}$$

$$\text{or} \quad \left.\begin{matrix} x + 3 = \pm 2 \\ 3 - y = \pm 3 \end{matrix}\right\} \text{or} \left.\begin{matrix} x + 3 = \pm 1 \\ 3 - y = \pm 6 \end{matrix}\right\}$$

neglecting $(0, 1)$ and $(-1, 0)$

$\therefore$ 6 points are possible.

Revisal Problems

1. (a) We know that, circumcentre, centroid and orthocentre of a triangle lie on a line, i.e. collinear and G divide OH in the ratio 1 : 2.

i.e. $OG : GH = 1 : 2$

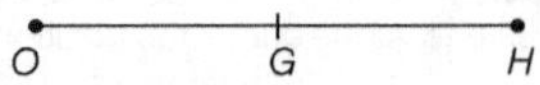

Let coordinates of H be (x, y).

By section formula, we have

$\therefore \quad 3 = \frac{x+12}{3}$

$\Rightarrow \quad x + 12 = 9$

$\Rightarrow \quad x = -3$

and $\quad 3 = \frac{y+4}{3}$

$\Rightarrow \quad y + 4 = 9$

$\Rightarrow \quad y = 5$

Hence, orthocentre is $(-3, 5)$.

2. (b) Since, origin is shifted to the point (1, 2).

$\therefore$ Replacing x by $(x+1)$ and y by $(y+2)$ in given equation of curve, we get

$$(y+2)^2 - 8(x+1) - 4(y+2) + 12 = 0$$

$$\Rightarrow \quad y^2 + 4 + 4y - 8x - 8 - 4y - 8 + 12 = 0$$

$$\Rightarrow \quad y^2 - 8x = 0$$

$$\Rightarrow \quad y^2 = 8x \quad \text{...(i)}$$

But it is given that $y^2 - 8x - 4y + 12 = 0$ becomes

$$y^2 = 4ax \quad \text{...(ii)}$$

$\therefore$ On comparing Eqs. (i) and (ii), we have

$$4a = 8$$
$$a = 2$$

3. (a) Since, orthocentre is the point of intersection of altitudes.

$\therefore$ It is clear that, orthocentre of ΔOAC is point B, i.e. (5, 4).

4. (c) Let $A(x_1, y_1)$ and $B(x_2, y_2)$ be two points and a point $R(x_1 + t(x_2 - x_1), y_1 + t(y_2 - y_1))$ divides the join of AB internally.

$\therefore$ The coordinates of the points of division are

$$\left(\frac{(1-t)x_1 + t\,x_2}{(1-t)+t}, \frac{(1-t)y_1 + t\,y_2}{(1-t)+t}\right)$$

Clearly, these are the coordinates of the points dividing the join of (x_1, y_1) and (x_2, y_2) in the ratio $t : (1-t)$.

For the internal division, we must have

$$t > 0$$

and $\quad 1 - t > 0$

$\Rightarrow \quad 0 < t < 1$

5. (b) We may suppose that, coordinates of two fixed points A and B are $(a, 0), (-a, 0)$ respectively and coordinates of variable point C are (h, k).

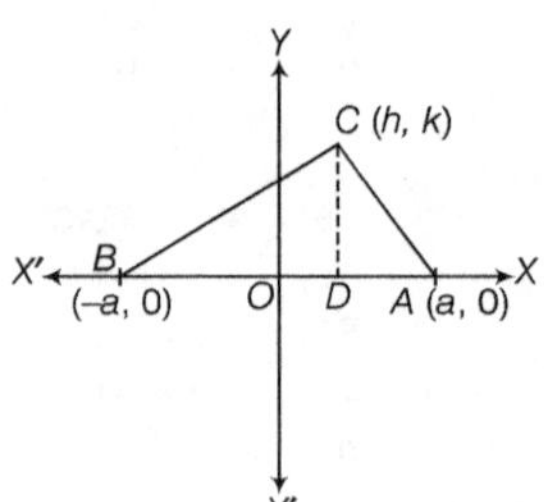

Now, in ΔCDA,

$$\cot A = \frac{DA}{CD} = \frac{a-h}{k}$$

and in ΔDBC,

$$\cot B = \frac{BD}{CD} = \frac{a+h}{k}$$

But $\quad \cot A + \cot B = \text{Constant}, C$

$$\cot A + \cot B = \lambda \quad \text{(say)}$$

$$\frac{a-h}{k} + \frac{a+h}{k} = \lambda$$

$\Rightarrow \quad \frac{2a}{k} = \lambda$

$\Rightarrow \quad 2a = k\lambda$

or $\quad 2a = y\lambda$

$\Rightarrow \quad y = \frac{2a}{\lambda}$

which is a line parallel to X-axis, i.e. AB.

6. (b) Let $A(5, -1)$, $B(-2, 3)$ and $C(h, k)$ be the vertices of a triangle and $O(0, 0)$ is orthocentre.

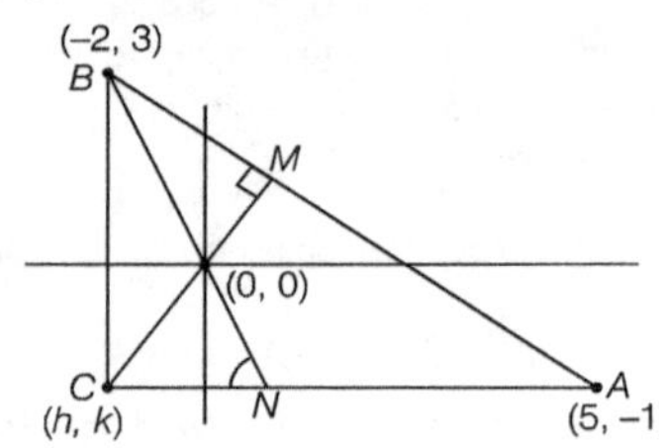

Now, draw $CM \perp AB$ and $BN \perp AC$.

$\therefore \quad \text{Slope } CM = \frac{-1}{\text{Slope } AB} = \frac{7}{4}$

$$\text{Slope } BN = \frac{-1}{\text{Slope } AC} = \frac{-(h-5)}{(k+1)}$$

$$\text{Equation of } CM = (y - k) = \frac{7}{4}(x - h)$$

$$\Rightarrow \quad y = \frac{7}{4}x - \frac{7h}{4} + k \quad \text{...(i)}$$

$$\text{Equation of } BN = y - 3 = -\frac{(h-5)}{k+1}(x+2)$$

$$\Rightarrow \quad y = \frac{-(h-5)}{k+1}x - \frac{2(h-5)}{k+1} + 3 \quad \text{...(ii)}$$

Since, O is the intersection point of CM and BN. Therefore, it satisfies Eqs. (i) and (ii).

From Eq. (i), we have

$$-\frac{7h}{4} + k = 0$$

$$\Rightarrow \quad 7h - 4k = 0 \quad \text{...(iii)}$$

and from Eq. (ii), we have

$$\frac{-2(h-5)}{k+1} + 3 = 0$$

$$-2h + 10 + 3k + 3 = 0$$

or $\quad 2h - 3k - 13 = 0 \quad$ ...(iv)

On solving Eqs. (iii) and (iv), we have

$$h = -4, k = -7$$

7. (b)

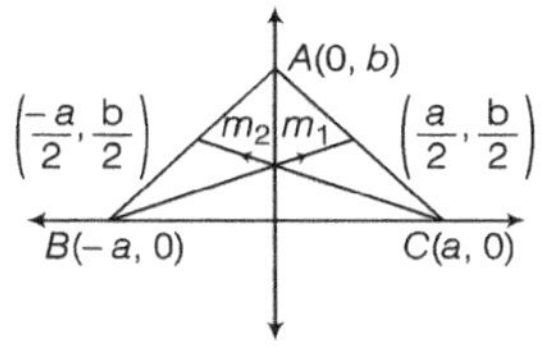

Here, $$m_1 = \frac{\frac{b}{2}}{\frac{a}{2}+a} = \frac{b}{3a}$$

and $$m_2 = \frac{\frac{b}{2}}{\frac{-3a}{2}} = -\frac{b}{3a}$$

$\therefore$ $m_1 + m_2 = 0$

8. (b) According to the figure,

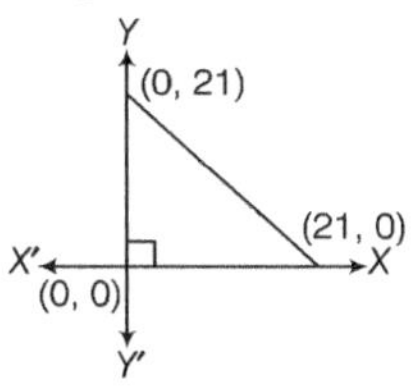

$$x + y < 21$$

where, $0 < x < 21$

$0 < y < 21$

$\Rightarrow$ $x + y + z = 21$

where, $0 < x < 21$

$0 < y < 21$

$0 < z < 21$

$\therefore$ Total number of integral points

$= \text{Coefficient } x^{21} \text{ in } (1 + x + x^2 + x^3 + \cdots + x^{21})^3$

$= {}^{21-1}C_{3-1} = {}^{20}C_2 = \frac{20!}{2!\,18!} = \frac{20\times 19}{2} = 190$

Explanations for Q. Nos. (9-11)

Since, $A = 90°$

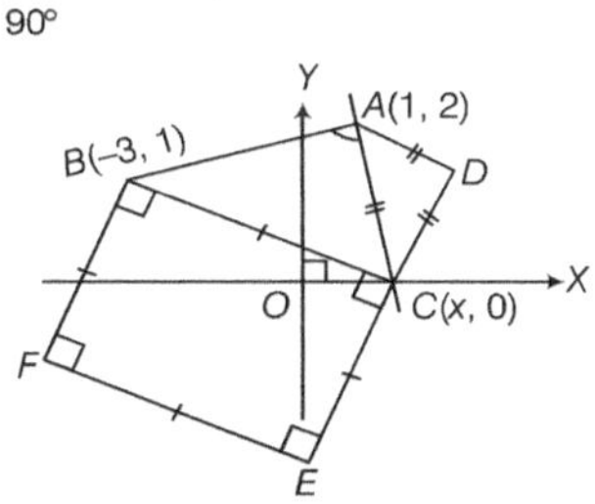

$\therefore$ $m_{AC} \cdot m_{AB} = -1$

$\Rightarrow$ $\frac{2}{1-x} \times \frac{1}{4} = -1$

$\Rightarrow$ $x = \frac{3}{2}$

$\Rightarrow$ $C\left(\frac{3}{2}, 0\right)$

Also, by rotation

$$F\left(-4, \frac{-7}{2}\right), E\left(\frac{1}{2}, \frac{-9}{2}\right) \text{ and } D\left(\frac{5+4\sqrt{3}}{4}, \frac{4+\sqrt{3}}{4}\right).$$

9. (d) Slope of $AF = \frac{2+\frac{7}{2}}{1+4} = \frac{11}{10}$

10. (c) Abscissa of centroid of ΔBCE,

$$= \left(\frac{-3+\frac{3}{2}+\frac{1}{2}}{3}\right) = \frac{-1}{3}$$

11. (b) $D = (\alpha, \beta)$

Then, $\beta = \frac{4+\sqrt{3}}{4}$ $\Rightarrow (4\beta - 4)^2 = 3$

12. (a) Since, $\angle A$ of ΔABC is obtuse.

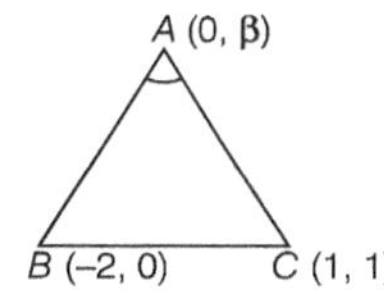

$$\therefore \quad \cos(\angle BAC) = \frac{(AB)^2 + (AC)^2 - (BC)^2}{2\cdot AB\cdot AC} < 0$$

$$= \frac{(\beta^2+4) + (1-\beta)^2 + 1^2 - 10}{2\cdot(\beta^2+4)(\beta^2-2\beta+2)} < 0$$

$$= \frac{\beta^2+4+1+\beta^2-2\beta+1-10}{2(\beta^2+4)(\beta^2-2\beta+2)} < 0$$

$$= \frac{2\beta^2-2\beta-4}{2(\beta^2+4)(\beta^2-2\beta+2)} < 0$$

$$= \frac{\beta^2-\beta-2}{(\beta^2+4)(\beta^2-2\beta+2)} < 0$$

$$= (\beta-2)(\beta+1) < 0$$

$[\because \beta^2 + 4 > 0 \text{ and } \beta^2 - 2\beta + 2 > 0]$

$\Rightarrow$ $\beta \in (-1, 2)$

13. (b) Whenever A is obtuse it is the largest angle also. But there may be more values of β for which A is largest but is not obtuse.

$\therefore$ l_1 is subset of l_2.

14. (c) $\angle A$ will be largest, if $a > b, a > c$, A, B and C are not collinear.

$\Rightarrow$ $a^2 > b^2, a^2 > c^2, \beta \neq \frac{2}{3}$

$\Rightarrow$ $10 > 1 + (\beta+1)^2, 10 > 4 + \beta^2$

$\Rightarrow$ $\beta^2 + 2\beta - 8 < 0$ and $\beta^2 < 6$

$\Rightarrow$ $-2 < \beta < 4$ and $-\sqrt{6} < \beta < \sqrt{6}$

$\therefore$ $\beta \in \left(-2, \frac{2}{3}\right) \cup \left(\frac{2}{3}, \sqrt{6}\right)$

15. (a) The line passes through fix point $(1, -1)$. If $\lambda = \frac{7}{5}$, then A, C and B are collinear

$\therefore$ $\lambda \neq \frac{7}{5}$

16. (a) Centroid of Δ is

$$h = \frac{1+3+\lambda}{3}$$

and $$k = \frac{4-1+0}{3} = 1$$

$\therefore$ $k = 1$

$\Rightarrow$ $y = 1$ is required locus.

17. (b) $x^2 = 4a(y-k)$ ...(i)

Locus is $y = 1$ ...(ii)

On solving Eqs. (i) and (ii), we get

$$x^2 = 4a(1-k), \text{as } x^2 > 0$$

$\Rightarrow \quad 4a(1-k) > 0$

$\therefore \quad (a < 0, k < 1)$ or $(a > 0, k > 1)$

18. (1) $\because$ Lines $x + ay + a = 0$; $bx + y + b = 0$ and $cx + cy + 1 = 0$ are concurrent.

$$\therefore \quad \begin{vmatrix} 1 & a & a \\ b & 1 & b \\ c & c & 1 \end{vmatrix} = 0$$

On applying $C_1 \to C_1 - C_2$, we get

$$\Rightarrow \quad \begin{vmatrix} 1-a & a & a \\ b-1 & 1 & b \\ 0 & c & 1 \end{vmatrix} = 0$$

On applying $C_2 \to C_2 - C_3$, we get

$$\Rightarrow \quad \begin{vmatrix} 1-a & 0 & a \\ b-1 & 1-b & b \\ 0 & c-1 & 1 \end{vmatrix} = 0$$

$\Rightarrow \quad (1-a)[(1-b) - b(c-1)] + a[(b-1)(c-1)] = 0$

$\Rightarrow (1-a)(1-b) - b(1-a)(c-1) + a(b-1)(c-1) = 0$

$\Rightarrow (1-a)(1-b) + b(a-1)(c-1) + a(b-1)(c-1) = 0$

Dividing by $(a-1)(b-1)(c-1)$, we get

$$\Rightarrow \quad \frac{a}{a-1} + \frac{b}{b-1} + \frac{1}{c-1} = 0$$

$$\Rightarrow \quad \frac{a}{a-1} + \frac{b}{b-1} + \frac{c}{c-1} = 1$$

[adding 1 on both sides]

19. (0)

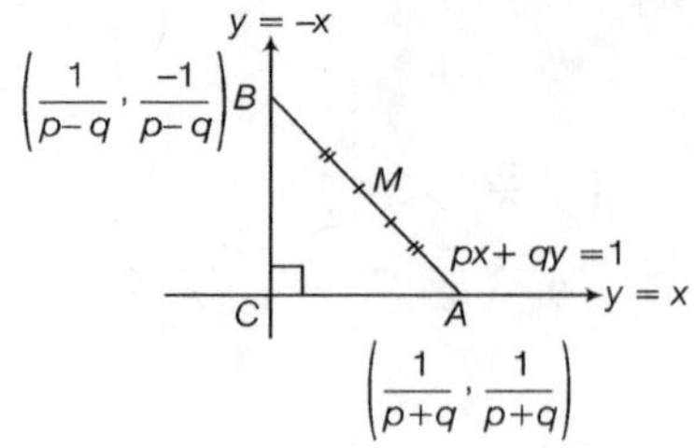

$$M\left(\frac{\frac{1}{p-q}+\frac{1}{p+q}}{2}, \frac{\frac{1}{p+q}-\frac{1}{p-q}}{2}\right) = \left(\frac{p}{p^2-q^2}, \frac{-q}{p^2-q^2}\right)$$

$$\therefore \quad x = \frac{p}{p^2-q^2}, y = \frac{-q}{p^2-q^2}$$

$$\Rightarrow \quad x^2 + y^2 = \frac{p^2+q^2}{(p^2-q^2)^2}$$

Since, $\quad p^2 + q^2 = 1$

$$\therefore \quad x^2 + y^2 = \frac{1}{(p^2-q^2)^2} \quad \text{and} \quad x^2 - y^2 = \frac{1}{(p^2-q^2)}$$

Given, $\quad x^2 + y^2 = k(x^2 - y^2)^2$

$$\Rightarrow \quad \frac{1}{(p^2-q^2)^2} = k \cdot \frac{1}{(p^2-q^2)^2}$$

$\therefore \quad k = 1 \Rightarrow k - 1 = 0$

20. (7) We have,

$$OA_n = n \cdot A_{n-1} = n(n-1)OA_{n-1} \cdots$$

$$= n(n-1) \cdots 2(OA_1) = n!\, OA_1 = n! \qquad [\because OA_1 = 1]$$

$\Rightarrow \quad \sqrt{2}(2520\sqrt{2}) = n! \quad \Rightarrow \quad 2520 \times 2 = n!$

$\Rightarrow \quad 7 \times 6 \times 5 \times 4 \times 3 \times 2 = n!$

$\therefore \quad n = 7$

JEE Main & AIEEE Archive

1. (b) Given, mid-points of a triangle are (0, 1), (1, 1) and (1, 0). Plot these points on a graph paper and make a triangle.

So, the sides of a triangle will be 2, 2 and $\sqrt{2^2 + 2^2} = 2\sqrt{2}$

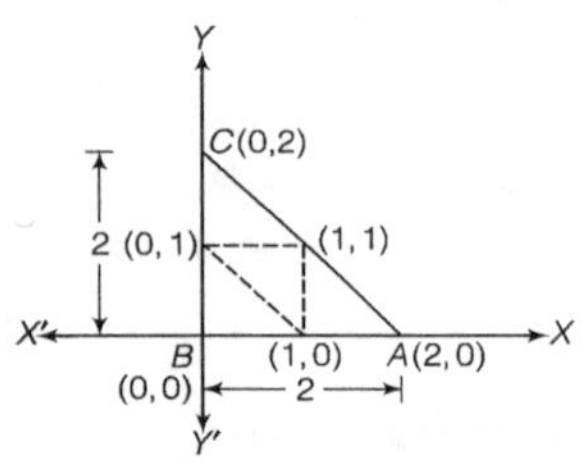

x-coordinate of incentre

$$= \frac{2 \times 0 + 2\sqrt{2} \cdot 0 + 2 \cdot 2}{2 + 2 + 2\sqrt{2}}$$

$$= \frac{2}{2+\sqrt{2}} \times \frac{2-\sqrt{2}}{2-\sqrt{2}} = 2 - \sqrt{2}$$

2. (c) Given line $L: 2x + y = k$ passes through point (say P) which divides a line segment (say AB) in the ratio 3 : 2, where $A(1, 1)$ and $B(2, 4)$.

Using section formula, the coordinates of the point P which divides AB internally in the ratio 3 : 2 are

$$P\left(\frac{3 \times 2 + 2 \times 1}{3+2}, \frac{3 \times 4 + 2 \times 1}{3+2}\right) \equiv P\left(\frac{8}{5}, \frac{14}{5}\right)$$

Also, since the line L passes through P, hence substituting the coordinates of $P\left(\frac{8}{5}, \frac{14}{5}\right)$ in the equation of $L: 2x + y = k$, we get

$$2\left(\frac{8}{5}\right) + \left(\frac{14}{5}\right) = k$$

$\therefore \quad k = 6$

3. (c) Given, third vertex lies on $2x + 3y = 9$, i.e. $\left(x, \frac{9-2x}{3}\right)$

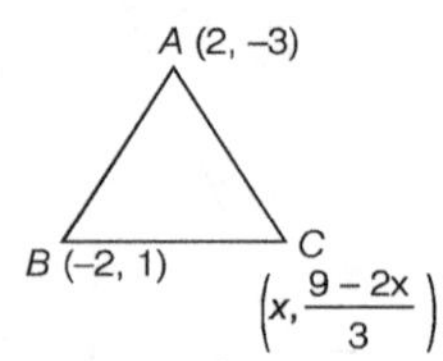

$\therefore$ Locus of centroid is

$$\left(\frac{2-2+x}{3}, \frac{-3+\frac{9-2x}{3}+1}{3}\right) = (h, k)$$

$$\therefore \quad h = \frac{x}{3}, k = \frac{3-2x}{9}$$

$\Rightarrow \quad 9k = 3 - 2(3h)$

$\Rightarrow \quad 9k = 3 - 6h$

$\Rightarrow \quad 2h + 3k = 1$ or $2x + 3y = 1$

4. (a) Let (x, y) denotes the coordinates in A, B and C plane.

Then, $$\frac{(x-1)^2+y^2}{(x+1)^2+y^2}=\frac{1}{9}$$

$$\Rightarrow \quad 9x^2+9y^2-18x+9=x^2+y^2+2x+1$$

$$\Rightarrow \quad 8x^2+8y^2-20x+8=0$$

$$\Rightarrow \quad x^2+y^2-\frac{5}{2}x+1=0$$

Hence, A, B and C lie on a circle with $C\left(\frac{5}{4}, 0\right)$.

5. (c) Since, $A(h, k)$, $B(1, 1)$ and $C(2, 1)$ are the vertices of a right angled ΔABC.

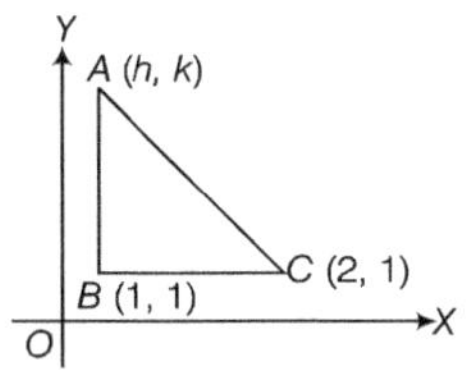

Now, $$AB=\sqrt{(1-h)^2+(1-k)^2}$$

$$BC=\sqrt{(2-1)^2+(1-1)^2}=1$$

and $$CA=\sqrt{(h-2)^2+(k-1)^2}$$

From Pythagoras theorem,

$$AC^2=AB^2+BC^2$$

$$\Rightarrow 4+h^2-4h+k^2+1-2k=h^2+1-2h+k^2+1-2k+1$$

$$\Rightarrow \quad 5-4h=3-2h$$

$$\therefore \quad h=1 \qquad \ldots\text{(i)}$$

Now, given that area of triangle is 1.

Then, area $(\Delta ABC)=\frac{1}{2}\times AB\times BC$

$$\Rightarrow \quad 1=\frac{1}{2}\times\sqrt{(1-h)^2+(1-k)^2}\times 1$$

$$\Rightarrow \quad 2=\sqrt{(1-h)^2+(1-k)^2} \qquad \ldots\text{(ii)}$$

$$\Rightarrow \quad 2=\sqrt{(k-1)^2} \qquad \text{[from Eq. (i)]}$$

$$\Rightarrow \quad 4=k^2+1-2k$$

$$\Rightarrow \quad k^2-2k-3=0$$

$$\Rightarrow \quad (k-3)(k+1)=0$$

$$\therefore \quad k=-1, 3$$

Thus, the set of values of k is $\{-1, 3\}$.

6. (b) Let D and E be the mid-points of AB and AC. The coordinates of B are

$$-1=\frac{1+x_2}{2} \quad \text{and} \quad 2=\frac{1+y_2}{2}$$

i.e. $$x_2=-3 \quad \text{and} \quad y_2=3$$

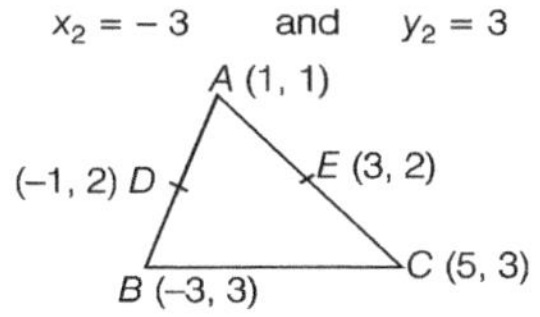

and coordinates of C are

$$3=\frac{1+x_3}{2} \quad \text{and} \quad 2=\frac{1+y_3}{2}$$

i.e. $$x_3=5 \quad \text{and} \quad y_3=3$$

$\therefore$ Centroid of triangle

$$=\left(\frac{x_1+x_2+x_3}{3}, \frac{y_1+y_2+y_3}{3}\right)$$

$$=\left(\frac{1-3+5}{3}, \frac{1+3+3}{3}\right)=\left(1, \frac{7}{3}\right)$$

7. (a) Let $P(\alpha, \beta)$ be the point which is equidistant from $A(a_1, b_1)$ and $B(a_2, b_2)$.

$$\therefore \quad PA=PB$$

$$\Rightarrow \quad (\alpha-a_1)^2+(\beta-b_1)^2=(\alpha-a_2)^2+(\beta-b_2)^2$$

$$\Rightarrow \alpha^2+a_1^2-2\alpha a_1+\beta^2+b_1^2-2\beta b_1$$
$$=\alpha^2+a_2^2-2\alpha a_2+\beta^2+b_2^2-2\beta b_2$$

$$\Rightarrow \quad 2(a_2-a_1)\alpha+2(b_2-b_1)\beta+(a_1^2+b_1^2-a_2^2-b_2^2)=0$$

Thus, the equation of locus (α, β) is

$$(a_2-a_1)x+(b_2-b_1)y+\frac{1}{2}(a_1^2+b_1^2-a_2^2-b_2^2)=0$$

But the given equation is

$$(a_2-a_1)x+(b_2-b_1)y-c=0$$

$$\therefore \quad c=-\frac{1}{2}(a_1^2+b_1^2-a_2^2-b_2^2)$$

$$=\frac{1}{2}(a_2^2+b_2^2-a_1^2-b_1^2)$$

8. (b) Given, a triangle, whose vertices are $(a\cos t, a\sin t)$, $(b\sin t, -b\cos t)$ and $(1, 0)$.

Let the coordinates of centroid be

$$x=\frac{a\cos t+b\sin t+1}{3}$$

$$\Rightarrow \quad 3x-1=a\cos t+b\sin t \qquad \ldots\text{(i)}$$

and $$y=\frac{a\sin t-b\cos t+0}{3}$$

$$\Rightarrow \quad 3y=a\sin t-b\cos t \qquad \ldots\text{(ii)}$$

On squaring and adding Eqs. (i) and (ii), we get

$$(3x-1)^2+(3y)^2=a^2(\cos^2 t+\sin^2 t)+b^2(\sin^2 t+\cos^2 t)$$

$$\Rightarrow \quad (3x-1)^2+(3y)^2=a^2+b^2 \qquad [\because \sin^2\theta+\cos^2\theta=1]$$

9. (a) Let the vertices of ΔABC be $A(4, 0)$, $B(-1, -1)$ and $C(3, 5)$.

Now, $AB=\sqrt{(-1-4)^2+(-1-0)^2}=\sqrt{25+1}=\sqrt{26}$

$$BC=\sqrt{(3+1)^2+(5+1)^2}=\sqrt{4^2+6^2}=\sqrt{16+36}=\sqrt{52}$$

and $$CA=\sqrt{(4-3)^2+(0-5)^2}=\sqrt{1+25}=\sqrt{26}$$

$$\therefore \quad CA^2+AB^2=(\sqrt{26})^2+(\sqrt{26})^2=26+26=52=BC^2$$

$$\Rightarrow CA^2+AB^2=BC^2$$

Thus, the triangle is isosceles and right angled triangle.

10. (d) Let the vertices of ΔABC be $A(1, \sqrt{3})$, $B(0, 0)$ and $C(2, 0)$.

Again, let $a=BC=\sqrt{(2-0)^2+(0-0)^2}=2$

$$b=AC=\sqrt{(2-1)^2+(0-\sqrt{3})^2}=2$$

and $$c=AB=\sqrt{(0-1)^2+(0-\sqrt{3})^2}=2$$

Since, all sides of a triangle are equal, then the triangle is an equilateral triangle.

Also, incentre is same as centroid of the triangle.

Hence, coordinates of incentre are

$$\left(\frac{1+0+2}{3}, \frac{\sqrt{3}+0+0}{3}\right), \text{ i.e. } \left(1, \frac{1}{\sqrt{3}}\right)$$

2. Straight Line

DPP-1 Slopes and Types of Straight Line

1. (d) We know that, the equation of a line passing through two given points (x_1, y_1) and (x_2, y_2), is

$$y - y_1 = \frac{y_2 - y_1}{x_2 - x_1}(x - x_1)$$

Here, $(x_1, y_1) \equiv (4, -2)$ and $(x_2, y_2) \equiv (-1, 3)$

$\therefore$ Equation of line

$$(y + 2) = \frac{3 + 2}{-1 - 4}(x - 4)$$

$$\Rightarrow \quad y + 2 = \frac{5}{-5}(x - 4)$$

$$\Rightarrow \quad y + 2 = -x + 4$$

$$\therefore \quad x + y - 2 = 0$$

2. (b) Given equation

$$y - y_1 = m(x - x_1)$$

$$\Rightarrow \quad y - y_1 = mx - mx_1$$

$$\Rightarrow \quad y = mx - mx_1 + y_1 \quad \ldots(i)$$

Given, m and x_1 are fixed i.e. mx is a constant.

Let $\quad mx_1 = c$

$$\therefore \quad y = mx - c + y_1$$

Here, y_1 is variable, for different values of y_1, we have set of lines having slope 'm'.

Therefore, these line are parallel.

3. (c) Given lines are

$$ax - y - b = 0 \quad \ldots(i)$$

$$x - 2by + a = 0 \quad \ldots(ii)$$

Lines (i) and (ii) meet the point (1, 2)

$$\therefore \quad a - 2 - b = 0 \Rightarrow a - b = 2 \quad \ldots(iii)$$

and $\quad 1 - 4b + a = 0 \Rightarrow a - 4b = -1 \quad \ldots(iv)$

On solving Eqs. (iii) and (iv), we get

$$a = 3, b = 1$$

4. (a) The equation of the line passing through the points $\left(at_1, \frac{a}{t_1}\right)$ and $\left(at_2, \frac{a}{t_2}\right)$, is

$$\left(y - \frac{a}{t_1}\right) = \frac{\frac{a}{t_2} - \frac{a}{t_1}}{at_2 - at_1}(x - at_1)$$

$$\Rightarrow \quad \left(y - \frac{a}{t_1}\right) = \frac{a\left(\frac{t_1 - t_2}{t_1 t_2}\right)}{a(t_2 - t_1)}(x - at_1)$$

$$\Rightarrow \quad y - \frac{a}{t_1} = -\frac{1}{t_1 t_2}(x - at_1)$$

$$\Rightarrow \quad t_1 t_2 y - t_2 a = -x + at_1$$

$$\therefore \quad t_1 t_2 y + x = a(t_1 + t_2)$$

5. (c) A line parallel to X-axis, passes through $(-3, 2)$ intersect Y-axis at a distance 2 unit above the origin is

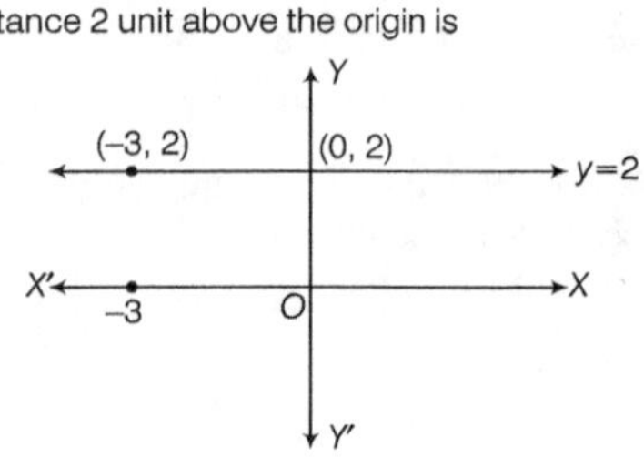

$$y = 2$$

$$\therefore \quad y - 2 = 0$$

6. (a) Given, vertices of rectangle are A (1, 3), C (5, 1) and slope of line BD is 2.

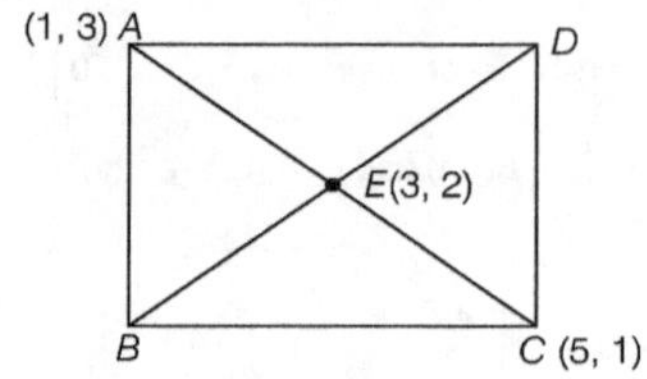

Let AC and BD intersect at point E.

$\therefore$ E will be mid-point of AC and BD, as diagonals of rectangle bisect each other.

$\therefore$ Coordinates of $E\left(\frac{1+5}{2}, \frac{3+1}{2}\right)$, i.e. E(3, 2).

$\therefore$ Equation of BD which is passes through (3, 2) and having slope 2.

$$(y - 2) = 2(x - 3)$$

$$\Rightarrow \quad y - 2 = 2x - 6$$

$$\therefore \quad 2x - y - 4 = 0$$

7. (b) Let the line intersect X-axis at $C(x_1, 0)$ and Y-axis at $D(0, y_1)$.

Given, A is mid-point of CD.

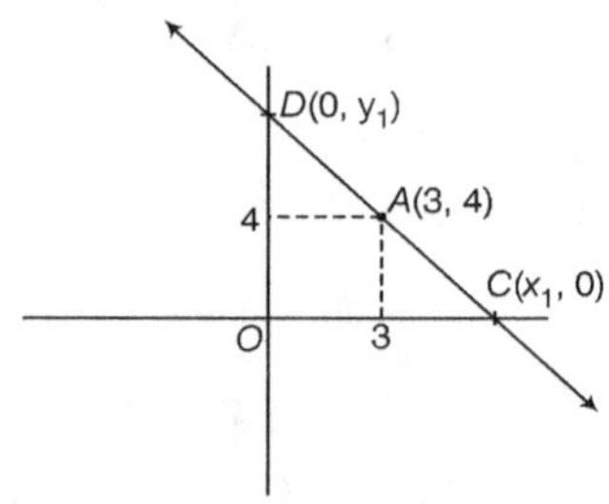

$$\therefore \quad \frac{x_1 + 0}{2} = 3 \Rightarrow x_1 = 6$$

$$\frac{y_1 + 0}{2} = 4 \Rightarrow y_1 = 8$$

Hence, coordinates of C and D are respectively (6, 0) and (0, 8).

Now, equation of CD is

$$\frac{x}{6} + \frac{y}{8} = 1 \quad \Rightarrow \quad \frac{4x + 3y}{24} = 1$$

$$\therefore \quad 4x + 3y = 24$$

8. (a) Let intercept made by the line on X-axis is k.

$\therefore$ Intercept on Y-axis is $2k$.

Now, equation of line in intercept form is

$$\frac{x}{k} + \frac{y}{2k} = 1 \quad \ldots(i)$$

Since, line (i) passes through the point (1, 2).

$$\therefore \quad \frac{1}{k} + \frac{2}{2k} = 1$$

$$\Rightarrow \quad \frac{2}{k} = 1$$

or $\quad k = 2$

$\therefore$ Required equation of line

$$\frac{x}{2} + \frac{y}{4} = 1$$

or $\quad 2x + y = 4$

9. (d) Let the line intersect X-axis at $(x_1, 0)$ and Y-axis at $(0, y_1)$.

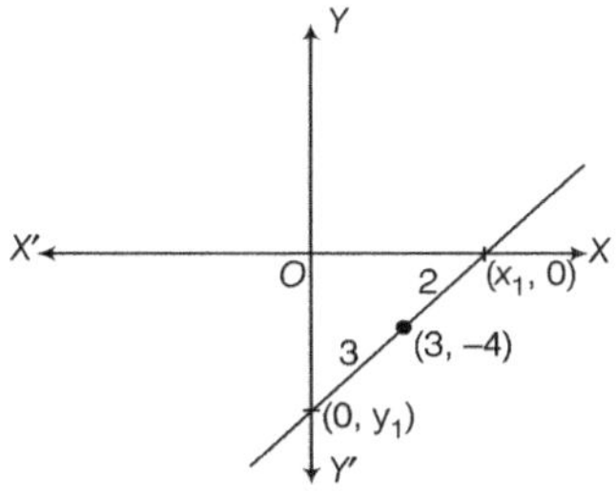

$\therefore \quad \frac{3x_1}{5} = 3 \quad \Rightarrow \quad x_1 = 5$

and $\quad \frac{2y_1}{5} = -4 \quad \Rightarrow \quad y_1 = -10$

$\therefore$ Equation of line

$$\frac{x}{5} + \frac{y}{-10} = 1$$

or $$\frac{x}{5} - \frac{y}{10} = 1$$

or $$2x - y = 10$$

10. (d) Let the line makes intercept a on X-axis and b on Y-axis.

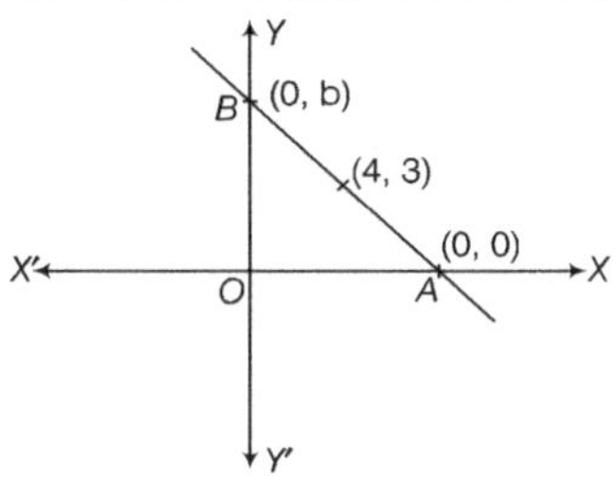

$\therefore$ Equation of line

$$\frac{x}{a} + \frac{y}{b} = 1 \quad \ldots \text{(i)}$$

It is given that

$$a + b = -1$$

$\Rightarrow \quad b = -(a + 1) \quad \ldots$(ii)

From Eqs. (i) and (ii), we have

$$\frac{x}{a} + \frac{y}{-(a+1)} = 1$$

or $$\frac{x}{a} - \frac{y}{a+1} = 1 \quad \ldots \text{(iii)}$$

Since, line (iii) passes through (4, 3).

$\therefore \quad \frac{4}{a} - \frac{3}{a+1} = 1$

$\Rightarrow \quad 4(a + 1) - 3a = a(a + 1)$

$\Rightarrow \quad 4a + 4 - 3a = a^2 + a$

$\Rightarrow \quad a + 4 = a^2 + a$

$\Rightarrow \quad a^2 - 4 = 0$

$\Rightarrow \quad a = \pm 2$

When $a = 2$, then $b = -3$

When $a = -2$, then $b = 1$

$\therefore$ Equation of such lines are

$$\frac{x}{2} - \frac{y}{3} = 1$$

or $$\frac{x}{-2} + \frac{y}{1} = 1$$

11. (c) Slope of the line,

$$m = \tan\theta = \tan 45^\circ = 1$$

$\therefore$ Equation of line having slope '1' is

$$y = 1 \cdot x + c$$

$\Rightarrow \quad y = x + c \quad \ldots$(i)

It is given that distance of origin from the line (i) is $\sqrt{2}$.

$\therefore \quad \sqrt{2} = \frac{|0 - 0 - c|}{\sqrt{2}}$

$\Rightarrow \quad c = 2 \quad \ldots$(ii)

From Eqs. (i) and (ii), we get

$$y = x + 2$$

$\therefore \quad y - x = 2$

12. (b) Let the equation of line in perpendicular form is

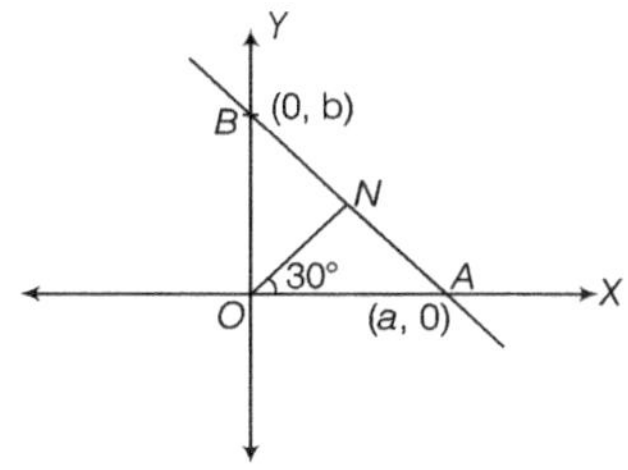

$$x\cos\alpha + y\sin\alpha = p$$

$\Rightarrow \quad x\cos 30^\circ + y\sin 30^\circ = p$

$$\frac{\sqrt{3}x}{2} + \frac{y}{2} = p$$

$\Rightarrow \quad \sqrt{3}x + y = 2p \quad \ldots$(i)

Again, let $OA = a$ and $OB = b$

Now, In ΔONA

$$\cos 30^\circ = \frac{p}{a}$$

$\Rightarrow \quad a = \frac{2p}{\sqrt{3}}$

In ΔONB,

$$\cos 60^\circ = \frac{p}{b}$$

$\Rightarrow \quad b = 2p$

Area of $\Delta ONA = \frac{50}{\sqrt{3}}$ (given)

$\Rightarrow \quad \frac{1}{2} \times OA \times OB = \frac{50}{\sqrt{3}}$

$\Rightarrow \quad \frac{1}{2} \times \frac{2p}{\sqrt{3}} \times 2p = \frac{50}{\sqrt{3}} \quad \left(\because NA = \frac{p}{\sqrt{3}}, ON = p\right)$

$\Rightarrow \quad p^2 = 25$

$\Rightarrow \quad p = \pm 5 \quad \ldots\ldots$(ii)

From Eqs. (i) and (ii), we have

$$\sqrt{3}x + y = \pm 10 \Rightarrow \sqrt{3}x + y \pm 10 = 0$$

13. (c) Equation of such line is

$$\frac{x-1}{\cos 60^\circ} = \frac{y-2}{\sin 60^\circ} = 3$$

$\Rightarrow \quad x = 1 + 3\cos 60^\circ$

and $\quad y = 2 + 3\sin 60^\circ$

$\Rightarrow \quad x = 1 + \frac{3}{2} = \frac{5}{2}$

and $\quad y = 2 + \frac{3\sqrt{3}}{2}$

$\therefore$ Coordinates of point are $\left(\frac{5}{2}, 2 + \frac{3\sqrt{3}}{2}\right)$.

14. (a) Slope of line, $m = \tan 45° = 1$

$\therefore$ Equation of line having slope '1' and passes through $P(1, 2)$, is

$$y - 2 = 1 \cdot (x - 1) \Rightarrow y - 2 = x - 1$$

$$\Rightarrow \quad x - y + 1 = 0 \quad \text{...(i)}$$

Line (i) meets with line

$$x + 2y - 7 = 0 \quad \text{...(ii)}$$

at point Q.

$\therefore$ Point of intersection of lines (i) and (ii) is $Q\left(\frac{5}{3}, \frac{8}{3}\right)$.

$$\therefore \quad PQ = \sqrt{\left(\frac{5}{3} - 1\right)^2 + \left(\frac{8}{3} - 2\right)^2} = \sqrt{\frac{4}{9} + \frac{4}{9}} = \frac{8}{9} = \frac{2\sqrt{2}}{3}$$

15. (c) Given equation of line,

$$x - y = 2 \quad \text{...(i)}$$

Since, the line (i) intersects X-axis at P.

$\therefore$ Coordinates of P are (2, 0).

Now, slope Eq. (i), $\quad m = 1$

Let θ be the angle subtended by line (i) with positive direction of X-axis.

$$\therefore \quad m = \tan\theta$$

$$\Rightarrow \quad \tan\theta = 1 \Rightarrow \theta = 45°$$

Since, the line (i) rotated through 45° in anti-clockwise direction.

$\therefore$ Slope of the line in new position,

$$m_1 = \tan 90° = \infty$$

Equation of line PQ in new position,

$$\frac{y - 0}{x - 2} = \infty \Rightarrow x = 2$$

16. (b) Given equation of line,

$$\sqrt{3}x + y = 0 \quad \text{...(i)}$$

and

$$\sqrt{3}x - y = 0 \quad \text{...(ii)}$$

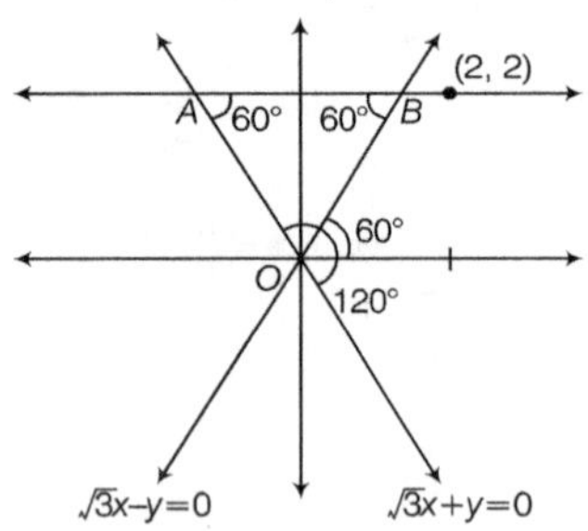

Slope of line (i), is

$$m_1 = -\sqrt{3} = \tan 120°$$

It makes 120° with X-axis.

Slope of line (ii), $\quad m_2 = \sqrt{3} = \tan 60°$

It makes 60° with X-axis.

A line passes through the point (2, 2) and intersect line (i) and (ii) at A and B, respectively such that ΔAOB is equilateral.

$\therefore$ This line will be parallel to X-axis.

Hence, equation of line is $y = 2$.

17. (d) Let $y = \cos x \cos(x + 2) - \cos^2(x + 1)$

$$= \cos(x + 1 - 1)\cos(x + 1 + 1) - \cos^2(x + 1)$$

$$= \cos^2(x + 1) - \sin^2 1 - \cos^2(x + 1)$$

$$[\because \cos^2 A - \sin^2 B = \cos(A + B)\cos(A - B)]$$

$$y = -\sin^2 1$$

This is a straight line which is parallel to X-axis.

Clearly, it passes through $\left(\frac{\pi}{2}, -\sin^2 1\right)$.

18. (a) Given lines are

$$3x + 4y = 9 \quad \text{...(i)}$$

and

$$y = mx + 1 \quad \text{...(ii)}$$

On solving Eqs. (i) and (ii), we get

$$x = \frac{5}{3 + 4m}, \; y = \frac{9m + 3}{4m + 3}$$

Now, for x to be an integer, 5 must be completely divisible by $(3 + 4m)$.

$$\therefore \quad 3 + 4m = \pm 5 \text{ or } \pm 1$$

If $\quad 3 + 4m = 5$ or -5

$$\Rightarrow \quad m = \frac{1}{2} \text{ or } -2$$

Now, $\quad m = -2$

If $\quad 3 + 4m = 1$ or -1

$$\Rightarrow \quad m = -\frac{1}{2} \text{ or } -1$$

$$\Rightarrow \quad m = -1$$

Hence, two values of m are possible.

19. (a) Let $A(1, 3)$, $B(5, 0)$ and $C(-1, 2)$ be three given points.

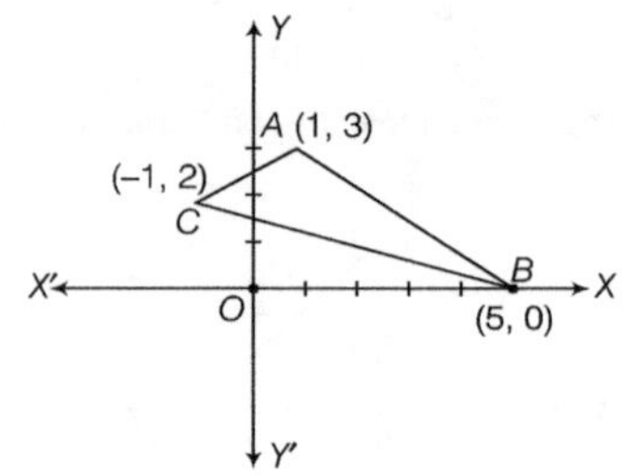

We have to find the inequality which is satisfied by all the points inside ΔABC.

$\therefore$ Among the given three inequalities which is satisfied by points A, B and C will be the correct answers.

By option (a), we have

$$3x + 2y \geq 0$$

$$A(1, 3): 3 \times 1 + 2 \times 3 = 9 \geq 0$$

$$B(5, 0): 15 + 0 = 15 \geq 0$$

$$C(-1, 2): -3 + 4 = 1 \geq 0$$

Hence, option (a) is correct.

20. (a) Let the line cut at X-axis at point $A(a, 0)$ and Y-axis at point $B(0, b)$.

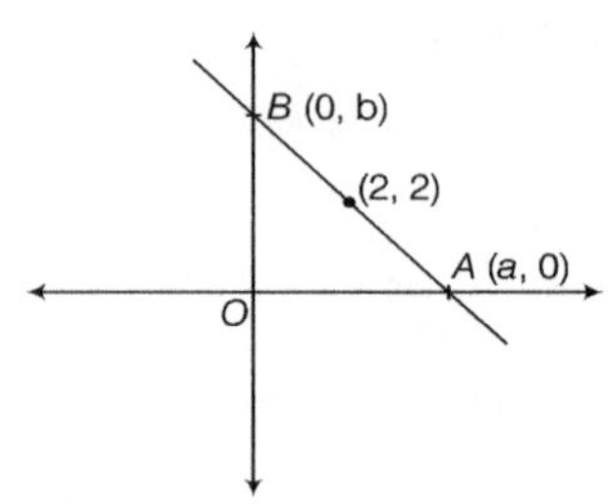

$\therefore$ Equation of line

$$\frac{x}{a} + \frac{y}{b} = 1 \quad \text{...(i)}$$

Since, line (i) passes through (2, 2)

$$\therefore \quad 2\left(\frac{1}{a} + \frac{1}{b}\right) = 1$$

$$\Rightarrow \quad \frac{2(a + b)}{ab} = 1 \quad \text{...(ii)}$$

Given, area $\Delta OAB = 9$ sq unit

$$\Rightarrow \quad \frac{1}{2}ab = 9$$

$$\Rightarrow \quad ab = 18 \quad \ldots\text{(iii)}$$

From Eqs. (ii) and (iii), we have

$$a + b = 9 \quad \ldots\text{(iv)}$$

Now, $(a-b)^2 = (a+b)^2 - 4ab = (9)^2 - 4 \times 18 = 9$

$$(a-b) = \pm 3 \quad \ldots\text{(v)}$$

On solving Eqs. (iv) and (v), we get

$$a = 6, b = 3 \quad \text{or} \quad a = 3, b = 6$$

Slope of line (i), is $\quad m = -\frac{b}{a}$

$$\Rightarrow \quad m = -\frac{1}{2} \quad \text{or} \quad -2$$

$$\text{Sum of slopes} = -\frac{1}{2} - 2 = -\frac{5}{2} = -2.5$$

21. (d) Since, diagonals of parallelogram $PQRS$ are perpendicular.

$\therefore$ $PQRS$ must be a rhombus.

22. (a) Given,

$$2(OA) + 3(OB) = 10$$

$$\Rightarrow \quad 2x_1 + 3y_1 = 10 \quad \ldots\text{(i)}$$

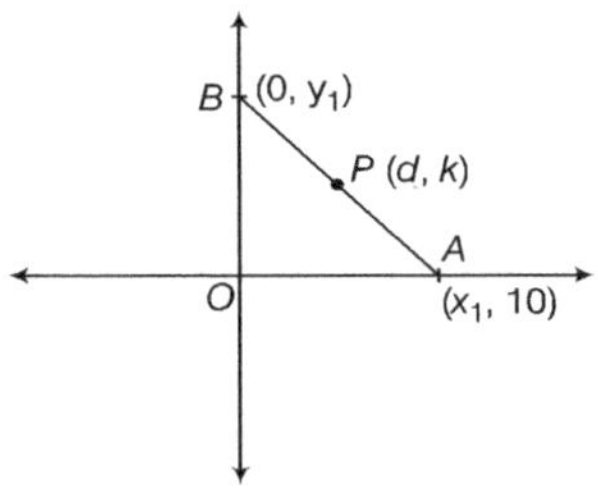

Let $P(h, k)$ be the mid-point of AB

$$\therefore \quad \frac{x_1}{2} = h \quad \Rightarrow \quad x_1 = 2h$$

$$\text{and} \quad \frac{y_1}{2} = k \quad \Rightarrow \quad y_1 = 2k$$

Substituting the values of x_1 and y_1 in Eq. (i), we get

$$2(2h) + 3(2k) = 10$$

$$\Rightarrow \quad 4h + 6k = 10$$

$$\Rightarrow \quad 2h + 3k = 5$$

$\therefore$ Required locus is

$$2x + 3y = 5$$

23. (a) Given lines are $\quad 3x + 4y = 0 \quad \ldots$(i)

$$4x + 3y = 0 \quad \ldots\text{(ii)}$$

and $\quad x = 3 \quad \ldots$(iii)

On taking mutually intersection of these lines, we get the vertices of triangle as

$$A(0, 0), B(3, -4) \text{ and } C\left(3, \frac{-9}{4}\right)$$

$$\therefore \quad AB = \sqrt{(3)^2 + (-4)^2} = 5 = c \quad \text{(say)}$$

$$BC = \sqrt{(3-3)^2 + \left(-\frac{9}{4} - 4\right)^2} = \frac{25}{4} = a \quad \text{(say)}$$

$$CA = \sqrt{(3-0)^2 + \left(\frac{-9}{4} - 0\right)^2} = \frac{15}{4} = b \quad \text{(say)}$$

$\therefore$ x-coordinate of incentre

$$= \frac{0 + 3 \times \frac{15}{4} + 3 \times 5}{\frac{25}{4} + \frac{15}{4} + 5} = \frac{\frac{105}{4}}{\frac{60}{4}} = \frac{105}{4} \times \frac{4}{60} = \frac{21}{12} = h$$

y-coordinate of incentre

$$= \frac{0 + \frac{15}{4} \times (-4) + 5 \times \left(\frac{-9}{4}\right)}{\frac{25}{4} + \frac{15}{4} + 5}$$

$$= \frac{-21}{12} = k$$

Now, $\quad h + k = \frac{21}{12} - \frac{21}{12} = 0$

24. (b) As, if $y = 0$

$$\Rightarrow \quad mx + b = 0$$

$$\therefore \quad x = -\frac{m}{b}, \quad \text{as } mb > 0$$

$$\Rightarrow \quad x < 0$$

Thus, (2009, 0) is not possible as x is +ve again, if $x = 0$.

$\Rightarrow y = b$. which can be +ve and – ve both.

Since independent of m.

25. (d) Given, equation of line,

$$y = mx + 2$$

$$\Rightarrow \quad \frac{x}{-\frac{2}{m}} + \frac{y}{2} = 1$$

It is given that x-intercept of the above line is greater than $\frac{1}{2}$.

$$\therefore \quad -\frac{2}{m} > \frac{1}{2}$$

$$\Rightarrow \quad -\frac{2}{m} - \frac{1}{2} > 0$$

$$\Rightarrow \quad -\left(\frac{2}{m} + \frac{1}{2}\right) > 0$$

$$\Rightarrow \quad -\left(\frac{4+m}{2m}\right) > 0$$

$$\Rightarrow \quad \frac{4+m}{2m} < 0$$

$$\Rightarrow \quad \text{Either } 4 + m > 0 \text{ and } 2m < 0$$

$$\text{or} \quad 4 + m < 0 \text{ and } 2m > 0$$

$$\Rightarrow \quad m > -4 \text{ and } m < 0$$

$$\therefore \quad m \in (-4, 0)$$

26. (c) Given,

$$4x^2 - y^2 + 2y - 1 = 0$$

$$\Rightarrow \quad 4x^2 - (y^2 - 2y + 1) = 0$$

$$\Rightarrow \quad 4x^2 - (y-1)^2 = 0$$

$$\Rightarrow \quad (2x + y - 1)(2x - y + 1) = 0$$

$\therefore$ Given equation represent pair of straight line, i.e.

$$2x + y - 1 = 0 \quad \ldots\text{(i)}$$

and $\quad 2x - y + 1 = 0 \quad \ldots$(ii)

Slope of line (i) is

$$m_1 = -2$$

Slope of line (ii) is

$$m_2 = 2$$

Hence, greatest slope is 2.

27. (c) Here, $\frac{x_1 + x_2}{2} = 5, \frac{y_1 + y_2}{2} = 6 \quad \ldots$(i)

Also, $\quad x_1 + 2y_1 = 6 \quad \ldots$(ii)

$$2x_2 + 3y_2 = 29 \quad \ldots\text{(iii)}$$

From Eqs. (ii) and (iii), we get

$$2(x_1 + x_2) + 4y_1 + 3y_2 = 41$$

$$\Rightarrow \quad 4y_1 + 3y_2 = 21 \quad \ldots\text{(iv)}$$

From Eqs. (i) and (iv), we get

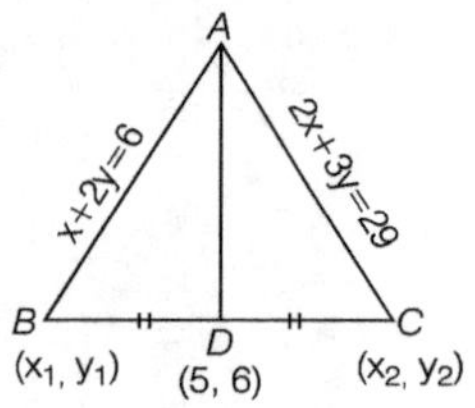

$$y_1 = -15 \quad \ldots(v)$$

From Eqs. (ii) and (v), we get

$$x_1 = 36 \quad \ldots(iv)$$

$\therefore$ $B(36, -15)$ and $D(5, 6)$

Hence, equation of BC is

$$21x + 31y = 291$$

28. (a) Since, P lies on X-axis, clearly Q also an X-axis and at the point, where $x = a$ intersect X-axis.

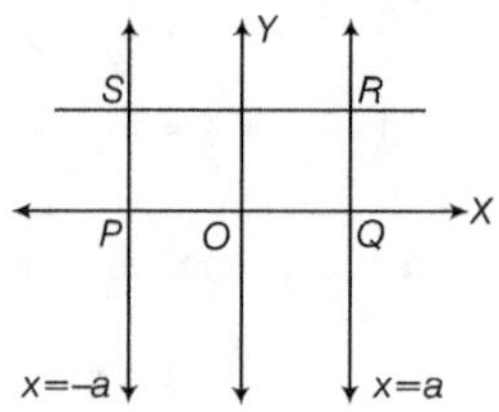

Similarly, S lies on the line $x = -a$, therefore R lies on $x = a$ (since, $SR \parallel PQ$).

Therefore, point R moving on the line $x = a$.

Hence, locus of R is a straight line.

29. (a)

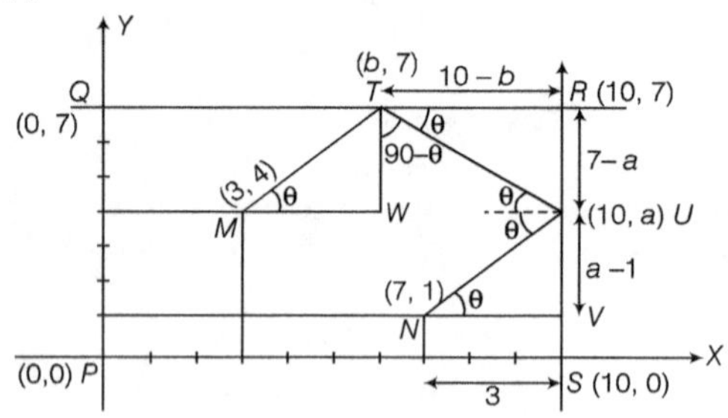

In ΔNVU,

$$\tan\theta = \frac{a-1}{3}$$

In ΔTRU, $\quad \tan\theta = \dfrac{7-a}{10-b}$

In ΔMWT, $\quad \tan\theta = \dfrac{7-4}{b-3} = \dfrac{3}{b-3}$

Hence, $\quad \dfrac{3}{b-3} = \dfrac{a-1}{3} = \dfrac{7-a}{10-b}$

From first two relations,

$$9 = ab - b - 3a + 3$$

$$\Rightarrow \quad 3a + 6 = ab - b \quad \ldots(i)$$

From last two relations,

$$10a - 10 - ab + b = 21 - 3a$$

$$ab - b = 13a - 31 \quad \ldots(ii)$$

From Eqs. (i) and (ii), we get

$$3a + 6 = 13a - 31$$

$$\Rightarrow \quad 10a = 37$$

$$\therefore \quad a = 3.7$$

30. (a) Equation of line AB,

$$\frac{x}{a} + \frac{y}{b} = 1 \quad \ldots(i)$$

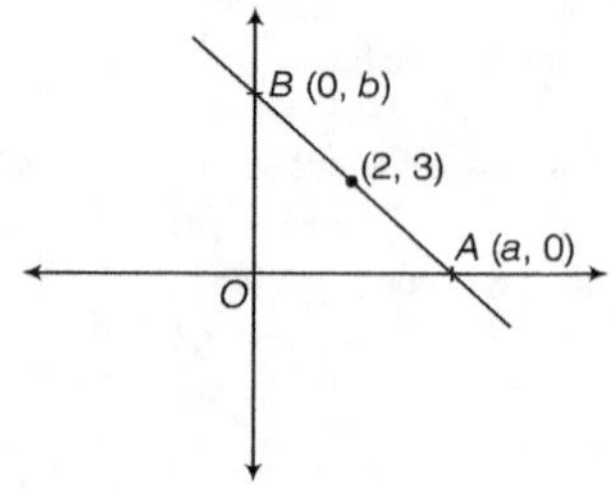

Since, point (2, 3) lies on line (i).

$$\therefore \quad \frac{2}{a} + \frac{3}{b} = 1$$

$$\Rightarrow \quad \frac{3a + 2b}{ab} = 1$$

$$\Rightarrow \quad 3a + 2b = ab$$

$$\Rightarrow \quad (3a + 2b)^2 = a^2b^2$$

$$\Rightarrow \quad 9a^2 + 4b^2 = a^2b^2 - 12ab \quad \ldots(ii)$$

Now, $\quad$ area of $\Delta AOB = 11$

$$\Rightarrow \quad \frac{1}{2}ab = 11$$

$$\Rightarrow \quad ab = 22 \quad \ldots(iii)$$

From Eqs. (ii) and (iii), we get

$$4b^2 + 9a^2 = (22)^2 - 12 \times 22$$

$$= 22\,(22 - 12)$$

$$= 22 \times 10 = 220$$

31. (a,b) Given, coordinates of point A and B are (3, 2) and (5, 4), respectively.

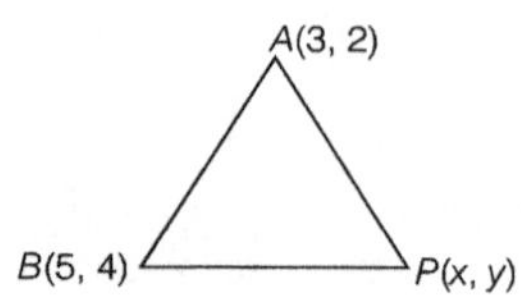

Since, ΔABP is equilateral.

$$\therefore \quad AB = AP = PB$$

Now, $\quad PA^2 = (x-3)^2 + (y-2)^2$

$$= x^2 + 9 - 6x + y^2 + 4 - 4y \quad \ldots(i)$$

$$PB^2 = (x-5)^2 + (y-4)^2$$

$$= x^2 + 25 - 10x + y^2 + 16 - 8y \quad \ldots(ii)$$

$$AB^2 = (5-3)^2 + (4-2)^2 = (2)^2 + (2)^2 = 8 \quad \ldots(iii)$$

$$\because \quad PA = PB$$

$$\Rightarrow \quad PA^2 = PB^2$$

$$\Rightarrow \quad x^2 + y^2 - 6x - 4y + 13 = x^2 + y^2 - 10x - 8y + 41$$

$$\Rightarrow \quad -6x + 10x - 4y + 8y = 41 - 13$$

$$\Rightarrow \quad 4x + 4y = 28$$

$$\Rightarrow \quad x + y = 7 \quad \ldots(iv)$$

$$\because \quad AB = AP$$

$$\Rightarrow \quad AB^2 = AP^2$$

$$\Rightarrow \quad 8 = x^2 + y^2 - 6x - 4y + 13$$

$$\Rightarrow \quad x^2 + y^2 - 6x - 4y + 5 = 0 \quad \ldots(v)$$

From Eqs. (iv) and (v), we get

$$x^2 + (7-x)^2 - 6x - 4(7-x) + 5 = 0$$

$$\Rightarrow \quad x^2 + 49 + x^2 - 14x - 6x - 28 + 4x + 5 = 0$$

$$\Rightarrow \quad 2x^2 - 16x + 26 = 0$$

$$x^2 - 8x + 13 = 0$$

$$\Rightarrow \quad x = \frac{8 \pm \sqrt{64 - 52}}{2} \quad \Rightarrow \quad x = \frac{8 \pm \sqrt{12}}{2}$$

$$\Rightarrow \quad x = 4 \pm \sqrt{3}$$

When $x = 4 + \sqrt{3} \Rightarrow y = 3 - \sqrt{3}$

When $x = 4 - \sqrt{3} \Rightarrow y = 3 + \sqrt{3}$

Therefore, coordinates of point P are $(4+\sqrt{3}, 3-\sqrt{3})$ or $(4-\sqrt{3}, 3+\sqrt{3})$.

32. (a,b) Given, equation of line are

$$2x + y = 5 \qquad \ldots(i)$$

$$x - 2y = 3 \qquad \ldots(ii)$$

The point of intersection of line (i) and (ii), is $A\left(\frac{13}{5}, \frac{-1}{5}\right)$

Given, $AB = AC = r$ (say)

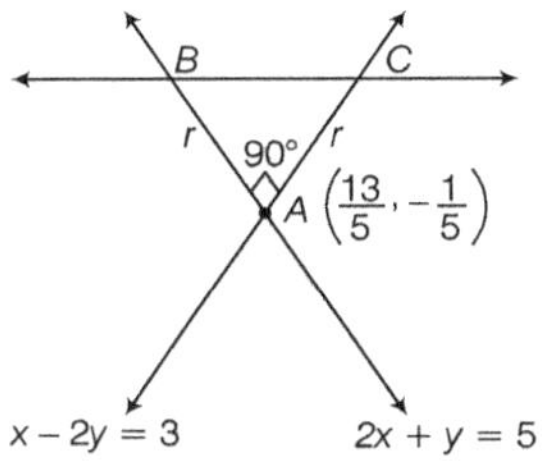

Coordinates of point C are $\left(\frac{13}{5} + \frac{2r}{\sqrt{5}}, -\frac{1}{5} + \frac{r}{\sqrt{5}}\right)$

and coordinates of point B are $\left(\frac{13}{5} - \frac{r}{\sqrt{5}}, \frac{-1}{5} + \frac{2r}{\sqrt{5}}\right)$.

$\therefore$ Equation of line BC is

$$\left(y + \frac{1}{5} - \frac{2r}{\sqrt{5}}\right) = \left(\frac{-\frac{1}{5} + \frac{r}{\sqrt{5}} + \frac{1}{5} - \frac{2r}{\sqrt{5}}}{\frac{13}{5} + \frac{2r}{\sqrt{5}} - \frac{13}{5} + \frac{r}{\sqrt{5}}}\right)\left(x - \frac{13}{5} + \frac{r}{\sqrt{5}}\right)$$

$$\Rightarrow \quad y + \frac{1}{5} - \frac{2r}{\sqrt{5}} = \frac{-r/\sqrt{5}}{3r/\sqrt{5}}\left(x - \frac{13}{5} + \frac{r}{\sqrt{5}}\right)$$

$$\Rightarrow \quad \left(y + \frac{1}{5} - \frac{2r}{\sqrt{5}}\right) = -\frac{1}{3}\left(x - \frac{13}{5} + \frac{r}{\sqrt{5}}\right)$$

$$\Rightarrow \quad 3y + \frac{3}{5} - \frac{6r}{\sqrt{5}} = -x + \frac{13}{5} - \frac{r}{\sqrt{5}}$$

$$\Rightarrow \quad x + 3y = \frac{13}{5} - \frac{3}{5} - \frac{r}{\sqrt{5}} + \frac{6r}{\sqrt{5}}$$

$$\Rightarrow \quad x + 3y = \frac{10}{5} + \frac{5r}{\sqrt{5}}$$

$$\Rightarrow \quad x + 3y = 2 + \sqrt{5}r \qquad \ldots(iii)$$

Since, line (iii) passes through (2, 3),

$$\therefore \quad 2 + 9 = 2 + \sqrt{5}r$$

$$\Rightarrow \quad r = \frac{9}{\sqrt{5}} \qquad \ldots(iv)$$

From Eqs. (iii) and (iv), we get

$$x + 3y = 2 + 9$$

$$\Rightarrow \quad x + 3y = 11$$

$$\Rightarrow \quad x + 3y - 11 = 0 \qquad \ldots(v)$$

Other position of line (v) will be perpendicular to the line (v) because line (i) and (ii) are perpendicular to each other.

$$\therefore \quad 3x - y = \lambda \qquad \ldots(vi)$$

Since, it passes through (2, 3)

$$\therefore \quad 3 \times 2 - 3 = \lambda \Rightarrow \lambda = 3$$

$\therefore$ Equation of line BC are

$$x + 3y - 11 = 0$$

$$3x - y - 3 = 0$$

33. (a,c) Given, equation of lines are

$$x + y = 0 \qquad \ldots(i)$$

$$3x + y - 4 = 0 \qquad \ldots(ii)$$

and

$$x + 3y - 4 = 0 \qquad \ldots(iii)$$

Taking mutual intersection of lines (i), (ii) and (iii), we get the vertices of triangles as (2, − 2), (1, 1) and (−2, 2).

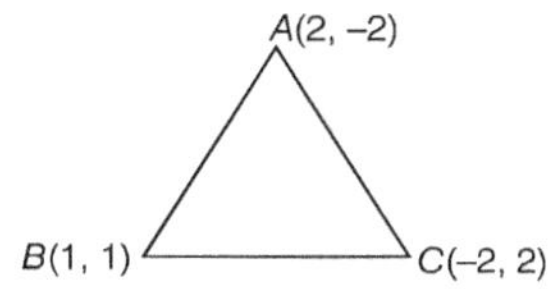

Now, $AB = \sqrt{(2-1)^2 + (-2-1)^2} = \sqrt{10}$

$$BC = \sqrt{(3)^2 + (-1)^2} = \sqrt{10}$$

$$AC = \sqrt{(2+2)^2 + (-2-2)^2} = \sqrt{32}$$

$$\because \quad AB = BC$$

$\Rightarrow$ ΔABC is isosceles

Hence, option (a) is correct.

Now, $\cos B = \frac{AB^2 + BC^2 - AC^2}{2AB \cdot BC}$

$$= \frac{10 + 10 - 32}{2 \times 10} = \frac{-12}{20} = -\frac{3}{5}$$

$$\cos B = \frac{-3}{5}$$

$$\Rightarrow \quad \angle B > 90° \qquad \left[\because \cos\theta < 0, \text{in } \theta \in \left(\frac{\pi}{2}, \frac{3\pi}{2}\right)\right]$$

$\Rightarrow$ ΔABC is obtuse angled triangle.

Hence, option (c) is correct.

34. (a,b) Given, $x^2 - 3|x| + 2 = 0 \qquad \ldots(i)$

$$y^2 - 3y + 2 = 0 \qquad \ldots(ii)$$

From Eq. (i), we have

$$|x|^2 - 3|x| + 2 = 0$$

$$\Rightarrow \quad |x| = 2, 1$$

If $|x| = 2$

$$\Rightarrow \quad x = \pm 2$$

and $|x| = 1$

$$\Rightarrow \quad x = \pm 1$$

From Eq. (ii), we have

$$y^2 - 3y + 2 = 0$$

$$\Rightarrow \quad y = 2, 1$$

$\therefore$ Coordinates of vertices of square are

(2, 2) (2, 1), (−2, 2), (−2, 1), (1, 2), (1, 1), (−1, 2) (−1. 1)

Hence, options (a) and (b) are correct.

35. (b,d) Given, coordinates of point $A(-5, 0)$, $B(3, 0)$ and point C lies on the line

$$x - y = 2 \qquad \ldots(i)$$

Let $C(h, k)$ be the third vertex of the ΔABC.

$$\therefore \quad h - k = 2 \qquad \ldots(ii)$$

Draw $CD \perp AB$

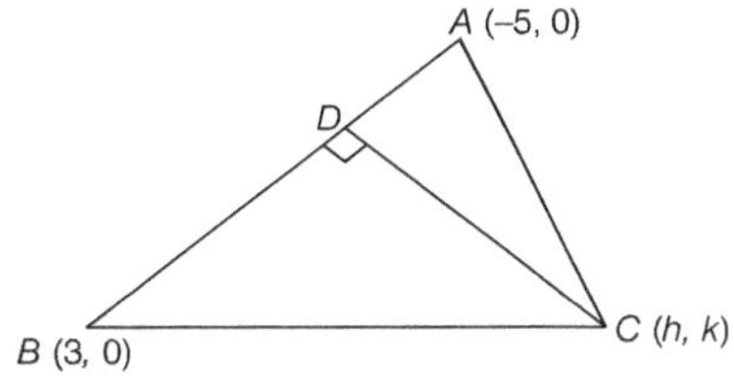

Now, $AB = \sqrt{(3+5)^2 + 0^2} = \sqrt{64} = 8$

Equation of AB,

$$y - 0 = \frac{0}{8}(x+5) \Rightarrow y = 0$$

$$CD = \left|\frac{k}{\sqrt{1}}\right| = k$$

Now, area of $\Delta ABC = 20$ (given)

$$\Rightarrow \left|\frac{1}{2} \times AB \times CD\right| = 20 \Rightarrow \frac{1}{2} \times 8 \times k = \pm 20$$

$$\Rightarrow k = \pm 5 \quad \ldots(\text{iii})$$

From Eqs. (ii) and (iii), we get

Coordinates of point C as $(7, 5)$ or $(-3, -5)$.

Hence, options (b) and (d) are correct.

36. (c)

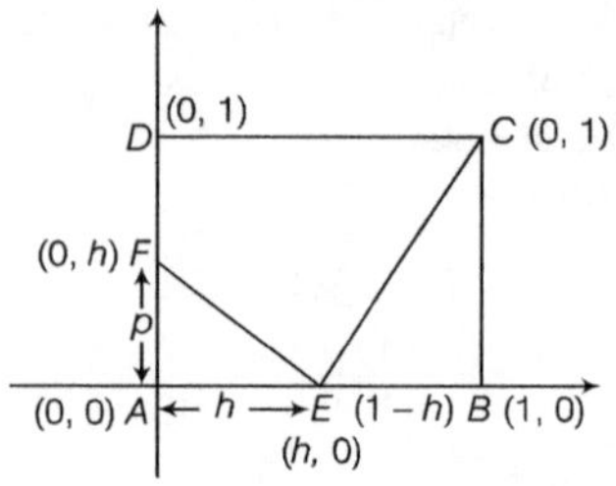

$$\text{Area of } CDFE = 1 - \frac{1}{2}h^2 - \frac{1}{2}(1-h)$$

$$= \frac{2 - h^2 - 1 + h}{2} = \frac{1 + h - h^2}{2}$$

At $h = \frac{1}{2}$, $\quad A_{\max} = \dfrac{1 + \frac{1}{2} - \frac{1}{4}}{2} = \dfrac{5}{8}$

37. (d) $PA^2 - PB^2 + PC^2 - PD^2$

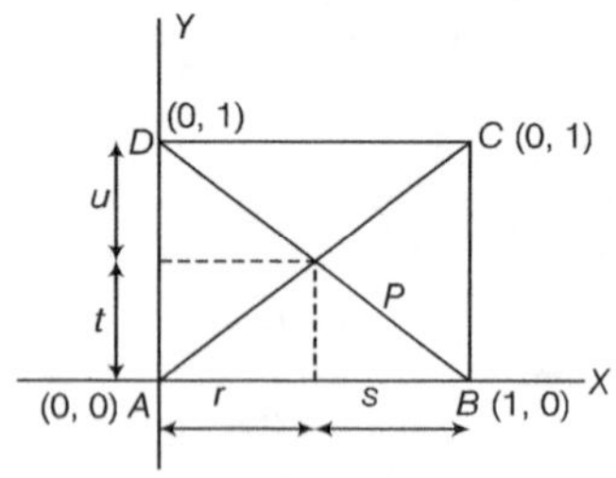

$$= (r^2 + t^2) - (t^2 + s^2) + (s^2 + u^2) - (u^2 + r^2)$$

$$= 0$$

38. (b) $\frac{1}{2} y \times 1 = \frac{1}{3} \times 1$

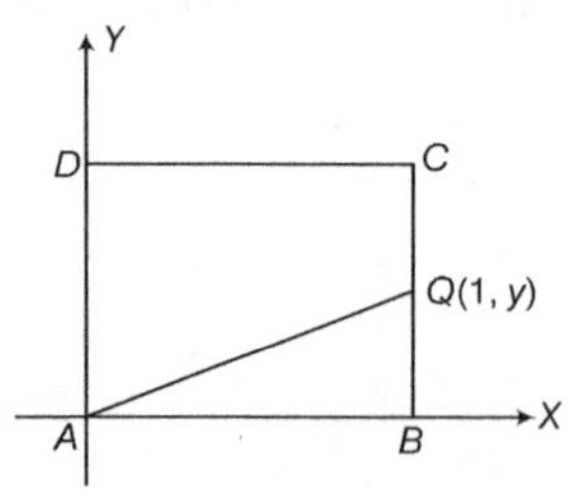

$$y = \frac{2}{3}$$

$$AQ = \sqrt{1^2 + y^2} = \sqrt{1 + \left(\frac{2}{3}\right)^2} = \sqrt{1 + \frac{4}{9}} = \sqrt{\frac{13}{9}} = \frac{\sqrt{13}}{3}$$

39. (0) The equation of a line through A. i.e. the point of intersection of AB and AC is

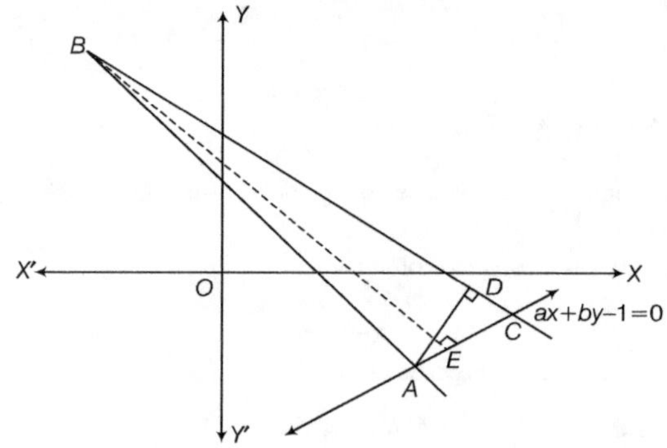

$$(2x + 3y - 1) + \lambda(ax + by - 1) = 0 \quad \ldots(\text{i})$$

It passes through $O(0, 0)$, then $\quad -1 - \lambda = 0$

$$\therefore \quad \lambda = -1$$

From Eq. (i), we get

$$2x + 3y - 1 - ax - by + 1 = 0$$

$$\Rightarrow (2-a)x + (3-b)y = 0$$

Since, $AD \perp BC$

$$\therefore \quad -\frac{(2-a)}{3-b} \times \left(-\frac{1}{2}\right) = -1$$

$$\Rightarrow \quad 2 - a = -6 + 2b$$

$$a + 2b = 8 \quad \ldots(\text{ii})$$

Similarly, $BE \perp AC$

$$\Rightarrow \quad a + b = 0 \quad \ldots(\text{iii})$$

Solving Eqs. (ii) and (iii), we get

$$b = 8, a = -8$$

$$\therefore \quad a + b = 0$$

40. (1) Given, lines are

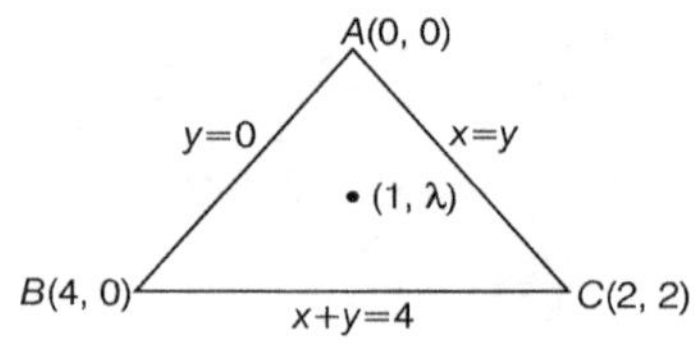

$$x = y \quad \ldots(\text{i})$$

$$y = 0 \quad \ldots(\text{ii})$$

$$x + y = 4 \quad \ldots(\text{iii})$$

The coordinates of vertices $A(0, 0)$, $B(4, 0)$ and $C(2, 2)$.

$\because$ Point $(1, \lambda)$ lies inside the ΔABC, then

(i) A and $(1, \lambda)$ must lies on the same side of BC.

(ii) B and $(1, \lambda)$ must lies on the same side of CA.

(iii) C and $(1, \lambda)$ must lies on the same side of AB.

Then,

(i) $\quad \dfrac{-4}{1 + \lambda - 4} > 0 \Rightarrow 1 + \lambda - 4 < 0$

$$\Rightarrow \quad \lambda < 3$$

(ii) $\quad \dfrac{4}{1 - \lambda} > 0 \Rightarrow 1 - \lambda > 0$

$$\Rightarrow \quad \lambda < 1$$

(iii) $\quad \dfrac{2}{\lambda} > 0 \Rightarrow \lambda > 0$

From Eqs. (i) and (ii), we get

$$\lambda \in (0, 1)$$

$$\Rightarrow \quad a + b = 0 + 1 = 1$$

DPP-2 Angle between Straight Lines, Image and Foot of Perpendicular

1. (i) Given, equations of lines are

$$x - y\sqrt{3} = 5 \quad \ldots\text{(i)}$$
$$\sqrt{3}x + y = 7 \quad \ldots\text{(ii)}$$

Slope of line (i), $m_1 = \frac{1}{\sqrt{3}}$

Slope of line (ii), $m_2 = -\sqrt{3}$

Let θ be the angle between the lines (i) and (ii),

$$\therefore \quad \tan\theta = \left|\frac{m_1 + m_2}{1 + m_1 m_2}\right| = \left|\frac{\frac{1}{\sqrt{3}} + \sqrt{3}}{1 - 1}\right|$$

$$= \infty \Rightarrow \theta = \frac{\pi}{2}$$

(ii) Given, equations of lines are

$$x - 4y = 3 \quad \text{and} \quad 6x - y = 11$$

Slope of line (i), $m_1 = \frac{1}{4}$

and slope of line (ii), $m_2 = 6$

Let θ be the angle between the lines, (i) and (ii)

$$\therefore \quad \tan\theta = \left|\frac{m_1 - m_2}{1 + m_1 m_2}\right| = \left|\frac{\frac{1}{4} - 6}{1 + \frac{6}{4}}\right| = \left|\frac{-\frac{23}{4}}{\frac{10}{4}}\right| = \left|-\frac{23}{10}\right|$$

$$\tan\theta = \frac{23}{10} \Rightarrow \theta = \tan^{-1}\left(\frac{23}{10}\right)$$

(iii) Given, equation of lines are

$$y = (2 - \sqrt{3})\,x + 5 \quad \ldots\text{(i)}$$

and $$y = (2 + \sqrt{3})x - 7 \quad \ldots\text{(ii)}$$

Slope of line (i), $m_1 = (2 - \sqrt{3})$

Slope of line (ii), $m_2 = (2 + \sqrt{3})$

Let θ be the angle between the lines (i) and (ii),

$$\therefore \quad \tan\theta = \left|\frac{m_1 - m_2}{1 + m_1 m_2}\right| = \left|\frac{2 - \sqrt{3} - 2 - \sqrt{3}}{1 + (2)^2 - (\sqrt{3})^2}\right|$$

$$= \left|\frac{-2\sqrt{3}}{2}\right| = \sqrt{3}$$

$$\theta = \tan^{-1}(\sqrt{3}) = \frac{\pi}{3}$$

(iv) Given, equation of lines are

$$y = 3x + 7 \quad \ldots\text{(i)}$$

and $$3y - x = 8 \quad \ldots\text{(ii)}$$

Slope of line (i), $m_1 = 3$

Slope of line (ii), $m_2 = 1/3$

Let θ be the angle between lines (i) and (ii)

$$\therefore \quad \tan\theta = \left|\frac{m_1 - m_2}{1 + m_1 m_2}\right| = \left|\frac{3 - \frac{1}{3}}{1 + 3 \times \frac{1}{3}}\right| = \left|\frac{\frac{8}{3}}{2}\right| = \frac{4}{3}$$

$$\theta = \tan^{-1}\left(\frac{4}{3}\right)$$

(v) Given, equation of lines are

$$(m^2 - mn)y = (mn + n^2)\,x + n^3 \quad \ldots\text{(i)}$$

and $$(mn + m^2)y = (mn - n^2)x + m^3 \quad \ldots\text{(ii)}$$

Slope of line (i), $m_1 = \left(\frac{mn + n^2}{m^2 - mn}\right)$

Slope of line (ii), $m_2 = \left(\frac{mn - n^2}{m^2 + mn}\right)$

Let θ be the angle between the (i) and (iii) lines

$$\therefore \quad \tan\theta = \left|\frac{m_1 - m_2}{1 + m_1 m_2}\right| = \left|\frac{\frac{mn + n^2}{m^2 - mn} - \frac{mn - n^2}{m^2 + mn}}{1 + \left(\frac{mn + n^2}{m^2 - mn}\right)\left(\frac{mn - n^2}{m^2 + mn}\right)}\right|$$

$$= \left|\frac{\frac{(m^2 + mn)(n^2 + mn) - (mn - n^2)(m^2 - mn)}{m^4 - (mn)^2}}{\frac{m^4 - (mn)^2 + (mn)^2 - (n)^4}{m^4 - (mn)^2}}\right|$$

$$= \left|\frac{m^2n^2 + m^3n + mn^3 + m^2n^2 - m^3n + m^2n^2 + m^2n^2 - mn^3}{m^4 - n^4}\right| = \left|\frac{4m^2n^2}{m^4 - n^4}\right|$$

$$\Rightarrow \quad \theta = \tan^{-1}\left|\frac{4m^2n^2}{m^4 - n^4}\right|$$

2. (i) Given lines are

$$x + y = 4 \quad \ldots\text{(i)}$$
$$4x + 3y = 10 \quad \ldots\text{(ii)}$$

Any point on the line (i) is (α, β).

$\therefore$ Distance of the point (α, β) from the line $4x + 3y = 10$ is 1

$$\Rightarrow \quad \left|\frac{4\alpha + 3\beta + 10}{5}\right| = 1$$

$$\Rightarrow \quad 4\alpha + 3\beta - 10 = \pm 5$$

$$4\alpha + 3\beta = 15 \quad \text{or} \quad 4\alpha + 3\beta = 5 \quad \ldots\text{(iii)}$$

Since, point (α, β) lies on line (i).

$$\therefore \quad \alpha + \beta = 4 \quad \ldots\text{(iv)}$$

On solving equations $4\alpha + 3\beta = 15$ and $\alpha + \beta = 4$, we get

$$\alpha = 3, \beta = 1$$

On solving equations $4\alpha + 3\beta = 5$ and $\alpha + \beta = 4$, we get

$$\alpha = -7, \beta = 11$$

$\therefore$ Sum of abscisse $= 3 - 7 = -4$

(ii) Since, PQ and RS is parallel to sides.

Thus, equations of PQ and RS are

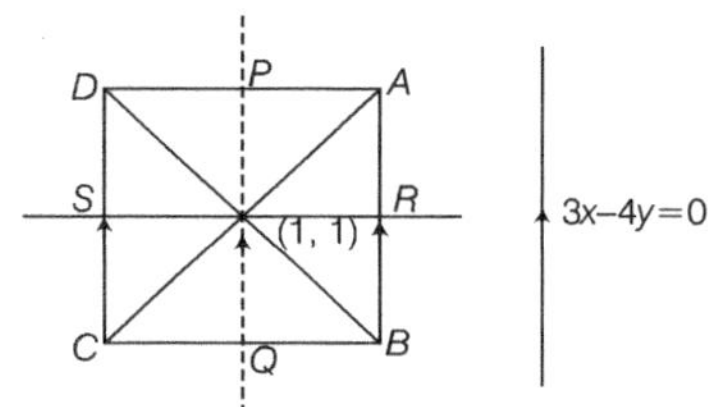

$3x - 4y = \lambda_1$ and $4x + 3y = \lambda_2$ passing through (1, 1)

$\therefore$ Equation of PQ is, $\quad 3x - 4y = -1$

and equation of RS is, $\quad 4x + 3y = 7$

Distance between $3x - 4y - 1 = 0$ and side $3x - 4y + \mu = 0$ is 2

$$\Rightarrow \quad \frac{|\mu + 1|}{\sqrt{3^2 + 4^2}} = 2 \quad \Rightarrow \quad \mu + 1 = \pm 10$$

$$\therefore \quad \mu = 11, -9$$

Now, AB and CD are

$$3x - 4y + 11 = 0$$

and $$3x - 4y - 9 = 0$$

Also, BC and AD are

$$4x + 3y + 3 = 0$$

and $$4x + 3y - 17 = 0$$

3. Consider the first situation when the line divide the join of $P(-3, 7)$ and $Q(5, -4)$ in the ratio 4 : 7 internally.

$\therefore$ Coordinates of point R are

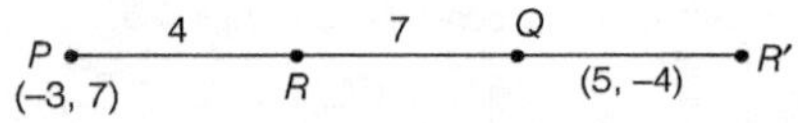

$$\left(\frac{4\times 5-7\times 3}{4+7}, \frac{-4\times 4+7\times 7}{4+7}\right) \text{ i.e. } \left(-\frac{1}{11}, 3\right)$$

$\therefore$ Equation of line passing through $\left(-\frac{1}{11}, 3\right)$ and perpendicular to PQ is

$$y-3=\frac{8}{11}\left(x+\frac{1}{11}\right)$$

$$\Rightarrow \quad 11y-33=8x+\frac{8}{11}$$

$$\Rightarrow \quad 121y-363=88x+8$$

$$\Rightarrow \quad 121y-88x=371$$

Again, consider the second situation when the line divides join of P and Q in the ratio 4 : 7 externally.

$\therefore$ Coordinates of point R' are

$$\left(\frac{4\times 5+7\times 3}{4-7}, \frac{-4\times 4-7\times 7}{4-7}\right) \text{ i.e. } \left(-\frac{41}{3}, \frac{65}{3}\right)$$

$\therefore$ Equation of line passing through $\left(-\frac{41}{3}, \frac{65}{3}\right)$ and perpendicular to PQ is

$$y-\frac{65}{3}=\frac{8}{11}\left(x+\frac{41}{3}\right)$$

$$\Rightarrow \quad 11y-\frac{715}{3}=8x+\frac{328}{3}$$

$$\Rightarrow \quad 33y-715=24x+328$$

$$\Rightarrow \quad 33y-24x=1043$$

4. (c) Given, points are $A(-1, 5)$, $B(0, 0)$ and $C(2, 2)$ and D is the mid-point of BC.

$\therefore$ Coordinates of D are $(1, 1)$.

Now, equation of line AD is

$$(y-5)=\frac{1-5}{1+1}(x+1)$$

$$\Rightarrow \quad y-5=-\frac{4}{2}(x+1)$$

$$\Rightarrow \quad y-5=-2(x+1)$$

$$\Rightarrow \quad 2x+y-3=0 \quad \ldots\text{(i)}$$

Slope of AD, $\quad m_1=-2$

$\therefore$ Slope of line perpendicular to AD,

$$m_2=\frac{1}{2}$$

Equation of a line perpendicular to AD and passes through point $(0, 0)$ is

$$(y-0)=\frac{1}{2}(x-0)$$

$$\Rightarrow \quad x-2y=0$$

5. (a) We know that, the equation of a line passes through point of intersection of two lines $a_1x+b_1y+c_1=0$ and $a_2x+b_2y+c_2=0$ is $(a_1x+b_1y+c_1)+\lambda(a_2x+b_2y+c_2)=0$

$\therefore$ Equation of a line passes through point of intersection of lines $y=x+7$ and $x+2y+1=0$ is

$$(x-y+7)+\lambda(x+2y+1)=0$$

$$\Rightarrow \quad (1+\lambda)x+(2\lambda-1)y+\lambda+7=0 \quad \ldots\text{(i)}$$

and $\quad 5x-2y+7=0 \quad \ldots$(ii)

Slope of line (i), $\quad m_1=\frac{-(1+\lambda)}{2\lambda-1}$

Slope of line (ii), $\quad m_2=\frac{5}{2}$

Since, lines (i) and (ii) are perpendicular to each other.

$$\therefore \quad m_1\cdot m_2=-1$$

$$-\frac{(1+\lambda)}{2\lambda-1}\times\frac{5}{2}=-1$$

$$\Rightarrow \quad 5+5\lambda=4\lambda-2 \Rightarrow \lambda=-7$$

Substituting the value of λ in line (i), we get

$$-6x-15y=0$$

$$\Rightarrow \quad 2x+5y=0$$

6. (d) Given, equation of line,

$$x\sec\theta+y\,\text{cosec}\,\theta=a \quad \ldots\text{(i)}$$

Slope of line (i), $\quad m_1=-\frac{\sec\theta}{\text{cosec}\,\theta}=-\tan\theta$

Let m_2 be the slope of required line.

$$\because \quad m_1\cdot m_2=-1$$

$$\therefore \quad m_2=-\frac{1}{-\tan\theta}=\cot\theta$$

Now, equation of a line passes through $(a\cos^3\theta, a\sin^3\theta)$ and perpendicular to the line (i) is

$$(y-a\sin^3\theta)=\cot\theta\,(x-a\cos^3\theta)$$

$$\Rightarrow \quad y\sin\theta-a\sin^4\theta=x\cos\theta-a\cos^4\theta$$

$$\Rightarrow \quad x\cos\theta-y\sin\theta=a(\sin^4\theta-\cos^4\theta)$$

$$\Rightarrow \quad x\cos\theta-y\sin\theta=a(\sin^2\theta+\cos^2\theta)\,(\sin^2\theta-\cos^2\theta)$$

$$\Rightarrow \quad x\cos\theta-y\sin\theta=a\cos2\theta \quad \left(\begin{array}{l}\because \sin^2\theta+\cos^2\theta=1,\\ \sin^2\theta-\cos^2\theta=\cos2\theta\end{array}\right)$$

7. (d) Let $A(2a, 0)$ and $B(0, a)$ be two ends of the base of isosceles ΔPAB.

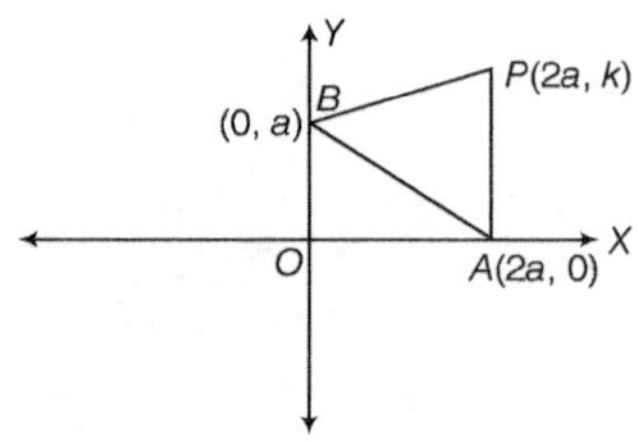

Since, side PA is parallel to Y-axis

$\therefore$ Coordinates of points P are $(2a, k)$

Now, $\quad PA=k$

$$PB=\sqrt{(2a-0)^2+(k-a)^2}=\sqrt{4a^2+(k-a)^2}$$

$$PA=PB \quad (\because \Delta PAB \text{ is isosceles})$$

$$\Rightarrow \quad PA^2=PB^2$$

$$\Rightarrow \quad k^2=4a^2+k^2+a^2-2ak$$

$$\Rightarrow \quad 5a^2-2ak=0$$

$$\Rightarrow \quad a(5a-2k)=0$$

$$\Rightarrow \quad 5a-2k=0 \quad (\because a\neq 0)$$

$$\Rightarrow \quad k=\frac{5a}{2}$$

$\therefore$ Equation of line PB is $\quad y-a=\frac{3a}{4a}(x-0)$

$$\Rightarrow \quad 3x-4y+4a=0$$

8. (a) Given, equations of line are

$$(\beta-\gamma)x+(\gamma-\alpha)y+(\alpha-\beta)=0 \quad \ldots\text{(i)}$$

and $\quad (\beta^3-\gamma^3)x+(\gamma^3-\alpha^3)y+(\alpha^3-\beta^3)=0 \quad \ldots$(ii)

Since, Eqs. (i) and (ii) represent same line.

$$\therefore \quad \frac{\beta-\gamma}{\beta^3-\gamma^3}=\frac{\gamma-\alpha}{\gamma^3-\alpha^3}=\frac{\alpha-\beta}{\alpha^3-\beta^3}$$

From first two relations, we have

$$\frac{\beta-\gamma}{\beta^3-\gamma^3}=\frac{\gamma-\alpha}{\gamma^3-\alpha^3}$$

$\Rightarrow \quad \dfrac{1}{\beta^2+\alpha^2+\alpha\beta}=\dfrac{1}{\gamma^2+\alpha^2+\alpha\gamma}$

$\Rightarrow \quad \beta^2+\alpha^2+\alpha\beta=\gamma^2+\alpha^2+\alpha\gamma$

$\Rightarrow \quad \beta^2+\alpha\beta-\gamma^2-\alpha\gamma=0$

$\Rightarrow \quad \beta^2-\gamma^2+\alpha(\beta-\gamma)=0$

$\Rightarrow \quad (\beta+\gamma)(\beta-\gamma)-\alpha(\beta-\gamma)=0$

$\Rightarrow \quad (\beta-\gamma)(\beta+\gamma+\alpha)=0$

Either $\quad \beta=\gamma$

or $\quad \alpha+\beta+\gamma=0$

$\therefore \quad \alpha+\beta+\gamma=0 \qquad (\because \beta\neq\gamma)$

9. (d) Given, equation of lines,

$3x-y=3 \qquad \dots(i)$

$3x-y=5 \qquad \dots(ii)$

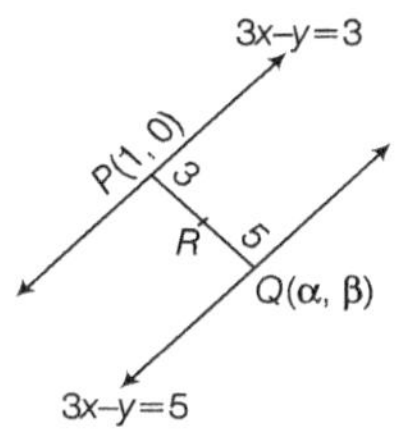

Put $y=0$ in line (i), we get

$3x=3$

$x=1$

$\therefore$ Point $(1, 0)$ lies on the line (i),

Let coordinates of point Q is (α, β), it satisfies Eq. (ii)

$\therefore \quad 3\alpha-\beta=5 \qquad \dots(iii)$

Now, slope of line (ii), $\quad m_1=3$

and slope of PQ is

$$m_2=\frac{\beta}{\alpha-1}$$

Since, PQ is perpendicular to $3x-y=5$.

$\therefore \quad m_1m_2=-1$

$\Rightarrow \quad \dfrac{\beta}{\alpha-1}=-\dfrac{1}{3} \Rightarrow 3\beta=1-\alpha$

$\Rightarrow \quad \alpha+3\beta=1 \qquad \dots(iv)$

On solving Eqs. (iii) and (v), we get

$$\alpha=\frac{8}{5}, \beta=\frac{-1}{5}$$

Let point $R(x, y)$ divides PQ in the ratio 3 : 5.

$\therefore \quad x=\dfrac{3\times\frac{8}{5}+5}{8}=\dfrac{49}{40}$

and $\quad y=\dfrac{3\times(-1/5)}{8}=-\dfrac{3}{40}$

$\therefore$ Any line parallel to lines (i) and (ii) is

$3x-y=\lambda \qquad \dots(v)$

Since, it passes through point $R\left(\dfrac{49}{40}, \dfrac{-3}{40}\right)$

$\therefore \quad 3\times\dfrac{49}{40}+\dfrac{3}{40}=\lambda \Rightarrow \lambda=\dfrac{15}{4}$

Substituting the value of λ in Eq. (v), we get

$$3x-y=\frac{15}{4}$$

$\Rightarrow \quad 12x-4y-15=0$

Alternate Method

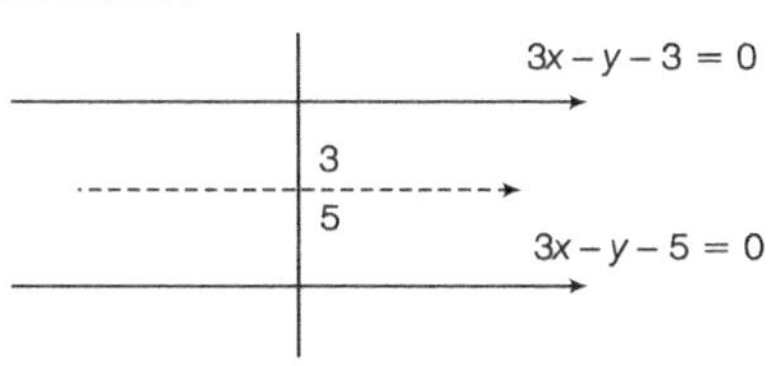

Required equation, is

$$\frac{3(3x-y-5)+5(3x-y-3)}{3+5}=0$$

$\Rightarrow \quad 24x-8y-30=0$

or $\quad 12x-4y-15=0$

10. (d) Given, equation of line is

$$\frac{x}{a}-\frac{y}{b}=1 \qquad \dots(i)$$

For point of intersection of line (i) with X-axis, put $y=0$ in line (i), we get

$$\frac{x}{a}=1 \Rightarrow x=a$$

$\therefore$ It intersect X-axis at $(a, 0)$.

Now, the equation of line perpendicular to line (i) is

$$\frac{x}{b}+\frac{y}{a}=\lambda; \forall \lambda\in R \qquad \dots(ii)$$

Line (ii) is passes through the point $(a, 0)$.

$\therefore \quad \dfrac{a}{b}=\lambda$

Substituting the value of λ in Eq. (ii), we get

$$\frac{x}{b}+\frac{y}{a}=\frac{a}{b}$$

11. (d) Slope of line segment $PQ=\dfrac{3-4}{k-1}=\dfrac{-1}{k-1}$

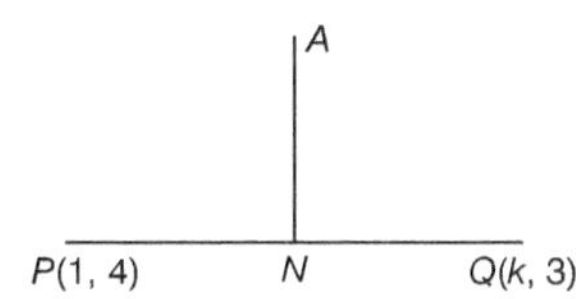

$\therefore$ Slope of $AN=(k-1)$

Now, equation of AN is

$$y-\frac{7}{2}=(k-1)\left[x-\left(\frac{k+1}{2}\right)\right]$$

For y-intercept, $x=0$, $y=-4$

$$-4-\frac{7}{2}=(k-1)\left[-\frac{(k+1)}{2}\right]$$

$\Rightarrow \quad -\dfrac{15}{2}=-\left(\dfrac{k^2-1}{2}\right)$

$\Rightarrow \quad k^2-1=15 \quad \Rightarrow k^2=16$

$\Rightarrow \quad k=\pm 4$

12. (a) We know that, lines perpendicular to the same line are parallel to each other. Therefore, slope of both the lines will be same.

i.e. $\quad p(p^2+1)=-(p^2+1) \quad \Rightarrow \quad p=-1$

$\therefore$ There is exactly one value of p.

13. (a) Equation of a line passing through the intersection of lines $ax+2by+3b=0$ and $bx-2ay-3a=0$ is

$$(ax+2by+3b)+\lambda(bx-2ay-3a)=0$$

$\Rightarrow \quad (a+\lambda b)x+(2b-2a\lambda)y+3b-3a\lambda=0 \qquad \dots(i)$

This line is parallel to X-axis

$\therefore$ Coefficient of x must be zero.

i.e. $\quad a+\lambda b=0 \quad \Rightarrow \quad \lambda=-\dfrac{a}{b}$

On putting this value in Eq. (i), we get

$$2(b^2+a^2)y+3(b^2+a^2)=0$$

$\Rightarrow \quad y=-\dfrac{3}{2}$

Negative sign shows that the line is below X-axis and at a distance 3/2 from it .

Alternate Method

Equation of given lines are

$$ax + 2by + 3b = 0 \quad \dots(i)$$

and $$bx - 2ay - 3a = 0 \quad \dots(ii)$$

On solving Eqs. (i) and (ii), we get the point of intersection as

$$x = 0,\ y = -\frac{3}{2}$$

Also, it is given that the required line is parallel to X-axis

$\therefore$ $m = 0$

Now, equation of a line passing through $\left(0, -\frac{3}{2}\right)$ having slope $m = 0$ is

$$(y + 3/2) = 0\,(x - 0)$$

$\Rightarrow$ $y = -3/2$

Thus, the required line is below X-axis at a distance $\frac{3}{2}$ from it.

14. (d) Given, equation of lines

$$L : \frac{x}{5} + \frac{y}{b} = 1 \quad \dots(i)$$

and $$K : \frac{x}{c} + \frac{y}{3} = 1 \quad \dots(ii)$$

Since, the line L passing through the point (13, 32)

$\therefore$ $\frac{13}{5} + \frac{32}{b} = 1 \Rightarrow b = -20$

The line 'K' is parallel to 'L', then its equation must be

$$\frac{x}{5} - \frac{y}{20} = \lambda$$

On comparing this equation with the given equation K, we get

$$20\lambda = -3$$

$\Rightarrow$ $\lambda = -\frac{3}{20}$

and $c = 5\lambda$

$\Rightarrow$ $c = 5 \times \left(-\frac{3}{20}\right)$

$\Rightarrow$ $c = -3/4$

Therefore, the distance between the lines L and K

$$= \left|\frac{\lambda - 1}{\sqrt{\frac{1}{25} + \frac{1}{400}}}\right| = \left|\frac{-\frac{3}{20} - 1}{\sqrt{\frac{17}{400}}}\right| = \frac{23}{\sqrt{17}}$$

15. (c) Given, equation of lines are

$$x + y = |a| \quad \dots(i)$$

$$ax - y = 1 \quad \dots(ii)$$

$\therefore$ Point of intersection of lines (i) and (ii) is

$$\left(\frac{1 + |a|}{1 + a}, \frac{a|a| - 1}{a + 1}\right)$$

Since, lines (i) and (ii) intersect in first quadrant.

$\therefore$ Both x and y coordinates are positive

i.e. $\frac{1 + |a|}{1 + a} \ge 0$

and $\frac{a|a| - 1}{a + 1} \ge 0$

$\Rightarrow$ $1 + |a| \ge 0$

and $a|a| - 1 \ge 0$

$\Rightarrow$ $a > -1$

and $a|a| \ge 1 \quad \dots(i)$

If $-1 < a < 0 \Rightarrow -a^2 > 1$ (not possible)

If $a \ge 0 \Rightarrow a^2 \ge 1 \Rightarrow a \ge 1$

$\therefore$ $a \in [1, \infty)$

16. (d) Given, a line through (1, 2) meets the coordinate axes at P and Q, then the area of ΔOPQ thus obtained is minimum.

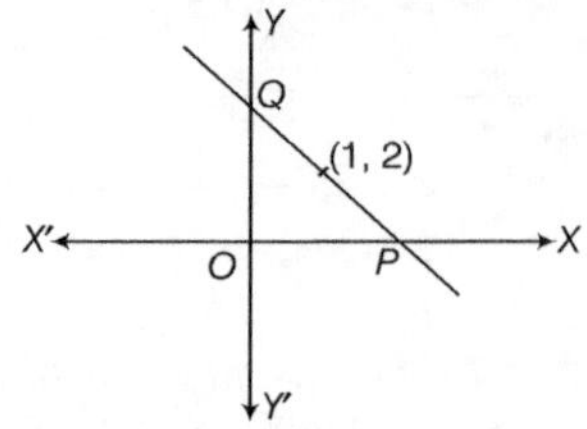

Let m be the slope of line PQ, then the equation of PQ is

$$y - 2 = m(x - 1)$$

Now, PQ meets X-axis at $P\left(1 - \frac{2}{m}, 0\right)$

and Y-axis at $Q(0, 2 - m)$

$\therefore$ $OP = 1 - \frac{2}{m}$ and $OQ = 2 - m$

Then, area of $\Delta OPQ = \frac{1}{2} OP \times OQ = \frac{1}{2} \times \left|\left(1 - \frac{2}{m}\right) \times (2 - m)\right|$

$$= \frac{1}{2}\left|2 - m - \frac{4}{m} + 2\right| = \frac{1}{2}\left|4 - \left(m + \frac{4}{m}\right)\right|$$

Let $f(m) = \frac{1}{2}\left[4 - \left(m + \frac{4}{m}\right)\right]$

$$f(m) = 2 - 2m - \frac{8}{m}$$

$\Rightarrow$ $f'(m) = -2 + \frac{8}{m^2}$

For maximum or minimum,

$f'(m) = 0 \Rightarrow -2 + \frac{8}{m^2} = 0$

$\Rightarrow$ $m^2 = 4 \Rightarrow m = \pm 2$

$$f''(m) = \frac{-16}{m^3}$$

$f''(2) = \frac{-16}{8} - 2 < 0$ (maximum)

$f''(-2) = -\frac{-16}{-8} = 2 > 0$ (minimum)

$\therefore$ Area is minimum or least, when $m = -2$.

17. (c) If $x - 3y = p$ and $ax + 2y = q$ are perpendicular, then

$\frac{1}{3} \cdot \left(-\frac{a}{2}\right) = -1$ i.e. $a = 6 \quad \dots(i)$

If $x - 3y = p$ and $ax + y = r$ are perpendicular, then

$\frac{1}{3} \cdot (-a) = -1$ i.e. $a = 3 \quad \dots(ii)$

$ax + 2y = q$ and $ax + y = r$ cannot be perpendicular, as

$-\frac{a}{2} \cdot (-a) = -1 \Rightarrow a^2 = -2$ (not possible)

Thus, $a = 6$ or $a = 3$

$\Rightarrow$ $(a - 6)(a - 3) = 0$

or $a^2 - 9a + 18 = 0$

18. (d) Given equation of line, $3x - 4y - 5 = 0 \quad \dots(i)$

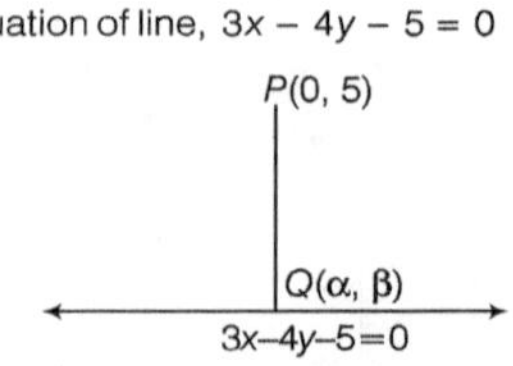

Let coordinates of foot of perpendicular are $Q(\alpha, \beta)$.

$\therefore$ $3\alpha - 4\beta = 5 \quad \dots(ii)$

Slope of line (i), $m_1 = \frac{3}{4}$

Slope of PQ, $m_2 = \frac{\beta - 5}{\alpha}$

Since, PQ is perpendicular on the line $3x - 4y - 5 = 0$

$\therefore$ $m_1 \cdot m_2 = -1$

$\Rightarrow$ $\frac{3}{4}\left(\frac{\beta - 5}{\alpha}\right) = -1$

$\Rightarrow$ $3(\beta - 5) = -4\alpha$

$\Rightarrow$ $5\alpha + 3\beta = 15$ ...(iii)

On solving Eqs. (i) and (iii), we get

$$\alpha = 3, \beta = 1$$

Hence, coordinates of foot of perpendicular are (3, 1).

19. (d) Given, lines are $2x + 3y = 7$ ...(i)

$$2x + 3y = 12 \quad \text{...(ii)}$$

Length of perpendicular from $A(3, -5)$ to the line (i) is

$$\left|\frac{2 \times 3 - 3 \times 5 - 7}{\sqrt{13}}\right| = \frac{16}{\sqrt{13}} = p_1 \quad \text{(say)}$$

Again, length of perpendicular from $A(3, -5)$ to line (ii) is

$$\left|\frac{2 \times 3 - 3 \times -12}{\sqrt{13}}\right| = \frac{21}{\sqrt{3}} = p_2 \quad \text{(say)}$$

Now, distance between lines (i) and (ii) is

$$d = \left|\frac{7 - 12}{\sqrt{13}}\right| = \frac{5}{\sqrt{13}}$$

Also, $p_1 + p_2 = \frac{16}{\sqrt{13}} + \frac{21}{\sqrt{13}} = \frac{37}{\sqrt{13}} \neq d$

$\therefore$ A does not lies between the lines and sum of perpendicular from A to the line $\neq \frac{5}{\sqrt{13}}$.

Hence, (d) is correct option.

20. (b) Given, equation of lines, are

$$3x + 4y + 2 = 0 \quad \text{...(i)}$$
$$3x + 4y + 5 = 0 \quad \text{...(ii)}$$
$$3x + 4y - 5 = 0 \quad \text{...(iii)}$$

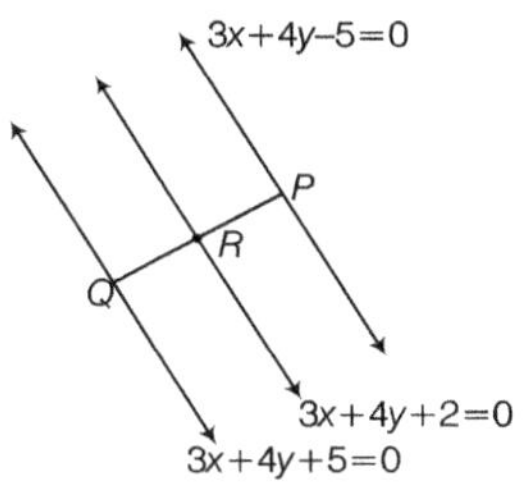

Any point P on the line can be find by putting $y = 0$ in

$$3x + 4y - 5 = 0$$

$\Rightarrow$ $3x + 4 \times 0 - 5 = 0$

$\Rightarrow$ $x = \frac{5}{3}$

$\therefore P\left(\frac{5}{3}, 0\right)$ is a point on the line

$$3x + 4y - 5 = 0$$

Now, slope of line $3x + 4y + 5 = 0$

$$m_1 = -\frac{3}{4}$$

Let m_2 be the slope of line PQ.

$\because$ $m_1 \cdot m_2 = -1$

($\because$ PQ is perpendicular to the given lines)

$\therefore$ $m_2 = \frac{4}{3}$

Equation of PQ, passing through (5/3, 0) having slope 4/3 is

$$y - 0 = \frac{4}{3}\left(x - \frac{5}{3}\right)$$

$$y = \frac{y}{3}x - \frac{20}{9}$$

$$2x - 3y = 20 \quad \text{...(iv)}$$

On solving equations $3x + 4y + 2 = 0$ and $12x - 9y = 20$, we get coordinates of point R,

i.e. $\left(\frac{62}{75}, \frac{-28}{25}\right)$

On solving equation $3x + 4y + 5 = 0$ and $12x - 9y = 20$, we get the coordinates of point Q,

i.e. $\left(\frac{35}{75}, -\frac{40}{25}\right)$

Let point R divides joint of PQ in the ratio $\lambda : 1$.

$\therefore$ $\frac{62}{75} = \frac{\frac{5}{3} \times \lambda + \frac{35}{75}}{\lambda + 1}$

$$\frac{62}{75} = \frac{125\lambda + 35}{75(\lambda + 1)}$$

$$(\lambda + 1)\,62 = 125\lambda + 35$$

$$\lambda = \frac{27}{63} = 3 : 7$$

Alternate Method

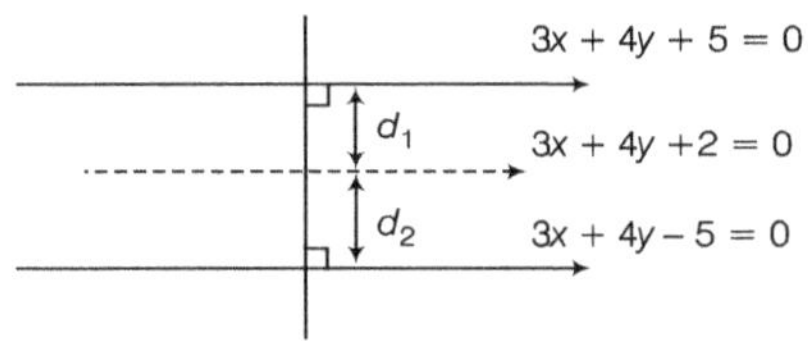

Here, $\frac{d_1}{d_2} = \frac{\left|\frac{5-2}{\sqrt{3^2 + 4^2}}\right|}{\left|\frac{2+5}{\sqrt{3^2 + 4^2}}\right|} = \frac{3}{7}$

21. (d) Let p_1 and p_2 be the length of perpendiculars drawn from $\left(\sqrt{a^2 - b^2}, 0\right)$ and $\left(-\sqrt{a^2 - b^2}, 0\right)$ to the line $\frac{x}{a}\cos\theta + \frac{y}{b}\sin\theta = 1$

$\therefore$ $p_1 = \left|\frac{\frac{\sqrt{a^2 - b^2}}{a}\cos\theta - 1}{\sqrt{\frac{\cos^2\theta}{a^2} + \frac{\sin^2\theta}{b^2}}}\right|$ ...(i)

and $p_2 = \left|\frac{-\frac{\sqrt{a^2 - b^2}}{a}\cos\theta - 1}{\sqrt{\frac{\cos^2\theta}{a^2} + \frac{\sin^2\theta}{b^2}}}\right|$ ...(ii)

Now, $p_1 \times p_2 = \left|\frac{-\left(\frac{\sqrt{a^2 - b^2}}{a}\cos\theta - 1\right)\left(\frac{\sqrt{a^2 - b^2}}{a}\cos\theta + 1\right)}{\frac{\cos^2\theta}{a^2} + \frac{\sin^2\theta}{b^2}}\right|$

$$= \left|\frac{\left(\frac{a^2 - b^2}{a^2}\cos^2\theta + 1\right)}{\frac{b^2\cos^2\theta + a^2\sin^2\theta}{a^2b^2}}\right|$$

$$= \left|\frac{(a^2 - b^2)\cos^2\theta - a^2}{b^2\cos^2\theta + a^2\sin^2\theta}\right| b^2 = \left|\frac{-b^2\cos^2\theta - a^2\sin^2\theta}{b^2\cos^2\theta + a^2\sin^2\theta}\right| b^2 = b^2$$

22. (a) Given, lines are

$$x + y = 4 \quad \ldots(i)$$
$$4x + 3y = 10 \quad \ldots(ii)$$

Let (α, β) be the point on the line

$$x + y = 4$$
$$\therefore \quad \alpha + \beta = 4 \quad \ldots(iii)$$

Now, distance of the point (α, β) from the line $4x + 3y = 10$ is

$$\Rightarrow \left|\frac{4\alpha + 3\beta - 10}{5}\right| = 1 \quad \text{(given)}$$
$$\Rightarrow |4\alpha + 3\beta - 10| = 5$$
$$\Rightarrow 4\alpha + 3\beta - 10 = \pm 5$$
$$4\alpha + 3\beta = 15 \quad \ldots(iv)$$
$$4\alpha + 3\beta = 5 \quad \ldots(v)$$

On solving Eqs. (iii) and (v), we get

$$\alpha = 3, \beta = 1$$

On solving Eqs. (iii) and (v), we get

$$\alpha = -7, \beta = 11$$

Hence, points are (3, 1), (–7, 11).

23. (d) Given, equation of line,

$$L: 2x + y = 2 \quad \ldots(i)$$

If the axes are rotated through an angle 45°, then

$$\left.\begin{aligned} x &= X\cos 45° - Y\sin 45° = \frac{X - Y}{\sqrt{2}} \\ y &= X\sin 45° + Y\cos 45° = \frac{X + Y}{\sqrt{2}} \end{aligned}\right\} \quad \ldots(ii)$$

Here, X and Y are coordinates of a point with respect to new axes. Let p and q be the intercept of line L on the new axes.

$\therefore$ Equation of line L is

$$\frac{X}{p} + \frac{Y}{q} = 1 \quad \ldots(iii)$$

Substituting the values of x and y in Eq. (i), we get

$$2\left(\frac{X - Y}{\sqrt{2}}\right) + \frac{X + Y}{\sqrt{2}} = 2$$
$$\Rightarrow \sqrt{2}(X - Y) + \frac{X + Y}{\sqrt{2}} = 2$$
$$\Rightarrow 2X - 2Y + X + Y = 2\sqrt{2}$$
$$\Rightarrow 3X - Y = 2\sqrt{2}$$
$$\Rightarrow \frac{X}{2\sqrt{2}/\sqrt{3}} - \frac{y}{2\sqrt{2}} = 1 \quad \ldots(iv)$$

Clearly, Eqs. (iii) and (iv) represent same line.

$$\therefore \quad p = \frac{2\sqrt{2}}{3}, q = 2\sqrt{2}$$

24. (b) Let reflection of point (3, 8) with respect to the line $x + 3y = 7$ is (x_1, y_1).

$$\therefore \quad \frac{x_1 - 3}{1} = \frac{y_1 - 8}{3} = \frac{-2(3 + 3 \times 8 - 7)}{10}$$

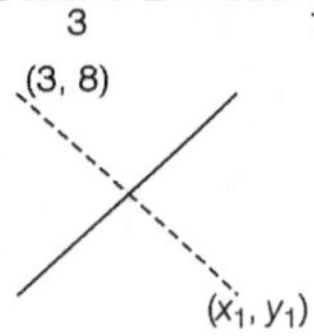

$$\Rightarrow x_1 - 3 = \frac{y_1 - 8}{3} = -4$$
$$\Rightarrow x_1 - 3 = -4$$
$$\Rightarrow x_1 = -1$$
$$\text{and} \quad y_1 = 8 = -4 \times 3$$
$$\Rightarrow y_1 = -4$$

$\therefore$ Reflection of point (3, 8) is (–1, – 4).

25. (c) The given line is $\frac{x}{a} + \frac{y}{b} = 1$

$$\Rightarrow bx + ay - ab = 0 \quad \ldots(i)$$

Given, p is the length of perpendicular from origin to the line (i).

$$\therefore \quad p = \left|\frac{b(0) + (a, 0) - ab}{\sqrt{b^2 + a^2}}\right|$$
$$\Rightarrow p = \frac{ab}{\sqrt{b^2 + a^2}}$$
$$\Rightarrow p^2 = \frac{a^2b^2}{b^2 + a^2}$$
$$\Rightarrow \frac{1}{p^2} = \frac{b^2 + a^2}{a^2b^2}$$
$$\Rightarrow \frac{1}{p^2} = \frac{1}{a^2} + \frac{1}{b^2}$$

26. (d) Given, equation of base of an equilateral ΔABC is

$$x + y = 2$$

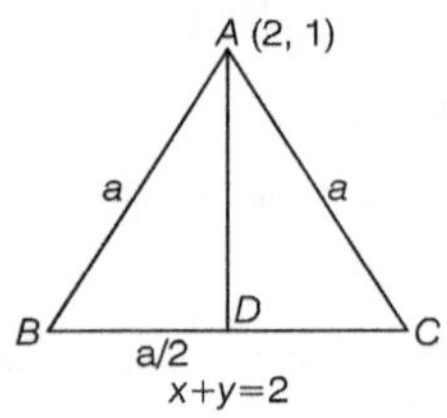

Draw $AD \perp BC$

$$\therefore \quad AD = \left|\frac{2 + 1 - 2}{\sqrt{2}}\right|$$
$$AD = \frac{1}{\sqrt{2}} \quad \ldots(i)$$

Let the side of an equilateral ΔABC be a.

$$\therefore \quad AD = \sqrt{a^2 - \frac{a^2}{4}} = \frac{\sqrt{3}}{2}a \quad \ldots(ii)$$

From Eqs. (i) and (ii), we get

$$\frac{\sqrt{3}}{2}a = \frac{1}{\sqrt{2}}$$
$$\Rightarrow \frac{\sqrt{3}}{\sqrt{2}}a = 1$$
$$\Rightarrow a = \sqrt{2/3}$$

27. (b) The equation of line with reference to the old axis is

$$\frac{x}{a} + \frac{y}{b} = 1 \quad \ldots(i)$$

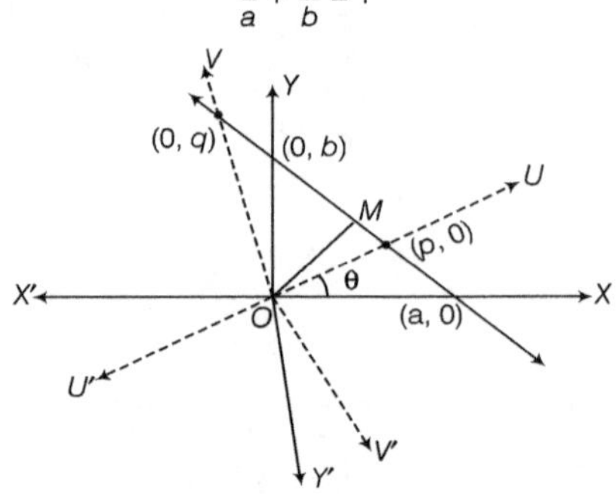

OM = Length of the perpendicular from $O(0, 0)$ on the line (i).

$$\Rightarrow OM = \frac{1}{\sqrt{\frac{1}{a^2} + \frac{1}{b^2}}} \quad \ldots(ii)$$

Let the axes be rotated through an angle θ with reference to the new axes, then equation of line is

$$\frac{x}{p} + \frac{y}{q} = 1 \quad \ldots(iii)$$

$\therefore OM$ = Length of perpendicular from $O(0, 0)$ on the line (iii)

$$OM = \frac{1}{\sqrt{p^2 + q^2}} \quad \ldots(iv)$$

From Eqs. (ii) and (iv), we get

$$\frac{1}{\sqrt{\frac{1}{a^2}+\frac{1}{b^2}}} = \frac{1}{\sqrt{\frac{1}{p^2}+\frac{1}{q^2}}}$$

$\Rightarrow$ $$\frac{1}{a^2}+\frac{1}{b^2} = \frac{1}{p^2}+\frac{1}{q^2}$$

or $$a^{-2}+b^{-2} = p^{-2}+q^{-2}$$

This is required relation.

Alternate Method

The equation of the line with reference to the old axes is

$$\frac{x}{a}+\frac{y}{b}=1 \quad \ldots(i)$$

If the axes are rotated through an angle θ, then

$$x = X\cos\theta - Y\sin\theta,\ y = X\sin\theta + Y\cos\theta$$

Here, X and Y are coordinates of a point $P(x, y)$ with respect to new axes.

The equation of line with respect to new axes is

$$\frac{X}{p}+\frac{Y}{q}=1 \quad \ldots(ii)$$

Substituting the values of x and y in line (i), we get

$$\left(\frac{\cos\theta}{a}+\frac{\sin\theta}{b}\right)X + \left(\frac{-\sin\theta}{a}+\frac{\cos\theta}{b}\right)Y = 1 \quad \ldots(iii)$$

Clearly, Eqs. (ii) and (iii) represent the same line. Therefore,

$$\frac{1}{p} = \frac{\cos\theta}{a}+\frac{\sin\theta}{b}$$

and $$\frac{1}{q} = \frac{-\sin\theta}{a}+\frac{\cos\theta}{b}$$

$\Rightarrow$ $$\frac{1}{p^2}+\frac{1}{q^2} = \frac{1}{a^2}+\frac{1}{b^2}$$

which is the required relation.

28. (b) The given lines are

$$4x + 2y = 9 \quad \ldots(i)$$
$$2x + y = -6 \quad \ldots(ii)$$

Sign of constant on RHS shows that two lines lie on opposite side of origin. Let any line through origin meet these lines at P and Q, respectively. Then, required ratio is $OP : OQ$.

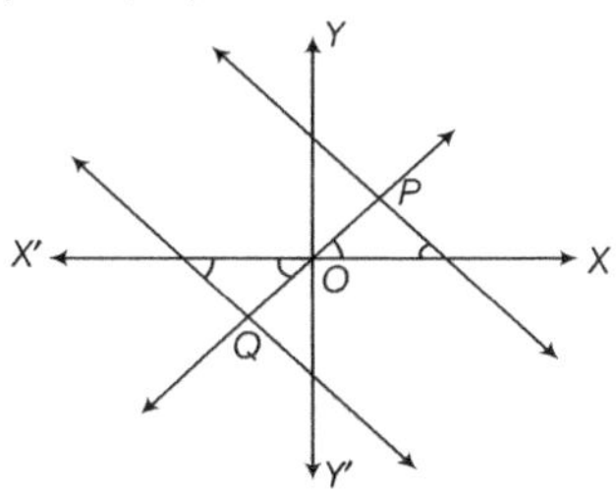

In particular, OP can be taken perpendicular to the two parallel lines. Then, required ratio is equal to the ratio of perpendicular distances of lines from origin.

Now, distance of origin from $4x + 2y - 9 = 0$ is

$$OP = \left|\frac{-9}{\sqrt{4^2+2^2}}\right| = \frac{9}{\sqrt{20}}$$

and distance of origin from $2x + y + 6 = 0$

$$OQ = \left|\frac{6}{\sqrt{2^2+1^2}}\right| = \frac{6}{\sqrt{5}}$$

Hence, the required ratio $\dfrac{OP}{OQ} = \dfrac{\frac{9}{\sqrt{20}}}{\frac{6}{\sqrt{5}}} = \dfrac{3}{4} = 3:4$

29. (b) Slope $QR = \dfrac{1+3}{4-2} = 2 = m_1$ (say)

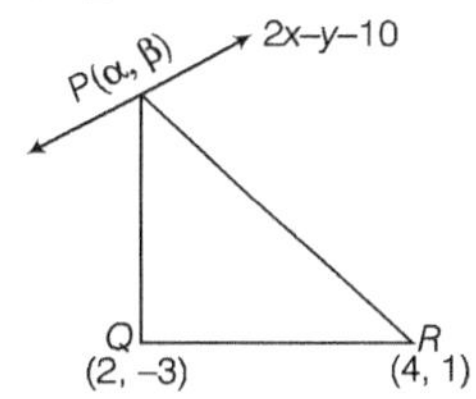

Slope $PQ = \dfrac{\beta+3}{\alpha-2} = m_2$ (say)

$\because$ $PQ \perp QR$

$\therefore$ $m_1 \cdot m_2 = -1$

$\Rightarrow$ $2\left(\dfrac{\beta+3}{\alpha-2}\right) = -1$

$\Rightarrow$ $2\beta + 6 = 2 - \alpha$

$\Rightarrow$ $\alpha + 2\beta = -4 \quad \ldots(i)$

Since, $P(\alpha, \beta)$ lies on $2x - y - 10 = 0$

$\therefore$ $2\alpha - \beta - 10 = 0 \quad \ldots(ii)$

On solving Eqs. (i) and (ii), we get

$$\alpha = \frac{16}{5},\ \beta = \frac{-18}{5}$$

Now, $$QR = \sqrt{(2-4)^2+(-3-1)^2} = \sqrt{20}$$

$$PQ = \sqrt{\left(\frac{16}{5}-2\right)^2+\frac{-18}{5}} = \sqrt{\frac{36}{25}+\frac{9}{25}} = \sqrt{\frac{45}{25}} = \frac{3}{5}$$

$\therefore$ Area of $\Delta PQR = \dfrac{1}{2} \times QR \times PQ$

$$= \frac{1}{2} \times \sqrt{20} \times \frac{3}{\sqrt{5}} = 3 \text{ sq units.}$$

30. (c) Given, equation of line

$$3x + 5y = 15 \quad \ldots(i)$$

Since, point $P(x, y)$ lies on line (i) and equidistant from the coordinate axes.

$\therefore$ The numerical value of abscissa and ordinate of point P are same.

Put $x = y$ in Eq. (i), we get

$$8x = 15$$
$$x = \frac{15}{8}$$

$\Rightarrow$ $$y = \frac{15}{8}$$

$\therefore$ Coordinates of point P are $\left(\dfrac{15}{8}, \dfrac{15}{8}\right)$.

$\Rightarrow$ P lies in Ist quadrant.

Now, put $x = -y$ in Eq. (i), we get

$$2y = 15$$
$$y = \frac{15}{2} \quad \Rightarrow \quad x = -\frac{15}{2}$$

$\therefore$ Coordinates of point P are $\left(-\dfrac{15}{2}, \dfrac{15}{2}\right)$.

Hence, P lies in IInd quadrant.

31. (a) Equation of line,

$$Ax + By + C = 0$$

$\Rightarrow$ $$\frac{x}{-\frac{C}{A}} + \frac{y}{-\frac{C}{B}} = 1$$

$\therefore$ x-intercept $= -\dfrac{C}{A}$

and y-intercept $= -\dfrac{C}{B}$

(i) x-intercept finite

$\Rightarrow \quad -\frac{C}{A} = \text{finite} \Rightarrow -\frac{C}{A} = k$ (say)

$\Rightarrow \quad C = -kA$

$\Rightarrow \quad A \neq 0 \text{ and } C \neq 0$

y-intercept infinite

$\Rightarrow \quad -\frac{C}{B} = \infty \Rightarrow B = 0$

(i) → (s)

(ii) x-intercept infinite

$\Rightarrow \quad -\frac{C}{A} = \infty \quad \Rightarrow \quad A = 0$

y-intercept finite

$-\frac{C}{B} = k$ (say)

$\Rightarrow \quad B \neq 0, C \neq 0$

∴ **(ii) → (p)**

(iii) Both x and y-intercepts are zero.

$\Rightarrow \quad -\frac{C}{A} = 0 \text{ and } -\frac{C}{B} = 0$

$\Rightarrow \quad C = 0 \text{ and } A, B \neq 0$

∴ **(iii) → (q)**

(iv) Both x and y-intercept are infinite.

$\Rightarrow \quad -\frac{C}{A} = \infty \text{ and } -\frac{C}{B} = \infty$

$\Rightarrow \quad C \neq 0, A = 0 \text{ and } C \neq 0, B = 0$

$\Rightarrow \quad C \neq 0 \text{ and } A, B = 0$

∴ **(iv) → (r)**

32. (b)

(i) Equation of line perpendicular to $4x - 3y = 10$ is

$$3x + 4y = \lambda \quad \text{...(i)}$$

Since, line (i) passes through (2, 3).

$\therefore \quad 3 \times 2 + 4 \times 3 = \lambda$

$\Rightarrow \quad \lambda = 18$

∴ Required equation of line

$$3x + 4y = 18$$

∴ **(i) → (q)**

(ii) Equation of line perpendicular to $7x + 8y = 5$ is

$$8x - 7y = \lambda \quad \text{...(i)}$$

Since, line (i) passes through point (−6, 10).

$\therefore \quad 8 \times -6 - 7 \times 10 = \lambda$

$\Rightarrow \quad \lambda = -118$

∴ Required equation of line

$$8x - 7y = -118$$

$\Rightarrow \quad 7y - 8x = 118$

∴ **(ii) → (s)**

(iii) Equation of line passing through (5, 7) and (−6, 3) is

$$y - 7 = \frac{3-7}{-6-5}(x - 5)$$

$\Rightarrow \quad y - 7 = \frac{-4}{-11}(x - 5)$

$\Rightarrow \quad 11y - 77 = 4x - 20$

$\Rightarrow \quad 4x - 11y = -57 \quad \text{...(i)}$

Equation of line perpendicular to Eq. (i) and passing through (2, − 3) is

$$y + 3 = -\frac{11}{4}(x - 2)$$

$\Rightarrow \quad 4y + 12 = -11x + 22$

$\Rightarrow \quad 4y + 11x = 22 - 12$

$\Rightarrow \quad 11x + 4y = 10$

∴ **(iii) → (p)**

(iv) Equation of line joining points (1, 3) and (2, 7) is

$$y - 3 = \frac{4}{1}(x - 1)$$

$\Rightarrow \quad y - 3 = 4x - 4$

$\Rightarrow \quad 4x - y = 1 \quad \text{...(i)}$

Equation of line perpendicular to Eq. (i) and passing through (−4, − 3) is

$$y + 3 = -\frac{1}{4}(x + 4)$$

$\Rightarrow \quad 4y + 12 = -x - 4$

$\Rightarrow \quad x + 4y + 16 = 0$

∴ **(iv) → (r)**

33. (c) Equation of line *AB*,

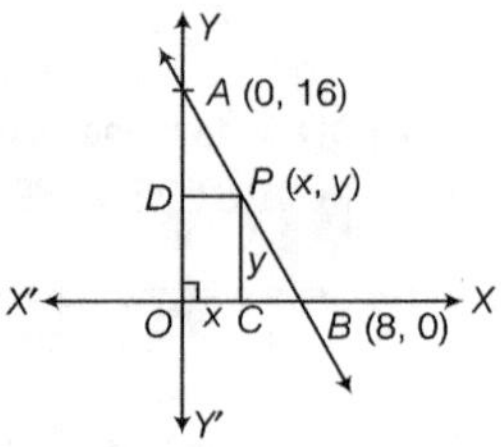

$$\frac{x}{8} + \frac{y}{16} = 1 \Rightarrow \frac{2x + y}{16} = 1$$

$\Rightarrow \quad 2x + y = 16 \quad \text{...(i)}$

Now, perpendicular distance of point (2, 2) from line (i) is

$$p = \frac{|2 \times 2 + 2 - 16|}{\sqrt{2^2 + 1^2}} = \frac{10}{\sqrt{5}} = \sqrt{20}$$

34. (a) Let the coordinates of point *P* be (*x*, *y*).

$\therefore \quad OC = DP = x \text{ and } OD = CP = y$

If *PDOC* is a square, then $OC = OD \Rightarrow x = y$

From Eq. (i), we get

$$2x + x = 16$$

$\Rightarrow \quad x = \frac{16}{3} = y$

$\therefore \quad x + y = \frac{16}{3} + \frac{16}{3} = \frac{32}{3}$

35. (b) Given, area of rectangle *PDOC* = 30 sq units

$\Rightarrow \quad x \cdot y = 30$

$\Rightarrow \quad y = \frac{30}{x} \quad \text{...(ii)}$

All positions of point *P* is/are the values of *x* and *y* which satisfy Eqs. (i) and (ii) simultaneously.

From Eqs. (i) and (ii), we get

$$2x + \frac{30}{x} = 16$$

$\Rightarrow \quad 2x^2 + 30 = 16x$

$\Rightarrow \quad 2x^2 - 16x + 30 = 0$

$\Rightarrow \quad x^2 - 8x + 15 = 0$

$\Rightarrow \quad x = \frac{8 \pm \sqrt{64 - 60}}{2} = 5, 3$

When $x = 5, y = 6$

and $x = 3, y = 10$

∴ Position of point *P* is (5, 6) or (3, 10).

DPP-3 Equation of Straight Line making an Angle 'α' with Given Straight Line

1. (a) Let $A(2, 3)$ be one vertex and $x + y = 2$ be the opposite side of an equilateral triangle. Clearly, remaining two sides pass through the point $A(2, 3)$ and make an angle of 60° with $x + y = 2$.

Let m be the slope of $x + y = 2$. Then, $m = -1$. So, the equations of the other two sides are

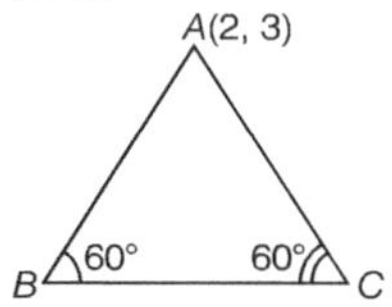

$$y - 3 = \frac{-1 - \tan 60°}{1 - \tan 60°}(x - 2)$$

and $$y - 3 = \frac{-1 + \tan 60°}{1 + \tan 60°}(x - 2)$$

$\Rightarrow$ $$y - 3 = \frac{-(1 + \sqrt{3})}{1 - \sqrt{3}}(x - 2)$$

$$= (2 + \sqrt{3})(x - 2)$$

and $$y - 3 = \frac{\sqrt{3} - 1}{\sqrt{3} + 1}(x - 2)$$

$$= (2 - \sqrt{3})(x - 2)$$

2. (c) Let $A(1, 2)$ be one vertex of square and diagonal BD is along X-axis.

$\therefore$ $$\angle ABD = \angle ADB = 45°$$

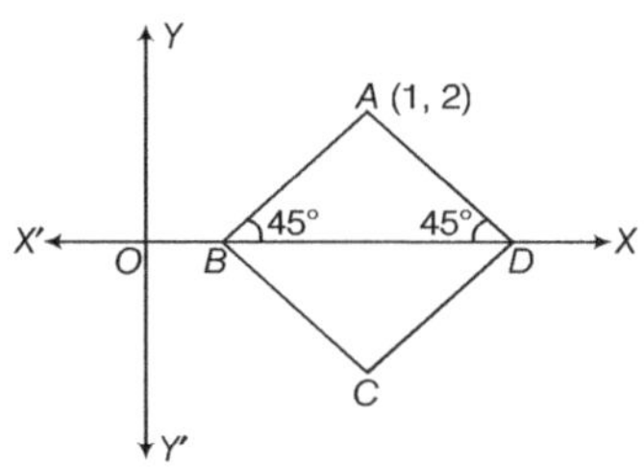

Let m be the slope of line BD, then $m = 0$.

$\therefore$ Equation of two sides of square is

$$y - 2 = \frac{0 \pm \tan 45°}{1 \mp \tan 45° \times 0}(x - 1)$$

$$y - 2 = \pm \tan 45° \, (x - 1)$$

$$y - 2 = \pm (x - 1)$$

$\Rightarrow$ $$y - 2 = (x - 1)$$

and $$y - 2 = -(x - 1)$$

$\Rightarrow$ $$x - y + 1 = 0$$

and $$x + y - 3 = 0$$

3. (b) Given, equation of line

$$y = mx + c \quad \text{...(i)}$$

Slope of line (i) is m.

Since, the required lines inclined at an angle $\tan^{-1} m$ to the line (i).

$\therefore$ Slope of required line $= \tan(\tan^{-1} m) = m$

Equation of lines

$$(y - 0) = \left(\frac{m \pm m}{1 \mp m^2}\right)(x - 0)$$

$\Rightarrow$ $$y = \frac{2m}{1 - m^2}(x - 0) \text{ and } y = 0$$

$\Rightarrow$ $$y(1 - m^2) = 2mx \text{ and } y = 0$$

$\Rightarrow$ $$2mx - (m^2 - 1)\,y = 0 \text{ and } y = 0$$

4. (a) Given, equation of line is

$$x - 2y = 0 \quad \text{...(i)}$$

Slope of line (i), $$m = \frac{1}{2}$$

We know that, diagonal of a square bisect vertex angle.

$\therefore$ Required lines are inclined at an angle of 45° with the diagonal.

Now, equation of lines

$$y - 0 = \frac{\frac{1}{2} \pm \tan 45°}{1 \pm \frac{1}{2}\tan 45°}(x - 3)$$

$\Rightarrow$ $$y = \frac{\frac{1}{2} \pm 1}{1 \mp \frac{1}{2}}(x - 3)$$

$\Rightarrow$ $$y = \frac{\frac{1}{2} + 1}{1 - \frac{1}{2}}(x - 3)$$

and $$y = \frac{\frac{1}{2} - 1}{1 + \frac{1}{2}}(x - 3)$$

$\Rightarrow$ $$y = 3(x - 3) \text{ and } y = -\frac{1}{3}(x - 3)$$

$\Rightarrow$ $$3x - y - 9 = 0$$

and $$x + 3y - 3 = 0$$

5. (b) Given, equation of line

$$\sqrt{3}x + y = 1 \quad \text{...(i)}$$

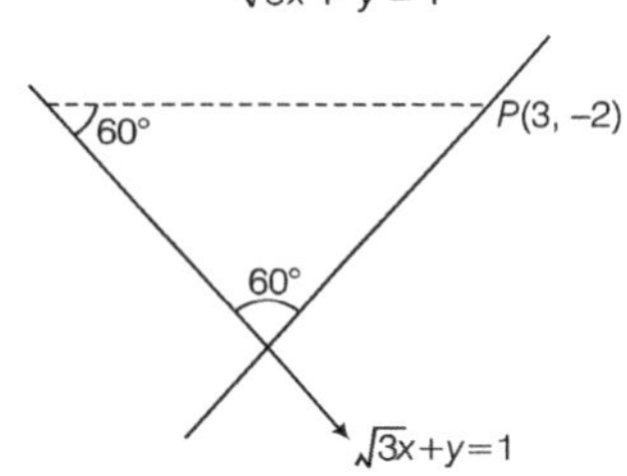

A straight line passing through P and making an angle $\alpha = 60°$ is given by

$$\frac{y - y_1}{x - x_1} = \tan(\theta \pm \alpha)$$

$\Rightarrow$ $$y = -\sqrt{3}x + 1$$

Then, $$\tan\theta = -\sqrt{3}$$

$\Rightarrow$ $$\frac{y + 2}{x - 3} = \frac{\tan\theta \pm \tan\alpha}{1 \mp \tan\alpha\tan\theta}$$

$\Rightarrow$ $$\frac{y + 2}{x - 3} = \frac{-\sqrt{3} + \sqrt{3}}{1 - (-\sqrt{3})\sqrt{3}}$$

and $$\frac{y + 2}{x - 3} = \frac{-\sqrt{3} - \sqrt{3}}{1 + (-\sqrt{3})\sqrt{3}}$$

$\Rightarrow$ $$y + 2 = 0$$

and $$y + 2 = \frac{-2\sqrt{3}}{-2}(x - 3)$$

$\Rightarrow$ $$y + 2 = 0$$

and $$y - \sqrt{3}x + 2 + 3\sqrt{3} = 0$$

Neglecting $y + 2 = 0$, as it does not intersect X-axis.

6. (a) The coordinates of A and B are shown in figure given below :

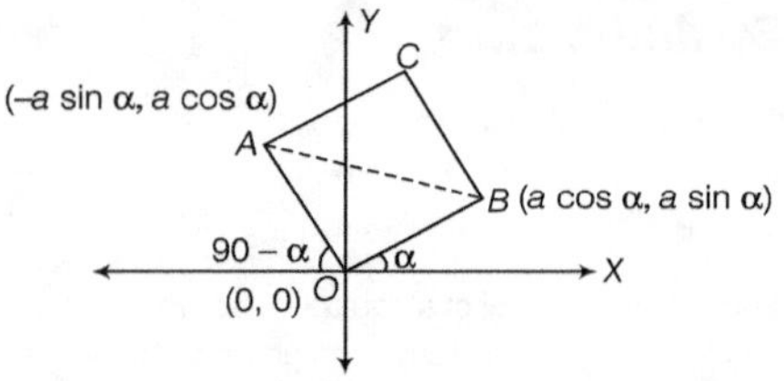

The equation of the diagonal AB is

$$(y - a\sin\alpha) = \frac{a\cos\alpha - a\sin\alpha}{-a\sin\alpha - a\cos\alpha}(x - a\cos\alpha)$$

$$\Rightarrow y(\cos\alpha + \sin\alpha) - a(\sin\alpha\cos\alpha + \sin^2\alpha)$$

$$= -(\cos\alpha - \sin\alpha)\,x + a\cos\alpha(\cos\alpha - \sin\alpha)$$

$$\Rightarrow y(\cos\alpha + \sin\alpha) + x(\cos\alpha - \sin\alpha) = a$$

7. (d) In the argand plane, P is represented by $e^{i\theta}$ and Q is represented by $e^{i(\alpha - \theta)}$.

Now, rotation about a line with angle α is given by $e^{i\theta} \to e^{i(2\alpha - \theta)}$.

Therefore, θ is obtained from P by reflection in the line making an angle $\frac{\alpha}{2}$.

8. (d) Given, equation of lines are

$$y = 3x + 1 \quad \text{...(i)}$$

$$2y = x + 3 \quad \text{...(ii)}$$

$$y = mx + 4 \quad \text{...(iii)}$$

Slope of line (i), $m_1 = 3$

Slope of line (ii), $m_2 = \frac{1}{2}$

and slope of line (iii) is m.

Since, lines (i) and (ii) are equally inclined to the line (iii).

$$\therefore \quad \left(\frac{m_1 - m}{1 + m_1 \cdot m}\right) = -\left(\frac{m_2 - m}{1 + m_2 \cdot m}\right)$$

$$\Rightarrow \left(\frac{3 - m}{1 + 3m}\right) = -\left(\frac{\frac{1}{2} - m}{1 + \frac{m}{2}}\right) \Rightarrow \left(\frac{3 - m}{1 + 3m}\right) = -\left(\frac{1 - 2m}{2 + m}\right)$$

$$\Rightarrow (3 - m)(2 + m) = -(1 + 3m)(1 - 2m)$$

$$6 + 3m - 2m - m^2 = -1 + 2m - 3m + 6m^2$$

$$\Rightarrow 7m^2 - 2m - 7 = 0$$

$$\Rightarrow m = \frac{2 \pm \sqrt{4 + 196}}{2 \times 7}$$

$$\Rightarrow m = \frac{2 \pm 10\sqrt{2}}{2 \times 7} = \frac{1 \pm 5\sqrt{2}}{7}$$

9. (c) Here, equation of diagonals of parallelogram are parallel to

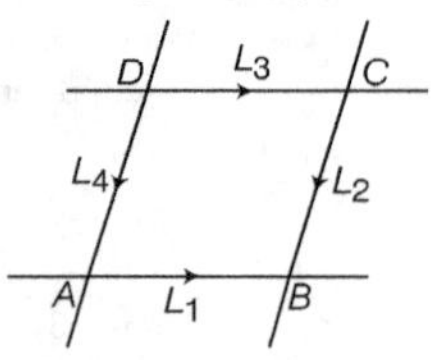

$$L_1 + L_2 = 0 \quad \text{and} \quad L_1 - L_2 = 0$$

i.e. $(l + m)x + (m + l)y + 2n = 0$ ...(i)

and $(l - m)x + (m - l)y = 0$ ...(ii)

Having slopes –1 and 1.

$\therefore$ Slope of diagonals are $m_{AC} = 1$ and $m_{BD} = -1$

$\Rightarrow \quad m_{AC} \cdot m_{BD} = -1$

Thus, angle between diagonals is 90°.

Explanations for Q. Nos. (10-12)

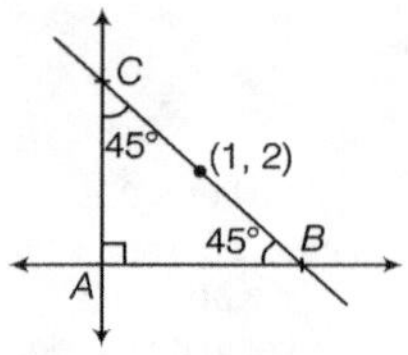

Given, equation of lines

$$3x + 4y = 5 \quad \text{...(i)}$$

$$4x - 3y = 15 \quad \text{...(ii)}$$

It is given that, ΔABC is right angled isosceles triangles in which $AB = AC$.

$\Rightarrow \quad \angle ABC = \angle ACB = 45°$

Therefore, BC is equally inclined with AB and AC.

Slope of line (i), $m_1 = -\frac{3}{4}$

Slope of line (ii), $m_2 = \frac{4}{3}$

Let m be the slope of line BC.

$$\therefore \quad \left(\frac{m_1 - m}{1 + m_1 m}\right) = -\left(\frac{m_2 - m}{1 + m_2 m}\right)$$

$$\Rightarrow \left(\frac{-\frac{3}{4} - m}{1 - \frac{3m}{4}}\right) = -\left(\frac{\frac{4}{3} - m}{1 + \frac{4}{3}m}\right)$$

$$\Rightarrow \left(\frac{\frac{-3 - 4m}{4}}{\frac{4 - 3m}{4}}\right) = -\left(\frac{\frac{4 - 3m}{3}}{\frac{3 + 4m}{3}}\right)$$

$$\Rightarrow \left(\frac{-3 - 4m}{4 - 3m}\right) = -\left(\frac{4 - 3m}{3 + 4m}\right)$$

$$\Rightarrow \frac{3 + 4m}{4 - 3m} = \frac{4 - 3m}{3 + 4m}$$

$$\Rightarrow (3 + 4m)^2 = (4 - 3m)^2$$

$$\Rightarrow 9 + 16m^2 + 24m = 16 + 9m^2 - 24m$$

$$\Rightarrow 7m^2 + 48m - 7 = 0$$

$$\Rightarrow m = \frac{-48 \pm \sqrt{2500}}{14}$$

$$\Rightarrow m = \frac{-48 \pm 50}{14} \quad \Rightarrow \quad m = \frac{1}{7}, -7$$

$\therefore$ Lines are $y - 2 = \frac{1}{7}(x - 1)$

and $y - 2 = -7(x - 1) \quad \Rightarrow \quad 7y - 14 = x - 1$

and $y - 2 = -7x + 7$

$\Rightarrow \quad x - 7y + 13 = 0$

and $7x + y - 9 = 0$

$\therefore \quad L_1 : x - 7y + 13 = 0$ ...(iii)

$L_2 : 7x + y - 9 = 0$ ...(iv)

10. (d) $l_1 : ax + by + c = 0$ and $l_2 : dx + ey + f = 0$

On comparing these lines with Eqs. (iii) and (iv), we get

$$a = 1, b = -7, c = 13, \quad d = 7, e = +1, f = -9$$

$$\therefore \quad c + f = 13 - 9 = 4$$

11. (a) Coordinates of point P are $(c + f - 1)$ i.e. $P(2, 3)$

[$\because c + f = 4$ units]

$\therefore$ Equation of line passing through $P(2, 3)$ and making an angle $(90 + 45°)$ i.e. 150° with positive X-axis.

$$\therefore \quad \frac{x - 2}{\cos 150°} = \frac{y - 3}{\sin 150°} = r$$

Coordinates of any point on it which is at distance r from (2, 3) is

$$x = 2 + r\cos 150°, \quad y = 3 + r\sin 150°$$

Here, $r = 4$

$$\therefore \quad x = 2 + 4\cos 150° = 2 - 4 \times \frac{\sqrt{3}}{2} = 2 - 2\sqrt{3}$$

$$y = 3 + 4\sin 150° = 3 + 4 \times \frac{1}{2} = 5$$

12. (d)

$|b - 2a - 1| = |-7 - 2 - 1|$

$OD = 10$

$\because \quad OA = OB$

$\Rightarrow \angle OBA = \angle OAB = 45°$

$OD \perp AB \Rightarrow \angle ODA = 90°$

$\therefore \angle AOD = 180° - (90° + 45°)$

$= 45° = \alpha$ (say)

$\therefore$ Equation of AB,

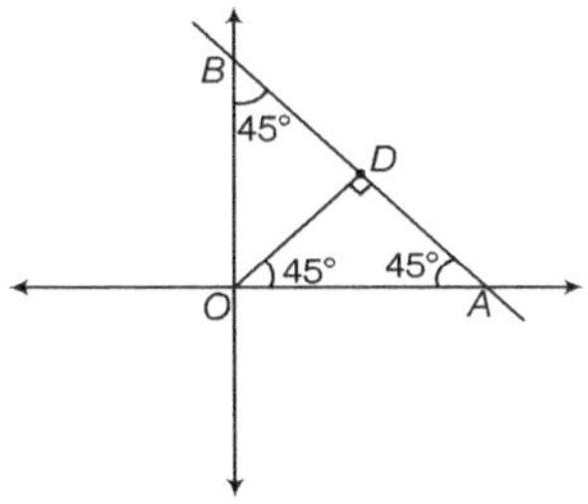

$x\cos\alpha + y\sin\alpha = p$

$$x\cos 45° + y\sin 45° = 10$$

$$\Rightarrow \quad \frac{x}{\sqrt{2}} + \frac{y}{\sqrt{2}} = 10$$

$$\therefore \quad x + y = 10\sqrt{2}$$

13. (a) Given, equation of line

$$x + y = 1 \qquad \text{...(i)}$$

Let $AB = BC = AC = a$

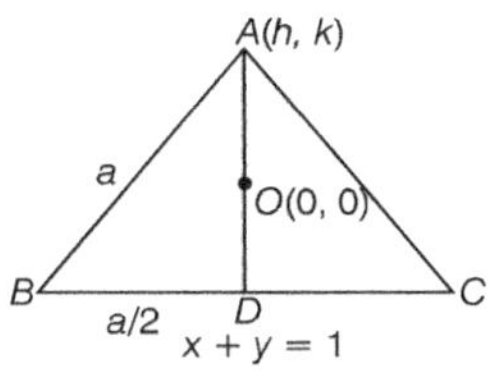

Draw $AD \perp BC$, AD is median of ΔABC

Equation of AD, $\quad y - 0 = 1(x - 0) \quad \Rightarrow \quad y = x$

$$\Rightarrow \quad x - y = 0 \qquad \text{...(ii)}$$

D is the point of intersection of line $x + y = 1$ and $x = y$.

On solving Eqs. (i) and (ii), we get

$$x = \frac{1}{2}, y = \frac{1}{2}$$

$\therefore$ Coordinates of point D are $\left(\frac{1}{2}, \frac{1}{2}\right)$.

We know that, centroid divides the median AD in the ratio 2 : 1.

$\therefore$ Coordinates of point A

$$0 = \frac{2 \times \frac{1}{2} + 1 \times h}{3} \Rightarrow h = -1$$

and
$$0 = \frac{2 \times \frac{1}{2} + 1 \times k}{3} \Rightarrow k = -1$$

Hence, point A is $(-1, -1)$.

Now,
$$AD = \sqrt{\left(-1 - \frac{1}{2}\right)^2 + \left(-1 - \frac{1}{2}\right)^2}$$

$$= \sqrt{\frac{9}{4} + \frac{9}{4}} = \sqrt{\frac{18}{4}} = \frac{3\sqrt{2}}{2}$$

Also, in ΔADB

$$AB^2 = AD^2 + BD^2$$

$$a^2 = AD^2 + \left(\frac{a}{2}\right)^2 \quad \Rightarrow \quad AD = \frac{\sqrt{3}}{2}a$$

$$\Rightarrow \quad \frac{\sqrt{3}}{2}a = \frac{3\sqrt{2}}{2} \quad \Rightarrow \quad a = \sqrt{6}$$

Now, area of $\Delta ABC = \frac{\sqrt{3}}{4} \times a^2 = \frac{\sqrt{3}}{4} \times 6$

$$= \sqrt{3} \times \frac{3}{2} = \frac{3\sqrt{3}}{2} \text{ sq units}$$

14. (d) Gradient of other two lines are

$$\frac{-1 + \tan 60°}{1 - (-1)\tan 60°} \text{ and } \frac{-1 - \tan 60°}{1 + (-1)\tan 60°}$$

$$\Rightarrow \quad \frac{-1 + \sqrt{3}}{1 + \sqrt{3}} \text{ and } \frac{-1 - \sqrt{3}}{1 - \sqrt{3}}$$

$$\Rightarrow \quad 2 - \sqrt{3} \text{ and } 2 + \sqrt{3}$$

15. (a,b) Given, equation of line

$$x - y + 1 = 0 \qquad \text{...(i)}$$

Slope of line (i), $m = 1$

The equation of lines making an angle $\frac{\pi}{4}$ with the line (i) and passing through origin is

$$y - 0 = \frac{1 \pm \tan\frac{\pi}{4}}{1 \mp 1 \times \tan\frac{\pi}{4}}(x - 0)$$

$$\Rightarrow \quad y = \frac{1 + 1}{1 - 1}(x)$$

and
$$y = \frac{1 - 1}{1 + 1}(x)$$

$$\Rightarrow \quad x = 0 \text{ and } y = 0$$

16. (b,c) Given, equation of line

$$y + 2x = 5 \qquad \text{...(i)}$$

Let the required line inclined at θ with the line (i).

$$\therefore \quad \theta = \text{arc} \tan\left(\frac{1}{2}\right)$$

$$\theta = \tan^{-1}\left(\frac{1}{2}\right)$$

$$\Rightarrow \quad \tan\theta = \frac{1}{2}$$

and slope of line (i), $\quad m = -2$

Equation of lines

$$y - 3 = \frac{-2 \pm \frac{1}{2}}{1 \mp (-2) \times \frac{1}{2}}(x - 2)$$

$$\Rightarrow \quad y - 3 = \frac{-2 + \frac{1}{2}}{1 - \frac{(-2)}{2}}(x - 2)$$

and
$$y - 3 = \frac{-2 - \frac{1}{2}}{1 + \frac{(-2)}{2}}(x - 2)$$

$$\Rightarrow \quad y - 3 = \frac{-3/2}{2}(x - 2) \text{ and } x = 2$$

$$\Rightarrow \quad 3x + 4y - 18 = 0 \text{ and } x = 2$$

17. (a,b,c,d) Given, equation of line

$$x - \sqrt{3}y = 0 \qquad \text{...(i)}$$

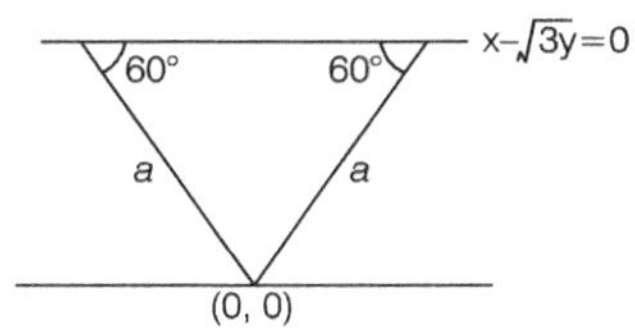

Slope of line (i), $\quad m = \frac{1}{\sqrt{3}}$

We know that, each angle of an equilateral triangle is 60°.

$\therefore$ Other two sides of equilateral triangle makes 60° angle with the line $x - \sqrt{3}y = 0$

Now, equation of other two sides of equilateral triangle which are passes through origin

$$y - 0 = \frac{\frac{1}{\sqrt{3}} \pm \tan 60^\circ}{1 \mp \frac{1}{\sqrt{3}} \times \tan 60^\circ}(x - 0)$$

$$\Rightarrow \quad y = \frac{\frac{1}{\sqrt{3}} \pm \sqrt{3}}{1 \mp \frac{\sqrt{3}}{\sqrt{3}}}(x)$$

$$\Rightarrow \quad y = \frac{\frac{1}{\sqrt{3}} + \sqrt{3}}{1 - 1} x$$

and
$$y = \frac{\frac{1}{\sqrt{3}} - \sqrt{3}}{1 + 1} x$$

$$\Rightarrow \quad x = 0 \text{ and } 2y = -\frac{2}{\sqrt{3}} x$$

$$\Rightarrow \quad x = 0 \quad \text{...(ii)}$$

and
$$\sqrt{3}y + x = 0 \quad \text{...(iii)}$$

Any point on the line (ii) at a distance a from the point (0, 0) is

$$x = a \cos \frac{\pi}{2}$$

$$y = a \sin\left(\pm \frac{\pi}{2}\right)$$

Slope of line $x = 0$, is

$$\tan \theta = \infty \Rightarrow \theta = \pm \frac{\pi}{2}$$

$$\Rightarrow \quad x = 0 \text{ and } y = \pm a$$

$\therefore$ Coordinates of second vertex are $(0, a)$ or $(0, -a)$. Again any point on the line (iii) at a distance a from the point (0, 0) is

$$x = a \cos \frac{5\pi}{6} \quad \text{or} \quad x = a \cos\left(-\frac{\pi}{6}\right)$$

$$y = a \sin \frac{5\pi}{6}$$

or
$$y = a \sin\left(-\frac{\pi}{6}\right)$$

Slope of line $\sqrt{3}y + x = 0$ is $-\frac{1}{\sqrt{3}}$

$$\tan \theta = -\frac{1}{\sqrt{3}} \Rightarrow \theta = \frac{5\pi}{6} \text{ or } \left(-\frac{\pi}{6}\right)$$

$$x = -\frac{\sqrt{3}}{2} a \text{ or } x = \frac{\sqrt{3}}{2} a$$

$$y = \frac{a}{2} \text{ or } y = -\frac{a}{2}$$

$\therefore$ Coordinates of third vertex are

$$\left(-\frac{\sqrt{3}}{2}a, \frac{a}{2}\right) \text{or} \left(\frac{\sqrt{3}}{2}a, -\frac{a}{2}\right).$$

18. (a,d) Here, $m_{AC} = \tan\theta = \frac{5 - 3}{7 - 1} = \frac{1}{3}$

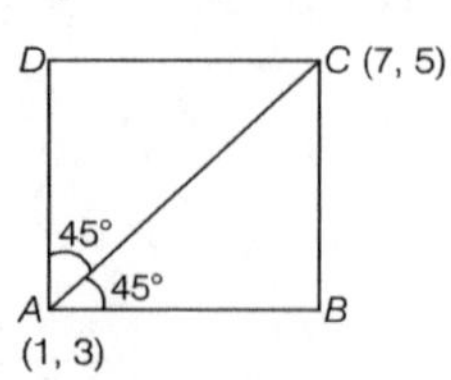

Thus, equation of AB and AD are

$$\frac{y - 3}{x - 1} = \tan(\theta \pm 45^\circ)$$

$$\Rightarrow \quad \frac{y - 3}{x - 1} = \frac{\tan\theta \pm 1}{1 \mp \tan\theta}$$

$$\therefore \quad \frac{y - 3}{x - 1} = \frac{\frac{1}{3} + 1}{1 - \frac{1}{3}} \quad \text{and} \quad \frac{y - 3}{x - 1} = \frac{\frac{1}{3} - 1}{1 + \frac{1}{3}}$$

$$\Rightarrow \quad 2x - y + 1 = 0 \quad \text{and} \quad 2y + x - 7 = 0$$

19. (0) The equation of line, which passes through (4, – 5) having slope m, is

$$y + 5 = m(x - 4)$$

$$\Rightarrow \quad y + 5 = mx - 4m$$

$$\Rightarrow \quad mx - y - 4m - 5 = 0 \quad \text{...(i)}$$

It is given that the distance of point (–2, 3) from line (i) is 12 units.

$$\therefore \quad \frac{|-2m - 3 - 4m - 5|}{\sqrt{m^2 + 1}} = 12$$

$$\Rightarrow \quad \frac{|-6m - 8|}{\sqrt{m^2 + 1}} = 12$$

$$\Rightarrow \quad \frac{-6m - 8}{\sqrt{m^2 + 1}} = \pm 12$$

Taking positive sign, we get

$$-6m - 8 = 12\left(\sqrt{m^2 + 1}\right)$$

$$(6m + 8)^2 = 144(m^2 + 1)$$

$$\Rightarrow \quad 36m^2 + 64 + 96m = 144m^2 + 144$$

$$\Rightarrow \quad 108m^2 - 96m + 80 = 0$$

$$\Rightarrow \quad 27m^2 - 24m + 20 = 0$$

$\Rightarrow$ Since, $D < 0$. Therefore real value of m is not possible.

Hence, no lines are possible.

20. (5) Given, equation of line

$$2x + y = 5 \quad \text{...(i)}$$

Slope of line (i), $m = -2$

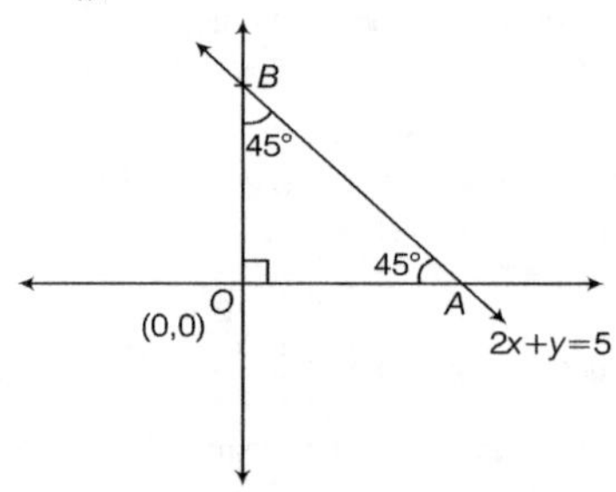

$\therefore$ Equation of lines passing through (0, 0) and making an angle 45° with the line AB, is

$$y - 0 = \frac{-2 \pm \tan 45^\circ}{1 \mp (-2)\tan 45^\circ}(x - 0)$$

$$\Rightarrow \quad y = \frac{-2 \pm 1}{1 \mp (-2)} x \quad \Rightarrow \quad y = \frac{-2 + 1}{1 - (-2)} x$$

and
$$y = \frac{-2 - 1}{1 + (-2)} x$$

$$\Rightarrow \quad y = -\frac{1}{3} x \text{ and } y = \frac{-3}{-1} x$$

$$\Rightarrow \quad x + 3y = 0 \text{ and } 3x + y = 0$$

Now, to get the coordinates of point A, on solving equation $2x + y = 5$ and $x + 3y = 0$, we get

$$x = 3, y = -1$$

$\therefore$ Coordinates of point A are (3, – 1).

Now, $OA = \sqrt{(3 - 0)^2 + (-1 - 0)^2}$

$$= \sqrt{9 + 1} = \sqrt{10}$$

Since, ΔOAB is isosceles triangle, right angled at O.

$$\therefore \quad OA = OB = \sqrt{10} \text{ and } \angle AOB = 90^\circ$$

Hence, area of $\Delta AOB = \frac{1}{2} \times OA \times OB$

$$= \frac{1}{2} \times \sqrt{10} \times \sqrt{10} = 5 \text{ sq units}$$

DPP-4 **Angle Bisector and Family of Concurrent Lines**

1. (b) Given lines are $3x - 4y + 7 = 0$ ...(i)

$12x + 5y - 2 = 0$ ...(ii)

Eq. (ii) can be written as

$-12x - 5y + 2 = 0$ ...(iii)

Now, $a_1a_2 + b_1b_2 = 3 \times (-12) + (-4) \times (-5)$

$\Rightarrow \quad = -36 + 20$

$= -16 < 0$

Equation of acute angle bisector

$$\left(\frac{3x - 4y + 7}{\sqrt{(3)^2 + (-4)^2}}\right) = \pm\left(\frac{-12x - 5y + 2}{\sqrt{(-12)^2 + (-5)^2}}\right)$$

[$\because a_1a_2 + b_1b_2 < 0$, origin lies in acute angle, $\therefore$ we take +ve sign]

$$\Rightarrow \quad \frac{3x - 4y + 7}{\sqrt{25}} = \frac{-12x - 5y + 2}{\sqrt{169}}$$

$$\Rightarrow \quad \frac{3x - 4y + 7}{5} = \frac{-12x - 5y + 2}{13}$$

$\Rightarrow \quad 39x - 52y + 91 = -60x - 25y + 10$

$\Rightarrow \quad 99x - 27y + 81 = 0$

$\Rightarrow \quad 11x - 3y + 9 = 0$

2. (a) Given lines are $x - 2y + 4 = 0$ ...(i)

$4x - 3y + 2 = 0$ ...(ii)

Now, $a_1a_2 + b_1b_2 = 1 \times 4 + (-2)(-3)$

$= 4 + 6 = 10 > 0$

Equation of bisector

$$\frac{x - 2y + 4}{\sqrt{1^2 + (-2)^2}} = \pm\left(\frac{4x - 3y + 2}{\sqrt{4^2 + (-3)^2}}\right)$$

$$\Rightarrow \quad \frac{x - 2y + 4}{\sqrt{1 + 4}} = \frac{4x - 3y + 2}{\sqrt{16 + 9}}$$

$$\Rightarrow \quad \frac{x - 2y + 4}{\sqrt{5}} = \frac{4x - 3y + 2}{5}$$

$\Rightarrow \quad \sqrt{5}x - 2\sqrt{5}y + 4\sqrt{5} = 4x - 3y + 2$

$\Rightarrow \quad (\sqrt{5} - 4)x + (3 - 2\sqrt{5})y + 4\sqrt{5} - 2 = 0$

$\Rightarrow \quad (4 - \sqrt{5})x - (3 - 2\sqrt{5})y + (2 - 4\sqrt{5}) = 0$

3. (a) Given lines are $3x - 4y + 7 = 0$ ...(i)

$12x - 5y - 8 = 0$ ...(ii)

Equation of bisector

$$\frac{3x - 4y + 7}{\sqrt{3^2 + (-4)^2}} = \pm\left(\frac{12x - 5y - 8}{\sqrt{(12)^2 + (-5)^2}}\right)$$

$$\Rightarrow \quad \frac{3x - 4y + 7}{5} = \pm\left(\frac{12x - 5y - 8}{13}\right)$$

$\Rightarrow \quad 39x - 52y + 91 = \pm(60x - 25y - 40)$

On taking +ve sign, we have

$39x - 52y + 91 = 60x - 25y - 40$

$\Rightarrow \quad -21x - 27y + 131 = 0$

$\Rightarrow \quad 21x + 27y - 131 = 0$

On taking –ve sign, we have

$39x - 52y + 91 = -60x + 25y + 40$

$\Rightarrow \quad 99x - 77y + 51 = 0$

4. (a) Let the slope of straight line equally inclined to given lines be m.

$$\therefore \quad \frac{m - \frac{3}{4}}{1 + \frac{3}{4}m} = -\left(\frac{m - 12/5}{1 + \frac{12}{5}m}\right)$$

$\Rightarrow \quad (4m - 3)(5 + 12m) = -(5m - 12)(4 + 3m)$

$\Rightarrow \quad 20m - 15 + 48m^2 - 36m = -20m + 48 - 15m^2 + 36m$

$\Rightarrow \quad 63m^2 - 32m - 63 = 0$

$\Rightarrow \quad (9m + 7)(7m - 9) = 0$

$\therefore \quad m = \frac{9}{7}$ and $m = \frac{-7}{9}$

Thus, required straight lines are

$$\frac{y - 5}{x - 4} = \frac{9}{7} \text{ and } \frac{y - 5}{x - 4} = -\frac{7}{9}$$

$\Rightarrow \quad 9x - 7y = 1$ and $7x + 9y = 73$

5. (b) The graph of the equations

$x - 2y = 0$

and $3x - y = 0$

is as shown in the figure given.

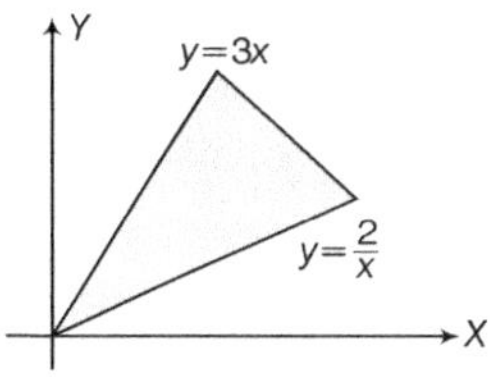

Since, given point (a, a^2) lies in the shaded region, then

$a^2 - \frac{a}{2} > 0$ and $a^2 - 3a < 0$

$\Rightarrow a \in (-\infty, 0) \cup \left(\frac{1}{2}, \infty\right)$ and $a \in (0, 3)$

$\Rightarrow \quad a \in (1/2, 3)$

6. (c) Slope of $QR = \frac{3\sqrt{3} - 0}{3 - 0} = \sqrt{3} = \tan\theta$

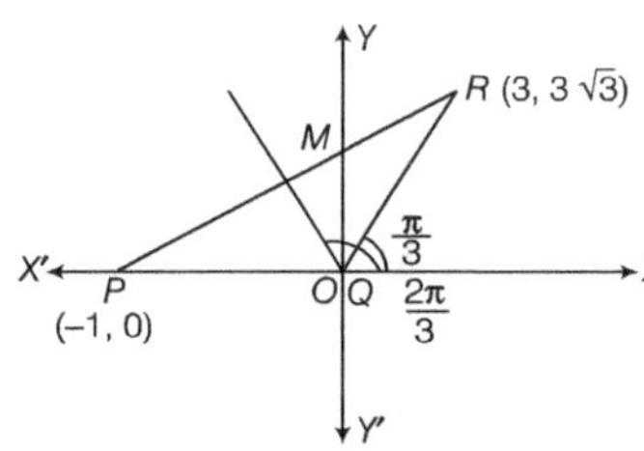

$\Rightarrow \quad \theta = \frac{\pi}{3}$

$\therefore$ The angle between PQR is $\frac{2\pi}{3}$.

So, the line QM makes an angle $\frac{2\pi}{3}$ from positive direction of X-axis.

Slope of line $QM = \tan\frac{2\pi}{3} = -\sqrt{3}$

Hence, equation of line QM is $y = -\sqrt{3}x$

$\therefore \quad \sqrt{3}x + y = 0$

7. (c) The equation of line passing through the point of intersection of lines $4x - 3y - 1 = 0$ and $5x - 2y - 3 = 0$ is

$(4x - 3y - 1) + \lambda(5x - 2y - 3) = 0$

$\Rightarrow \quad (4 + 5\lambda)x - (3 + 2\lambda)y - 3\lambda - 1 = 0$ ...(i)

Slope of line (i) is $m_1 = \frac{4 + 5\lambda}{3 + 2\lambda}$

Given, line (i) is parallel to the line $2x - 3y + 2 = 0$ ...(ii)

$\therefore$ Slope of line (ii) is $m_2 = \frac{2}{3} \quad \Rightarrow \quad m_1 = m_2$

$$\Rightarrow \quad \frac{4 + 5\lambda}{3 + 2\lambda} = \frac{2}{3}$$

$\Rightarrow \quad 12 + 15\lambda = 6 + 4\lambda$

$\Rightarrow \quad 11\lambda = -6 \quad \Rightarrow \quad \lambda = \frac{-6}{11}$

Substituting the value of λ in Eq. (i), we get

$14x - 21y + 7 = 0$

$\therefore \quad 2x - 3y + 1 = 0$

8. (b) Given equation of line

$$(2\cos\theta + 3\sin\theta)x + (3\cos\theta - 5\sin\theta)y - (5\cos\theta - 2\sin\theta) = 0 \quad \ldots(i)$$

Consider option (b) (1, 1), put $x = 1$ and $y = 1$ in Eq. (i), we get

$$2\cos\theta + 3\sin\theta + 3\cos\theta - 5\sin\theta - 5\cos\theta + 2\sin\theta = 0$$

Hence, point is (1, 1).

9. (b) Given, $\quad 3a + 2b + 4c = 0$

$$\Rightarrow \quad \frac{3}{4}a + \frac{b}{2} + c = 0 \quad \ldots(i)$$

and family of straight line is

$$ax + by + c = 0 \quad \ldots(ii)$$

Subtracting Eq. (i) from Eq. (ii), we get

$$a\left(x - \frac{3}{4}\right) + b\left(y - \frac{1}{2}\right) = 0$$

Which is equation of line passing through the point of intersection of the line

$$x - \frac{3}{4} = 0 \text{ and } y - \frac{1}{2} = 0$$

$$\therefore \quad x = \frac{3}{4} \text{ and } y = \frac{1}{2}$$

10. (a) The equation of straight line passing through the point of intersection of $ax + by + c = 0$ and $a'x + b'y + c' = 0$ is

$$(ax + by + c) + \lambda(a'x + b'y + c') = 0$$

$$\Rightarrow \quad (a + \lambda a')x + (b + \lambda b')y + (c + \lambda c') = 0 \quad \ldots(i)$$

It is given that line (i) is parallel to Y-axis.

$\therefore$ Coefficient of $y = 0$

$$\Rightarrow \quad b + \lambda b' = 0$$

$$\Rightarrow \quad \lambda = -\frac{b}{b'}$$

Substituting the value of λ in Eq. (i), we get

$$\left(a - \frac{a}{b'}a'\right)x + \left(b - \frac{b}{b'}b'\right)y + \left(c - \frac{b}{b'}c'\right) = 0$$

$$\Rightarrow \quad (ab' - ba')x + (b'c - bc') = 0$$

11. (b) Given equation of line,

$$(3x - y + 5) + \lambda(2x - 3y - 4) = 0$$

$$\Rightarrow \quad (3 + 2\lambda)x - (3\lambda + 1)y + (5 - 4\lambda) = 0 \quad \ldots(i)$$

Since, line (i) is parallel to Y-axis.

$\therefore$ Coefficient of $y = 0 \Rightarrow -(3\lambda + 1) = 0$

$$\Rightarrow \quad \lambda = -\frac{1}{3}$$

12. (b) Given equation, $\quad (2 + k)x + (1 + k)y = 5 + 7k$

$$\Rightarrow \quad (2x + y - 5) + k(x + y - 7) = 0$$

which is equation of lines passing through point of intersection of lines

$$2x + y - 5 = 0 \quad \ldots(i)$$

and $\quad x + y - 7 = 0 \quad \ldots(ii)$

On solving Eqs. (i) and (ii), we get

$$x = -2, y = 9$$

13. (a) Given equation of line, $\quad ax + by + c = 0 \quad \ldots(i)$

and a, b, c are in an AP.

$$2b = a + c \quad \ldots(ii)$$

Eq. (i) can be written as $2ax + 2by + 2c = 0$

$$\Rightarrow \quad 2ax + (a + c)y + 2c = 0 \quad \text{[using Eq. (ii)]}$$

$$\Rightarrow \quad 2ax + ay + cy + 2c = 0$$

$$\Rightarrow \quad a(2x + y) + c(y + 2) = 0$$

$$\Rightarrow \quad 2x + y = 0 \text{ and } y + 2 = 0$$

$$\Rightarrow \quad 2x = -y \text{ and } y = -2$$

$$\Rightarrow \quad x = 1 \text{ and } y = -2$$

Hence, line (i) passes through the point (1, −2).

14. (c) The given equation can be written as

$$a(x + y - 1) + b(2x + 3y - 1) = 0$$

which is equation of a line passing through the point of intersection of the lines

$$x + y - 1 = 0 \text{ and } 2x + 3y - 1 = 0$$

The point of intersection of these lines is (2, −1). Hence, the given family of lines passes through the point (2, − 1) for all values of a and b.

15. (b) Given equation of line

$$(y - m_1x - c_1) + \lambda(y - m_2x - c_2) = 0 \quad \ldots(i)$$

Clearly, Eq. (i) passes through the point of intersection of lines

$$y - m_1x - c_1 = 0 \quad \ldots(ii)$$

and $\quad y - m_2x - c_2 = 0 \quad \ldots(iii)$

On solving Eqs. (ii) and (iii), we get

$$x = \frac{c_2 - c_1}{m_1 - m_2}$$

and $\quad y = \dfrac{m_1c_2 - m_2c_1}{m_1 - m_2}$

16. (a) Let the equation of variable line be

$$ax + by + c = 0 \quad \ldots(i)$$

Let p_1, p_2 and p_3 be the length of perpendicular from points (2, 0) (0, 2) and (1, 1) to the line (i), respectively.

$$\therefore \quad p_1 = \frac{|2a + c|}{\sqrt{a^2 + b^2}}, \; p_2 = \frac{|2b + c|}{\sqrt{a^2 + b^2}},$$

$$p_3 = \frac{|a + b + c|}{\sqrt{a^2 + b^2}}$$

Given, $\quad p_1 + p_2 + p_3 = 0$

$$\Rightarrow \quad \frac{1}{\sqrt{a^2 + b^2}}[|2a + c| + |2b + c| + |a + b + c|] = 0$$

$$\Rightarrow \quad |3a + 3b + 3c| = 0$$

$$\Rightarrow \quad a + b + c = 0$$

$$\Rightarrow \quad b = -a - c \quad \ldots(ii)$$

From Eqs. (i) and (ii), we get

$$ax - ay - cy + c = 0$$

$$\Rightarrow \quad a(x - y) - c(y - 1) = 0$$

$$\Rightarrow \quad x - y = 0 \text{ and } y - 1 = 0$$

$$\Rightarrow \quad x = y \text{ and } y = 1$$

$$\Rightarrow \quad x = 1 \text{ and } y = 1$$

17. (b) From the following figure,

$$y + \sqrt{3}x = 2, x > 0$$

$$y - \sqrt{3}x = 2, x < 0$$

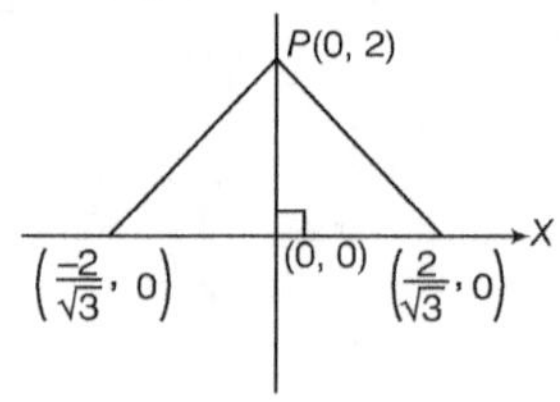

$\therefore$ Foot of perpendicular is (0, 0).

18. (b) Since, a, b, c are in HP, then $\frac{1}{a}, \frac{1}{b}, \frac{1}{c}$ are in an AP.

$$\therefore \quad \frac{2}{b} = \frac{1}{a} + \frac{1}{c}$$

$$\Rightarrow \quad \frac{1}{a} - \frac{2}{b} + \frac{1}{c} = 0$$

Hence, straight line $\frac{x}{a} + \frac{y}{b} + \frac{1}{c} = 0$ is always passes through a fixed point (1, −2).

19. (d) Given lines are
$$x + y + 1 = 0 \quad \ldots(i)$$
$$4x + 3y + 4 = 0 \quad \ldots(ii)$$
$$x + \alpha y + \beta = 0 \quad \ldots(iii)$$
It is given that lines are concurrent, then
$$\begin{vmatrix} 1 & 1 & 1 \\ 4 & 3 & 4 \\ 1 & \alpha & \beta \end{vmatrix} = 0$$
Applying $C_2 \to C_2 - C_1$ and $C_3 \to C_3 - C_1$, we get
$$\Rightarrow \begin{vmatrix} 1 & 0 & 0 \\ 4 & -1 & 0 \\ 1 & \alpha - 1 & \beta - 1 \end{vmatrix} = 0$$
$$\Rightarrow (\beta - 1)(-1 - 0) = 0 \quad \text{(expanding along } c_3)$$
$$\Rightarrow \beta = 1$$
Again, $\alpha^2 + \beta^2 = 2 \Rightarrow \alpha^2 + 1^2 = 2$
$$\Rightarrow \alpha^2 = 2 - 1 \Rightarrow \alpha^2 = 1$$
$$\therefore \alpha = \pm 1$$

20. (a) $a(3x + 4y + 6) + b(x + y + 2) = 0$ intersect at $(-2, 0)$.

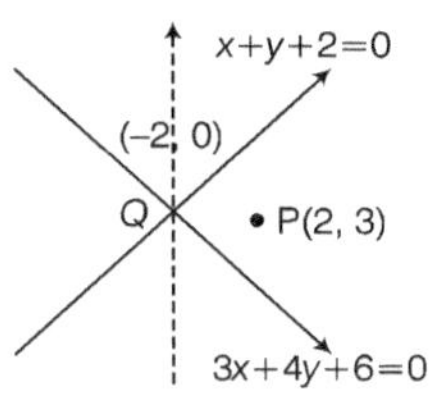

Thus, equation of straight line passing through $Q(-2, 0)$ having maximum distance from $P(2, 3)$ would be perpendicular to PQ.
$$\therefore m_{PQ} = \frac{3}{4}$$
Now, required straight line, is
$$\frac{y - 0}{x + 2} = \frac{-4}{3}$$
$$\Rightarrow 3y = -4x - 8$$
$$\therefore 4x + 3y + 8 = 0$$

21. (b) Equation of altitude BD
$$7x - 10y + 1 = 0 \quad \ldots(i)$$
Equation of bisector of $\angle B$, i.e. BF
$$3x - 2y + 5 = 0 \quad \ldots(ii)$$

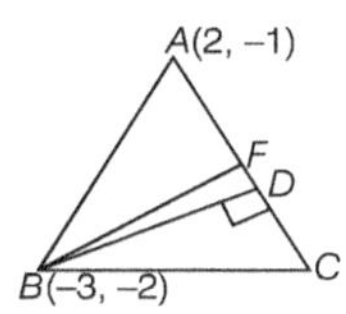

On solving Eqs. (i) and (ii), we get

The coordinates of point B as $\left(-3, \frac{11}{5}\right)$.
$$\text{Slope of } AB, m_1 = \frac{-2 + 1}{-3 - 2}$$
$$m_1 = \frac{1}{5}$$
$$\text{Slope of line (ii)}, \quad m = \frac{3}{2}$$
Let the slope of line BC is m_2,
$$\therefore \frac{m_1 - m}{1 + m_1 m} = -\left(\frac{m_2 - m}{1 + m_2 m}\right) \quad \left(\because BF \text{ is equally inclined with } AB \text{ and } BC\right)$$
$$\Rightarrow \frac{\frac{1}{5} - \frac{3}{2}}{1 + \frac{1}{5} \times \frac{3}{2}} = -\left(\frac{m_2 - \frac{3}{2}}{1 + \frac{3m_2}{2}}\right)$$
$$\Rightarrow -1 = -\left(\frac{2m_2 - 3}{2 + 3m_2}\right)$$
$$\Rightarrow 2 + 3m_2 = 2m_2 - 3$$
$$\Rightarrow m_2 = -5$$
Now, equation of BC
$$y + 2 = -5(x + 3)$$
$$\Rightarrow y + 2 = -5x - 15$$
$$\Rightarrow 5x + y = -17$$
$$\Rightarrow 5x + y + 17 = 0$$

22. (a) Given equation of line is
$$(k + 1)^2 x + ky - 2k^2 - 2 = 0 \quad \ldots(i)$$
$$\Rightarrow (k^2 + 1 + 2k)x + ky - 2k^2 - 2 = 0$$
$$\Rightarrow (k^2 + 1)x + 2kx + ky - 2(k^2 + 1) = 0$$
$$\Rightarrow (k^2 + 1)x - 2(k^2 + 1) + k(2x + y) = 0$$
$$\Rightarrow (k^2 + 1)x - 2(k^2 + 1) = 0$$
and $2x + y = 0$
$$\Rightarrow x = 2 \text{ and } 2x + y = 0$$
$$\Rightarrow x = 2 \text{ and } y = -4$$
Hence, line (i) passes through point $(2, -4)$.

Now, equation of line having slope 2 and passes through $(2, -4)$, is
$$y + 4 = 2(x - 2)$$
$$\Rightarrow y + 4 = 2x - 4$$
$$\Rightarrow y = 2x - 8$$

23. (b) $A(3, 0)$, $B(4, 0)$

Let l be $(y - 2) = m(x - 2)$.
$$\therefore C\left(2 - \frac{2}{m}, 0\right)$$
Now, $3\left(2 - \frac{2}{m}\right) = 4^2 = 16$
$$\Rightarrow 6m - 6 = 16m$$
$$\Rightarrow m = \frac{-3}{5}$$
$\therefore$ The line is, $(y - 2) = \frac{-3}{5}(x - 2)$
$$\Rightarrow 3x + 5y - 16 = 0$$

24. (a) $AB = \sqrt{8^2 + 19^2} = \sqrt{425}$
$$AC = \sqrt{16^2 + 13^2} = \sqrt{425}$$
$\therefore \Delta$ is isosceles.

Thus, Statement I is correct and Statement II is correct explanation of Statement I.

25. (b) Let (h, k) be the point on the line
$$L = 3x + y + 4 = 0$$
Now, $PA = AB$
$$\sqrt{(h + 5)^2 + (k - 6)^2} = \sqrt{(h - 3)^2 + (k - 2)^2}$$
$$\Rightarrow h^2 + 25 + 10h + k^2 + 36 - 12k = h^2 + 9 - 6h + k^2 + 4 - 4k$$
$$\Rightarrow 10h - 12k + 61 = 13 - 6h - 4k$$
$$\Rightarrow 16h - 8k + 48 = 0 \quad \ldots(i)$$
and (h, k) lies on line L.
$$\therefore 3h + k + 4 = 0 \quad \ldots(ii)$$
On solving Eqs. (i) and (ii), we get
$$h = -2, k = 2$$
Hence, only one pair $(-2, 2)$ on line L which is equidistant from A and B.

$\therefore$ Statement I is true.

Statement II Point $(-2, 2)$ is equidistant from A and B, then by sign conversion, points A and B lie on opposite sides.

26. (a,c,d) The coordinates of triangle are $A(1, -2)$, $B(k, -2k)$ $C(l, l - 3)$

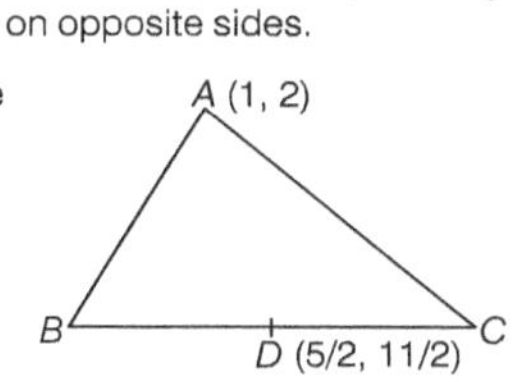

$$\therefore \frac{k + 1 + l}{3} = 2$$
and $\frac{-2 - 2k + l - 3}{3} = 3$
$$\Rightarrow k = -3, l = 8$$
$$\therefore B(-3, 6) \text{ and } C(8, 5)$$

Equation of BC is $\quad x + 11y = 63$

$\Rightarrow \quad p + q = 74$

Medium through A, is $\quad 5x - y - 7 = 0$

Altitude through A, is $\quad 11x - y - 13 = 0$

27. (a,b) Let the length of the sides of the rhombus be λ.

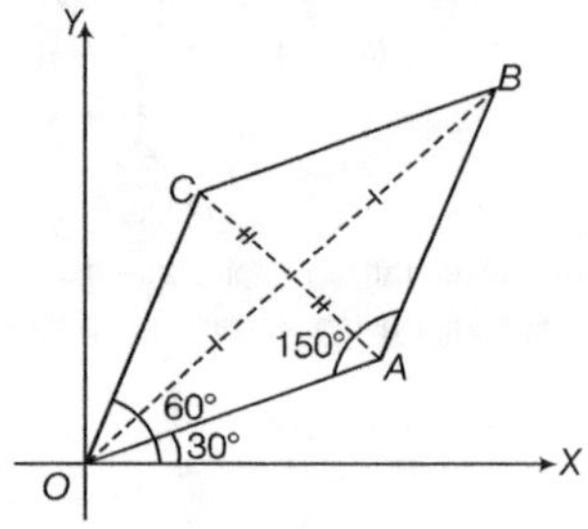

$\Rightarrow \quad$ Area $= \lambda^2 \cdot \sin 30° = 2$

$\Rightarrow \quad \lambda = 2$

$\therefore \quad OA = OC = 2$

$\Rightarrow \quad OB^2 = OA^2 + AB^2 - 2\, OA \cdot AB \cdot \cos 150°$

$= 4 + 4 - 2(4) \cdot \left(-\frac{\sqrt{3}}{2}\right)$

$= 4(2 + \sqrt{3})$

$\therefore \quad OB = 2\sqrt{2 + \sqrt{3}}$

$\Rightarrow \quad B(\sqrt{4 + 2\sqrt{3}}, \sqrt{4 + 2\sqrt{3}})$

$= (1 + \sqrt{3}, 1 + \sqrt{3}) \in I$ quadrant

or $\quad B(-1 - 1\sqrt{3}, -1 - \sqrt{3}) \in$ III quadrant

28. (a,d) Given lines are

$$x - 2y - 6 = 0 \quad \ldots(i)$$

$$3x + y - 4 = 0 \quad \ldots(ii)$$

$$\lambda x + 4y + \lambda^2 = 0 \quad \ldots(iii)$$

It is given that the line are concurrent.

$$\Rightarrow \begin{vmatrix} 1 & 3 & \lambda \\ -2 & 1 & 4 \\ -6 & -4 & \lambda^2 \end{vmatrix} = 0$$

$\Rightarrow \quad 1(\lambda^2 + 16) - 3(-2\lambda^2 + 24) + \lambda(8 + 6) = 0$

$\Rightarrow \quad \lambda^2 + 16 + 6\lambda^2 - 72 + 14\lambda = 0$

$\Rightarrow \quad 7\lambda^2 + 14\lambda - 56 = 0$

$\Rightarrow \quad \lambda^2 + 2\lambda - 8 = 0$

$\therefore \quad \lambda = \frac{-2 \pm \sqrt{4 + 32}}{2} = \frac{-2 \pm \sqrt{36}}{2} = \frac{-2 \pm 6}{2} = \frac{4}{2}, \frac{-8}{2}$

$\therefore \quad = 2, -4$

29. (a,c,d) Given lines are $\quad ax + by + c = 0$

$bx + cy + a = 0$

$cx + ay + b = 0$

Since, lines are concurrent.

$$\therefore \begin{vmatrix} a & b & c \\ b & c & a \\ c & a & b \end{vmatrix} = 0$$

Applying $R_1 \to R_1 + R_2 + R_3$, we get

$$\Rightarrow \begin{vmatrix} a+b+c & a+b+c & a+b+c \\ b & c & a \\ c & a & b \end{vmatrix} = 0$$

$$\Rightarrow (a + b + c)\begin{vmatrix} 1 & 1 & 1 \\ b & c & a \\ c & a & b \end{vmatrix} = 0$$

Applying $C_2 \to C_2 - C_1$ and $C_3 \to C_3 - C_1$, we get

$$\Rightarrow (a + b + c)\begin{vmatrix} 1 & 0 & 0 \\ b & c-b & a-b \\ c & a-c & b-c \end{vmatrix} = 0$$

$\Rightarrow \quad (a + b + c)\{(c - b)(b - c) - (a - b)(a - c)\} = 0$

$\Rightarrow \quad (a + b + c)\{(bc + ca + ab) - (a^2 + b^2 + c^2)\} = 0$

$\Rightarrow \quad (a + b + c)[a^2 + b^2 + c^2 - ab - bc - ca] = 0$

$\Rightarrow \quad a^2 + b^2 + c^2 - ab - bc - ca = 0 \quad (\because a + b + c \neq 0)$

$\Rightarrow \quad a^2 + b^2 + c^2 = ab + bc + ca$

$\Rightarrow \quad a = b = c$

Also, $\quad (a + b + c)(a^2 + b^2 + c^2 - ab - bc - ca) = 0$

$\Rightarrow \quad a^3 + b^3 + c^3 - 3abc = 0$

30. (b,d) The coordinates of vertices are

$A\left(\frac{5}{4}, \frac{7}{8}\right)$, $B(-7, 5)$ and $C\left(\frac{1}{3}, \frac{1}{9}\right)$

$\because P(\alpha, \alpha^2)$ lies inside the ΔABC, then

(1) A and P must lie on the same side BC.

(2) B and P must lie on the same side of CA.

(3) C and P must lie on the same side of AB

Then,

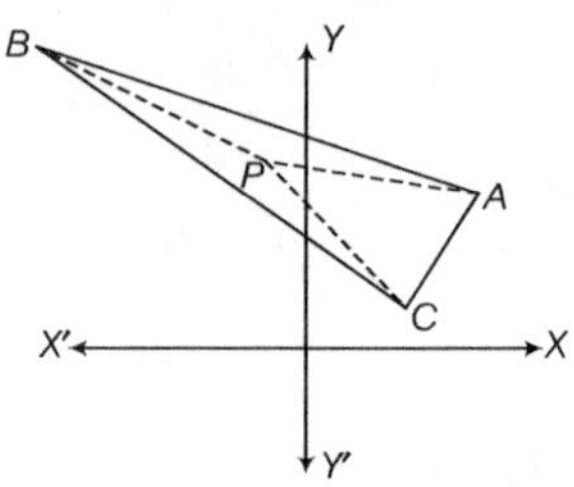

$$\frac{\frac{5}{2} + \frac{21}{8} - 1}{2\alpha + 3\alpha^2 - 1} > 0$$

$\Rightarrow \quad \frac{33}{2\alpha + 3\alpha^2 - 1} > 0$

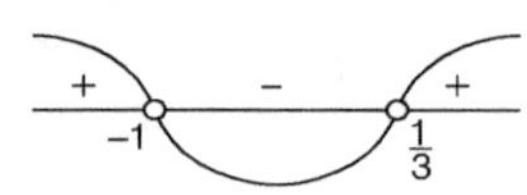

$3\alpha^2 + 2\alpha - 1 > 0$

$(\alpha + 1)\left(\alpha - \frac{1}{3}\right) > 0$

$\Rightarrow \quad \alpha \in (-\infty, -1) \cup \left(\frac{1}{3}, \infty\right) \quad \ldots(i)$

and $\quad \frac{-35 - 30 - 1}{5\alpha - 6\alpha^2 - 1} > 0$

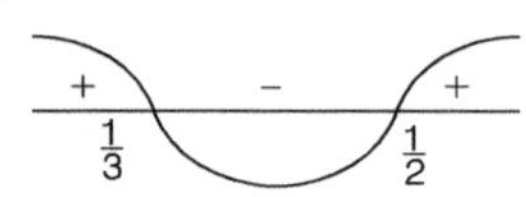

$\Rightarrow \quad 5\alpha - 6\alpha^2 - 1 < 0$

$\Rightarrow \quad \left(\alpha - \frac{1}{3}\right)\left(\alpha - \frac{1}{2}\right) > 0$

$\Rightarrow \quad \alpha \in \left(-\infty, \frac{1}{3}\right) \cup \left(\frac{1}{2}, \infty\right) \quad \ldots(ii)$

and $\quad \frac{\frac{1}{3} + \frac{2}{9} - 3}{\alpha + 2\alpha^2 - 3} > 0$

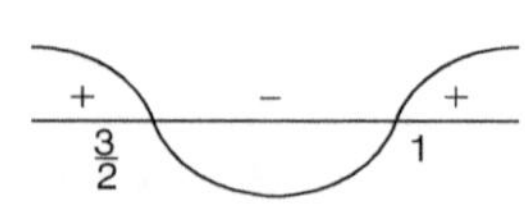

$\Rightarrow \quad \alpha + 2\alpha^2 - 3 < 0$

$\Rightarrow \quad (2\alpha + 3)(\alpha - 1) < 0$

$\alpha \in \left(\frac{-3}{2}, 1\right) \quad \ldots(iii)$

From Eqs. (i), (ii) and (iii), we get

$$\alpha \in \left(\frac{-3}{2}, -1\right) \cup \left(\frac{1}{2}, 1\right)$$

31. (a,b) Given family of lines, $(1+2\lambda)x-(1+\lambda)y+1=0$

$$\Rightarrow \quad (x-y+1)+\lambda(2x-y)=0$$

$\therefore$ Family of lines passing through point of intersection of lines

$$x-y+1=0 \quad \text{and} \quad 2x-y=0$$

i.e. through the point (1, 2).

Since, the family of lines form an equilateral triangle with the line $x+y=5$.

$$\therefore \quad y-2=\frac{-1\pm\tan 60^\circ}{1\mp(-1)\tan 60^\circ}(x-1)$$

$$y-2=\frac{-1\pm\sqrt{3}}{1\mp(-\sqrt{3})}(x-1) \Rightarrow y-2=\frac{-1+\sqrt{3}}{1+\sqrt{3}}(x-1)$$

$$\text{and} \quad y-2=\frac{-1-\sqrt{3}}{1-\sqrt{3}}(x-1) \Rightarrow y-2=(2-\sqrt{3})(x-1)$$

$$\text{and} \quad y-2=(2+\sqrt{3})(x-1)$$

32. (a,c) Find the angular bisector of $x-\sqrt{3}y+2\sqrt{3}=0$ and $\sqrt{3}x-y+3=0$. This will be parallel to the diagonal which is passing through (1, 2).

33. (a, c) Let $C(h, k)$ be the third vertex of ΔABC which are lies on the line $y=x+3$.

$$\therefore \quad k=h+3 \quad \ldots(\text{i})$$

Area of ΔABC,

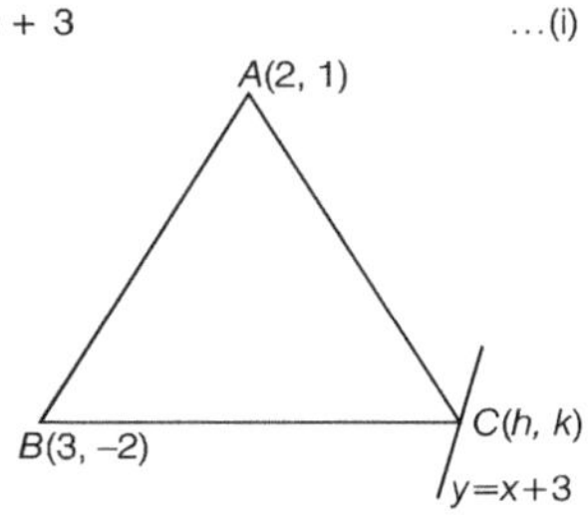

$$\Rightarrow \quad \frac{1}{2}\begin{vmatrix} 2 & 1 & 1 \\ 3 & -2 & 1 \\ h & k & 1 \end{vmatrix}=\pm 5$$

$$\Rightarrow \quad \begin{vmatrix} 2 & 1 & 1 \\ 3 & -2 & 1 \\ h & k & 1 \end{vmatrix}=\pm 10$$

$$\Rightarrow \quad \begin{vmatrix} 2 & 1 & 1 \\ 1 & -3 & 0 \\ h-2 & k-1 & 0 \end{vmatrix}=\pm 10$$

Applying $R_2 \to R_2-R_1$ and $R_3 \to R_3 \to R_1$

$$\Rightarrow \quad (k-1)+3(h-2)=\pm 10$$

$$\Rightarrow \quad k-1+3h-6=\pm 10$$

$$\Rightarrow \quad 3h+k-7=\pm 10$$

$$\Rightarrow \quad 3h+k-7=10$$

$$\text{or} \quad 3h+k-7=-10$$

$$\Rightarrow \quad 3h+k=17 \quad \text{or} \quad 3h+k=-3$$

$$\Rightarrow \quad 3h+k=17 \quad \ldots(\text{ii})$$

$$\text{and} \quad 3h+k=-3 \quad \ldots(\text{iii})$$

On solving Eqs. (i) and (ii), we get

$$h=\frac{7}{2}, \quad k=\frac{13}{2}$$

On solving Eqs. (i) and (iii), we get

$$h=-\frac{3}{2}, k=\frac{3}{2}$$

Hence, coordinates of third vertex are $\left(\frac{7}{2},\frac{13}{2}\right)$ and $\left(\frac{-3}{2},\frac{3}{2}\right)$.

34. (d) Let mirror image of (4, 3) about the line $(2-3m)x+(1+4m)y+(2-m)=0$ be (h, k).

$$\therefore \quad \frac{h-4}{2-3m}=\frac{k-3}{1+4m}$$

$$=\frac{-2[(2-3m)\cdot(4)+(1+4m)\cdot(3)+(2-m)]}{(2-3m)^2+(1+4m)^2}$$

$$\Rightarrow \quad \frac{h-4}{2-3m}=\frac{k-3}{1+4m}=\frac{-2(13-m)}{(2-3m)^2+(1+4m)^2}$$

Eliminating m, the locus passes through $\left(\frac{-62}{11},\frac{-41}{11}\right)$.

35. (d) Let the slope of BC be m.

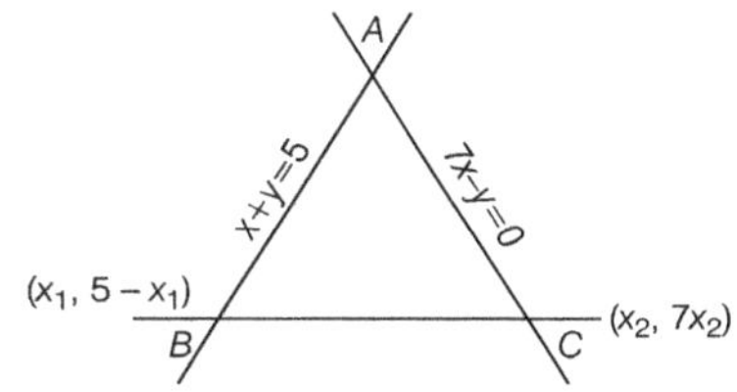

$$\angle ABC=\angle ACB$$

$$\Rightarrow \quad \frac{m-7}{1+7m}=\frac{-m-1}{1-m}$$

$$\Rightarrow \quad m=-3, \frac{1}{3}$$

$$\therefore \quad \text{Slope of } BC=\frac{7x_2-(5-x_1)}{x_2-x_1}=-3 \text{ or } \frac{1}{3}$$

$$\Rightarrow \quad 5x_2-x_1=4$$

$$\text{or} \quad 5x_2+x_1=6$$

Hence, option (d) is not possible.

36. (c) Given, $x_1=\frac{7}{2}$

$$\Rightarrow \quad x_2=\frac{4+\frac{7}{2}}{5} \quad \text{or} \quad \frac{6-\frac{7}{2}}{5}$$

$$\therefore \quad x_2=\frac{3}{2} \quad \text{or} \quad \frac{1}{2}$$

Thus, option (c) is not possible.

37. (b) Given, area = 5 sq units

Since, there are two possible positions of B and C. Thus, possible number of ΔABC is 4.

38. (b) Let $C(h, k)$, then

$$h-4k+10=0 \quad \ldots(\text{i})$$

and $E\left(\frac{h+3}{2},\frac{k-1}{2}\right)$ lies on

$$6x+10y-59=0$$

$$\Rightarrow \quad 3h+5k=55 \quad \ldots(\text{ii})$$

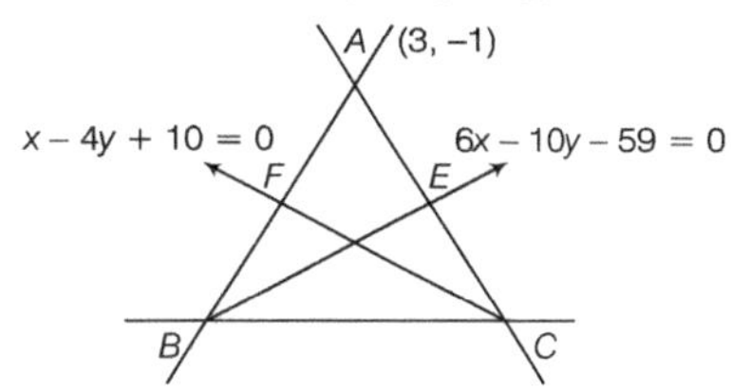

From Eqs. (i) and (ii), we get

$$h=10, k=5$$

$$\therefore \quad \text{Slope of } AC \text{ is } \left(\frac{6}{7}\right)$$

$$\text{and} \quad \text{Slope of } BC \text{ is } \left(-\frac{2}{9}\right).$$

39. (c) Equation of BC, is

$$2x+9y-65=0$$

Now, the point B can be found by solving BC and BE.

Hence, equation of AB can be determined as

$$18x+13y=41$$

40. (c) Distance, $AC=\sqrt{85}$ units

DPP-5 Locus

1. (c) Given equation of lines are

$$\frac{x}{a}+\frac{y}{b}=1 \quad \ldots(i)$$

and $$a+b=10 \quad \ldots(ii)$$

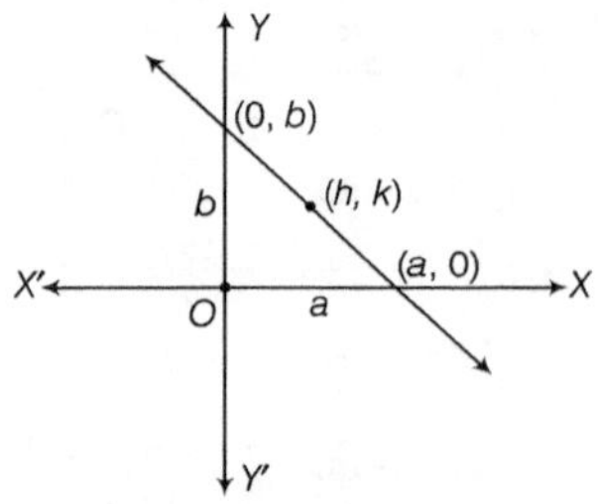

Let the locus of mid-point of the intercept between the axes is (h, k).

$\therefore$ $$h=\frac{a+0}{2} \Rightarrow a=2h$$

and $$k=\frac{0+b}{2} \Rightarrow b=2k$$

Now, $a+b=10$

$\Rightarrow$ $2h+2k=10$

$\Rightarrow$ $h+k=5$

or $x+y=5$

2. (c) Let the line makes intercepts a and b on X and Y-axes respectively.

$\therefore$ $$a+b=c \quad \text{(given)}$$

Let the locus of mid-point of intercept between the axes be (h, k).

$\therefore$ $$h=\frac{a+0}{2} \Rightarrow h=\frac{a}{2}$$

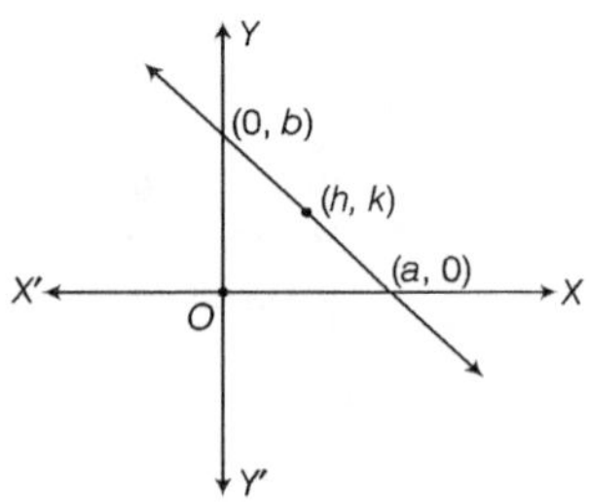

$$k=\frac{0+b}{2} \Rightarrow k=\frac{b}{2}$$

Now, $a+b=c$

$\Rightarrow$ $2h+2k=c$

or $2(x+y)=c$

3. (b) Let the equation of variable line be

$$\frac{x}{h}+\frac{y}{k}=1 \quad \ldots(i)$$

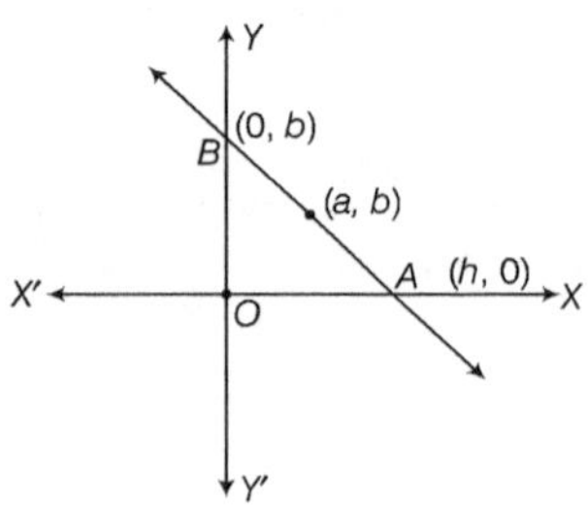

Equation of line parallel to Y-axis and passing through point A is

$$x=h \quad \ldots(ii)$$

Equation of line parallel to X-axis and passing through point B is

$$y=k \quad \ldots(iii)$$

Now, line (i) is also passes through (a, b).

$\therefore$ $$\frac{a}{h}+\frac{b}{k}=1 \Rightarrow \frac{a}{x}+\frac{b}{y}=1$$

which is required locus.

4. (d) Given, equation of lines

$$x\cos\alpha+y\sin\alpha=a \quad \ldots(i)$$

$$x\sin\alpha-y\cos\alpha=b \quad \ldots(ii)$$

On squaring and adding Eqs. (i) and (ii), we get

$$x^2(\cos^2\alpha+\sin^2\alpha)+y^2(\sin^2\alpha+\cos^2\alpha)$$
$$+2xy\cos\alpha\sin\alpha-2xy\sin\alpha\cos\alpha=a^2+b^2$$

$\Rightarrow$ $$x^2+y^2=a^2+b^2 \quad (\because \sin^2\alpha+\cos^2\alpha=1)$$

5. (a) Let (h, k) be the point.

$\therefore$ Distance between (h, k) and $(3, -2)$ is

$$=\sqrt{(h-3)^2+(k+2)^2}$$

and distance of point (h, k) from the line $5x-12y=13$ is

$$=\frac{|5h-12k-13|}{\sqrt{25+144}}=\frac{|5h-12k-13|}{\sqrt{169}}$$

$$=\frac{|5h-12k-13|}{13}$$

According to the question,

$$(h-3)^2+(k+2)^2=\frac{|5h-12k-13|}{13}$$

$\Rightarrow$ $$h^2+9-6h+k^2+4+4k=\frac{|5h-12k-13|}{13}$$

$\Rightarrow$ $$13(h^2+k^2-6h+4k+13)=\pm(5h-12k-13)$$

On taking +ve sign, wen have

$$13h^2+13k^2-78h+52k+169=5h-12k-13$$

$\Rightarrow$ $$13h^2+13k^2-83h+64k+182=0$$

$\Rightarrow$ $$13x^2+13y^2-83x+64y+182=0$$

On taking –ve sign, we have

$$13h^2+13k^2-78h+52k+169=-5h+12k+13$$

$\Rightarrow$ $$13h^2+13k^2-73h+40k+156=0$$

$\Rightarrow$ $$13x^2+13y^2-73x+40y+156=0$$

6. (d) Let the equation of line segment be

$$\frac{x}{a}+\frac{y}{b}=1 \quad \ldots(i)$$

Its ends are at point A on X-axis and at point B on Y-axis.

Let $P(h, k)$ divides the join of AB in the ratio 1 : 2.

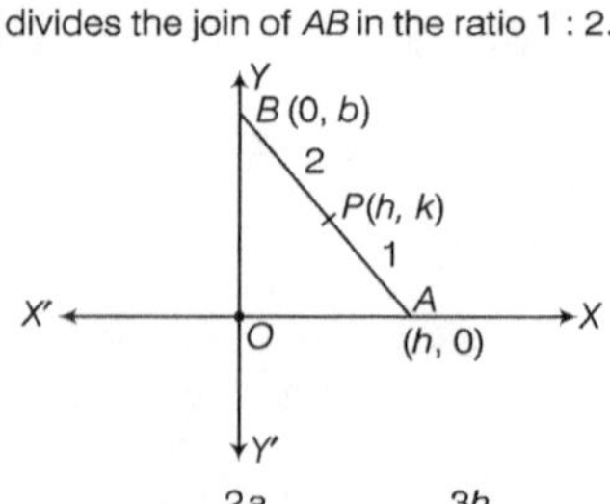

$\therefore$ $$h=\frac{2a}{3} \Rightarrow a=\frac{3h}{2}$$

$$k=\frac{b}{3} \Rightarrow b=3k$$

Now, in ΔOAB,

$$OA^2 + OB^2 = AB^2$$

$\Rightarrow$ $a^2 + b^2 = l^2$ $(\because OA = a, OB = b, AB = l)$

$\Rightarrow$ $\left(\frac{3h}{2}\right)^2 + (3k)^2 = l^2$

$\Rightarrow$ $\frac{9h^2}{4} + 9k^2 = l^2$

$\Rightarrow$ $9h^2 + 36k^2 = 4l^2$

$\Rightarrow$ $9x^2 + 36y^2 = 4l^2$

7. (a) We know that, the foot of perpendicular is (x_2, y_2), drawn from point (x_1, y_1) to the line $ax + by + c = 0$.

Then, $\frac{x_2 - x_1}{a} = \frac{y_2 - y_1}{b} = \frac{-(ax_1 + by_1 + c)}{a^2 + b^2}$

Let foot of perpendicular be (α, β).

$\therefore$ $\frac{\alpha - 0}{\frac{1}{a}} = \frac{\beta - 0}{\frac{1}{b}} = -\left[\frac{(-1)}{\frac{1}{a^2} + \frac{1}{b^2}}\right]$

$\Rightarrow$ $a\alpha = b\beta = c^2$

$\Rightarrow$ $\alpha = \frac{c^2}{a}, \beta = \frac{c^2}{b}$

$\Rightarrow$ $a^2 + \beta^2 = \frac{c^4}{a^2} + \frac{c^4}{b^2}$

$\Rightarrow$ $\alpha^2 + \beta^2 = c^4\left[\frac{1}{a^2} + \frac{1}{b^2}\right]$

$\Rightarrow$ $\alpha^2 + \beta^2 = c^4 \times \frac{1}{c^2}$

$\Rightarrow$ $\alpha^2 + \beta^2 = c^2$

$\Rightarrow$ $x^2 + y^2 = c^2$

8. (c) Line $x\cos\alpha + y\sin\alpha = p$ intersect at X and Y-axes at point $\left(\frac{p}{\cos\alpha}, 0\right)$ and $\left(0, \frac{p}{\sin\alpha}\right)$, respectively.

Let $P(h, k)$ be the mid-point of intercept between axes.

$\therefore$ $h = \frac{p}{2\cos\alpha}$

$k = \frac{p}{2\sin\alpha}$

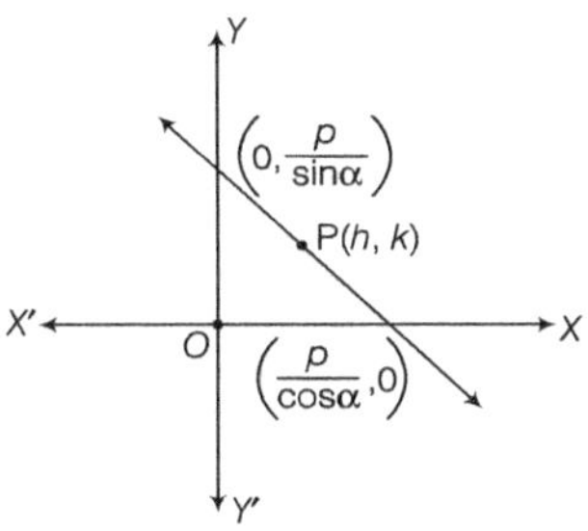

Now, $\frac{1}{h^2} + \frac{1}{k^2} = \frac{4}{p^2}[\cos^2\alpha + \sin^2\alpha]$

$\Rightarrow$ $\frac{1}{h^2} + \frac{1}{k^2} = \frac{4}{p^2}$

$\Rightarrow$ $\frac{1}{x^2} + \frac{1}{y^2} = \frac{4}{p^2}$

9. (c) Let locus of C be (h, k).

Now, centroid of ΔABC,

$$x = \frac{2 - 2 + h}{3} = \frac{h}{3}$$

and $y = \frac{-3 + 1 + k}{3} = \frac{k - 2}{3}$, i.e. $\left(\frac{h}{3}, \frac{k - 2}{3}\right)$

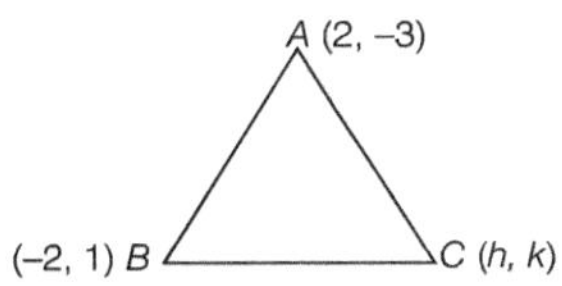

Since, centroid lies on the line

$$2x + 3y = 1$$

$\therefore$ $\frac{2h}{3} + \frac{3(k - 2)}{3} = 1$

$\Rightarrow$ $\frac{2h}{3} + (k - 2) = 1$

$\Rightarrow$ $2h + 3k - 6 = 3$

$\Rightarrow$ $2h + 3k = 9$

$\Rightarrow$ $2x + 3y = 9$

10. (a) Let the two perpendicular lines be taken as the coordinate axes. If (h, k) is any point on the locus, then according to the given condition $|h| + |k| = 1$.

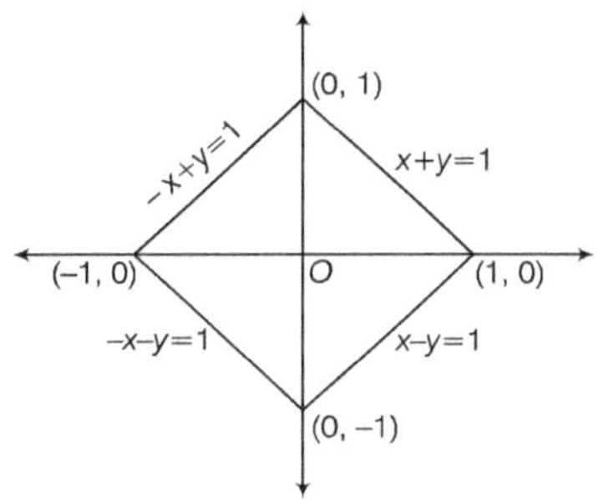

Hence, the locus of (h, k) is $|x| + |y| = 1$. This consist of four line segments enclosing a square as shown in the figure.

11. (a) Equation of line parallel to X-axis and passing through $P(h, k)$ is

$$y = k \quad \ldots(\text{i})$$

and given equation of lines, $y = x$ ...(ii)

$$x + y = 2 \quad \ldots(\text{iii})$$

Let ABC be the triangle formed by the points of intersection of the lines (i), (ii) and (iii) as shown in the figure,

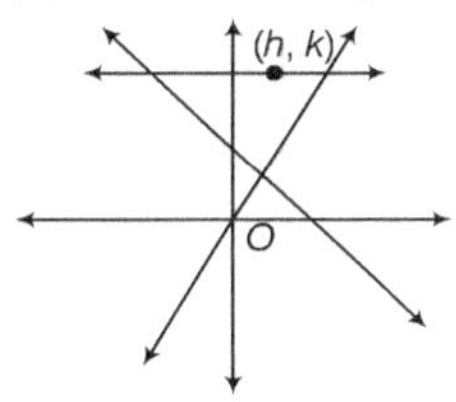

Then, $A(k, k), B(1, 1), C(2 - k, k)$

$\therefore$ Area of $\Delta = \frac{1}{2}\begin{vmatrix} k & k & 1 \\ 1 & 1 & 1 \\ 2 - k & k & 1 \end{vmatrix} = 4h^2$

Applying $C_1 \to C_1 - C_2$, we get

$$\frac{1}{2}\begin{vmatrix} 0 & k & 1 \\ 0 & 1 & 1 \\ 2-2k & k & 1 \end{vmatrix} = 4h^2$$

$$\Rightarrow \quad \frac{1}{2}|(2-2k)(k-1)| = 4h^2$$

$$\Rightarrow \quad (k-1)^2 = 4h^2$$

$$\Rightarrow \quad k = -1 = 2h \quad \text{or} \quad k - 1 = -2h$$

$$\Rightarrow \quad k = 2h + 1 \quad \text{or} \quad k = -2h + 1$$

Hence, locus of (h, k) is $y = 2x + 1$ or $y = -2x + 1$.

12. (a) Equation of line which intersect coordinate axes at $A(h, 0)$ and $B(0, k)$ is

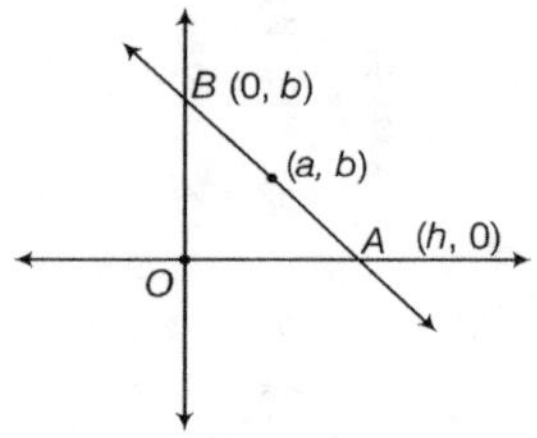

$$\frac{x}{h} + \frac{y}{k} = 1 \qquad \ldots(i)$$

Now, coordinates of centroid of ΔOAB are

$$x = \frac{0 + h + 0}{3} \Rightarrow x = \frac{h}{3}$$

$$y = \frac{0 + 0 + k}{3} \Rightarrow y = \frac{k}{3}$$

Also, line (i) passes through point (a, b).

$$\therefore \quad \frac{a}{h} + \frac{b}{k} = 1$$

$$\Rightarrow \quad ka + bh = hk$$

$$\Rightarrow \quad 3ya + 3bx = 9xy \qquad (\because h = 3x, k = 3y)$$

$$\Rightarrow \quad bx + ay = 3xy$$

$$\Rightarrow \quad bx + ay - 3xy = 0$$

13. (a) Let $A(\alpha, 0)$ and $B(0, \beta)$ be two points on the X and Y-axes respectively.

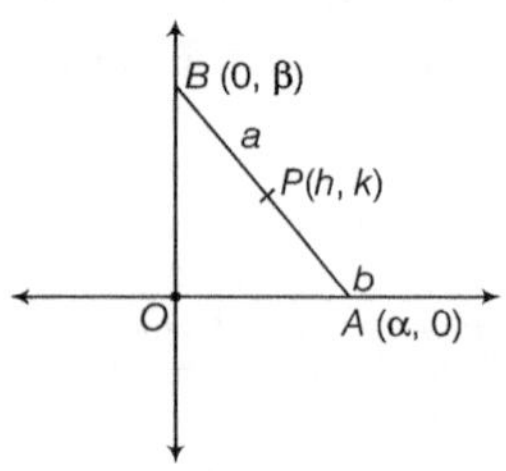

$$\therefore \quad h = \frac{a \cdot \alpha}{a + b} \Rightarrow \alpha = \left(\frac{a+b}{b}\right)h$$

$$k = \frac{b \cdot \beta}{a + b} \Rightarrow \beta = \left(\frac{a+b}{b}\right)k$$

Now, $\quad OA^2 + OB^2 = AB^2$

$$\Rightarrow \quad \alpha^2 + \beta^2 = (a + b)^2$$

$$\Rightarrow \quad \left(\frac{a+b}{a}\right)^2 h^2 + \left(\frac{a+b}{b}\right)^2 k^2 = (a+b)^2$$

$$\Rightarrow \quad \frac{h^2}{a^2} + \frac{k^2}{b^2} = \frac{(a+b)^2}{(a+b)^2}$$

$$\Rightarrow \quad \frac{h^2}{a^2} + \frac{k^2}{b^2} = 1 \Rightarrow \quad \frac{x^2}{a^2} + \frac{y^2}{b^2} = 1$$

which is required locus.

14. (b) $AB = a$ (constant)

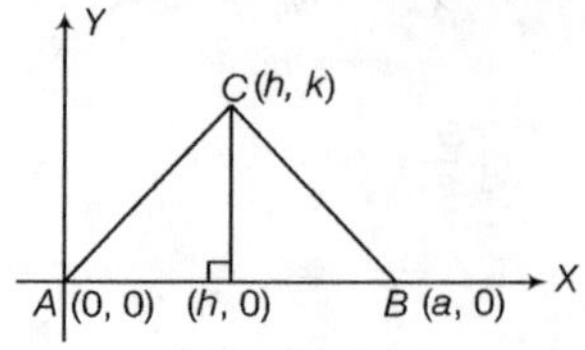

$$\cot A = \frac{h}{k} \quad \text{and} \quad \cot B = \frac{a - h}{k}$$

Since, $\quad \cot A + \cot B = \lambda$

$$\Rightarrow \quad \frac{h}{k} + \frac{a - h}{k} = \lambda$$

$$\Rightarrow \quad k = \frac{a}{\lambda} \quad \text{or} \quad y = \frac{a}{\lambda}$$

i.e. parallel to X-axis.

15. (d) Coordinates of mid-point M of AB is $(t, 2)$.

Equation of MR,

$$y - 2 = \frac{t}{2}(x - t)$$

$$\Rightarrow \quad 2(y - 2) = t(x - t)$$

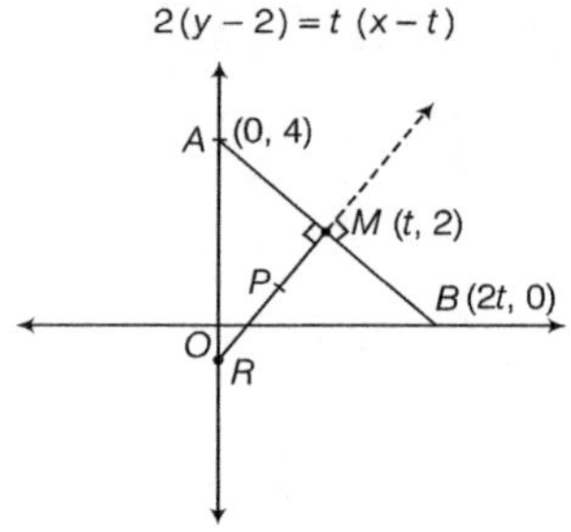

$$\Rightarrow \quad 2y - 4 = tx - t^2$$

$$\Rightarrow \quad tx - 2y + 4 - t^2 = 0 \qquad \ldots(i)$$

It intersect Y-axis at $R\left(0, 2 - \frac{t^2}{2}\right)$.

Let $P(h, k)$ be the mid-point of MR.

$$\therefore \quad h = \frac{t}{2} \Rightarrow t = 2h$$

$$\Rightarrow \quad k = \frac{2 + 2 - \frac{t^2}{2}}{2} = 2 - \frac{t^2}{4}$$

$$\Rightarrow \quad k = 2 - \frac{(2h)^2}{4} \Rightarrow k = 2 - \frac{4h^2}{4}$$

$$\Rightarrow \quad k = 2 - h^2$$

$$\Rightarrow \quad y = 2 - x^2$$

16. (d) Distance of all the points from (0, 0) are 5 units. That means circumcentre of the triangle formed by the given point is (0, 0).

If $G(h, k)$ be the centroid of triangle, then

$$3h = 3 + 5(\cos\theta + \sin\theta)$$

and $\quad 3k = 4 + 5(\sin\theta - \cos\theta)$

If $H(\alpha, \beta)$ is the orthocentre, then

$$OG : GH = 1 : 2$$

$$\Rightarrow \quad \alpha = 3h, \beta = 3k$$

$$\Rightarrow \quad \cos\theta + \sin\theta = \frac{\alpha - 3}{5}, \sin\theta - \cos\theta = \frac{\beta - 4}{5}$$

$$\Rightarrow \quad \sin\theta = \frac{\alpha + \beta - 7}{10}, \cos\theta = \frac{\alpha - \beta + 1}{10}$$

Thus, the locus of (α, β) is

$$(x + y - 7)^2 + (x - y + 1)^2 = 100$$

17. (b) $d(x, y) = \max\{|x|, |y|\}$ (given)

where, $d(x, y) = 1$

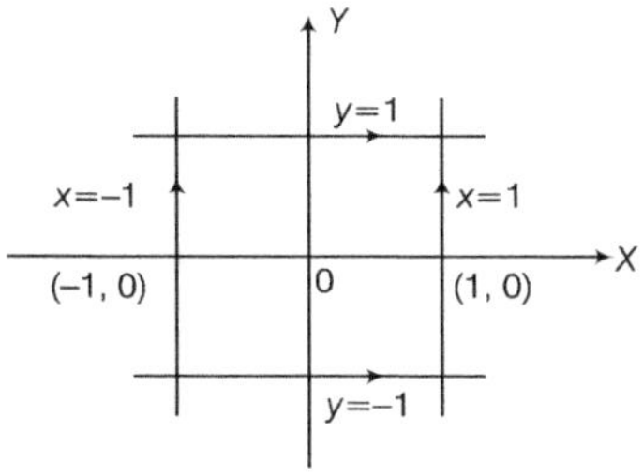

$\Rightarrow \quad |x| = 1$ or $|y| = 1$

i.e. a square of area 4 sq units.

18. (d) We have, $P(1, 0), Q(-1, 0), R(2, 0)$

Let $S(x, y)$

Given,

$$SQ^2 + SR^2 = 2SP^2$$

Hence, $(x+1)^2 + y^2 + (x-2)^2 + y^2 = 2[(x-1)^2 + y^2]$

$\Rightarrow \quad (x+1)^2 + y^2 + (x-2)^2 + y^2 = 2x^2 + 2y^2 - 4x + 2$

$\Rightarrow \quad 2x^2 + 2y^2 - 2x + 5 = 2x^2 + 2y^2 - 4x + 2$

$\Rightarrow \quad 2x + 3 = 0$

$\Rightarrow \quad x = \frac{-3}{2}$

which is a straight line parallel to Y-axis.

19. (d) Given equation is

$$z^2 - xz + (x - y)^2 = 0$$

Since, the quadratic equation has equal roots

$\therefore \quad D = 0$

$\Rightarrow \quad (-x)^2 - 4 \times 1 \times (x-y)^2 = 0$

$\Rightarrow \quad x^2 - 4(x-y)^2 = 0$

$\Rightarrow \quad x^2 - 4(x^2 + y^2 - 2xy) = 0$

$\Rightarrow \quad x^2 - 4x^2 - 4y^2 - 8xy = 0$

$\Rightarrow \quad -3x^2 - 4y^2 - 8xy = 0$

$\Rightarrow \quad 3x^2 + 4y^2 + 8xy = 0$

which represent a pair of straight lines passes through origin with slope $\frac{3}{2}$ and $\frac{1}{2}$.

20. (a) Let $P(h, k)$ be the mid-point of BC.

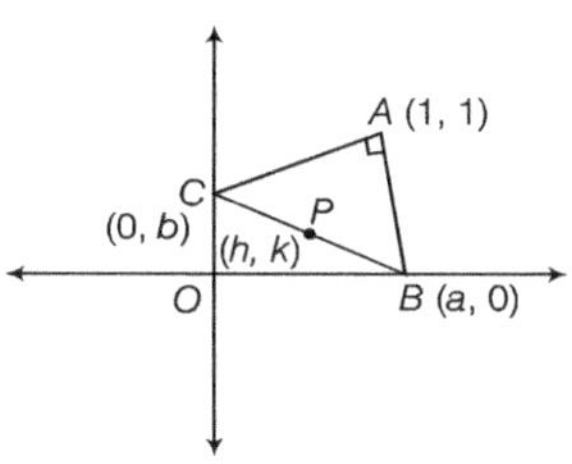

$\therefore \quad a = 2h$

$b = 2k$

Now, $AB^2 = (a-1)^2 + 1^2$

$= a^2 - 2a + 2$

$AC^2 = b^2 - 2b + 2$

and $BC^2 = a^2 + b^2$

In ΔABC,

$$AB^2 + AC^2 = BC^2$$

$\Rightarrow \quad a^2 - 2a + 2 + b^2 - 2b + 2 = a^2 + b^2$

$\Rightarrow \quad a + b = 2$

$\Rightarrow \quad 2h + 2k = 2$

$\Rightarrow \quad h + k = 1$

$\Rightarrow \quad x + y = 1$

which is required locus.

21. (a) Let (h, k) be point of intersection, then

$$\frac{h}{a} + \frac{k}{b} = 1$$

Also, $ah + bk = 1$

(given, $a^2 + b^2 = ab, \Rightarrow \frac{a}{b} + \frac{b}{a} = 1$)

Multiplying $h^2 + k^2 + hk\left(\frac{b}{a} + \frac{a}{b}\right) = 1$

$\therefore \quad h^2 + k^2 + hk = 1$

or $\quad x^2 + y^2 + xy - 1 = 0$

22. (a) Coordinates of circumcentre $\left(\frac{l}{l^2 - m^2}, \frac{m}{m^2 - l^2}\right)$.

$\therefore \quad h = \frac{l}{l^2 - m^2}, k = \frac{m}{m^2 - l^2}$

Also, $l^2 + m^2 = 1$

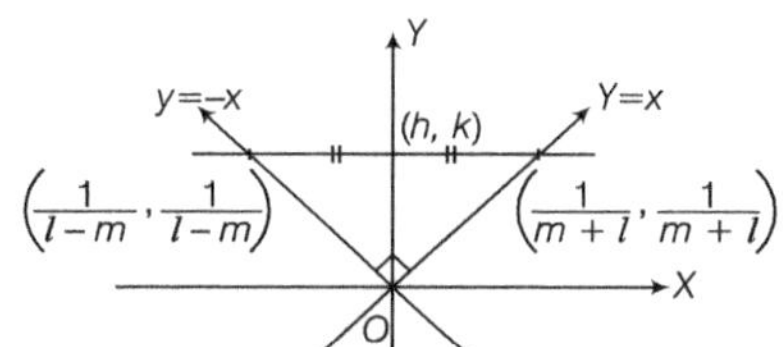

Equating and adding and then subtracting, we get

$$h^2 + k^2 = \frac{l^2 + m^2}{(l^2 - m^2)^2} = \frac{1}{(l^2 - m^2)^2} \quad \text{...(i)}$$

$$h^2 - k^2 = \frac{l^2 - m^2}{(l^2 - m^2)^2} = \frac{1}{(l^2 - m^2)} \quad \text{...(ii)}$$

$\therefore \quad (h^2 + k^2) = (h^2 - k^2)^2$

$\Rightarrow$ Locus is $x^2 + y^2 = (x^2 - y^2)^2$

23. (a) Let, $R(h, k)$ be any point on OM.

$$\text{Area of } \Delta OPR = \frac{1}{2}\begin{vmatrix} x_1 & y_1 & 1 \\ h & k & 1 \\ 0 & 0 & 1 \end{vmatrix}$$

$$= \frac{1}{2}|kx_1 - hy_1|$$

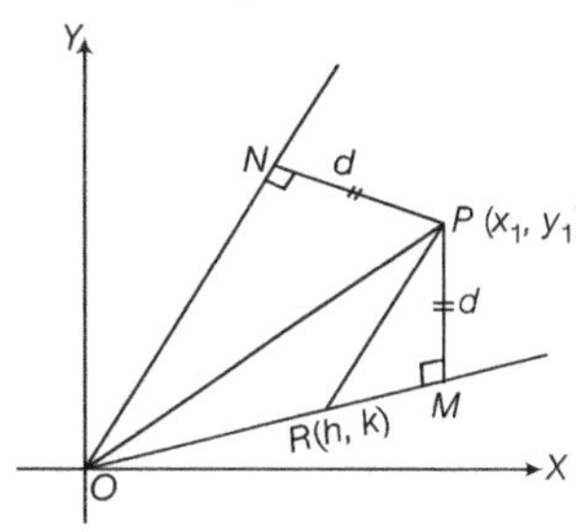

Also, area of $\Delta OPR = \frac{\sqrt{h^2 + k^2 \cdot d}}{2}$

$\therefore \quad \frac{1}{2}|kx_1 - hy_1| = \frac{\sqrt{h^2 + k^2}}{2} \cdot d$

Locus of (h, k), is

$$(xy_1 - yx_1)^2 = (x^2 + y^2) \cdot d^2$$

24. (c) Assume $C(a, b)$

$\Rightarrow \quad 9a + 7b + 4 = 0 \quad \ldots(i)$

Also, $h = \frac{a+6}{3}$

and $k = \frac{b-6}{3} \quad \ldots(ii)$

From Eqs. (i) and (ii), we get

$$27h - 54 + 21k + 42 + 4 = 0$$

$\therefore \quad 27h + 21k - 8 = 0$

Locus is $27x + 21y - 8 = 0$

or $9x + 7y - \frac{8}{3} = 0$

i.e. parallel to L.

25. (d) Let the point $C(x, y)$.

and $AC = 3BC$

$\Rightarrow \quad \sqrt{(x-1)^2 + (y-2)^2} = 3\sqrt{(x-5)^2 + (y-6)^2}$

$\Rightarrow \quad 8(x^2 + y^2) - 88x - 104y + 544 = 0$

$\therefore$ Locus of C is a circle, thus infinite number of triangle's is possible.

26. (a) $y = ax^2 + 2x + 3$ has maxima/minima occurs at $\left(-\frac{1}{a}, 3 - \frac{1}{a}\right)$.

$\therefore \quad h = \frac{-1}{a}$

$k = 3 - \frac{1}{a}$

or $k = 3 + h$

i.e. $y = x + 3$

which is a straight line with slope 1 and Y intercept 3.

27. (d) The graph of the equation,

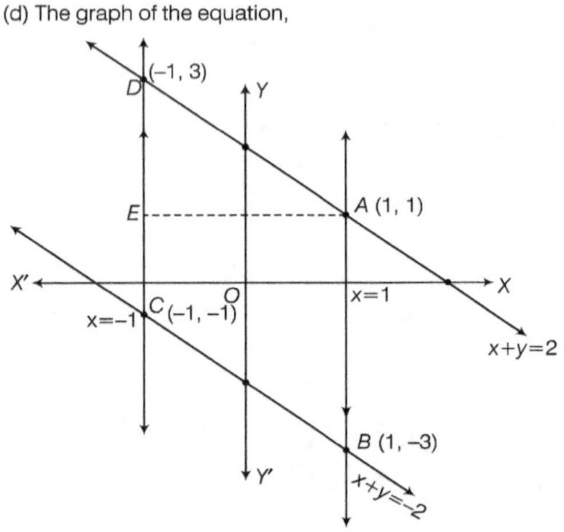

$|x| = 1 \quad \ldots(i)$

and $|x + y| = 2 \quad \ldots(ii)$

is given above.

Clearly, it represent a parallelogram.

Now, area of parallelogram = Base × Height

$= AB \times AE$

$= 4 \times 2 = 8$ sq units

Explanations for Q. Nos. (28-30)

Let $A(x_1, y_1)$, $B(x_2, y_2)$ and $C(x_3, y_3)$ be the vertices of triangle ABC.

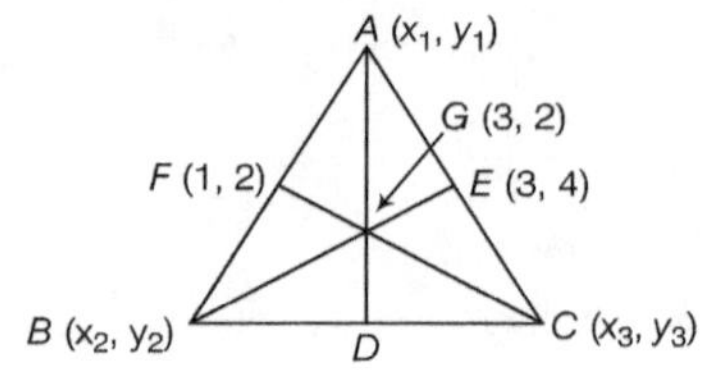

$\therefore$ Centroid

$$x_1 + x_2 + x_3 = 3 \times 3$$

$x_1 + x_2 + x_3 = 9 \quad \ldots(i)$

and $y_1 + y_2 + y_3 = 6 \quad \ldots(ii)$

Since, F is mid-point of AB,

$\therefore \quad x_1 + x_2 = 2 \quad \ldots(iii)$

and $y_1 + y_2 = 4 \quad \ldots(iv)$

Similarly, E is mid-point of AC

$\therefore \quad x_1 + x_3 = 6 \quad \ldots(v)$

$y_1 + y_3 = 8 \quad \ldots(vi)$

On solving Eqs. (i), (iii) and (ii), (iv), we get

$$x_3 = 7, y_3 = 2$$

On solving Eqs. (i), (v) and (ii), (vi), we get

$$x_2 = 3 \quad y_2 = -2$$

$\therefore$ Coordinates of B and C are $(3, -2)$ and $(7, 2)$, respectively.

28. (a) Equation of AB,

$$y - 2 = \frac{-4}{2}(x - 1)$$

$\Rightarrow \quad y - 2 = -2(x - 1)$

$\Rightarrow \quad y - 2 = -2x + 2$

$\Rightarrow \quad 2x + y = 4$

29. (b) Coordinates of point D are $\left(\frac{3+7}{2}, \frac{-2+2}{2}\right)$ i.e. $(5, 0)$

30. (c) Substituting the value of x_2, x_3 and y_2, y_3 in Eqs. (i) and (ii), we get

$$x_1 = -1, y_1 = 6$$

$\therefore$ Coordinates of point A are $(-1, 6)$.

Equation of BC,

$$y + 2 = \frac{2+2}{7-3}(x - 3)$$

$\Rightarrow \quad y + 2 = \frac{4}{4}(x - 3)$

$\Rightarrow \quad y + 2 = x - 3$

$\Rightarrow \quad x - y - 5 = 0$

$\therefore$ Perpendicular distance from A to the line BC is

$$= \frac{|-1 - 6 - 5|}{\sqrt{1^2 + 1^2}}$$

$$= \frac{|-12|}{\sqrt{2}} = \frac{12}{\sqrt{2}} = 6\sqrt{2}$$

Revisal Problems (JEE Main)

1. (a) Equation of line passing through the intersection of lines $ax + 2by + 3b = 0$ and $bx - 2ay - 3a = 0$ is

$$(ax + 2by + 3b) + \lambda (bx - 2ay - 3a) = 0 \quad \text{...(i)}$$

Now, this line is parallel to X-axis, so coefficient of x should be zero.

i.e. $$a + \lambda b = 0 \Rightarrow \lambda = -\frac{a}{b}$$

On putting this value in Eq. (i), we get

$$b(ax + 2by + 3b) - a(bx - 2ay - 3a) = 0$$

$$\Rightarrow \quad 2b^2y + 3b^2 + 2a^2y + 3a^2 = 0$$

$$\Rightarrow \quad 2(b^2 + a^2)y + 3(b^2 + a^2) = 0$$

$$y = -\frac{3}{2}$$

The negative sign shows that the line is below X-axis and at a distance $\frac{3}{2}$ from it.

Alternate Method

Equation of given lines are

$$ax + 2by + 3b = 0 \quad \text{...(i)}$$

and $$bx - 2ay - 3a = 0 \quad \text{...(ii)}$$

On solving Eqs. (i) and (ii), we get

The point of intersection $x = 0, y = -\frac{3}{2}$

Also, required line is parallel to X-axis

$\therefore$ $$m = 0$$

Now, equation of line which is passing through $\left(0, -\frac{3}{2}\right)$ having slope $m = 0$, is

$$\left(y + \frac{3}{2}\right) = 0\,(x - 0), \; y = -\frac{3}{2}$$

Thus, the required line is below X-axis at a distance of $\frac{3}{2}$ from it.

2. (a) Intercept made by the line on X-axis is $= \frac{-32}{3}$

and on Y-axis $= \frac{24}{5}$

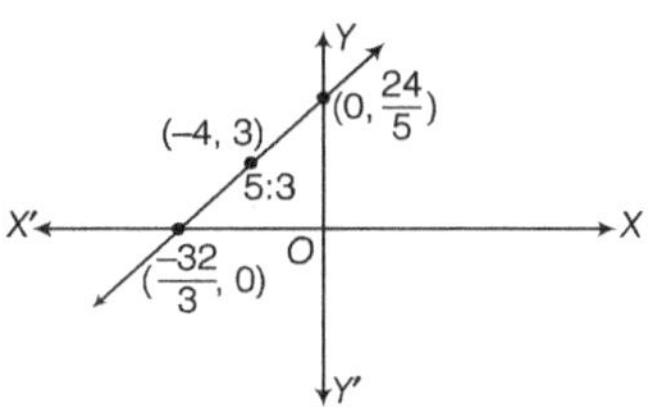

Equation of line,

$$y - 3 = \frac{\frac{24}{5}}{\frac{32}{3}}(x + 4)$$

$$y - 3 = \frac{72}{160}(x + 4)$$

$$y - 3 = \frac{9}{20}(x + 4)$$

$$20(y - 3) = 9(x + 4)$$

$$\Rightarrow \quad 9x - 20y + 96 = 0$$

3. (a) From the figure, it is clear that point B is the mirror image of the point A in the line $x - y + 5 = 0$. Similarly, point C is the mirror image of point A in the line $x + 2y = 0$

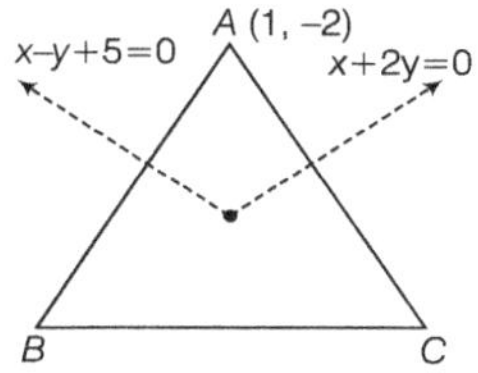

Let point B be (α, β).

$$\therefore \quad \frac{\alpha - 1}{1} = \frac{\beta + 2}{-1} = \frac{-2(1 + 2 + 5)}{2}$$

$$\Rightarrow \quad \alpha - 1 = \frac{\beta + 2}{-1} = -8$$

$$\Rightarrow \quad \alpha = -7, \beta = 6$$

$\therefore$ Coordinates of point B are $(-7, 6)$.

Again, let (γ, δ) be the coordinates of point C

$$\therefore \quad \frac{\gamma - 1}{1} = \frac{\delta + 2}{2} = \frac{-2(1 - 4)}{5}$$

$$\Rightarrow \quad \gamma - 1 = \frac{\delta + 2}{2} = \frac{6}{5}$$

$$\Rightarrow \quad \gamma = \frac{11}{5}, \delta = \frac{2}{5}$$

$\therefore$ Point C is $\left(\frac{11}{5}, \frac{2}{5}\right)$.

Now, equation of BC

$$y - 6 = \frac{\frac{2}{5} - 6}{\frac{11}{5} + 7}(x + 7) \Rightarrow y - 6 = \frac{-\frac{28}{5}}{\frac{46}{5}}(x + 7)$$

$$\Rightarrow \quad (y - 6) = \frac{-28}{46}(x + 7)$$

$$\Rightarrow \quad y - 6 = \frac{-14}{23}(x + 7)$$

$$\Rightarrow \quad 23y - 138 = -14x - 98$$

$$\Rightarrow \quad 14x + 23y - 40 = 0$$

4. (b) Here, $m_{PQ} = 1$ (since, PQ is parallel to $x - y + 1 = 0$)

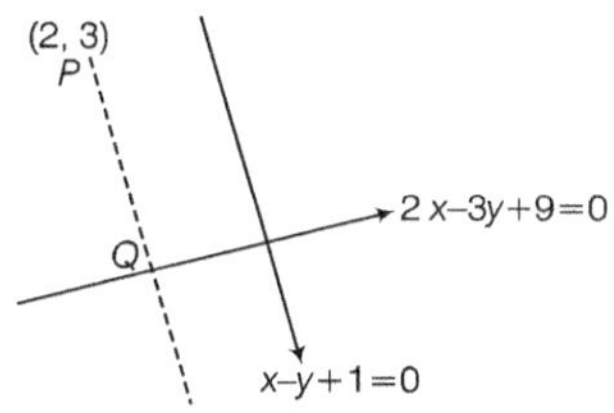

$\therefore$ $Q(2 + r\cos 45°, 3 + r\sin 45°)$ lie on

$$2x - 3y + 9 = 0$$

$$\Rightarrow \quad 2\left(2 + \frac{r}{\sqrt{2}}\right) - 3\left(3 + \frac{r}{\sqrt{2}}\right) + 9 = 0$$

$$\Rightarrow \quad 4 + \frac{2r}{\sqrt{2}} - 9 - \frac{3r}{\sqrt{2}} + 9 = 0$$

$$r = 4\sqrt{2}$$

5. (d) We know that, if point (x_1, y_1) and (x_2, y_2) lie on the same side of $ax + by + c = 0$, then $ax_1 + by_1 + c = 0$ and $ax_2 + by_2 + c = 0$ have same sign.

$\therefore$ Among the given options (d) is the correct answer as

$$3 \times (-1) - 8(-1) - 7 = -3 + 8 - 7$$
$$= -2 < 0$$

and $$3 \times 3 - 8 \times 7 - 7 = 9 - 56 - 7 = -54 < 0$$

6. (a) Let reflection of point $(4, -13)$ be (α, β) in the line $5x + y + 6 = 0$.

$$\therefore \quad \frac{\alpha - 4}{5} = \frac{\beta + 13}{1} = \frac{-2(5 \times 4 - 13 + 6)}{5^2 + 1^2}$$

$$\Rightarrow \quad \frac{\alpha - 4}{5} = \frac{\beta + 13}{1} = \frac{-26}{26}$$

$$\Rightarrow \quad \frac{\alpha - 4}{5} = \frac{\beta + 13}{1} = -1$$

$$\Rightarrow \quad \alpha = -1 \text{ and } \beta = -14$$

$\therefore$ Reflection is $(-1, -14)$.

7. (a) Given lines are $$3x - 4y + 7 = 0 \quad \text{...(i)}$$

$$12x + 5y - 2 = 0$$

or $$-12x - 5y + 2 = 0 \quad \text{...(ii)}$$

Now, $$a_1a_2 + b_1b_2 = 3(-12) + (-4)(-5)$$
$$= -36 + 20 = -16 < 0$$

$\therefore$ Acute angle bisector is

$$\frac{3x-4y+7}{\sqrt{9+16}} = \frac{-12x-5y+2}{\sqrt{144+25}}$$

$$\Rightarrow \quad \frac{3x-4y+7}{\sqrt{25}} = \frac{-12x-5y+2}{\sqrt{169}}$$

$$\Rightarrow \quad 13(3x-4y+7) = 5(-12x-5y+2)$$

$$\Rightarrow \quad 39x-52y+91 = -60x-25y+10$$

$$\Rightarrow \quad 99x-27y+81 = 0$$

$$\therefore \quad 11x-3y+9 = 0$$

8. (a) Any point on the line $x+y=4$ is $(x, 4-x)$.

Now, distance of the point $(x, 4-x)$ from the line $4x+3y=10$ is

$$\frac{|4x+3(4-x)-10|}{\sqrt{4^2+3^2}} = 1$$

$$\Rightarrow \quad \frac{|4x+12-3x-10|}{5} = 1$$

$$\Rightarrow \quad x+2 = \pm 5 \quad \Rightarrow \quad x = 3, -7$$

When $x=3$, $\quad y=1$

When $x=-7$, $\quad y=11$

$\therefore$ Points are (3, 1) and (−7, 11).

9. (c) Since, $BA \perp OA$

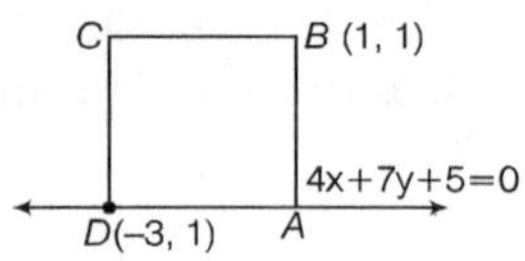

$$\therefore \text{Slope of } AB = -\frac{1}{\text{Slope } DA} = -\frac{1}{-\frac{4}{7}} = \frac{7}{4}$$

Equation of AB, $\quad y-1 = \frac{7}{4}(x-1) \quad \Rightarrow \quad 4y-4 = 7x-7$

$$7x-4y-3=0$$

Since, $CD \parallel AB$.

$\therefore$ Equation of CD, $\quad y-1 = \frac{7}{4}(x+3)$

$$4y-4 = 7x+21$$

$$7x-4y+25=0$$

Now, since $BC \parallel AD$.

Equation of BD, $\quad 4x+7y=\lambda$

This is passes through (1, 1).

$\therefore \quad 4\times 1 + 7\times 1 = \lambda \quad \Rightarrow \quad \lambda = 11$

Now, equation of BC, $\quad 4x+7y=11$

10. (b)

$$\frac{x}{a}+\frac{y}{b}=1 \quad \ldots\text{(i)}$$

$$\frac{x}{a'}+\frac{y}{b'}=-1 \quad \ldots\text{(ii)}$$

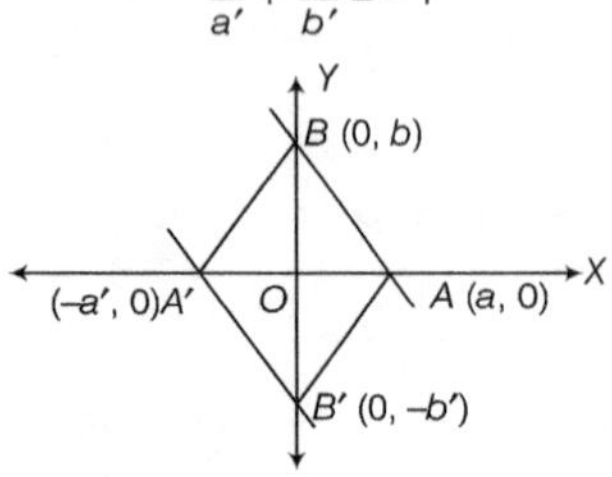

Since, A, B, A', B' are concyclic.

$$\therefore \quad -aa' = -bb' \quad \Rightarrow \quad aa' = bb' \quad \ldots\text{(iii)}$$

Equation of altitude through A' is $y-0 = \frac{a}{b}(x+a')$

It intersect the altitude $x=0$ at $y = \frac{aa'}{b}$

$\therefore$ Orthocenter is $\left(0, \frac{aa'}{b}\right)$.

Which is same as $(0, b')$ by Eq. (iii).

Hence, (b) is the correct answer.

11. (d) Given, equation of lines are

$$3x+4y=7a \quad \ldots\text{(i)}$$

$$3x+4y=7b \quad \ldots\text{(ii)}$$

$$4x+3y=7c \quad \ldots\text{(iii)}$$

$$4x+3y=7d \quad \ldots\text{(iv)}$$

Here, $\quad a_1=3, b_1=4, c_1=-7a, d_1=-7b$

$a_2=4, b_2=3, c_2=-7c, d_2=-7d$

$\therefore$ Area of parallelogram

$$= \left|\frac{(c_1-d_1)(c_2-d_2)}{(a_1b_2-a_2b_1)}\right| = \frac{|(-7a+7b)(-7c+7d)|}{9-16}$$

$$= \frac{|(7b-7a)(7d-7c)|}{-7} = 7|(a-b)(c-d)|$$

12. (a) Since, x_1, y_1 are the roots of the equation

$$x^2+8x-20=0$$

$$\Rightarrow \quad x^2+10x-2x-20=0$$

$$\Rightarrow \quad x(x+10)-2(x+10)=0$$

$$\Rightarrow \quad (x+10)(x-2)=0$$

$$\Rightarrow \quad x = 2, -10$$

$$\therefore \quad x_1 = 2, y_1 = -10$$

Similarly, x_2, y_2 are the roots of the equation

$$4x^2+32x-57=0$$

$$\Rightarrow \quad 4x^2+38x-6x-57=0$$

$$\Rightarrow \quad 2x(2x+19)-3(2x+19)=0$$

$$(2x-3)(2x+19)=0 \quad \Rightarrow \quad x = \frac{3}{2}, \frac{-19}{2}$$

$$\therefore \quad x_2 = \frac{3}{2}, y_2 = \frac{-19}{2}$$

and x_3, y_3 are the roots of the equation

$$9x^2+72x-112=0$$

$$\Rightarrow \quad 9x^2+84x-12x-112=0$$

$$\Rightarrow \quad 3x(3x+28)-4(3x+28)=0$$

$$\Rightarrow \quad (3x+28)(3x-4)=0$$

$$\Rightarrow \quad x = \frac{-28}{3}, \frac{4}{3}$$

$$\therefore \quad x_3 = \frac{-28}{3}, y_3 = \frac{4}{3}$$

$$\therefore \quad \begin{vmatrix} 2 & -10 & 1 \\ \frac{3}{2} & \frac{-19}{2} & 1 \\ \frac{-28}{3} & \frac{4}{3} & 1 \end{vmatrix}$$

Applying $R_2 \to R_2 - R_1$ and $R_3 \to R_3 - R_1$, we get

$$= \begin{vmatrix} 2 & -10 & 1 \\ \frac{-1}{2} & \frac{1}{2} & 0 \\ \frac{-34}{3} & \frac{34}{3} & 0 \end{vmatrix} = -\frac{1}{2}\times\frac{34}{3} + \frac{34}{3}\times\frac{1}{2} = 0$$

$\therefore (x_1, y_1), (x_2, y_2)$ and (x_3, y_3) are collinear.

13. (b) Given lines are $\lambda x + (\sin\alpha)y + \cos\alpha = 0 \quad \ldots\text{(i)}$

$$x + (\cos\alpha)y + \sin\alpha = 0 \quad \ldots\text{(ii)}$$

$$x - (\sin\alpha)y + \cos\alpha = 0 \quad \ldots\text{(iii)}$$

Since, lines are concurrent.

$$\therefore \quad \begin{vmatrix} \lambda & \sin\alpha & \cos\alpha \\ 1 & \cos\alpha & \sin\alpha \\ 1 & -\sin\alpha & \cos\alpha \end{vmatrix} = 0$$

Applying $R_3 \to R_3 - R_1$ and expanding, we get

$$\lambda = \cos\alpha + \sin 2\alpha$$

$$\therefore \quad \alpha \in [-\sqrt{2}, \sqrt{2}]$$

14. (b) Given, family of lines

$$(\cos\theta + \sin\theta)x + (\cos\theta - \sin\theta)y - 3(3\cos\theta + \sin\theta) = 0$$

$$\Rightarrow \quad \cos\theta(x + y - a) + \sin\theta(x - y - 3) = 0$$

$$(x + y - 9) + \tan\theta(x - y - 3) = 0$$

$$\Rightarrow \quad x + y - 9 = 0 \quad \text{and} \quad x - y - 3 = 0$$

On solving these equation, we have

$$x = 6,\ y = 3$$

∴ Point of intersection of lines is (6, 3).

i.e. $M(6, 3)$

Let reflection of M in the line $y - x = 0$ is (α, β).

$$\therefore \quad \frac{\alpha - 6}{-1} = \frac{\beta - 3}{1} = \frac{-2(3 - 6)}{2}$$

$$\Rightarrow \quad \frac{\alpha - 6}{-1} = \frac{\beta - 3}{1} = 3$$

$$\Rightarrow \quad \alpha = 3, \beta = 6$$

∴ Reflection of M is (3, 6).

15. (a) Given lines are $|x| + |y| = 2$

$$x + y = 2,\ x > 0,\ y > 0$$
$$-x + y = 2,\ x < 0,\ y > 0$$
$$-x - y = 2,\ x < 0,\ y < 0$$
$$x - y = 2,\ x > 0,\ y < 0$$

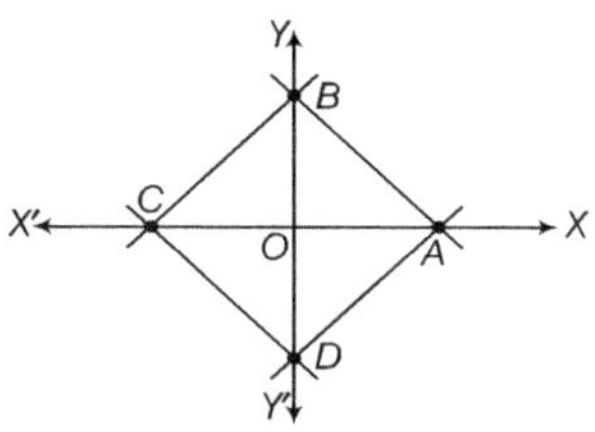

$$\text{Area} = \frac{1}{2} \times AC \times BD = \frac{1}{2} \times 4 \times 4 = 8 \text{ sq units}$$

16. (d) Equation of diagonal not passing through origin, is

$$(2x + 3y)(3x - 4y) = [(2x + 3y) - 5][(3x - 4y) - 3]$$

$$\Rightarrow \quad (2x + 3y)(3x - 4y) = (2x + 3y)(3x - 4y) - 3(2x + 3y) - 5(3x - 4y) + 15$$

$$\Rightarrow \quad 21x - 11y - 15 = 0$$

17. (d) Given lines are,

$$ax + amy + 1 = 0 \quad \text{...(i)}$$
$$bx + (m + 1)by + 1 = 0 \quad \text{...(ii)}$$
and
$$cx + (m + 2)cy + 1 = 0 \quad \text{...(iii)}$$

It is given that lines (i), (ii) and (iii) are concurrent.

$$\therefore \quad \begin{vmatrix} a & am & 1 \\ b & (m + 1)b & 1 \\ c & (m + 2)c & 1 \end{vmatrix} = 0$$

Applying $R_2 \to R_2 - R_1$ and $R_3 \to R_3 - R_1$, we get

$$\begin{vmatrix} a & am & 1 \\ b - a & m(b - a) + b & 0 \\ c - a & m(c - a) + 2c & 0 \end{vmatrix} = 0$$

Expanding along C_3, we get

$$1[(b - a)(m(c - a) + 2c) - (c - a)(m(b - a) + b)] = 0$$

$$\Rightarrow \quad (b - a)(c - a)m + 2c(b - a) - (c - a)(b - a)m - b(c - a) = 0$$

$$\Rightarrow \quad 2c(b - a) - b(c - a) = 0$$

$$\Rightarrow \quad 2bc - 2ac - bc + ab = 0$$

$$\Rightarrow \quad bc - 2ac + ab = 0$$

$$\Rightarrow \quad b(a + c) = 2ac \quad \Rightarrow \quad b = \frac{2ac}{a + c}$$

∴ a, b, c are in HP.

18. (d) Given equation of sides of triangle,

$$2x - 3y - 6 = 0 \quad \text{...(i)}$$
$$3x - y + 3 = 0 \quad \text{...(ii)}$$
$$3x + 4y - 12 = 0 \quad \text{...(iii)}$$

First draw the exact diagram of triangle.

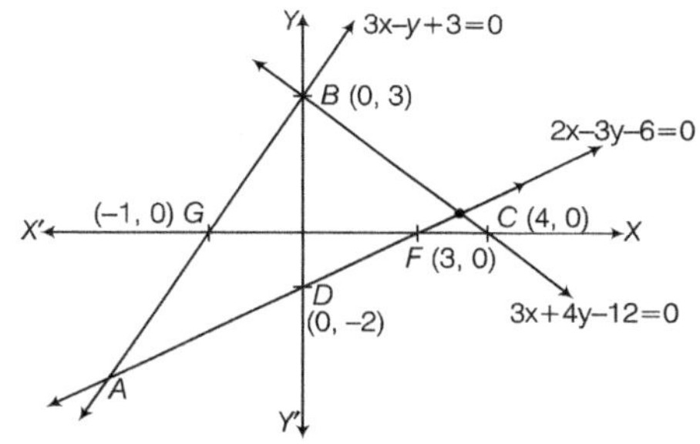

The point $P(\alpha, 0)$ move on the line $y = 0$ for all α and the point $Q(0, \beta)$ move on the line $x = 0$ for all β.

Now, G and F are the intersection points of $3x - y + 3 = 0$, $y = 0$ and $2x - 3y - 6 = 0$, $y = 0$ respectively.

$$\therefore \quad F \equiv (3, 0),\ G \equiv (-1, 0)$$

Thus, the points on the line $y = 0$, whose x-coordinate lies between -1 and 3 lie on or inside the ΔABC.

Hence, $-1 \le \alpha \le 3$

i.e. $\alpha \in [-1, 3]$

Similarly,

The point $(0, \beta)$ lies on the line DB.

Since, B and D are the intersection points of $3x - y + 3 = 0$, $x = 0$ and $2x - 3y - 6 = 0$, $x = 0$

$$\therefore \quad B \equiv (0, 3) \quad \text{and} \quad D \equiv (0, -2)$$

Thus, the points on the line $x = 0$, whose y-coordinate lies between -2 to 3 lie on or inside the ΔABC.

Hence, $-2 \le \beta \le 3$

i.e. $\beta \in [-2, 3]$

19. (b) Let

$$L(x, y) = 2x - y + 5 \quad \text{...(i)}$$

$$\therefore \quad L(4, -2) = 2 \times 4 + 2 + 5 = 15$$

and

$$L(2, -4) = 2 \times 2 + 4 + 5 = 13$$

∴ A and B lie on the same side of the line

$$3x + 2y + 10 = 0$$

Equation of line AB is

$$(y + 2) = \frac{-4 + 2}{2 - 4}(x - 4)$$

$$\Rightarrow \quad y + 2 = \frac{-2}{-2}(x - 4)$$

$$\Rightarrow \quad y + 2 = x - 4$$

$$\Rightarrow \quad x - y - 6 = 0 \quad \text{...(ii)}$$

Line (ii) meet line (i) at P.

On solving Eqs. (i) and (ii), we get

$$\therefore \quad P(-11, -17)$$

20. (a) Let $C'(\alpha, \beta)$ be the image of point $C(5, 3)$ about the line $y = 0$

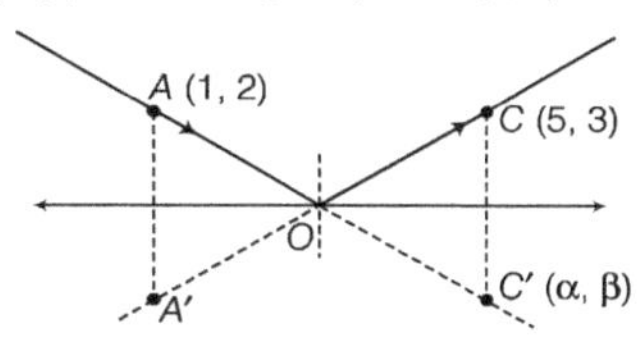

$$\therefore \quad \frac{\alpha - 5}{0} = \frac{\beta - 3}{1} = \frac{-2(3)}{1}$$

$$\Rightarrow \quad \alpha = 5, \beta = -3$$

Now, equation of AC', i.e. AB is

$$y-2=\frac{-3-2}{5-1}(x-1)$$

$$\Rightarrow \quad y-2=\frac{-5}{4}(x-1)$$

$$\Rightarrow \quad 4(y-2)=-5x+5$$

$$\Rightarrow \quad 4y-8=-5x+5$$

$$\therefore \quad 5x+4y-13=0$$

21. (b) The equation of bisector of the angle containing origin is

$$\frac{3x-y+1}{\sqrt{10}}=\frac{-x-2y+5}{\sqrt{5}}$$

$$\Rightarrow \quad 3x-y+1=\sqrt{2}(-x-2y+5)$$

$$\Rightarrow \quad 3x-y+1=-\sqrt{2}x-2\sqrt{2}y+5\sqrt{2}$$

$$\Rightarrow \quad (3+\sqrt{2})x+(2\sqrt{2}-1)y+1-5\sqrt{2}=0$$

This line is also containing the point $(a^2, a+1)$.

$$\therefore \quad (3+\sqrt{2})a^2+(2\sqrt{2}-1)(a+1)+1-5\sqrt{2}=0$$

$$\Rightarrow \quad (3+\sqrt{2})a^2+(2\sqrt{2}-1)a-3\sqrt{2}=0$$

$$\Rightarrow \quad (3a^2-a)+\sqrt{2}(a^2+2a-3)=0$$

$$\Rightarrow \quad 3a^2-a=0 \text{ and } a^2+2a-3=0$$

$$\Rightarrow \quad a(3a-1)=0 \text{ and } (a-1)(a+3)=0$$

$$\Rightarrow \quad a=0, \frac{1}{3} \text{ and } a=1,-3$$

$$\Rightarrow \quad a \in (-3, 0) \cup \left(\frac{1}{3}, 1\right)$$

22. (c) Equation of lines passing through origin and equally inclined with the $2x+y=a$ is ...(i)

$$y-y_1=\tan(\theta \pm \alpha)(x-x_1)$$

$$\Rightarrow \quad y-0=\frac{\tan\theta \pm \tan\alpha}{1 \pm \tan\theta\tan\alpha}(x-0)$$

$$\Rightarrow \quad y=\frac{-2 \pm 1}{1 \mp (-2)(1)}x$$

$$\Rightarrow \quad y=\frac{-2+1}{1+2}x \quad \text{or} \quad y=\frac{-2-1}{1-2}x$$

$$\Rightarrow \quad y=\frac{-1}{3}x \quad \text{or} \quad y=\frac{-3}{-1}x$$

$$\Rightarrow \quad 3y=-x \quad \text{or} \quad y=3x$$

$$\Rightarrow \quad x+3y=0 \quad \text{or} \quad 3x-y=0$$

$$\Rightarrow \quad x+3y=0 \quad \text{...(ii)}$$

$$\Rightarrow \quad 3x-y=0 \quad \text{...(iii)}$$

On solving Eq. (i) and (ii), we get

$$x=\frac{3a}{5}, y=\frac{-a}{5}$$

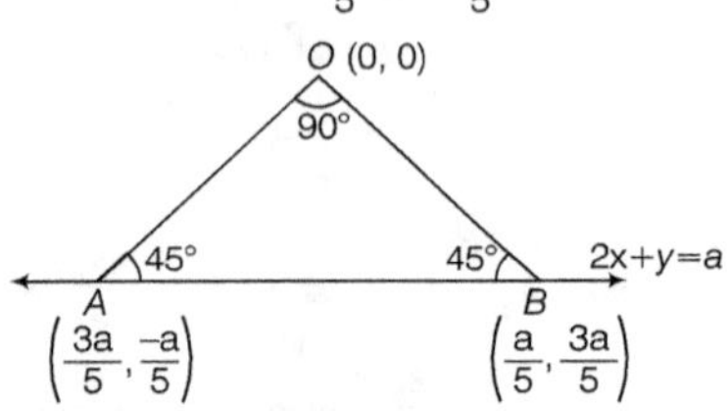

Now, $\quad OA=\sqrt{\frac{9a^2}{25}+\frac{a^2}{25}}=\sqrt{\frac{10a^2}{25}}$

$$=\frac{a}{5}\sqrt{10}=OB$$

$$\therefore \text{ Area } \Delta OAB=\frac{1}{2}\times OA\times OB$$

$$=\frac{1}{2}\times OA^2 \qquad (\because OA=OB)$$

$$=\frac{1}{2}\times\frac{10a^2}{25}=\frac{a^2}{5} \text{ sq units}$$

23. (c) Let (x_1, y_1) be the image of the point $(1, 2)$ about the line $y=x$.

Then, $\quad x_1=2, y_1=1 \quad$ ...(i)

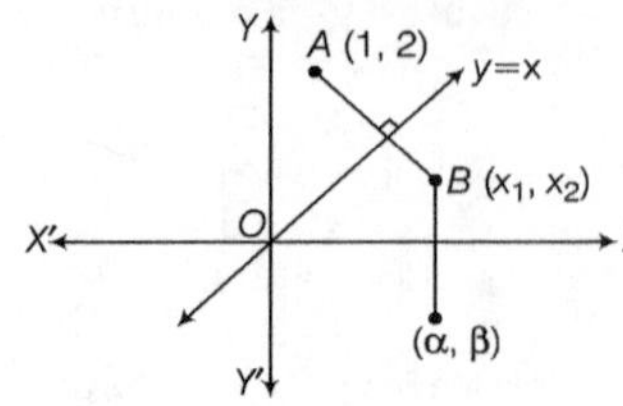

Also, given image of (x_1, y_1) by the mirror line $y=0$ is (α, β). Then, $\alpha=x_1=2$

and $\quad \beta=-y_1=-1 \quad$ (from Eq. (i))

Hence, $\quad \alpha=2, \beta=-1$

24. (d) Given lines are $\quad x+3y-2=0 \quad$ ...(i)

$$x-7y+5=0 \quad \text{...(ii)}$$

Slope of line (i), $\quad m=-\frac{1}{3}$

Slope of line (ii), $\quad m_1=\frac{1}{7}$

Let the slope of required line be m_2.

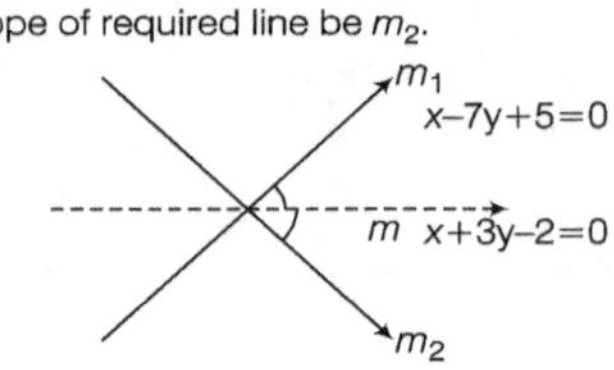

Now, $\left(\frac{m_1-m}{1+m_1m}\right)=-\left(\frac{m_2-m}{1+m_2m}\right) \Rightarrow \left(\frac{m_1-m}{1+m_1m}\right)=\left(\frac{m-m_2}{1+m_2m}\right)$

$$\Rightarrow \quad \frac{\frac{1}{7}+\frac{1}{3}}{1+\frac{1}{7}\times\frac{1}{3}}=\frac{-\frac{1}{3}-m_2}{1-\frac{m_2}{3}}$$

$$\Rightarrow \quad \frac{10}{22}=\frac{-1-3m_2}{3-m_2}$$

$$\Rightarrow \quad 30-10m_2=-22-66m_2$$

$$\Rightarrow \quad 56m_2=-52$$

$$\Rightarrow \quad m_2=-\frac{52}{56}=-\frac{13}{14}$$

Point of intersection of lines (i) and (ii) is $\left(-\frac{1}{10}, \frac{7}{10}\right)$

Equation of required line is

$$\left(y-\frac{7}{10}\right)=-\frac{13}{4}\left(x+\frac{1}{10}\right)$$

$$\Rightarrow \quad (10y-7)=\frac{-13}{14}(10x+1)$$

$$\Rightarrow \quad 140y-98=-130x-13$$

$$\therefore \quad 130x+140y-85=0$$

or $\quad 26x+28y-17=0$

25. (c) Equation of line passing through point E is

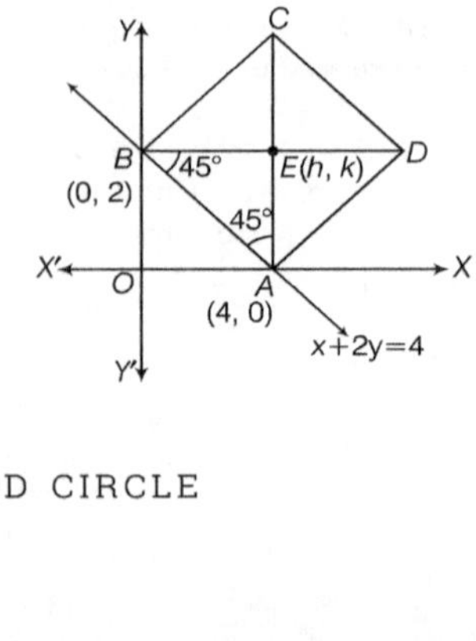

$$y - k = \frac{-\frac{1}{2} \pm 1}{1 \mp \left(-\frac{1}{2}\right) \cdot 1}(x - h)$$

$\Rightarrow$ $$y - k = \frac{-\frac{1}{2} + 1}{1 + \frac{1}{2}}(x - h)$$

and $$(y - k) = \frac{-\frac{1}{2} - 1}{1 - \frac{1}{2}}(x - h)$$

$\Rightarrow$ $$y - k = \frac{1/2}{3/2}(x - h)$$

and $$y - k = \frac{-3/2}{1/2}(x - h)$$

$\Rightarrow$ $$y - k = \frac{1}{3}(x - h)$$

and $$y - k = -3(x - h)$$

$\Rightarrow$ $$3y - 3k - x + h = 0$$

and $$3x + y - k - 3h = 0$$

$\Rightarrow$ $$x - 3y + 3k - h = 0 \quad \text{...(i)}$$

and $$3x + y - k - 3h = 0 \quad \text{...(ii)}$$

Line (i) passes through point (0, 2)

$\therefore$ $$3k - h = 6 \quad \text{...(iii)}$$

and line (ii) passes through point (4, 0)

$\therefore$ $$k + 3h = 12 \quad \text{...(iv)}$$

On solving Eqs. (iii) and (iv), we get

$$h = 3, k = 3$$

26. (b) Let image of $A(3, 10)$ about the line $2x + y - 6 = 0$ is $A'(\alpha, \beta)$.

Then, $$\frac{\alpha - 3}{2} = \frac{\beta - 10}{1} = \frac{-2(6 + 10 - 6)}{2^2 + 1^2}$$

$\therefore$ $$\alpha = -5, \beta = 6$$

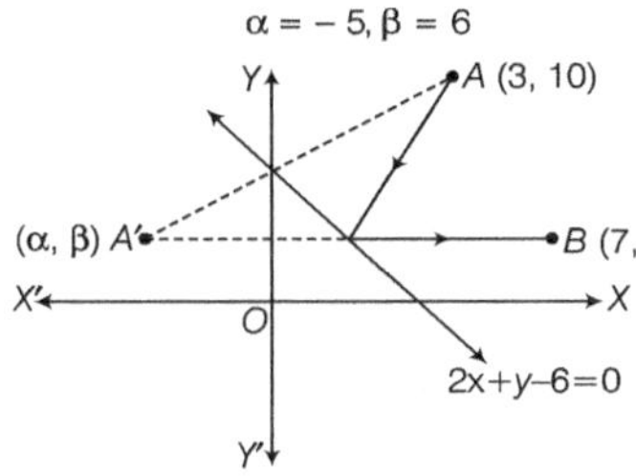

$\therefore$ $$A' = (-5, 6)$$

Now, equation of reflected ray $A'B$ is

$$y - 6 = \frac{2 - 6}{7 + 5}(x + 5)$$

$\Rightarrow$ $$y - 6 = -\frac{1}{3}(x + 5)$$

$\therefore$ $$x + 3y - 13 = 0$$

27. (a) Value of $3x - 4y + 12 = 0$ at $(-1, 4)$ is -7 (negative). Value of $12x - 5y + 7$ at $(-1, 4)$ is -25 (negative). Equation of bisectors is

$$\frac{3x - 4y + 12}{5} = \pm\left(\frac{12x - 5y + 7}{13}\right)$$

$\Rightarrow$ $$13(3x - 4y + 12) = \pm 5(12x - 5y + 7)$$

Taking –ve sign,

$$39x - 52y + 156 = 60x - 25y + 35$$

$\Rightarrow$ $$-21x - 27y + 121 = 0$$

$\Rightarrow$ $$21x + 27y - 121 = 0$$

28. (a)

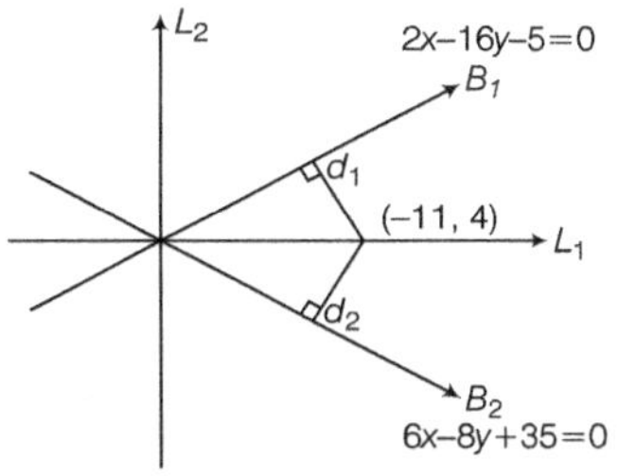

$$d_1 = \frac{|-22 - 64 - 5|}{\sqrt{4 + 256}} = \frac{91}{2\sqrt{65}} \sim 5.6 \quad \text{...(i)}$$

$$d_2 = \frac{|-704 + 32 + 35|}{\sqrt{(64)^2 + 8^2}} = \frac{637}{63.49} \sim 10.03 \quad \text{...(ii)}$$

Since, $d_1 < d_2$

$\therefore$ B_1 is acute angle bisector and B_2 is obtuse angle bisector.

29. (b) Let the variable line be $ax + by + c = 0$. ...(i)

Where algebraic sum of perpendiculars from (2, 0) (0, 2) and (1, 1) is zero.

$\therefore$ $$\frac{2a + 0 + c}{\sqrt{a^2 + b^2}} + \frac{0 + 2b + c}{\sqrt{a^2 + b^2}} + \frac{a + b + c}{\sqrt{a^2 + b^2}} = 0$$

$\Rightarrow$ $$3a + 3b + 3c = 0$$

$\Rightarrow$ $$a + b + c = 0 \quad \text{...(ii)}$$

From Eqs. (i) and (ii), we get

$ax + by + c = 0$ always passes through a fixed point (1, 1).

30. (b) Suppose a line passing through (1, 2) makes an angle θ with X-axis. Then, its equation is

$$\frac{x - 1}{\cos\theta} = \frac{y - 2}{\sin\theta}$$

The coordinates of a point on this line at a distance $\frac{\sqrt{6}}{3} = \sqrt{\frac{2}{3}}$ from (1, 2) is given by

$$\frac{x - 1}{\cos\theta} = \frac{y - 2}{\sin\theta} = \sqrt{\frac{2}{3}}$$

or $$\left(1 + \sqrt{\frac{2}{3}}\cos\theta, 2 + \sqrt{\frac{2}{3}}\sin\theta\right)$$

This point lie on the line $x + y = 4$.

$\therefore$ $$1 + \sqrt{\frac{2}{3}}\cos\theta + 2 + \sqrt{\frac{2}{3}}\sin\theta = 4$$

$\Rightarrow$ $$\sin\theta + \cos\theta = \sqrt{\frac{3}{2}}$$

$\Rightarrow$ $$\sin\left(\theta + \frac{\pi}{4}\right) = \frac{\sqrt{3}}{4}$$

$\Rightarrow$ $$\theta + \frac{\pi}{4} = \frac{\pi}{3}, \frac{2\pi}{3}$$

$\therefore$ $$\theta = \frac{\pi}{12} \text{ and } \frac{5\pi}{12} \text{ or } 75^\circ$$

31. (b) The straight lines $x\cos\alpha + y\sin\alpha = c$ and $x\sin\alpha - y\cos\alpha = 0$ are mutually perpendicular.

Line $x\cos\alpha + y\sin\alpha = c$ is in normal form, so perpendicular from the origin as this line is of length c and makes angle $\frac{\pi}{4}$ with OX.

The line $x\sin\alpha - y\cos\alpha = 0$ passes through origin and makes angle α with X-axis. So, it is along the perpendicular from the origin on $x\cos\alpha + y\sin\alpha = c$.

Line $ax + by + c = 0$ makes angle $\frac{\pi}{4}$ with $x\cos\alpha + y\sin\alpha = c$, $x\cos\alpha + y\sin\alpha = c$ and $x\sin\alpha - y\cos\alpha = 0$ are mutually perpendicular. So, $ax + by + c = 0$ is a bisector of the angle between them.

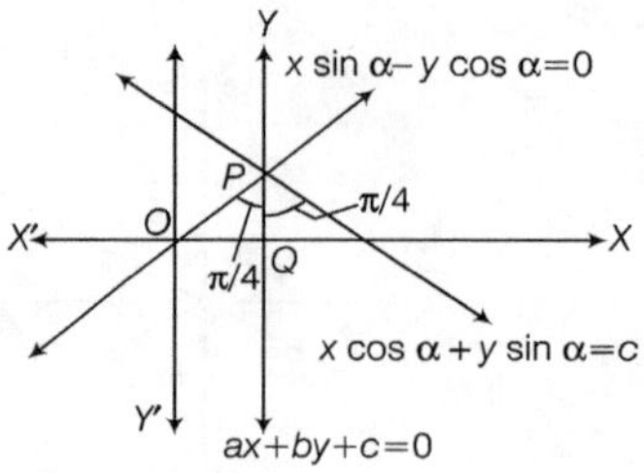

In ΔOPQ, we have

$$OQ = OP\sin\frac{\pi}{4} = \frac{c}{\sqrt{2}}$$

Also, OQ = Length of perpendicular from (0, 0) on $ax + by + c = 0$.

$$\Rightarrow \quad OQ = \frac{c}{\sqrt{a^2 + b^2}} \quad \Rightarrow \quad \frac{c}{\sqrt{2}} = \frac{c}{\sqrt{a^2 + b^2}}$$

$$\therefore \quad a^2 + b^2 = 2$$

32. (b) If point $P(\alpha, \alpha^2 - 2)$ lies inside the ΔABC, then

(a) A and P must lie on the same side of BC, whose equation is

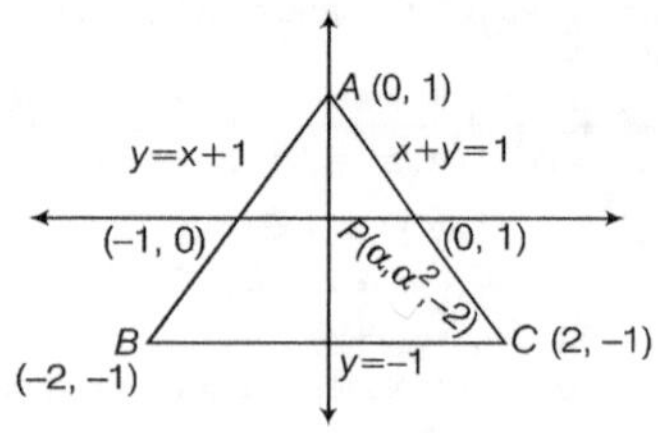

$$y + 1 = 0$$

$$\therefore \quad (1 + 1)(\alpha^2 - 2 + 1) > 0$$

$$\Rightarrow \quad \alpha^2 - 1 > 0 \Rightarrow \alpha \in (-\infty, -1) \cup (1, \infty) \quad \ldots\text{(i)}$$

(b) B and P must lie on the same side of AC.

$$\therefore (-2 - 1 - 1)(\alpha + \alpha^2 - 2 - 1) > 0$$

$$\Rightarrow \quad \alpha^2 + \alpha - 3 < 0$$

$$\Rightarrow \quad \frac{-1 - \sqrt{13}}{2} < \alpha < \frac{-1 + \sqrt{13}}{2} \quad \ldots\text{(ii)}$$

(c) C and P must lie on the same side of AB.

$$\therefore \quad (2 + 1 + 1)(\alpha - \alpha^2 + 2 + 1) > 0$$

$$\Rightarrow \quad \alpha^2 - \alpha - 3 < 0 \Rightarrow \frac{1 - \sqrt{13}}{2} < \alpha < \frac{1 + \sqrt{13}}{2} \quad \ldots\text{(iii)}$$

From Eqs. (i), (ii) and (iii), we have

$$\alpha \in \left(\frac{1 - \sqrt{3}}{2}, -1\right) \cup \left(1, \frac{-1 + \sqrt{13}}{2}\right)$$

33. (a) If point P lies between the given parallel lines, then

$$\left(1 + \frac{t}{\sqrt{2}} + 4 + \frac{2t}{\sqrt{2}} - 1\right)\left(2 + \frac{2t}{\sqrt{2}} + 8 + \frac{4t}{\sqrt{2}} - 15\right) < 0$$

$$\Rightarrow \quad \left(4 + \frac{3t}{\sqrt{2}}\right)\left(-5 + \frac{6t}{\sqrt{2}}\right) < 0$$

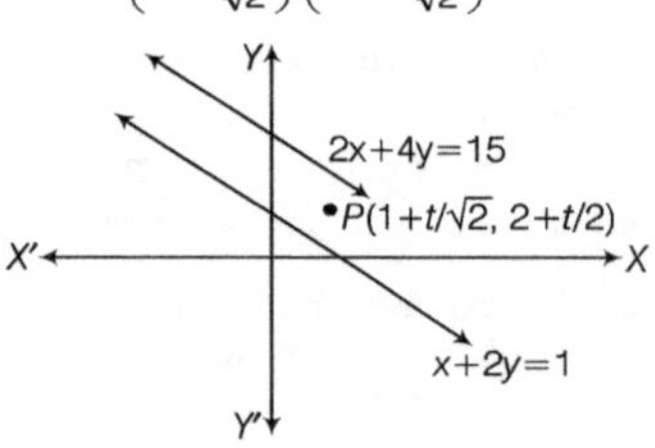

$$\Rightarrow \quad \frac{-4\sqrt{2}}{3} < t < \frac{5\sqrt{2}}{6} \quad \Rightarrow \quad t \in \left(\frac{-4\sqrt{2}}{5}, \frac{5\sqrt{2}}{6}\right)$$

34. (d) The equation of line through $P(2, 4)$ and parallel to $y - x - 1 = 0$ is

$$\frac{x - 2}{\cos\frac{\pi}{4}} = \frac{y - 4}{\sin\frac{\pi}{4}}$$

The coordinates of Q are given by

$$\frac{x - 2}{\cos\frac{\pi}{4}} = \frac{y - 4}{\sin\frac{\pi}{4}} = -3\sqrt{2}$$

$$\Rightarrow \quad x = -1, y = 1$$

Thus, the coordinates of Q are $(-1, 1)$.

Therefore, coordinates of R are given by

$$\frac{x + 1}{-1} = \frac{y - 1}{1} = -2\,\frac{(1 + 1 - 1)}{(-1)^2 + (1)^2}$$

$$\Rightarrow \quad \frac{x + 1}{-1} = \frac{y - 1}{1} = -1$$

$$\Rightarrow \quad x = 0, y = 0$$

Hence, coordinates of R are (0, 0).

35. (c) Given, a, c, b are in GP.

$$\therefore \quad c^2 = ab$$

$$c = \sqrt{ab}$$

$$\text{Now,} \quad ax + by + c = 0 \quad \ldots\text{(i)}$$

$$\Rightarrow \quad ax + by + \sqrt{ab} = 0$$

$$\Rightarrow \quad ax + by = -\sqrt{ab}$$

$$\Rightarrow \quad \frac{x}{-\sqrt{\frac{b}{a}}} + \frac{y}{-\sqrt{\frac{a}{b}}} = 1$$

$\therefore$ Line (i) making intercept on the axes, $\sqrt{\frac{b}{a}}$ and $\sqrt{\frac{a}{b}}$ respectively.

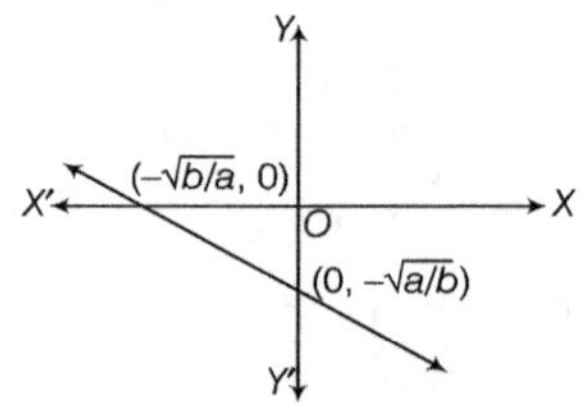

$\therefore$ This line forms a triangle with the axes whose area is constant.

36. (c) We have, $x + y + \lambda(2x - y + 1) = 0$

Clearly, it represent a family of lines passing through the intersection of the lines $x + y = 0$ and $2x - y + 1 = 0$, i.e. the point $\left(-\frac{1}{3}, \frac{1}{3}\right)$.

The required line passes through $\left(-\frac{1}{3}, \frac{1}{3}\right)$ and is perpendicular to the line joining (1, 4) and $\left(\frac{-1}{3}, \frac{1}{3}\right)$. So, its equation is

$$y - \frac{1}{3} = \frac{-4}{11}\left(x + \frac{1}{3}\right)$$

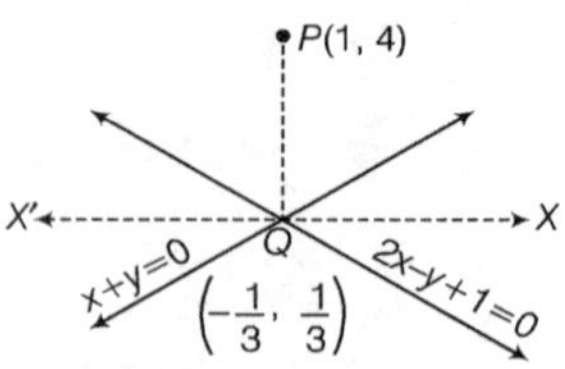

$$\Rightarrow \quad 12x + 33y = 7$$

37. (c) Given lines are,

$$L_1 : \frac{x}{3} + \frac{y}{4} = 1$$

$$L_2 : \frac{x}{4} + \frac{y}{3} = 1$$

$$L_3 : \frac{x}{3} + \frac{y}{4} = 2$$

$$L_4 : \frac{x}{4} + \frac{y}{3} = 2$$

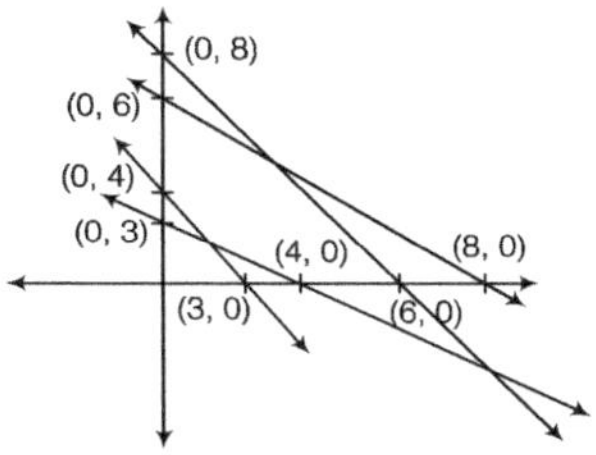

From the figure, Statement I is correct. But statement II is incorrect.

38. (a) $AB = \sqrt{425}$ and $AC = \sqrt{425}$

$\therefore$ ΔABC is isosceles

Hence, centroid lie on angle bisector.

39. (a)
$$(2\cos\theta_1\cos\theta_2 + 2\cos\theta_2\cos\theta_3 + 2\cos\theta_3\cos\theta_1 + \cos^2\theta_1 + \cos^2\theta_2 + \cos^2\theta_3) + (\sin^2\theta_1 + \sin^2\theta_2 + \sin^2\theta_3 + 2\sin\theta_1\sin^2\theta_2 + 2\sin\theta_2\sin\theta_3 + 2\sin\theta_3\sin\theta_1) = 0$$

$$\Rightarrow \quad (\cos\theta_1 + \cos\theta_2 + \cos\theta_3)^2 + (\sin\theta_1 + \sin\theta_2 + \sin\theta_3)^2 = 0$$

$$\Rightarrow \quad \cos\theta_1 + \cos\theta_2 + \cos\theta_3 = 0$$

and $\sin\theta_1 + \sin\theta_2 + \sin\theta_3 = 0$

$\therefore$ Centroid and circumcentre of ΔABC is at origin.

$\Rightarrow \Delta ABC$ is equilateral.

$\Rightarrow$ Orthocentre is also at origin.

40. (a) We know that, the area of triangle is invariant w.r.t. to the translation of axes.

$\therefore$ Statement II is true.

Consider points $P(0, 0)$, $Q(1, 2)$ and $R(2, 1)$.

Now, shift the origin at (20, 22).

$\therefore$ Coordinates of point P are (20, 220), i.e. A,

Coordinates of point Q are (21, 24), i.e. B

and coordinates of point R are (22, 23), i.e. C.

Hence, (ar ΔABC) = ar (ΔPQR).

41. (b) Equation of bisector of $\angle A$ is

$$\frac{3x + 4y}{5} = \frac{-y + 15}{1}$$

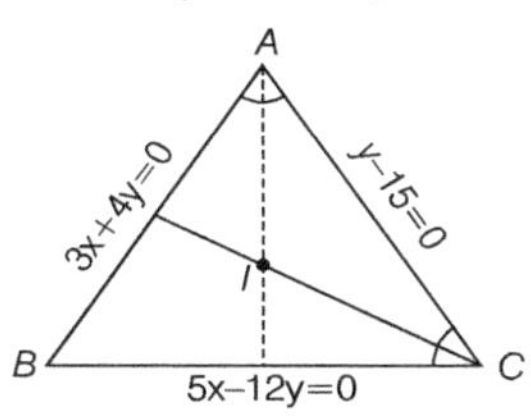

On taking +ve sign because $a_1a_2 + b_1b_2 < 0$

$$3x + 4y = -5y + 75$$

$$3x + 9y - 75 = 0 \quad \ldots\text{(i)}$$

Equation of bisector of $\angle C$

$$\frac{5x - 12y}{13} = -\frac{(-y + 15)}{1}$$

On taking –ve sign because $a_1a_2 + b_1b_2 > 0$

$$\Rightarrow \quad 5x - 12y = +13y - 195$$

$$\Rightarrow \quad 5x - 25y + 195 = 0 \quad \ldots\text{(ii)}$$

Point of intersection of lines (i) and (ii) is incenter.

$\therefore$ $I(1, 8)$

Statement I is correct.

Clearly, Statement II is correct but not the correct explanation of Statement I.

42. (a) Since, θ is symmetric to $P(4, -3)$ w.r.t. bisector of first and third quadrant.

$$\therefore \quad PQ = 2 \text{ (distance of } P(4, -3) \text{ w.r.t. } x - y = 0)$$

$$= 2\frac{|4 + 3|}{\sqrt{2}} = 7\sqrt{2}$$

43. (b) Given,

Equation of sides of quadrilateral are

$$3x + 2y + 1 = 0 \quad \ldots\text{(i)}$$

$$3x + 2y + 2 = 0 \quad \ldots\text{(ii)}$$

$$2x + 3y + 1 = 0 \quad \ldots\text{(iii)}$$

$$2x + 3y + 2 = 0 \quad \ldots\text{(iv)}$$

Distance between lines (i) and (ii) is

$$d = \frac{|2 - 1|}{\sqrt{13}} = \frac{1}{\sqrt{13}}$$

and distance between the lines (iii) and (iv) is

$$d = \frac{|2 - 1|}{\sqrt{13}} = \frac{1}{\sqrt{13}}$$

Since, distance between the two set of parallel lines is same.

$\therefore$ It is a rhombus.

Hence, angle between diagonals is $\frac{\pi}{2}$.

Statement I is correct.

Clearly, Statement II is correct but not the correct explanation of Statement I.

44. (d) Bisector of the given lines are

$$\frac{(3x + 4y - 12)}{5} = \pm\frac{(4x + 3y - 12)}{5}$$

$$\Rightarrow \quad 3x + 4y - 12 = \pm(4x + 3y - 12)$$

On taking +ve sign, we have

$$3x + 4y - 12 = 4x + 3y - 12$$

$$-x + y = 0$$

On taking –ve sign, we have

$$3x + 4y - 12 = -4x - 3y + 12$$

$$\Rightarrow \quad 7x + 7y - 24 = 0$$

$\therefore$ Statement I is incorrect.

Also, any point on the bisector is always equidistant from the given lines. Statement II is correct.

45. (c) Here, $5 = \frac{a}{2}$ and $\frac{b}{2} = 6$

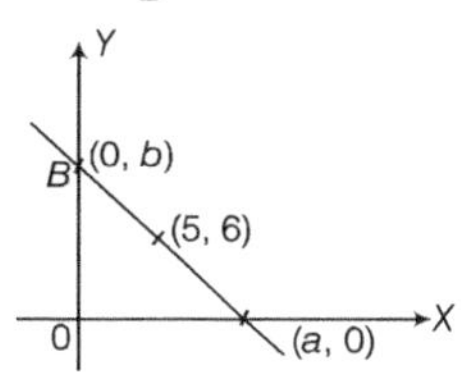

$$\Rightarrow \quad a = 10, b = 12$$

$$\therefore \quad \text{Area of } \Delta AOB = \frac{1}{2} \times 10 \times 12 = 60 \text{ sq units}$$

Statement I is coorrect.

Statement II is incorrect.

Revisal Problems (JEE Advanced)

1. (d) Let coordinates of vertex A be (x_1, y_1) and centroid be (h, k).

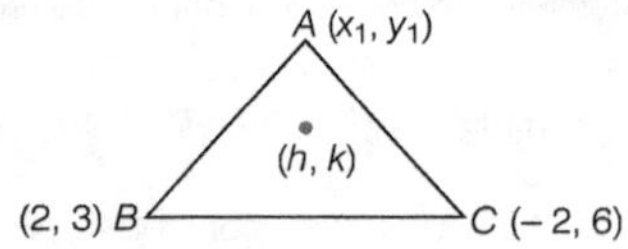

$\therefore \quad h = \dfrac{x_1 + 2 - 2}{3} \quad \Rightarrow \quad x_1 = 3h$

and $\quad k = \dfrac{y_1 + 3 + 6}{3} \quad \Rightarrow \quad y_1 = 3k - 9$

Now, $BC = \sqrt{16 + 9} = 5$

$AC = \sqrt{(3h + 2)^2 + (3k - 9 - 6)^2} = \sqrt{(3h + 2)^2 + 3k - 15)^2}$

$AB = \sqrt{(3h - 2)^2 + (3k - 9 - 3)^2} = \sqrt{(3h - 2)^2 + (3k - 12)^2}$

Perimeter = 14

$AB + BC + AC = 14$

$\Rightarrow \quad AB + AC = 9$

Which represent an ellipse.

2. (d) There exists exactly 4 triangles ABC, APC, APB and BPC satisfying the given condition of which three triangles will be obtuse angle.

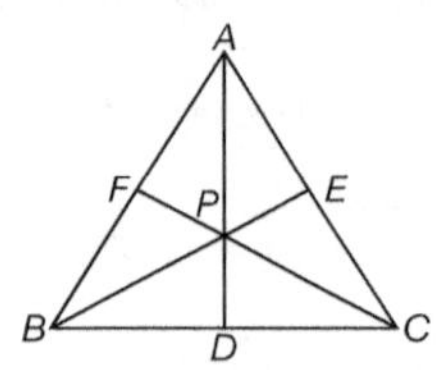

3. (c) Since, $ax^2 + bx + 10 = 0$ has imaginary roots.

Therefore, $f(x) = ax^2 + bx + 10 \geq 0$ or ≤ 0 for all $x \in R$.

We have, $\quad f(0) = 10 \geq 0$

$\therefore \quad f(x) = ax^2 + bx + 10 \geq 0$ for all $x \in R$

$\Rightarrow \quad f(5) = 25a + 5b + 10 \geq 0$

$\Rightarrow \quad 5a + b + 2 \geq 0$

$\Rightarrow \quad 5a + b \geq -2$

$\Rightarrow$ Minimum value of $5a + b$ is -2.

$\Rightarrow \quad 5\alpha + \beta = -2$

[$\because \alpha$ and β are the values of a and b for which $5a + b$ is minimum]

$\Rightarrow \quad \beta = -5\alpha - 2$

Now, $\quad \alpha(4x + 2y + 3) + \beta(x - y - 1) = 0$

$\Rightarrow \quad \alpha(4x + 2y + 3) - (5\alpha + 2)(x - y - 1) = 0$

$\Rightarrow \quad \alpha(-x + 7y + 8) - 2(x - y - 1) = 0$

Clearly, it represent a family of lines passing through the intersection of the lines $-x + 7y + 8 = 0$ and $x - y - 1 = 0$, i.e., the point $\left(-\dfrac{1}{6}, -\dfrac{7}{6}\right)$.

4. (b) Equation of four sides of quadrilateral are

$xy(x - 2)(y - 3) = 0$

$\Rightarrow \quad x = 0$

$y = 0$

$x = 2$

$y = 3$

$\therefore$ Required equation is

$x - 4y - 5 = 0$.

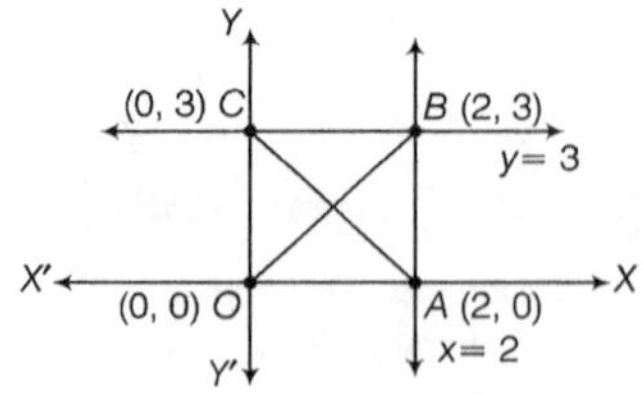

5. (a) Clearly, $AC + BC$ will be minimum when points B, C and D (image of A in X-axis) are collinear. In that case, C is the intersection of BD with X-axis.

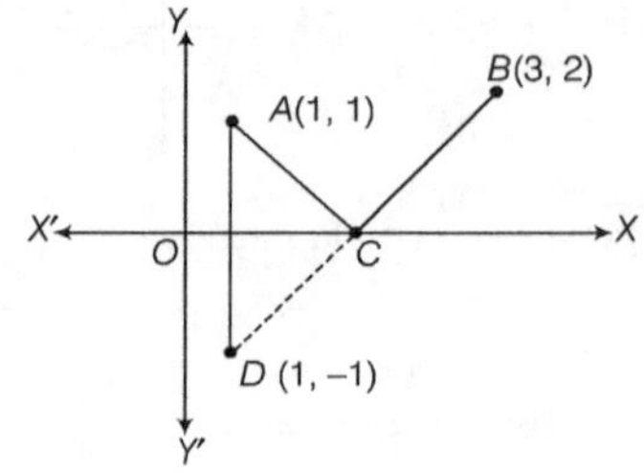

The equation of BD is

$$y + 1 = \frac{2 + 1}{3 - 1}(x - 1)$$

$\Rightarrow \quad 3x - 2y - 5 = 0$

It cut X-axis at $\left(\dfrac{5}{3}, 0\right)$.

Hence, the coordinates of the required point are $\left(\dfrac{5}{3}, 0\right)$.

6. (a) Let the coordinates of A be (h, k). Then,

$d(A, 0) = 1$

$\Rightarrow \quad \max\{|h|, |k|\} = 1$

$\Rightarrow \quad |h| \leq 1$ and $|k| \leq 1$

$\Rightarrow$ Locus of $A(h, k)$ is $|x| \leq 1, |y| \leq 1$

Clearly, $|x| \leq 1$ and $|y| \leq 1$ represent the region enclosed by the square shown in the figure given below :

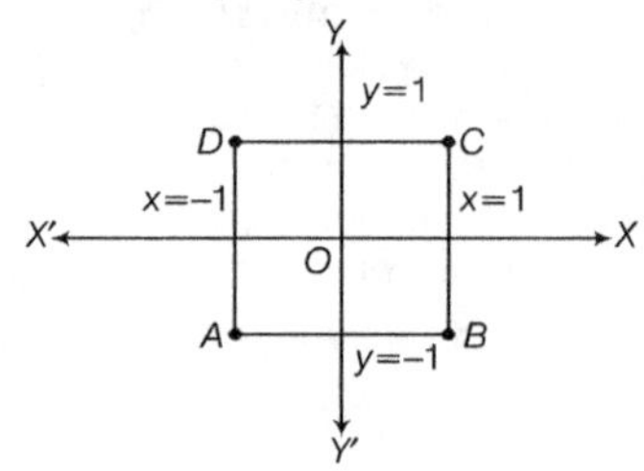

7. (b) If the axes are rotated about the origin in anti-clockwise direction, the relations relating old coordinates (x, y) and new coordinates (X, Y) are

$x = X\cos 60° - Y\sin 60°$

and $\quad y = X\sin 60° + Y\cos 60°$.

$\Rightarrow \quad x = \dfrac{X - \sqrt{3}Y}{2}$

and $\quad y = \dfrac{\sqrt{3}X + Y}{2}$

The equation of the line $x + y = 1$ with reference to new coordinate axes is

$$\frac{X - \sqrt{3}Y}{2} + \frac{\sqrt{3}X - Y}{2} = 1$$

$\Rightarrow \quad (\sqrt{3} + 1)X + (1 - \sqrt{3})Y = 2$

This intercepts length p and q on the new coordinate axes.

$\therefore \quad p = \dfrac{2}{\sqrt{3} + 1}$ and $q = \dfrac{2}{1 - \sqrt{3}}$

$\Rightarrow \quad \dfrac{1}{p^2} + \dfrac{1}{q^2} = \dfrac{(\sqrt{3} + 1)^2 + (1 - \sqrt{3})^2}{4} = 2$

8. (a) $P_1Q \perp P_2Q$

$$\Rightarrow \quad \left(\frac{k+5}{h-1}\right)\cdot\left(\frac{k-1}{h-2}\right) = -1$$

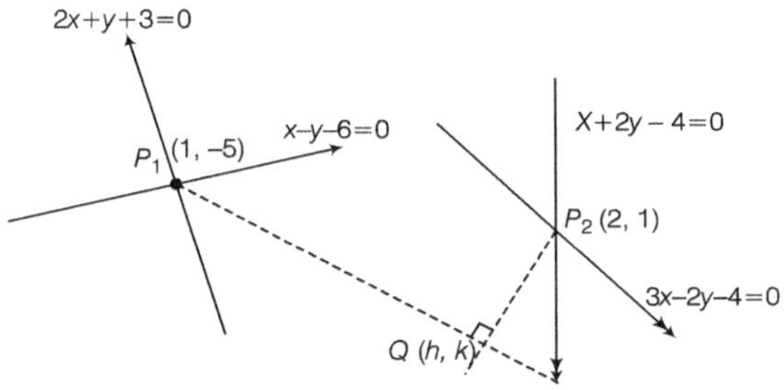

$\therefore$ Locus of (h, k) is $x^2 + y^2 - 3x + 4y - 3 = 0$.

9. (d) $A = \frac{1}{2}\left|\begin{vmatrix} 1 & p^2 & 1 \\ 0 & 1 & 1 \\ p & 0 & 1 \end{vmatrix}\right|$

$$= \frac{1}{2}\left|[1(1-0) + p(p^2-1)]\right| \qquad \text{(expanding along } C_1\text{)}$$

$$= \frac{1}{2}\left|(p^3 - p + 1\right|$$

$$\therefore \quad A = \frac{1}{2}\left|p^3 - b + 1\right|$$

Now, minimum value of modulus is zero.

Since, $A(p)$ is cubic, it must vanish for some p other than given in (a), (b), (c).

10. (b) Area of rectangle $BCDE$ is $4mn$.

Area of $\Delta ABE = \frac{2m(m-n)}{2} = m^2 - mn$

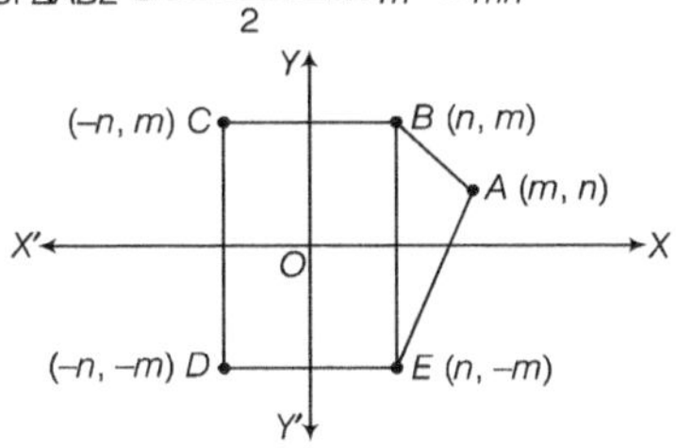

$\therefore$ Area of pentagon $ABCDE = 4mn + m^2 - mn$

$$= m^2 + 3mn = m(m + 3n)$$

11. (b)

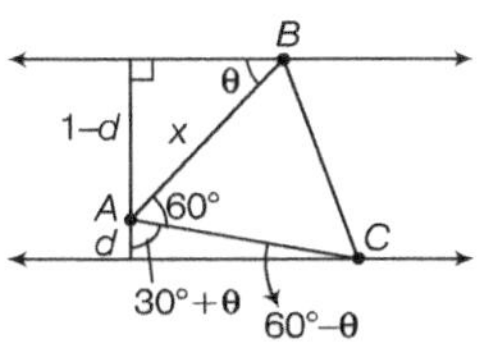

From the figure, $\quad x\cos(\theta + 30^\circ) = d \quad$ ...(i)

$$x\sin\theta = 1 - d \quad \text{...(ii)}$$

On dividing Eq. (i) by Eq. (ii), we get

$$\sqrt{3}\cot\theta = \frac{1+d}{1-d}$$

On squaring Eq. (ii) and putting the value of $\cot\theta$, we have

$$x^2 = \frac{1}{3}(4d^2 - 4d + 4)$$

$$\Rightarrow \quad x = 2\sqrt{\frac{d^2 - d + 1}{3}}$$

12. (c) $\frac{y-x}{y+x} = k \qquad (k > 1)$

$$\Rightarrow \quad (y - x) = k(y + x)$$

$$\Rightarrow \quad y(1-k) = x(1+k)$$

$$\Rightarrow \quad y = \frac{(1+k)x}{(1-k)}, \qquad \text{where } \frac{1+k}{1-k} < -1$$

13. (d) $\frac{49}{\Delta} = \frac{q^2}{a^2}$ or $\frac{7}{\sqrt{\Delta}} = \frac{q}{a}$ ...(i)

Similarly, $\quad \frac{2}{\sqrt{\Delta}} = \frac{p}{a}$ ...(ii)

and $\quad \frac{3}{\sqrt{\Delta}} = \frac{r}{a}$ ...(iii)

From Eqs. (i), (ii) and (iii), we get

$$\frac{12}{\sqrt{\Delta}} = \frac{p+q+r}{a} = 1 \Rightarrow \Delta = 144 \text{ sq units}$$

14. (a) x coordinate $= r\cos\theta$

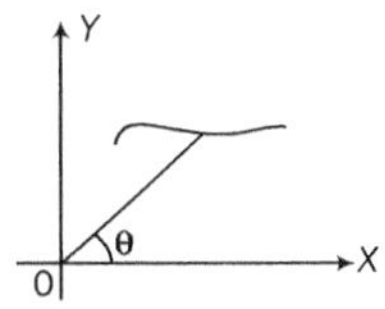

where, $r(\theta) = \cos\theta + \frac{1}{2}$

$$\therefore \quad x \text{ coordinate} = \left(\cos\theta + \frac{1}{2}\right)\cdot\cos\theta$$

$$= \cos^2\theta + \frac{1}{2}\cos\theta = \left(\cos\theta + \frac{1}{4}\right)^2 - \frac{1}{16}$$

Hence, smallest x-coordinate is $\left(-\frac{1}{16}\right)$.

15. (a) P', P'' are images of P w.r.t. $y = x$, $y = 0$, respectively.

$$\therefore \quad P'(3, 5) \text{ and } P''(5, -3)$$

Equation of $P'P''$, is

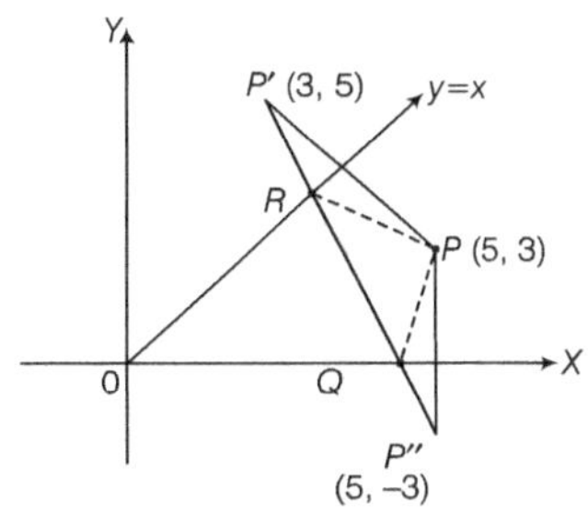

$$y + 3 = \frac{5+3}{5-3}(x - 5)$$

$$\therefore \quad Q = 5 + 3\left(-\frac{1}{4}\right) = \frac{17}{4}$$

$$\therefore \quad = \left(\frac{17}{4}, 0\right)$$

16. (d) As, (1, 1) satisfies $x = y$, it must satisfy exactly one other equation, i.e. either $x + ay + 2 = 0$ or $ax + y + 2 = 0$.

As (1, 1) either simultaneously satisfies both or none, all the lines must be concurrent.

$\therefore$ Circumcircle is of zero radius.

Hence, option (d) is correct.

17. (c) Let $PM = k$

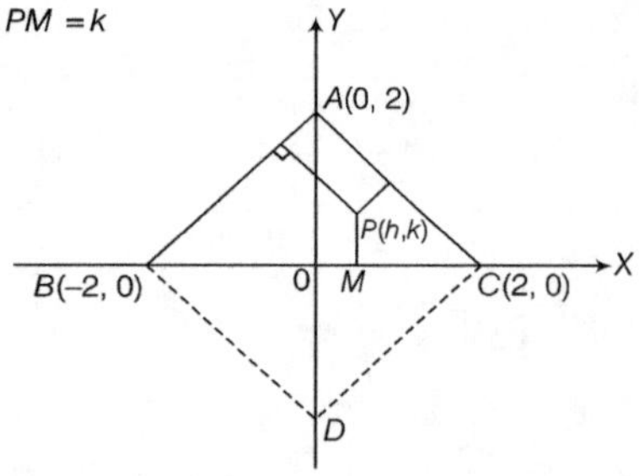

Equation of AB, $-x + y = 2$
Equation of AC, $x + y = 2$
According to the question,

$$\left(\frac{2-h-k}{\sqrt{2}}\right)\left(\frac{2+h-k}{\sqrt{2}}\right) = 2k^2$$

$\Rightarrow \quad h^2 + 3k^2 + 4k = 4$

$\Rightarrow \quad h^2 + 3\left(k^2 + \frac{4}{3}k + \frac{4}{9}\right) = 4 + \frac{4}{3}$

$\Rightarrow \quad h^2 + 3\left(k + \frac{2}{3}\right)^2 = \frac{16}{3}$

$\therefore \quad \frac{h^2}{16/3} + \frac{(k + 2/3)^2}{16/9} = 1$

Hence, locus is $\frac{x^2}{16/3} + \frac{(y + 2/3)^2}{16/9} = 1$

$\Rightarrow$ Ellipse with $e = \sqrt{2/3}$ and $D\left(0, -\frac{2}{3}\right)$.

18. (c) PA is circumradius of ΔABC.

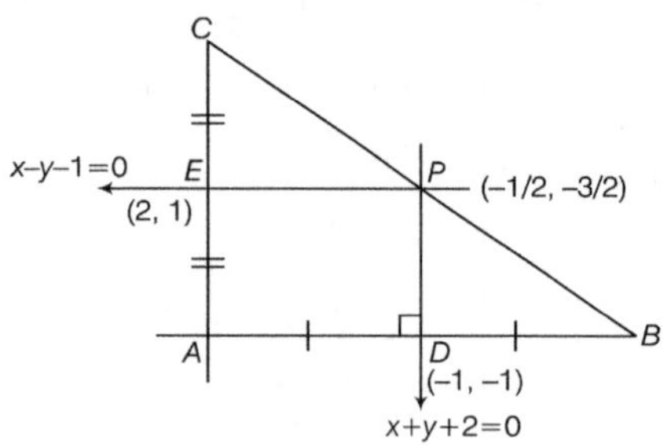

$$PA^2 = PD^2 + PE^2$$

$$= \left(-1 + \frac{1}{2}\right)^2 + \left(-1 + \frac{3}{2}\right)^2 + \left(\frac{5}{2}\right)^2 + \left(\frac{5}{2}\right)^2$$

$$= \frac{1}{4} + \frac{1}{4} + \frac{25}{4} + \frac{25}{4} = 13$$

$$PA = \sqrt{13}$$

19. (d) Given family of lines

$$(x + 2y + 1) + \lambda(x + y - 2) = 0 \quad \ldots(i)$$

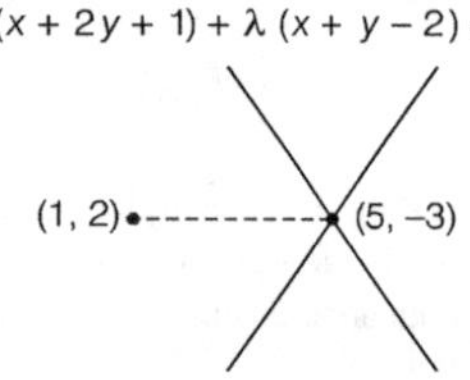

Points of intersection of these lines can be find by solving simultaneouly, the equations

$$x + 2y + 1 = 0$$
$$x + y - 2 = 0$$

Point of intersection is $(5, -3)$.

The required line is passes through $(5, -3)$ and is perpendicular to the line joining $(1, 2)$ and $(5, -3)$. So, its equation is

$$y + 3 = \frac{4}{5}(x - 5)$$

$\Rightarrow \quad 5y + 15 = 4x - 20$

$\Rightarrow \quad 4x - 5y = 35$

20. (a) Since, $(\lambda^2 + 1, \lambda)$ is mirror image of $(\lambda, \lambda + 1)$, then

$$\frac{\lambda^2 + 1 - \lambda}{1} = \frac{\lambda - \lambda - 1}{2} = \frac{-2(\lambda + 2\lambda + 2 - 1)}{5}$$

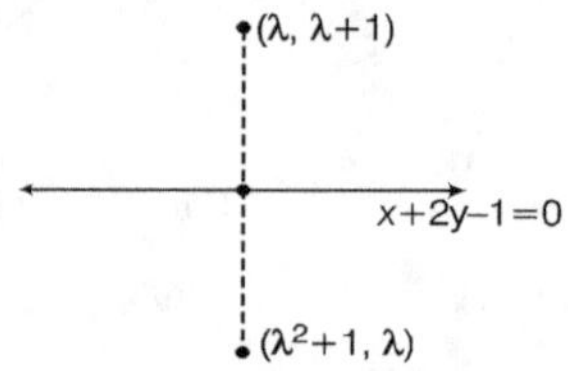

$\Rightarrow \quad \frac{\lambda^2 - \lambda + 1}{1} = -\frac{1}{2} = \frac{-2(3\lambda + 1)}{5}$

$\Rightarrow \quad \lambda^2 - \lambda + 1 = \frac{-6\lambda - 2}{5}$

and $\quad \frac{-2(3\lambda + 1)}{5} = -\frac{1}{2}$

$\Rightarrow \quad 5\lambda^2 - 5\lambda + 5 = -6\lambda - 2$

and $\quad -6\lambda - 2 = -\frac{5}{2}$

$\Rightarrow \quad 5\lambda^2 + \lambda + 7 = 0$

and $\quad -6\lambda = -\frac{5}{2} + 2$

$\Rightarrow \quad 5\lambda^2 + \lambda + 7 = 0$

and $\quad \lambda = +\frac{5}{12} - \frac{1}{3}$

$\Rightarrow \quad 5\lambda^2 + \lambda + 7 = 0$ and $\lambda = \frac{1}{12}$

Since, $\lambda = \frac{1}{12}$ does not satisfy $5\lambda^2 + \lambda + 7 = 0$

Hence, no value of λ is possible.

21. (d) Equation of BC is $y = 2$, which is parallel to x-axis.

$\therefore \quad \tan\frac{B}{2} = \frac{4}{3} \Rightarrow B > \frac{\pi}{2}$

$\quad \tan\frac{C}{2} = 2 \Rightarrow C > \frac{\pi}{2}$

In a triangle, two angles cannot be greater than 90° and hence, there is no such triangle.

22. (a) Slope of $AF = \frac{5 - (-4)}{-4 + 1} = -3$

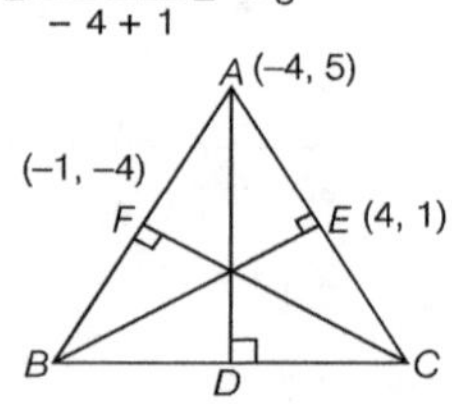

Slope of $CF = \frac{1}{3}$

Also, F lies on CF.

$\therefore$ Equation of $CF = x - 3y - 11 = 0 \quad \ldots(i)$

Slope of $AE = \frac{5 - 1}{-4 - 4} = -\frac{1}{2}$

Slope of $BC = \frac{-1}{-1/2} = 2$. Also, E lies on BE.

$\therefore$ Equation of $BE = 2x - y - 7 = 0 \quad \ldots(ii)$

On solving Eqs. (i) and (ii), we get point $O(2, -3)$.
Since, CB is passing through intersection of AC and FC.
Equation of $CB \Rightarrow (x + 2y - 6) + \lambda(x - 3y - 11) = 0$

or $\quad x(\lambda + 1) + y(2 - 3\lambda) - (6 + 11\lambda) = 0 \quad \ldots$(iii)

Slope of $CB = \frac{\lambda + 1}{3\lambda - 2}$; also slope of $AO = \frac{5 - (-3)}{-4 - 2} = -\frac{4}{3}$

Since, $CB \perp AO$; $\frac{\lambda + 1}{3\lambda - 2} \times -\frac{4}{3} = -1$ or $\lambda = 2$

$\therefore$ Equation of $BC \Rightarrow 3x - 4y - 28 = 0$

23. (c) Area of $\Delta = \frac{1}{2}\begin{vmatrix} x_1 & y_1 & 1 \\ x_2 & y_2 & 1 \\ x_3 & y_3 & 1 \end{vmatrix} = \frac{1}{2}\begin{vmatrix} m & n & 1 \\ 12 & 19 & 1 \\ 23 & 20 & 1 \end{vmatrix} = 70$

$\Rightarrow \quad -m + 11n = 197 \pm 140 \quad \ldots$(i)

Median through A has slope -5.

$\Rightarrow \quad 5m + n = 107 \quad \ldots$(ii)

From Eqs. (i) and (ii), we get $m = 20$, $n = 7$

$\Rightarrow \quad m + n = 27$

24. (a) Let (x_1, y_1) and (x_2, y_2) be end points of diameter. Thus, equation of circle is

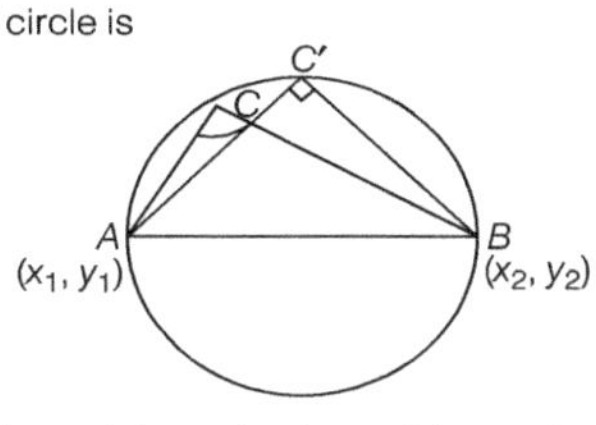

$$(x - x_1)(x - x_2) + (y - y_1)(y - y_2) = 0$$

If $C(x_3, y_3)$ makes an obtuse angle.
$\Rightarrow C$ lies inside the circle.

$\therefore \quad S_1 < 0$

or $\quad (x_3 - x_1)(x_3 - x_2) + (y_3 - y_1)(y_3 - y_2) < 0$

25. (d) First 3 equations are suggestive that (x_1, y_1) (x_2, y_2) and (x_3, y_3) are the vertices of ΔBCA.

Now, $\quad A = \frac{1}{2}\begin{vmatrix} x_1 & y_1 & 1 \\ x_2 & y_2 & 1 \\ x_3 & y_3 & 1 \end{vmatrix}, \quad 4A^2 = \begin{vmatrix} x_1 & y_1 & 1 \\ x_2 & y_2 & 1 \\ x_3 & y_3 & 1 \end{vmatrix}^2$

$\therefore \quad 4A^2 = 4s(s - a)(s - b)(s - c)$

$\Rightarrow \quad 16A^2 = 2s(2s - 2a)(2s - 2b)(2s - 2c) = \text{RHS}$

Thus, Statement I is incorrect.

26. (d) Here, $a(2x + 3y - 5) + b(x - 2y + 1) = 0 \quad \ldots$(i)

This is a family of lines passing through the point of intersection of $2x + 3y + 5 = 0$ and $x - 2y + 1 = 0$, i.e. $(1, 1)$.

Let $P(h, k)$ be the image of $A(0, 0)$ on the line passing through $B(1, 1)$.

$\Rightarrow \quad AB^2 = PB^2$

$\Rightarrow \quad 1 + 1 = (h - 1)^2 + (k + 1)^2$

or $\quad x^2 + y^2 - 2x - 2y = 0$ is locus of P.

27. (b) Since, origin is equidistant from all three straight lines. Thus, incentre is origin.

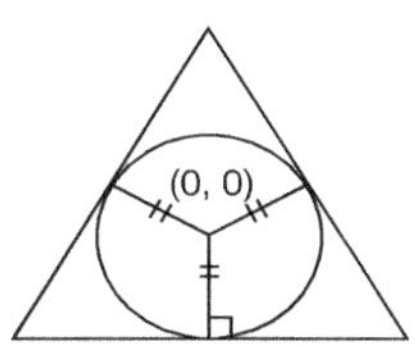

and inradius $(r) = 2$

28. (a) Here, $P = \left(\frac{2x_1 + x_2 + 3x_3}{6}, \frac{2y_1 + y_2 + y_3}{6}\right)$

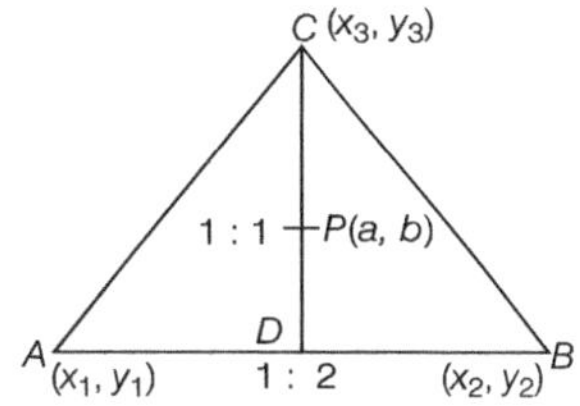

$$\left(\frac{2x_1 + x_2}{3}, \frac{2y_1 + y_2}{3}\right)$$

Hence, P lies inside the Δ.

$\therefore$ Ar of $\Delta PBC <$ Ar ΔABC

29. (b) Here, $\begin{vmatrix} a & a^2b & a^2 + c \\ b & b^3 & b^2 + c \\ c & c^2b & c^2 + c \end{vmatrix} - \begin{vmatrix} a & 1 & a^2 + c \\ b & 1 & b^2 + c \\ c & 1 & c^2 + c \end{vmatrix} = 0$

$\Rightarrow \quad b\begin{vmatrix} a & a^2 & a^2 + c \\ b & b^2 & b^2 + c \\ c & c^2 & c^2 + c \end{vmatrix} - \begin{vmatrix} a & 1 & a^2 \\ b & 1 & b^2 \\ c & 1 & c^2 \end{vmatrix} = 0$

$\Rightarrow \quad b\begin{vmatrix} a & a^2 & c \\ b & b^2 & c \\ c & c^2 & c \end{vmatrix} - \begin{vmatrix} a & 1 & a^2 \\ b & 1 & b^2 \\ c & 1 & c^2 \end{vmatrix} = 0$

$\Rightarrow \quad bc\begin{vmatrix} a & a^2 & 1 \\ b & b^2 & 1 \\ c & c^2 & 1 \end{vmatrix} + \begin{vmatrix} a & a^2 & 1 \\ b & b^2 & 1 \\ c & c^2 & 1 \end{vmatrix} = 0$

$\Rightarrow \quad (bc + 1)\begin{vmatrix} a & a^2 & 1 \\ b & b^2 & 1 \\ c & c^2 & 1 \end{vmatrix} = 0$

$\Rightarrow \quad (bc + 1)(a - b)(b - c)(c - a) = 0$

$\therefore$ Either $a = b$ or $b = c$ or $c = a$

Hence, triangle is isosceles.

30. (a) $\quad B'C' = 7 = a'$

$$A'B' = \sqrt{144 + 81} = 15 = c'$$
$$A'C' = \sqrt{144 + 256} = 20 = b'$$

where, $A'B'C'$ is the pedal Δ of ΔABC.

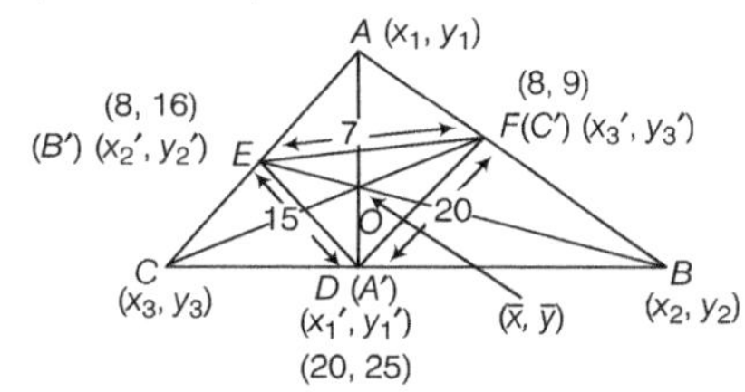

Note *Altitudes bisects each interior angle of the pedal Δ i.e. $\pi - 2A, \pi - 2B, \pi - 2C$.*

$O(\bar{x}, \bar{y})$ is the orthocentre of ΔABC and also the incentre of $\Delta A'B'C'$.

$$\bar{x} = \frac{a'x_1' + b'x_2' + c'x_3'}{a' + b' + c'},$$

where a', b', c' are the sides of pedal Δ.

$$= \frac{7(20) + 15(8) + 20(8)}{7 + 15 + 20} = 10$$

$$\bar{y} = \frac{a'y_1' + b'y_2' + c'y_3'}{a' + b' + c'} = \frac{7(25) + 15(9) + 20(16)}{7 + 15 + 20} = 15$$

$\therefore$ Orthocentre is $(10, 15)$. Hence, **(i) $\rightarrow$ (r)**

Now, slope of $OA' = \dfrac{25-15}{20-10} = 1$

$\Rightarrow \quad m_{BC} = -1$

Slope of $OB' = \dfrac{16-15}{8-10} = -\dfrac{1}{2}$

$\Rightarrow \quad m_{AC} = 2$

Slope of $OC' = \dfrac{9-15}{8-10} = 3$

$\Rightarrow \quad m_{AB} = -\dfrac{1}{3}$

Equation of AB, is

$$(y-9) = -\frac{1}{3}(x-8)$$

$\Rightarrow \quad 3y + x = 25 \quad \ldots\text{(i)}$

Equation of AC, is

$$(y-16) = 2(x-8)$$

$\Rightarrow \quad y + 2x = 0 \quad \ldots\text{(ii)}$

Equation of BC, is

$$(y-25) = -1(x-20)$$

$\Rightarrow \quad y + x = 45 \quad \ldots\text{(iii)}$

Hence, **(ii) → (p)**

Solving we get coordinates of

$$A(5, 10), B(50, -5), C(15, 30)$$

Centroid of ΔABC is $\left(\dfrac{70}{3}, \dfrac{35}{3}\right) = (a, b)$

$\therefore \quad a + b = 35$

Hence, **(iii) → (q)**

Now, orthocentre $O(10, 15)$ and centroid $\left(\dfrac{70}{3}, \dfrac{35}{3}\right)$

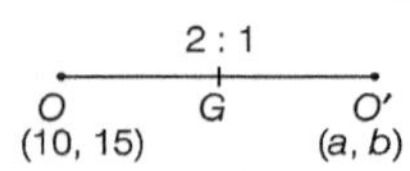

$\therefore \quad \dfrac{70}{3} = \dfrac{1(10) + 2(a)}{3} \Rightarrow a = 4$

$\dfrac{35}{3} = \dfrac{1(15) + 2(b)}{3} \Rightarrow b = 10$

∴ Circumcentre of ΔABC is $(4, 10)$.

Hence, **(iv) → (s)**

Explanations for Q. Nos. (31-33)

Equation of PM, is $(y-1) = \tan\theta(x-1)$

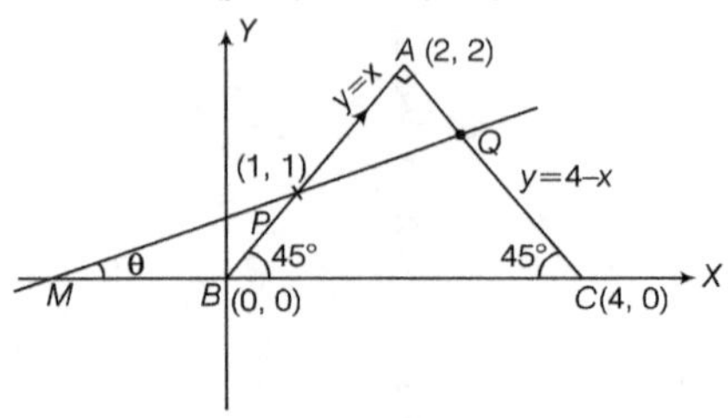

Intersection point Q of AC and MP,

$$4 - x - 1 = \tan\theta\,(x-1)$$

$\Rightarrow \quad Q\left(\dfrac{3+\tan\theta}{1+\tan\theta}, \dfrac{1+3\tan\theta}{1+\tan\theta}\right)$

$$\text{Area of } \Delta APQ = \frac{1}{2}\begin{vmatrix} 1 & 1 & 1 \\ 2 & 2 & 1 \\ \dfrac{3+\tan\theta}{1+\tan\theta} & \dfrac{1+3\tan\theta}{1+\tan\theta} & 1 \end{vmatrix}$$

$$= \frac{1-\tan\theta}{1+\tan\theta}$$

Area of quadrilateral $BPQC$

$$A = \frac{1}{2}(4 \times 2) - \left(\frac{1-\tan\theta}{1+\tan\theta}\right) = \frac{3+5\tan\theta}{1+\tan\theta}$$

$$= 5 - \frac{2}{1+\tan\theta}$$

note $\theta \in \left(0, \dfrac{\pi}{4}\right) \Rightarrow 1 + \tan\theta \in (1, 2)$

$$(PQ)^2 = \left(\frac{3+\tan\theta}{1+\tan\theta} - 1\right)^2 + \left(\frac{1+3\tan\theta}{1+\tan\theta} - 1\right)^2$$

$$= \frac{4}{(\cos\theta + \sin\theta)^2} = \frac{4}{1+\sin 2\theta}$$

$\Rightarrow \quad PQ \in (\sqrt{2}, 2)$

31. (d) **32.** (d) **33.** (c)

Explanations for Q. Nos. (34-36)

Given, $\quad (4a+3)x - (a+1)y - (2a+1) = 0$

$\Rightarrow \quad (3x - y - 1) + a(4x - y - 2) = 0$

Family of lines passing through the fixed point $P(1, 2)$.

34. (d) $\quad \dfrac{k}{h}\cdot\left(\dfrac{k-2}{h-1}\right) = -1$

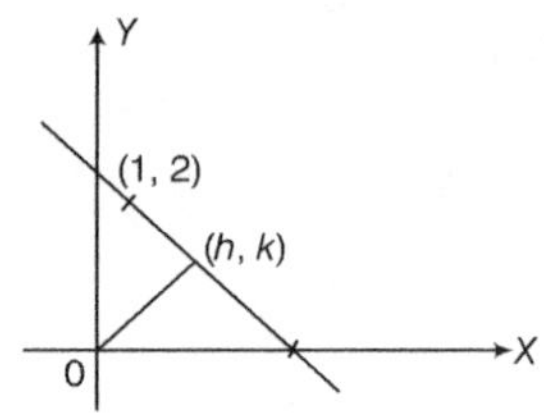

∴ Locus is $x(x-1) + y(y-2) = 0$

$\Rightarrow \quad x^2 + y^2 - 2y - x = 0$

$\Rightarrow \quad \left(x - \dfrac{1}{2}\right)^2 + (y-1)^2 = \dfrac{5}{4}$

or $\quad (2x-1)^2 + 4(y-1)^2 = 5$

35. (a) We have, $\quad (y-2) = m(x-1)$

This makes an angle of $\dfrac{\pi}{4}$ with $3x - 4y = 2$ with slope $3/4$.

$$\therefore \quad \left|\frac{m - 3/4}{1 + \frac{3}{4}m}\right| = 1$$

$\Rightarrow \quad m = 7, -\dfrac{1}{7} \quad$ (neglecting $-1/7$)

Hence, the line is $\quad y - 2 = 7(x-1)$

$\Rightarrow \quad 7x - y - 5 = 0$

36. (c) Again, $\quad (y-2) = m(x-1)$

$$x = 0, y = 2 - m;\ y = 0, x = 1 - \frac{2}{m}$$

$\therefore \quad 2A = (2-m)\left(1 - \dfrac{2}{m}\right) \quad (m < 0)$

$\Rightarrow \quad 2A = 2 - m - \dfrac{4}{m} + 2$

$$= 4 + \left(-m - \frac{4}{m}\right)$$

Let $\quad -m = M \quad (M > 0)$

$\Rightarrow \quad 2A = 4 + M + \dfrac{4}{M}$

$\therefore \quad (2A)_{min} = 4 + 4$

$(A)_{min} = 4$

Explanations for Q. Nos. (37-39)

On solving,

$$C\left(\frac{3}{2}, 0\right), F\left(-4, -\frac{7}{2}\right), E\left(\frac{1}{2}, -\frac{9}{2}\right), D\left(\frac{5+4\sqrt{3}}{4}, \frac{4+\sqrt{3}}{4}\right)$$

37. (d) Thus, slope of $AF = \frac{11}{10}$.

38. (c) Abscissa of centroid of $\Delta BCE = -\frac{1}{3}$.

39. (b) $D(\alpha, \beta) \Rightarrow (4\beta - 4)^2 = 3$

40. (d) Let $OA = r_1, OP = r, OB = r_2$

$$A(r_1 \cos\theta, r_1 \sin\theta)$$
$$B(r_2 \cos\theta, r_2 \sin\theta)$$
$$P(x, y) = (r\cos\theta, r\sin\theta)$$

A lie on $y - x - 20 = 0$

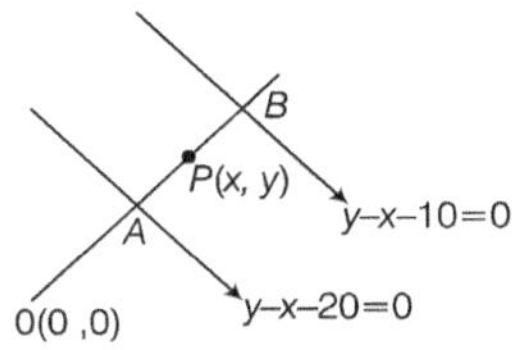

$\therefore \quad r_1 \sin\theta - r_1 \cos\theta - 20 = 0$

$$\frac{20}{r_1} = \sin\theta - \cos\theta$$

Also, $\frac{10}{r_2} = \sin\theta - \cos\theta$

$\therefore \quad \frac{2}{OP} = \frac{1}{OA} + \frac{1}{OB}$

$\Rightarrow \quad \frac{2}{r} = \frac{1}{r_1} + \frac{1}{r_2} = \frac{3\sin\theta - 3\cos\theta}{20}$

$\therefore 3(r\sin\theta) - 3(r\cos\theta) = 40 \quad \Rightarrow \quad 3y - 3x = 40$

41. (c) $\quad OP^2 = OA \cdot OB$

$$r^2 = \frac{10 \times 20}{(\sin\theta - \cos\theta)^2}$$

$\Rightarrow \quad (r\sin\theta - r\cos\theta)^2 = 200$

Hence, locus is $(y - x)^2 = 200$

42. (c) $OP = OA + OB$

$$r = \frac{10}{\sin\theta - \cos\theta} + \frac{20}{\sin\theta - \cos\theta}$$

$\Rightarrow \quad r = \frac{30}{\sin\theta - \cos\theta}$

$\Rightarrow \quad r\sin\theta - r\cos\theta = 30$

$\Rightarrow \quad y - x = 30$

43. (d) $\because \quad OR = AR$

$\Rightarrow \quad |x - 0| + |y - 0| = |x - 1| + |y - 2|$

$\Rightarrow \quad |x| + |y| = |x - 1| + |y - 2|$,

as $\quad 0 \le x < 1, 0 \le y < 2$

$\therefore \quad x + y = 1 - x + 2 - y$

$\Rightarrow \quad 2x + 2y = 3$

44. (d) $\because OS = OR$

$\Rightarrow |x| + |y| = |x - 2| + |y - 3|$

$\because \; x \ge 2$ and $0 \le y < 3$

$\Rightarrow \quad x + y = x - 2 + 3 - y$

$\Rightarrow \quad y = \frac{1}{2}$

45. (c) Centroid $G\left(\frac{7}{3}, \frac{8}{3}\right)$, orthocentre $O(1, 6)$ and circumcentre $C(3, 1)$.

$\therefore \quad OC = |1 - 3| + |6 - 1| = 7$ units

46. (a) A line passing through $(\sqrt{2}, \sqrt{3})$ is

$$y - \sqrt{3} = m(x - \sqrt{2})$$

$\Rightarrow \quad y - mx = \sqrt{3} - \sqrt{2}m \quad \ldots$(i)

Reflection of any point w.r.t. diameter lie on the circle. Thus, two rational points cannot exist on the circle as their mid-point should satisfy Eq. (i).

47. (b) $y - mx = -\sqrt{2}m - \sqrt{2}$

$\Rightarrow \quad y - mx = -\sqrt{2}(m + 1)$

For the above equation to have rational coefficient $m = -1$.

$\Rightarrow \quad y + x = 0$

Hence, reflection of (1, 2) w.r.t. above line lies on the circle.

Any other reflection point of (1, 2) will yield an irrational point.

48. (a) $(1, -1)$ lie on $y + x = 0$.

Hence, only one rational point.

Explanations for Q. Nos. (49–51)

Equation of ellipse satisfying $PA + PB = 2$.

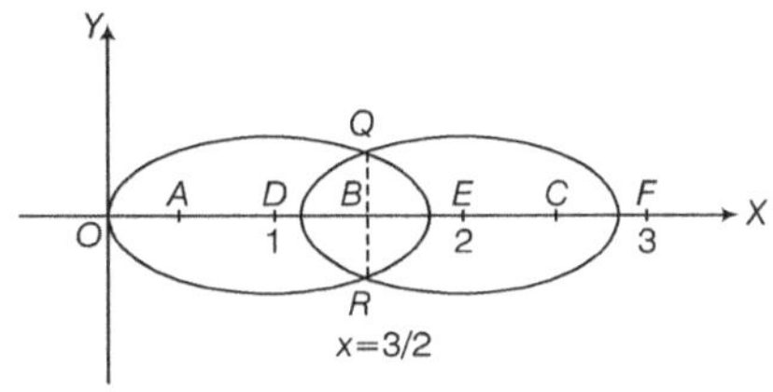

with major axis = 2 is $\quad \frac{(x-1)^2}{1} + \frac{y^2}{3/4} = 1$

Similarly, the equation of ellipse satisfying $PB + PC = 2$ is

$$\frac{(x-2)^2}{1} + \frac{y^2}{3/4} = 1$$

Required region is $DQERD$, $y = \frac{\sqrt{3}}{2}\sqrt{1 - (x-2)^2}$

$\therefore \quad$ Required area $= 4\int_1^{3/2} y\,dx = 2\sqrt{3}\int_1^{3/2}\sqrt{1 - (x-2)^2}\,dx$

$$= \sqrt{3}(\pi/3 - \sqrt{3}/4)$$

49. (d) **50.** (c) **51.** (b)

52. (c) Since, diagonal is member of both the families.

$\therefore$ it passes through (1, 0) and (– 1, 1).

Hence, eqaution of diagonal AC is $x + 2y - 1 = 0$.

53. (b) $\because$ (3, 2) does not lie on AC.

$\therefore$ It lies on BD. Hence, equation of BD is $2x - y = 4$.

54. (c) Point of intersection of diagonal AC and BD is $\left(\frac{9}{5}, -\frac{2}{5}\right)$.

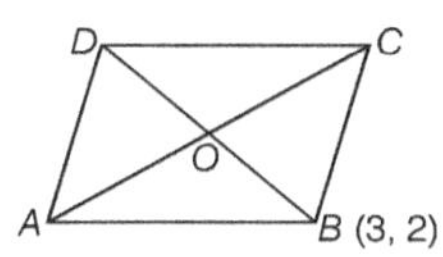

$\Rightarrow \quad$ Vertex $D\left(\frac{3}{5}, -\frac{14}{5}\right)$

Length of $BD = \frac{6\sqrt{5}}{5}$

$\Rightarrow \quad BD = \frac{12\sqrt{5}}{5} = \frac{12}{\sqrt{5}}$

Ar. (rhombus $ABCD$) $= \frac{1}{2} \times d_1 \times d_2 = 12\sqrt{5}$

$$d_2 = 10 \text{units}$$
$$AC = 10$$

Using parametric form SC and AC.

$$A\left(\frac{9}{5}-2\sqrt{5}, -\frac{2}{5}+\sqrt{5}\right) \text{ and } C\left(\frac{9}{5}+2\sqrt{5}, -\frac{2}{5}-\sqrt{5}\right)$$

$\therefore$ Centroid of $\Delta\, ACD$ is $\left(\frac{7}{5}, -\frac{6}{5}\right)$.

55. (6) For consistency,

$$\begin{vmatrix} a & 1 & b \\ b & 1 & a \\ a & b & ab \end{vmatrix} = 0$$

$\Rightarrow \quad (a-b)(a-b^2) = 0$

Either $a = b$ or $a = b^2$.

When $a = b$, then ordered pairs are (0, 0), (1, 1), (2, 2), (3, 3), (4, 4).

When $\quad a = b^2$, then (4, 2)

Hence, number of ordered pairs are 6.

56. (6) Given, $\quad x^2 - 4x + 1 = 0$

$\Rightarrow \quad x = 2 + \sqrt{3}$

From below figure, $I(a, b)$ lies on line $y = x$. So, $b = a$.

Also, from $\Delta\, IBD$.

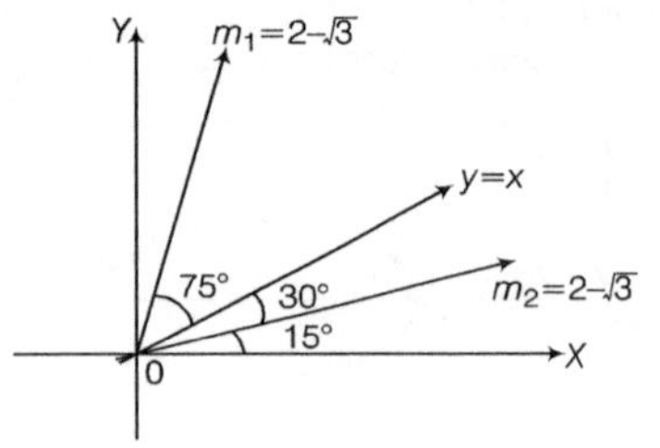

$$\frac{r}{BI} = \sin 30°$$

$\Rightarrow \quad BI = 2 \qquad (\because \text{ as } r = 1)$

$\therefore \quad BI = 2 = \sqrt{a^2 + a^2}$

$\Rightarrow \quad a = \sqrt{2}$

So, $\quad I(\sqrt{2}, \sqrt{2}) = (a, b)$

Hence, $\quad a^2 + b^4 = 6$

57. (4) Area $= \frac{1}{2}ab$

$AD: y = x + 3, \quad BE: y = 2x + 4$

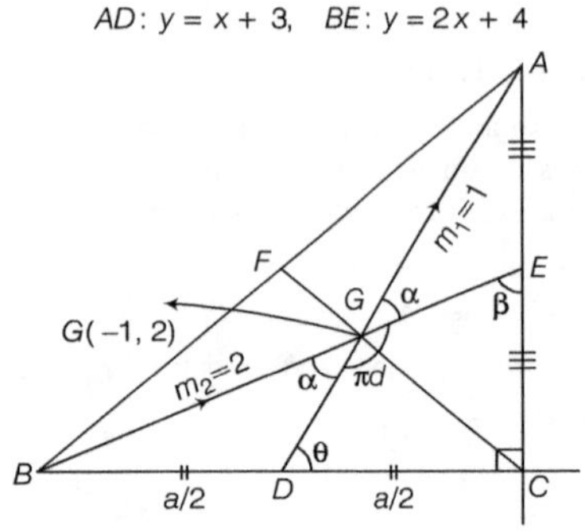

Solving $G(-1, 2)$.

Acute angle α between the medians is

$$\tan\alpha = \frac{m_1 - m_2}{1 + m_1 m_2}$$

$\Rightarrow \quad \tan\alpha = \frac{2-1}{1+2} = \frac{1}{3}$

Now, $(180 - \alpha) + 90° + \theta + \beta = 360°$

$\Rightarrow \quad \alpha = \theta + \beta - 90°$

$\cot\alpha = -\tan(\theta + \beta)$

$\Rightarrow \quad -3 = \dfrac{\tan\theta + \tan\beta}{1 - \tan\theta\cdot\tan\beta}$

$\Rightarrow \quad -3 = \dfrac{\frac{2b}{a} + \frac{2a}{b}}{1 - \frac{2b}{a}\cdot\frac{2a}{b}}$

$\Rightarrow \quad 9 = \dfrac{2(a^2 + b^2)}{ab}$

$\Rightarrow \quad 9ab = 2(3600)$

$\Rightarrow \quad \dfrac{ab}{2} = 400 = abc$

$\therefore \quad a + b + c = 4$

58. (1) $P\left(\dfrac{\alpha\cos\alpha - \sin\alpha}{\sin\alpha - \alpha}, \dfrac{\sin^2\alpha - \alpha^2\cos\alpha}{(\sin\alpha - \alpha)\cdot\sin\alpha\cos\alpha}\right)$

As limit, $\alpha \to 0$

$P(2, -1)$

$\therefore \quad (h + k) = 1$

59. (9) $\dfrac{AD}{DE} = \dfrac{BC}{DC} = \dfrac{AB}{AC}$

$\therefore$ Point $D\left(\frac{1}{3}, \frac{1}{3}\right)$ and $E(-2, 0)$.

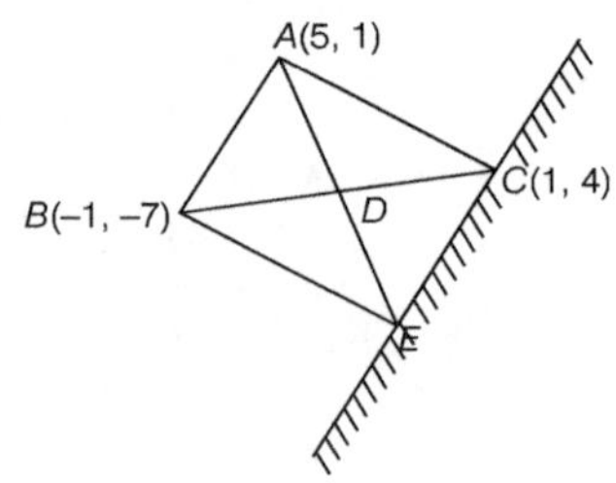

Area of ΔACE + area ΔABC = 2 area ΔACE

$$= 2 \times \frac{19}{2} = 19 = 10 + 9$$

$\therefore \quad a = 9$

60. (8) The line $y = \sqrt{3}\,x$ can be written as $x = r/2$, $y = \dfrac{r\sqrt{3}}{2}$. If this line cuts the given curve, then

$$\frac{r^4}{16} + \frac{ar^3\sqrt{3}}{8} + \frac{br^2\sqrt{3}}{4} + \frac{cr}{2} + \frac{dr\sqrt{3}}{2} + 6 = 0$$

$\therefore \quad \dfrac{OA\cdot OB\cdot OC\cdot OD}{12} = \dfrac{1}{12}|r_1 r_2 r_3 r_4| = 8$

JEE Main & AIEEE Archive

1. (b) Coordinates of

$$S = \left(\frac{7+6}{2}, \frac{3-1}{2}\right) = \left(\frac{13}{2}, 1\right)$$

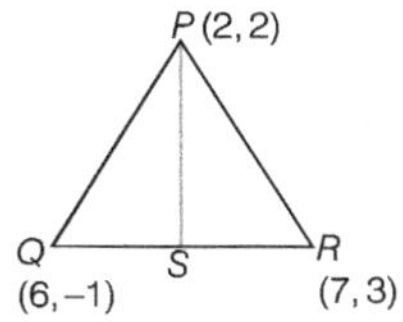

Slope of the line PS is $\frac{-2}{9}$.

Required equation passing through (1, – 1) is

$$y + 1 = \frac{-2}{9}(x - 1)$$

$\Rightarrow \quad 2x + 9y + 7 = 0$

2. (c) Let coordinates of the point be $(\alpha, -\alpha)$.

Since, $(\alpha, -\alpha)$ lie on $4ax + 2ay + c = 0$

and $\quad 5bx + 2by + d = 0$

$\therefore \quad 4a\alpha - 2a\alpha + c = 0$

$\Rightarrow \quad \alpha = \frac{-c}{2a} \quad$ ...(i)

Also, $\quad 5b\alpha - 2b\alpha + d = 0$

$\Rightarrow \quad \alpha = \frac{-d}{3b} \quad$ ...(ii)

From Eqs. (i) and (ii), $\quad \frac{-c}{2a} = \frac{-d}{3b}$

$\therefore \quad 3bc = 2ad$

3. (b) Take any point B (0, 1) on given line.

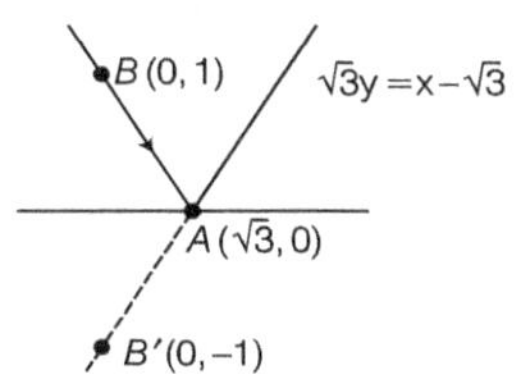

Equation of AB',

$$y - 0 = \frac{-1-0}{0-\sqrt{3}}(x - \sqrt{3})$$

$\Rightarrow \quad -\sqrt{3}y = -x + \sqrt{3} \Rightarrow x - \sqrt{3}y = \sqrt{3}$

$\Rightarrow \quad \sqrt{3}y = x - \sqrt{3}$

Alternate Method

Slope of given line is $-\frac{1}{\sqrt{3}}$.

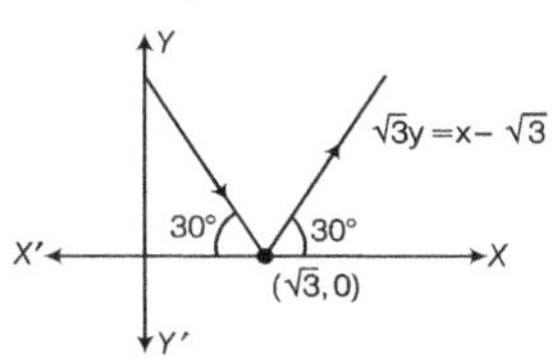

From figure, equation of reflected rays is

$$y = \frac{1}{\sqrt{3}}(x - \sqrt{3}) \Rightarrow \sqrt{3}y = x - \sqrt{3}$$

4. (c) Given

(i) A line through (1, 2) meets the coordinate axes at P and Q.

(ii) The area of ΔOPQ is minimum.

To find The slope of line PQ.

Let m be the slope of the line PQ, then the equation of PQ is

$$y - 2 = m(x - 1)$$

Now, PQ meets X-axis at $P\left(1 - \frac{2}{m}, 0\right)$

and Y-axis at $Q(0, 2 - m)$.

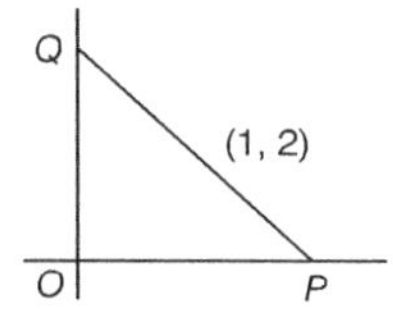

$\Rightarrow \quad OP = 1 - \frac{2}{m}$

and $\quad OQ = 2 - m$

Also, area of $\Delta OPQ = \frac{1}{2}(OP)(OQ)$

$$= \frac{1}{2}\left|\left(1 - \frac{2}{m}\right)(2 - m)\right|$$

$$= \frac{1}{2}\left|2 - m - \frac{4}{m} + 2\right| = \frac{1}{2}\left|4 - \left(m + \frac{4}{m}\right)\right|$$

Let $\quad f(m) = 4 - \left(m + \frac{4}{m}\right)$

$\Rightarrow \quad f'(m) = -1 + \frac{4}{m^2}$

Now, $\quad f'(m) = 0$

$\Rightarrow \quad m^2 = 4 \Rightarrow m = \pm 2$

$\Rightarrow \quad f(2) = 0$ and $f(-2) = 8$

Since, the area cannot be zero, hence the required value of m is -2.

5. (c) Here, $L_1 : y - x = 0$, $L_2 : 2x + y = 0$

and $\quad L_3 : y + 2 = 0$ shown as

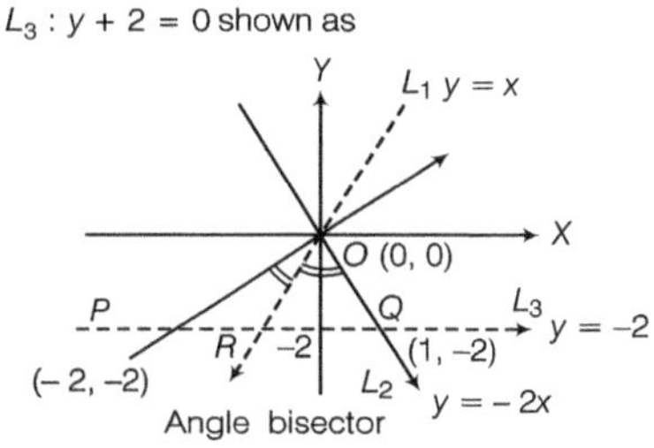

$$|PO| = \sqrt{4 + 4} = 2\sqrt{2}$$

$$|OQ| = \sqrt{1 + 4} = \sqrt{5}$$

Since, OR is angle bisector.

$\therefore \quad \frac{OP}{OQ} = \frac{PR}{RQ}$

In a triangle, angle bisector divides the opposite sides in the ratio of side containing the angle.

$\Rightarrow \quad \frac{PR}{RQ} = \frac{2\sqrt{2}}{\sqrt{5}}$

So, Statement I is correct.

But it does not divide the triangle in two similar triangles.

So, Statement II is incorrect.

6. (c)

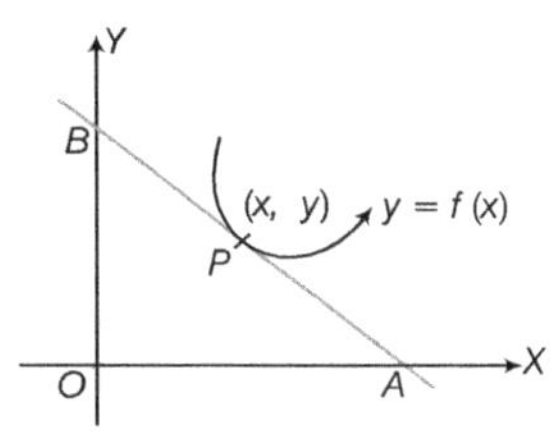

Equation of AB is $\dfrac{Y-y}{X-x} = \left(\dfrac{dy}{dx}\right)_{(x,\,y)}$ ...(i)

x-intercept is $\left(x - y\cdot\left(\dfrac{dx}{dy}\right), 0\right)$

y-intercept is $\left(0,\ y - x\dfrac{dy}{dx}\right)$

As P is mid-point of AB.

$\therefore \quad 2x = x - y\cdot\left(\dfrac{dx}{dy}\right) \quad \Rightarrow \quad y\dfrac{dx}{dy} = -x$

$\therefore \quad \dfrac{dx}{x} + \dfrac{dy}{y} = 0$

Integrating both sides, we get

$$\log x + \log y = \log C$$

$\Rightarrow \quad xy = C$, as it passes through (2, 3).

$\therefore \quad C = 6 \quad \Rightarrow \quad xy = 6$

7. (c) As, $x + y = |a|$ and $ax - y = 1$, intersect in I quadrant.

Therefore, x and y intercept are positive.

i.e. $\quad x = \dfrac{1+|a|}{1+a} \ge 0$

and $\quad y = \dfrac{a|a|-1}{a+1} \ge 0 \quad \Rightarrow \quad 1 + a \ge 0$

and $\quad a|a| - 1 \ge 0 \quad \Rightarrow \quad a \ge -1$

and $\quad a|a| \ge 1$ ...(i)

***Case* I** If $\quad -1 < a < 0$

$\Rightarrow \quad -a^2 > 1$ [not possible]

***Case* II** If $\quad a \ge 0 \Rightarrow a^2 \ge 1$

$\Rightarrow \quad a \ge 1$ ...(ii)

$\therefore \quad a \ge 1$ or $a \in [1, \infty)$

8. (d) Since, the line L is passing through the point (13, 32). Therefore,

$$\frac{13}{5} + \frac{32}{b} = 1 \quad \Rightarrow \quad \frac{32}{b} = -\frac{8}{5}$$

$\Rightarrow \quad b = -20$

The line K is parallel to the line L, its equation must be

$$\frac{x}{5} - \frac{y}{20} = a \text{ or } \frac{x}{5a} - \frac{y}{20a} = 1$$

On comparing with $\dfrac{x}{c} + \dfrac{y}{3} = 1$, we get

$$20a = -3, c = 5a = -\frac{3}{4}$$

Hence, the distance between lines

$$= \frac{|a-1|}{\sqrt{\frac{1}{25} + \frac{1}{400}}} = \frac{\left|\frac{-3}{20} - 1\right|}{\sqrt{\frac{17}{400}}} = \frac{23}{\sqrt{17}}$$

9. (a) Lines perpendicular to same line are parallel to each other.

$\therefore \quad -p(p^2+1) = p^2 + 1 \quad \Rightarrow p = -1$

Hence, there is exactly one value of p.

10. (a) Since, slope of $PQ = \dfrac{4-3}{1-k} = \dfrac{1}{1-k}$

Slope of $AM = (k-1)$

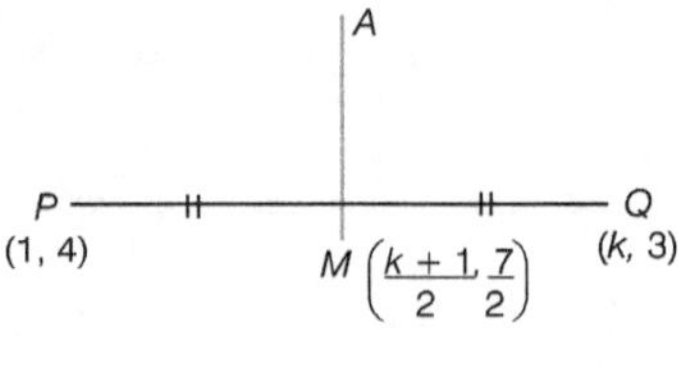

$\therefore$ Equation of AM is

$$y - \frac{7}{2} = (k-1)\left[x - \left(\frac{k+1}{2}\right)\right]$$

For y intercept, $x = 0$, $y = -4$

$$-4 - \frac{7}{2} = -(k-1)\left(\frac{k+1}{2}\right) \quad \Rightarrow \quad \frac{15}{2} = \frac{k^2-1}{2}$$

$\Rightarrow \quad k^2 - 1 = 15 \quad \Rightarrow \quad k^2 = 16$

$\therefore \quad k = \pm 4$

11. (a) Now, slope of $QR = \dfrac{3\sqrt{3}-0}{3-0} = \sqrt{3} = \tan\theta$

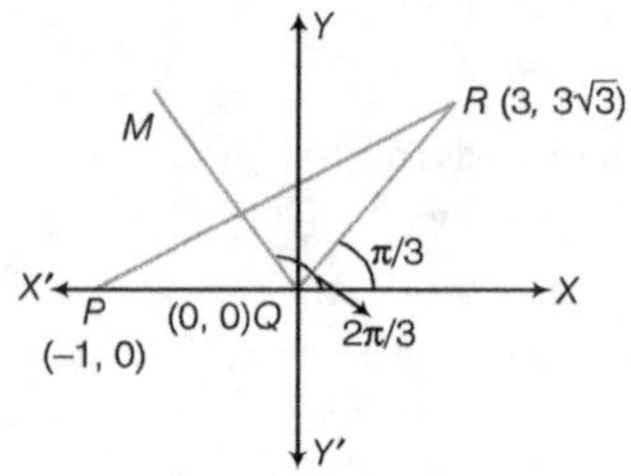

$\Rightarrow \quad \theta = \dfrac{\pi}{3}$

The angle between PQR is $\dfrac{2\pi}{3}$, so the line QM makes an angle $\dfrac{2\pi}{3}$ from positive direction of X-axis.

Slope of the line $QM = \tan\dfrac{2\pi}{3} = -\sqrt{3}$

Hence, equation of line QM is $y = -\sqrt{3}x$

or $\quad \sqrt{3}x + y = 0$

12. (b) Since, A is the mid-point of line PQ.

$\therefore \quad 3 = \dfrac{a+0}{2} \Rightarrow a = 6$

and $\quad 4 = \dfrac{0+b}{2} \Rightarrow b = 8$

Hence, the equation of line is

$$\frac{x}{6} + \frac{y}{8} = 1 \text{ or } 4x + 3y = 24$$

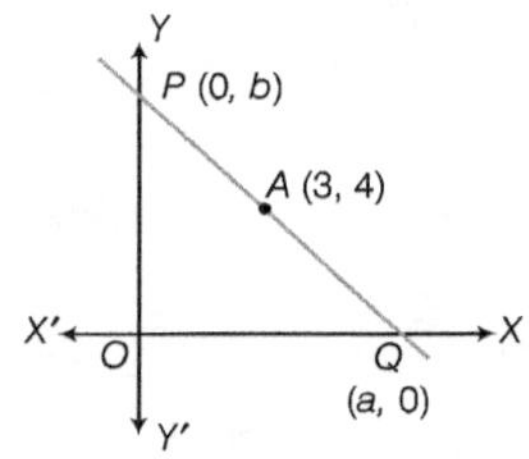

13. (b) The graph of equations $x - 2y = 0$ and $3x - y = 0$ is as shown in the figure. Since, given point (a, a^2) lies in the shaded region.

Then, $\quad a^2 - \dfrac{a}{2} > 0$ and $a^2 - 3a < 0$

$\Rightarrow \quad a(2a-1) > 0$ and $a(a-3) < 0$

$\Rightarrow \quad a \in (-\infty, 0) \cup \left(\dfrac{1}{2}, \infty\right)$

and $\quad 0 < a < 3$

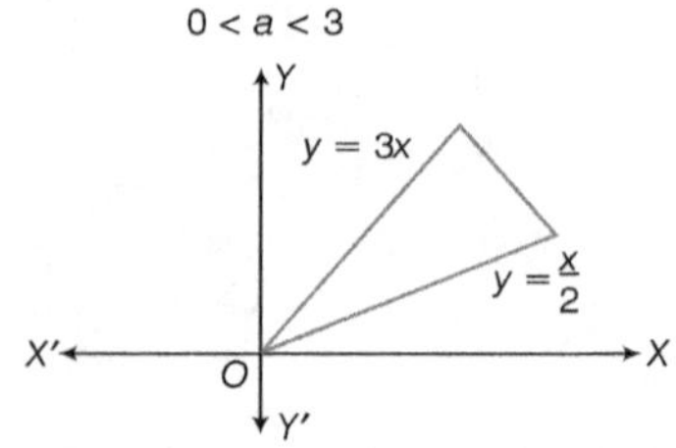

i.e. $\quad a \in (0,3) \quad \Rightarrow \quad a \in \left(\dfrac{1}{2}, 3\right)$

14. (d) Equation of a line passing through the intersection of lines $ax + 2by + 3b = 0$ and $bx - 2ay - 3a = 0$ is

$$(ax + 2by + 3b) + \lambda(bx - 2ay - 3a) = 0 \quad \ldots(i)$$

Now, this line is parallel to X-axis, so coefficient of x should be zero.

i.e. $$a + \lambda b = 0$$

$$\Rightarrow \lambda = -\frac{a}{b}$$

On putting this value in Eq. (i), we get

$$b(ax + 2by + 3b) - a(bx - 2ay - 3a) = 0$$

$$\Rightarrow 2b^2y + 3b^2 + 2a^2y + 3a^2 = 0$$

$$\Rightarrow 2(b^2 + a^2)y + 3(b^2 + a^2) = 0$$

$$\therefore y = -\frac{3}{2}$$

The negative sign shows that the line is below X-axis at a distance $\frac{3}{2}$ from it.

Alternate Method

Equation of given lines are

$$ax + 2by + 3b = 0 \quad \ldots(i)$$

and $$bx - 2ay - 3a = 0 \quad \ldots(ii)$$

On solving Eqs. (i) and (ii), we get the point of intersection

$$x = 0, \quad y = -\frac{3}{2}$$

Also, required line is parallel to X-axis.

$$\therefore m = 0$$

Now, equation of line which is passing through $\left(0, -\frac{3}{2}\right)$ having slope $m = 0$ is

$$\left(y + \frac{3}{2}\right) = 0(x - 0)$$

$$\Rightarrow y = -\frac{3}{2}$$

Thus, the required line is below X-axis at a distance of $\frac{3}{2}$ from it.

15. (b) Since, a, b and c are in HP. Then $\frac{1}{a}, \frac{1}{b}$ and $\frac{1}{c}$ are in AP.

$$\therefore \frac{2}{b} = \frac{1}{a} + \frac{1}{c}$$

$$\Rightarrow \frac{1}{a} - \frac{2}{b} + \frac{1}{c} = 0$$

Hence, straight line $\frac{x}{a} + \frac{y}{b} + \frac{1}{c} = 0$ is always passes through a fixed point $(1, -2)$.

16. (a) Let (x, y) be the coordinates of vertex C and (x_1, y_1) be the coordinates of centroid of the triangle.

$$\therefore x_1 = \frac{x + 2 - 2}{3} \text{ and } y_1 = \frac{y - 3 + 1}{3}$$

$$\Rightarrow x_1 = \frac{x}{3} \text{ and } y_1 = \frac{y - 2}{3} \quad \ldots(i)$$

Since, the centroid lies on the line $2x + 3y = 1$.

So, point (x_1, y_1) satisfied the equation of line.

$$\therefore 2x_1 + 3y_1 = 1$$

$$\Rightarrow \frac{2x}{3} + \frac{3(y - 2)}{3} = 1 \quad \text{[from Eq. (i)]}$$

$$\Rightarrow 2x + 3y - 6 = 3$$

$$\Rightarrow 2x + 3y = 9$$

Hence, this is the required equation of locus of the vertex C.

17. (d) Let the intercepts on the coordinate axes be a and b.

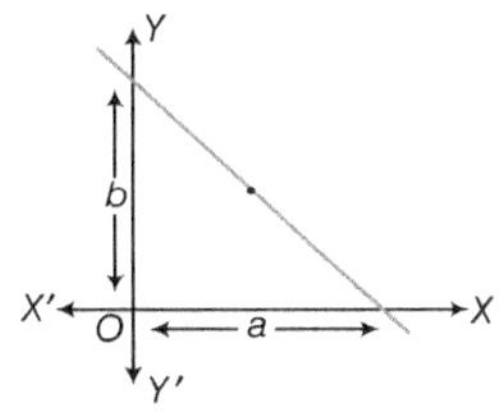

According to the given condition,

$$a + b = -1 \Rightarrow b = -a - 1 = -(a + 1)$$

Let equation of line be $\frac{x}{a} + \frac{y}{b} = 1$.

$$\Rightarrow \frac{x}{a} - \frac{y}{a + 1} = 1 \quad \ldots(i)$$

Since, this line passes through a point (4, 3).

$$\therefore \frac{4}{a} - \frac{3}{a + 1} = 1 \Rightarrow \frac{4a + 4 - 3a}{a(a + 1)} = 1$$

$$\Rightarrow a + 4 = a^2 + a \Rightarrow a^2 = 4$$

$$\therefore a = \pm 2$$

On putting the values of a in Eq. (i), we get

$$\frac{x}{2} - \frac{y}{3} = 1 \text{ and } \frac{x}{-2} + \frac{y}{1} = 1$$

18. (d) Since, line OA makes an angle α with X-axis and given $OA = a$, then coordinates of A are $(a \cos \alpha, a \sin \alpha)$.

Also, $OB \perp OA$, then OB makes an angle $(90° + \alpha)$ with X-axis, then coordinates of B are $[a \cos (90° + \alpha), a \sin (90° + \alpha)]$.

i.e. $$(-a \sin \alpha, a \cos \alpha)$$

Equation of the diagonal AB not passing through the origin is

$$(y - a \sin \alpha) = \frac{a\cos \alpha - a\sin \alpha}{-a \sin \alpha - a \cos \alpha}(x - a\cos \alpha)$$

$$\Rightarrow (\sin \alpha + \cos \alpha)(y - a\sin \alpha) = (\sin \alpha - \cos \alpha)(x - a\cos \alpha)$$

$$\Rightarrow y(\sin \alpha + \cos \alpha) + x(\cos \alpha - \sin \alpha)$$

$$= a \sin \alpha(\sin \alpha + \cos \alpha) - a\cos \alpha(\sin \alpha - \cos \alpha)$$

$$= a(\sin^2 \alpha + \sin \alpha \cos \alpha - \cos \alpha \sin \alpha + \cos^2 \alpha)$$

$$\therefore y(\sin \alpha + \cos \alpha) + x(\cos \alpha - \sin \alpha) = a$$

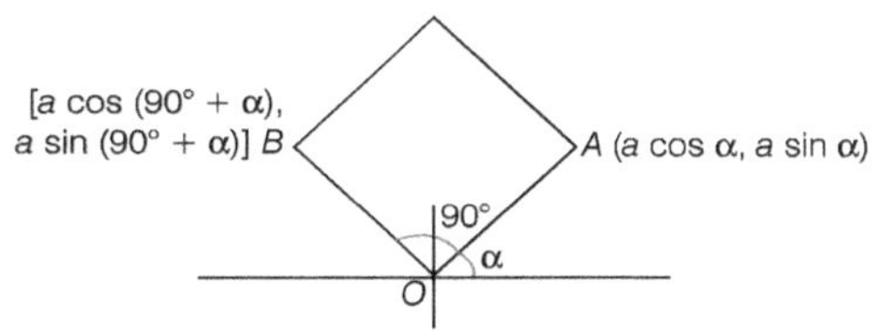

19. (c) The angle bisector for the given two lines $24x + 7y - 20 = 0$ and $4x - 3y - 2 = 0$, are

$$\frac{24x + 7y - 20}{25} = \pm \frac{4x - 3y - 2}{5}$$

Taking positive sign, we get

$$2x + 11y - 5 = 0$$

This equation of line is already given.

Therefore, the given three lines are concurrent with one line bisecting the angle between the other two.

20. (b) Given lines $\sqrt{3}x + y = 0$ makes an angle of 120° with OX and $\sqrt{3}x - y = 0$ makes an angle of 60° with OX. So, the required line is $y - 2 = 0$.

JEE Advanced & IIT JEE Archive

1. (a) **Concept involved** Application of inequality sum and differences, along with lengths of perpendicular. For this type of questions involving inequality, we should always check all options.

Situation analysis Check all the inequalities according to options and use length of perpendicular from the point (x_1, y_1) to $ax + by + c = 0$

i.e. $$\frac{|ax_1 + by_1 + c|}{\sqrt{a^2 + b^2}}$$

As, $a > b > c > 0$

$a - c > 0$ and $b > 0$

$\Rightarrow \quad a + b - c > 0 \quad \ldots(i)$

$a - b > 0$ and $c > 0$

$a + c - b > 0 \quad \ldots(ii)$

$\therefore$ options (a) and (c) are correct.

Also, the point of intersection for $ax + by + c = 0$ and $bx + ay + c = 0$ is

$$\left(\frac{-c}{a+b}, \frac{-c}{a+b}\right)$$

The distance between (1, 1) and $\left(\frac{-c}{a+b}, \frac{-c}{a+b}\right)$

i.e. less than $2\sqrt{2}$.

$$\therefore \quad \sqrt{\left(1 + \frac{c}{a+b}\right)^2 + \left(1 + \frac{c}{a+b}\right)^2} < 2\sqrt{2}$$

$$\Rightarrow \quad \left(\frac{a+b+c}{a+b}\right)\sqrt{2} < 2\sqrt{2}$$

$$\Rightarrow \quad a + b + c < 2a + 2b$$

or $\quad a + b - c > 0 \quad \ldots(iii)$

From Eqs. (ii) and (iii), option (c) is correct.

2. (b) A straight line passing through $P(3, -2)$ and making an angle of $\alpha = 60°$, is given by

$$\frac{y - y_1}{x - x_1} = \tan(\theta \pm \alpha)$$

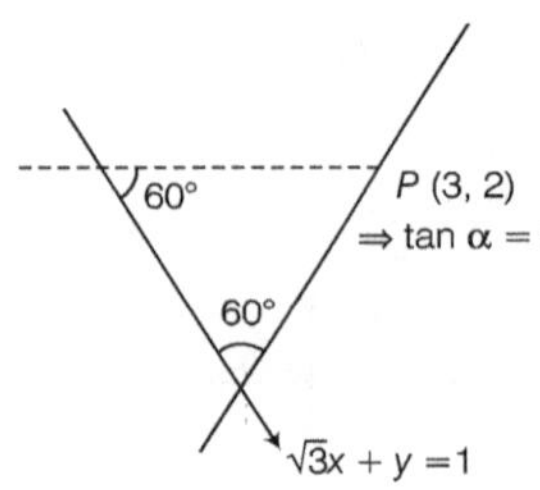

$\Rightarrow \quad \sqrt{3}x + y = 1$

$\Rightarrow \quad y = -\sqrt{3}x + 1$, then $\tan\theta = -\sqrt{3}$

$$\Rightarrow \quad \frac{y+2}{x-3} = \frac{\tan\theta \pm \tan\alpha}{1 \mp \tan\theta\tan\alpha}$$

$$\frac{y+2}{x-3} = \frac{-\sqrt{3} + \sqrt{3}}{1 - (-\sqrt{3})(\sqrt{3})}$$

and $$\frac{y+2}{x-3} = \frac{-\sqrt{3} - \sqrt{3}}{1 + (-\sqrt{3})(\sqrt{3})}$$

$$\Rightarrow \quad y + 2 = 0 \text{ and } \frac{y+2}{x-3} = \frac{-2\sqrt{3}}{1-3} = \sqrt{3}$$

$$y + 2 = \sqrt{3}x - 3\sqrt{3}$$

Neglecting, $y + 2 = 0$ as it does not intersect Y- axis.

3. (d) Given, lines are $(1 + p)x - py + p(1 + p) = 0 \quad \ldots(i)$

and $\quad (1+q)x - qy + q(1+q) = 0 \quad \ldots(ii)$

On solving Eqs. (i) and (ii), we get

$$C\{pq, (1+p)(1+q)\}$$

$\therefore$ Equation of altitude CM passing through C and perpendicular to AB is

$$x = pq \quad \ldots(iii)$$

$\because$ Slope of line (ii) is $\left(\frac{1+q}{q}\right)$.

$\therefore$ Slope of altitude BN (as shown in figure) is $\frac{-q}{1+q}$.

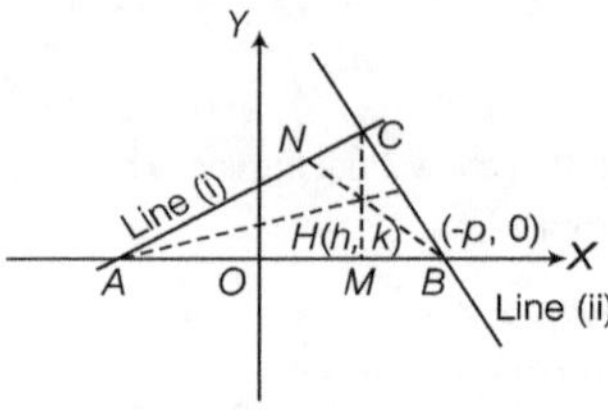

$\therefore \quad$ Equation of BN is $y - 0 = \frac{-q}{1+q}(x + p)$

$$\Rightarrow \quad y = \frac{-q}{(1+q)}(x + p) \quad \ldots(iv)$$

Let orthocentre of triangle be $H(h, k)$, which is the point of intersection of Eqs. (iii) and (iv).

On solving Eqs. (iii) and (iv), we get

$x = pq$ and $y = -pq$

$\Rightarrow \quad h = pq$ and $k = -pq$

$\therefore \quad h + k = 0$

Hence, locus of $H(h, k)$ is $x + y = 0$.

4. (a)

(i) Solving equations L_1 and L_3,

$$\frac{x}{-36 + 10} = \frac{y}{-25 + 12} = \frac{1}{2 - 15}$$

$\Rightarrow \quad x = 2, y = 1$

L_1, L_2, L_3 are concurrent, if point (2, 1) lies on L_2

$\therefore \quad 6 - k - 1 = 0 \Rightarrow k = 5$

(ii) Either L_1 is parallel to L_2 or L_3 is parallel to L_2, then

$$\frac{1}{3} = \frac{3}{-k} \quad \text{or} \quad \frac{3}{5} = \frac{-k}{2}$$

$$\Rightarrow \quad k = -9 \quad \text{or} \quad k = \frac{-6}{5}$$

(iii) L_1, L_2, L_3 form a triangle, if they are not concurrent, or not parallel.

$$\therefore \quad k \neq 5, -9, -\frac{6}{5} \Rightarrow k = \frac{5}{6}$$

(iv) L_1, L_2, L_3 do not form a triangle, if

$$k = 5, -9, -\frac{6}{5}$$

(i) → (s), (ii) → (p,q), (iii) → (r), (iv) → (p,q,s)

5. (c) It is not necessary that the bisector of an angle will divide the triangle into two similar triangles. Therefore, Statement II is false.

Now, we verify Statement I.

In ΔOPQ, OR is the internal bisector of $\angle POQ$.

$$\therefore \quad \frac{PR}{RQ} = \frac{OP}{OQ}$$

$$= \frac{\sqrt{2^2 + 2^2}}{\sqrt{1^2 + 2^2}} = \frac{2\sqrt{2}}{\sqrt{5}}$$

Hence, Statement I is correct.

3. Pair of Straight Line

DPP-1 Homogeneous Pair of Straight Lines

1. (d) It is given that

$$3x + 4y = 0$$

$\Rightarrow y = -\dfrac{3x}{4}$ is one of the line given by

$$6x^2 - xy + 4cy^2 = 0$$

$\therefore\ y = -\dfrac{3x}{4}$ satisfies it.

Now, $\quad 6x^2 + \dfrac{3}{4}x^2 + 4c \times \dfrac{9}{16}x^2 = -0$

$$\Rightarrow \quad \frac{27}{4} + \frac{9c}{4} = 0$$

$$\therefore \quad c = -3$$

2. (d) Given homogeneous equation

$$x^2 - 2cxy - 7y^2 = 0$$

$$\therefore \quad m_1 + m_2 = \frac{-(2c)}{-7} = \frac{2c}{7}$$

$$m_1 \cdot m_2 = -\frac{1}{7}$$

Given, $\quad m_1 + m_2 = 4m_1m_2$

$$\frac{2c}{7} = 4\left(-\frac{1}{7}\right)$$

$$\Rightarrow \quad \frac{2c}{7} = -\frac{4}{7}$$

$$\Rightarrow \quad 2c = -4$$

$$\therefore \quad c = -2$$

3. (c) Given, $ax^2 + 2hxy + by^2 = 0$ ...(i)

$$m_1 + m_2 = -\frac{2h}{b} \quad \text{...(ii)}$$

$$m_1 \cdot m_2 = \frac{a}{b} \quad \text{...(iii)}$$

$$m_1 : m_2 = 1 : 3$$

$$\frac{m_1}{m_2} = \frac{1}{3}$$

$$\Rightarrow \quad 3m_1 = m_2 \quad \text{...(iv)}$$

Using Eq.(iv) in Eqs. (ii) and (iii), we get

$$4m_1 = \frac{-2b}{b} \text{ and } 3m_1^2 = \frac{a}{b}$$

$$\Rightarrow \quad m_1 = \frac{-h}{2b} \text{ and } 3m_1^2 = \frac{a}{b}$$

$$\Rightarrow \quad 3\left(\frac{-h}{2b}\right)^2 = \frac{a}{b}$$

$$\Rightarrow \quad 3h^2 = 4ab$$

$$\therefore \quad \frac{h^2}{ab} = \frac{4}{3}$$

4. (b) Clearly, $y = x$ is one of the lines given by the equation

$$ax^2 + 2hxy + by^2 = 0$$

$$\Rightarrow \quad ax^2 + 2hx^2 + bx^2 = 0 \quad (\text{reset } y = x)$$

$$\therefore \quad a + b = -2h$$

5. (a) Clearly, lines given by $xy = 0$ are $x = 0, y = 0$, i.e. the coordinates axes and their bisectors are $y = \pm\ x$.

Putting $y = \pm\ x$ in $my^2 + (1 - m^2)xy - mx^2 = 0$, we get

$$1 - m^2 = 0$$

$$\Rightarrow \quad m = \pm 1$$

6. (a) Given pair of lines,

$$(x^2 + y^2)\sqrt{3} = 4xy$$

$$\Rightarrow \quad \sqrt{3}(x^2 + y^2) - 4xy = 0 \quad \text{...(i)}$$

Equation of bisector of the angles between the pair of lines (i) is

$$\frac{x^2 - y^2}{\sqrt{3} - \sqrt{3}} = \frac{xy}{-2}$$

$$\Rightarrow \quad -2(x^2 - y^2) = 0$$

$$\therefore \quad y^2 - x^2 = 0$$

7. (a) Given equations of pair of straight lines are

$$3x^2 - 7xy + 4y^2 = 0 \quad \text{...(i)}$$

and $\quad 6x^2 - 5xy + y^2 = 0$ ...(ii)

Let θ_1 and θ_2 be the angles between the pair of lines (i) and (ii).

$$\therefore \quad \tan\theta_1 = \left|\frac{2\sqrt{\frac{49}{4} - 12}}{3 + 4}\right| \Rightarrow \tan\theta_1 = \left|\frac{2\sqrt{\frac{1}{4}}}{7}\right|$$

$$\tan\theta_1 = \left|\frac{1}{7}\right| \quad \text{...(iii)}$$

and $\quad \tan\theta_2 = \left|\dfrac{2\sqrt{\frac{25}{4} - 6}}{6 + 1}\right|$

$$\tan\theta_2 = \left|\frac{2\sqrt{\frac{1}{4}}}{7}\right|$$

$$\tan\theta_2 = \left|\frac{1}{7}\right| \quad \text{...(iv)}$$

From Eqs. (iii) and (iv), we get

$$\tan\theta_1 = \tan\theta_2$$

$$\therefore \quad \theta_1 = \theta_2$$

8. (d) We know that, the equation of the pair of lines through the origin and perpendicular to the pair of lines given by $ax^2 + 2hxy + by^2 = 0$ can be obtained by interchange in the coefficient of x^2 and y^2 and changing the sign of the term containing xy.

$\because$ Given equation of pair of lines is

$$5x^2 - 7xy - 3y^2 = 0 \quad \text{...(i)}$$

Now, equation of line perpendicular to pair of lines (i) is

$$3x^2 + 7xy - 5y^2 = 0$$

9. (b) Given equation of pair of straight lines are

$$x^2 - 2mxy - y^2 = 0 \quad \text{...(i)}$$

$$x^2 - 2nxy - y^2 = 0 \quad \text{...(ii)}$$

Equation of bisector of pair of lines represented by (i) is

$$\frac{x^2 - y^2}{1 + 1} = \frac{xy}{-m}$$

$$\Rightarrow \quad -m(x^2 - y^2) = 2xy$$

$$\Rightarrow \quad my^2 - 2xy - mx^2 = 0 \quad \text{...(iii)}$$

Eqs. (ii) and (iii) are same.

$$\therefore \quad \frac{1}{m} = \frac{-2n}{-2} = \frac{-1}{-m}$$

$$\Rightarrow \quad mn - 1 = 0$$

10. (a) Given equation, $ab(x^2 - y^2) + (a^2 - b^2)xy = 0$ ...(i)

$\Rightarrow \quad abx^2 - aby^2 + a^2xy - b^2xy = 0$

$\Rightarrow \quad (abx^2 + a^2xy) - (aby^2 + b^2xy) = 0$

$\Rightarrow \quad ax(bx + ay) - by(ay + bx) = 0$

$\Rightarrow \quad (bx + ay)(ax - by) = 0$

$\Rightarrow \quad ax - by = 0$ and $bx + ay = 0$

11. (c) Given equation, $(x-5)^2 + (x-5)(y-6) - 2(y-6)^2 = 0$ ...(i)

Clearly, it passes through (5, 6).

Also, it is a general equation of second degree represent, pair of straight line.

12. (a) Any line parallel to the $y = m_1x + c_1$ and passes through origin is

$y = m_1x$ or $y - m_1x = 0$ ...(i)

Any line parallel to $y = m_2x + c_2$ and passes through origin is

$y = m_2x$

$\Rightarrow \quad y - m_2x = 0$ ...(ii)

Multiplying Eqs. (i) and (ii), we get

$(y - m_1x)(y - m_2x) = 0$

$\Rightarrow \quad y^2 + m_1m_2x^2 - m_1xy - m_2xy = 0$

$\Rightarrow \quad m_1m_2x^2 + y^2 - (m_1 + m_2)xy = 0$

13. (c) Given equation of pair of lines,

$ax^2(b - c) - xy(ab - bc) + cy^2(a - b) = 0$

$\Rightarrow \quad abx^2 - acx^2 - abxy + bcxy + acy^2 - bcy^2 = 0$

$\Rightarrow \quad x^2a(b - c) + bcxy - acxy - abxy + acxy + c(a - b)y^2 = 0$

$\Rightarrow \quad a(b - c)x^2 + c(a - b)y^2 - c(a - b)xy - a(b - c)xy = 0$

$\Rightarrow \quad x[a(b - c)x - c(a - b)y] - y[a(b - c)x - c(a - b)y] = 0$

$\Rightarrow \quad (x - y)[a(b - c)x - c(a - b)y] = 0$

$\Rightarrow \quad x - y = 0$

and $\quad a(b - c)x - c(a - b)y = 0$

14. (a) Given equation, $4x^2 + 12xy + 9y^2 = 0$

On comparing it with homogeneous equation of second degree, we get

$a = 4, b = 9, h = 6$

Now, $\quad h^2 - ab = (6)^2 - 4 \times 9 = 36 - 36 = 0$

$\therefore$ Lines are real and coincident.

15. (a) Given, $\quad a^2x^2 + bcy^2 = a(b + c)xy$

$a^2x^2 + bcy^2 - a(b + c)xy = 0$

On comparing it with homogeneous equation of second degree, we have

$$A = a^2, B = bc, H = \frac{-a(b + c)}{2}$$

Since, lines are coincident.

$\therefore \quad H^2 - AB = \dfrac{a^2(b + c)^2}{4} - a^2bc = 0$

$\Rightarrow \quad \dfrac{a^2(b^2 + c^2 + 2bc)}{4} - a^2bc = 0$

$\Rightarrow \quad \dfrac{a^2(b^2 + c^2 + 2bc) - 4a^2bc}{4} = 0$

$\Rightarrow \quad \dfrac{a^2b^2 + a^2c^2 + 2a^2bc - 4a^2bc}{4} = 0$

$\Rightarrow \quad \dfrac{a^2(b^2 + c^2 - 2bc)}{4} = 0$

$\Rightarrow \quad a^2(b - c)^2 = 0$

$\Rightarrow \quad a(b - c) = 0$

$\Rightarrow \quad a = 0$ or $b - c = 0$

$\therefore \quad a = 0$ or $b = c$

16. (d) Given, $\quad 2x^2 - 2hxy + 2y^2 = 0$ ...(i)

Eq. (i) represent two coincident lines passing through origin

$\therefore \quad h^2 - ab = 0 \Rightarrow h^2 - 2 \times 2 = 0$

$\Rightarrow \quad h^2 = 4$

$\therefore \quad h = \pm 2$

17. (c) Given, $\quad ax^2 + 2hxy + by^2 = 0$ ...(i)

and $\quad m_1 = \lambda m_2$ ...(ii)

$m_1 + m_2 = \dfrac{-2h}{a}$ ...(iii)

$m_1.m_2 = \dfrac{a}{b}$ ...(iv)

From Eqs. (ii) and (iii), we have

$(\lambda + 1)m_2 = \dfrac{-2h}{b}$ ...(v)

and from Eqs. (ii) and (iv), we have

$\lambda m_2^2 = \dfrac{a}{b}$

$m_2^2 = \dfrac{a}{b \cdot \lambda}$ ...(vi)

On squaring Eqs. (v) and using (vi), we have

$(\lambda + 1)^2 \dfrac{a}{b\lambda} = \dfrac{4h^2}{b^2} \Rightarrow \dfrac{(\lambda + 1)^2}{\lambda} \cdot a = \dfrac{4h^2}{b}$

$\therefore \quad 4\lambda h^2 = ab(\lambda + 1)^2$

18. (a) Let the equation of coincident line be $y = mx$. Then, it must satisfy the given both equations.

$\therefore \quad ax^2 + 2hx \cdot mx + bm^2x^2 = 0$

and $\quad a'x^2 + 2h'x \cdot mx + b'm^2x^2 = 0$

$\Rightarrow \quad bm^2 + 2hm + a = 0$

and $\quad b'm^2 + 2h'm + a' = 0$

$\Rightarrow \quad \dfrac{m^2}{2ha' - 2h'a} = \dfrac{m}{ab' - a'b} = \dfrac{1}{2h'b - 2hb'}$

$\Rightarrow \quad \dfrac{m^2}{2(ha' - h'a)} = \dfrac{m}{ab' - a'b} = \dfrac{1}{2(h'b - hb')}$

$\Rightarrow \quad m^2 = \dfrac{ha' - h'a}{h'b - hb'}$ and $m = \dfrac{1}{2}\dfrac{(ab' - a'b)}{(h'b - hb')}$

$\Rightarrow \quad \dfrac{1}{4}\dfrac{(ab' - a'b)^2}{(h'b - hb')^2} = \dfrac{(ha' - h'a)}{(h'b - hb')}$

$\Rightarrow \quad (ab' - a'b)^2 = 4(ha' - h'a)(h'b - hb')$

which is required condition.

19. (a) Given, $\quad ax^2 + (a + b)xy + by^2 + x + y = 0$

$\Rightarrow \quad ax^2 + axy + bxy + by^2 + x + y = 0$

$\Rightarrow \quad (ax^2 + bxy + x) + (axy + by^2 + y) = 0$

$\Rightarrow \quad x(ax + by + 1) + y(ax + by + 1) = 0$

$\Rightarrow \quad (ax + by + 1)(x + y) = 0$

$\Rightarrow \quad ax + by + 1 = 0$

$\therefore \quad x + y = 0$

20. (a) Given, $\quad y^2 - x^2 + 2x - 1 = 0$ ...(i)

On comparing it with general equation of second degree, we have

$a = -1, b = 1, c = -1, h = 0\ g = 1, f = 0$

$abc + 2fgh - af^2 - bg^2 - ch^2 = 1 + 0 - 0 - 1 - 0 = 0$

$\therefore$ Eq (i) represent pair of the straight lines.

Alternate Method

We have, $\quad y^2 - x^2 + 2x - 1 = 0$

$\Rightarrow \quad y^2 - (x - 1)^2 = 0 \Rightarrow (y + x - 1)(y - x + 1) = 0$

$\Rightarrow \quad y + x - 1 = 0, y - x + 1 = 0$

Hence, the given equation represent a pair of starigh lines.

21. (c) Given, $x^2 - 2xy\cot\theta - y^2 = 0$ ...(i)

It is a homogeneous equation of second degree which represent pair of straight lines passing through origin.

Let $y = mx$ be one of the line.

$\therefore \quad x^2 - 2x(mx)\cot\theta - m^2x^2 = 0$

$\Rightarrow \quad x^2 - 2x^2m\cot\theta - m^2x^2 = 0$

$\Rightarrow \quad 1 - 2m\cot\theta - m^2 = 0$

$\Rightarrow \quad m^2 + 2m\cot\theta - 1 = 0$

$$\Rightarrow m = \frac{-2\cot\theta \pm \sqrt{4\cot^2\theta + 4}}{2} = \frac{-2\cot\theta \pm 2\sqrt{\cot^2\theta + 1}}{2}$$

$= -\cot\theta \pm \operatorname{cosec}\theta$

$= (-\cot\theta + \operatorname{cosec}\theta)$ or $(-\cot\theta - \operatorname{cosec}\theta)$

$\therefore$ Lines are $\quad y = \dfrac{1-\cos\theta}{\sin\theta}x$

$\Rightarrow \quad y\sin\theta - (1-\cos\theta)x = 0$

$y = -\dfrac{(1+\cos\theta)}{\sin\theta}x \Rightarrow y\sin\theta + (1+\cos\theta)x = 0$

22. (b) The joint equation of pair of straight lines passing through the origin and perpendicular to the pair of lines $3x^2 - 8xy + 5y^2 = 0$ is $5x^2 + 8xy + 3y^2 = 0$

Shifting the origin at (1, 2), we get

$$5(x-1)^2 + 8(x-1)(y-2) + 3(y-2)^2 = 0$$

$\Rightarrow \quad (5x + 3y - 11)(x + y - 3) = 0$

This is the joint equation of pair of lines drawn through (1, 2) and perpendicular to pair of lines $3x^2 - 8xy + 5y^2 = 0$.

23. (a) The general quadratic equation

$f(x, y) = ax^2 + 2hxy + by^2 + 2gx + 2fy + c = 0$ ...(i)

Given, $\quad \Delta = 0$

$\Rightarrow$ Eq. (i) represent pair of straight lines

and $\quad h^2 = ab$

$\therefore$ Equation represents two parallel lines.

24. (b) Given, $x^2 + 2hxy + 2y^2 = 0$ ...(i)

and $\quad m_1 = 2m_2$ ...(ii)

Now, $\quad m_1 + m_2 = \dfrac{-2h}{2} = -h$ ...(iii)

and $\quad m_1 \cdot m_2 = \dfrac{1}{2}$ ...(iv)

From Eqs. (ii) and (iv), we have

$$2m_2^2 = \frac{1}{2}, m_2^2 = \frac{1}{4}$$

$$m_2 = \pm\frac{1}{2}$$

Using this in Eq. (ii), we get

$$m_1 = \pm 1$$

Now, $\quad h = -(m_1 + m_2) = -\left(\pm 1 \pm \dfrac{1}{2}\right) = -\dfrac{3}{2}$ or $\dfrac{3}{2}$

25. (b) Given, $4x^2 + 2hxy - 7y^2 = 0$ ...(i)

and $\quad m_1 + m_2 = m_1m_2$ ...(ii)

Now, $\quad m_1 + m_2 = \dfrac{-2h}{-7} = \dfrac{2h}{7}$

$m_1m_2 = \dfrac{4}{-7} \Rightarrow \dfrac{2h}{7} = -\dfrac{4}{7}$

$\Rightarrow \quad 2h = -4 \Rightarrow h = -2$

26. (b) Given, $\quad x^2 - 4y^2 = 0$

$x^2 - (2y)^2 = 0 \Rightarrow (x+2y)(x-2y) = 0$

$\Rightarrow \quad x + 2y = 0$ ...(i)

and $\quad x - 2y = 0$ ...(ii)

$x = a$ ...(iii)

Taking mutual intersection of lines (i), (ii) and (iii), we get the vertices of triangle as

$$(0, 0), \left(a, \frac{a}{2}\right) \text{ and } \left(a, -\frac{a}{2}\right)$$

$$\therefore \text{ Area of } \Delta = \frac{1}{2}\left|\begin{matrix} 0 & 0 & 1 \\ a & \frac{a}{2} & 1 \\ a & -\frac{a}{2} & 1 \end{matrix}\right| = \frac{1}{2}\left|1\left(-\frac{a^2}{2} - \frac{a^2}{2}\right)\right|$$

$$= \frac{1}{2}|-a^2| = \frac{a^2}{2} \text{ sq unit}$$

27. (a) Given, equation of pair of lines

$6x^2 + 41xy - 7y^2 = 0$ ...(i)

Since, lines represented by Eq. (i) make angles α and β with X-axis, then

$\tan\alpha = m_1$ (say), $\tan\beta = m_2$ (say)

From Eq. (i), we have

$$m_1 + m_2 = \frac{-41}{-7} = \frac{41}{7}$$

$$m_1 \cdot m_2 = -\frac{6}{7}$$

$\Rightarrow \quad \tan\alpha \cdot \tan\beta = -\dfrac{6}{7}$

28. (c) Given, equation of lines

$x^2(\sec^2\theta - \sin^2\theta) - 2xy\tan\theta + y^2\sin^2\theta = 0$ ...(i)

Let m_1 and m_2 be the slopes of lines represented by Eq. (i)

$\therefore \quad m_1 + m_2 = \dfrac{2\tan\theta}{\sin^2\theta}$

$m_1 \cdot m_2 = \dfrac{\sec^2\theta - \sin^2\theta}{\sin^2\theta}$

Now, $m_1 - m_2 = \sqrt{(m_1 + m_2)^2 - 4m_1m_2}$

$$= \sqrt{\frac{4\tan^2\theta}{\sin^4\theta} - 4\left(\frac{\sec^2\theta - \sin^2\theta}{\sin^2\theta}\right)}$$

$$= 2\sqrt{\frac{\tan^2\theta - \sin^2\theta\sec^2\theta + \sin^4\theta}{\sin^4\theta}}$$

$$= 2\sqrt{\frac{\tan^2\theta - \tan^2\theta + \sin^4\theta}{\sin^4\theta}}$$

$\therefore \quad m_1 - m_2 = 2$

29. (d) Given, two lines are $x - 6y = 0$ and $x - y = 0$.

We know that, $\quad \dfrac{0 + x_1 + x_2}{3} = 1$

and $\quad \dfrac{0 + y_1 + y_2}{3} = 0$

$\Rightarrow \quad x_1 + x_2 = 3$ and $y_1 + y_2 = 0$ ...(i)

Also, $\quad x_1 - 6y_1 = 0$ ...(ii)

and $\quad x_2 - y_2 = 0$ ...(iii)

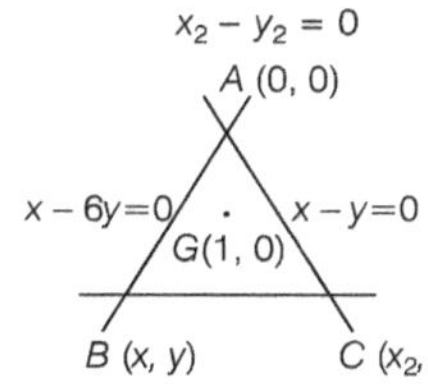

On solving, coordinates of B and C are $\left(\dfrac{18}{5}, \dfrac{3}{5}\right)$ and $\left(\dfrac{-3}{5}, \dfrac{-3}{5}\right)$, respectively.

Hence, the equation of third side, i.e. BC is

$$2x - 7y - 3 = 0$$

30. (b) Given, equation of pair of lines is

$$ax^2 - bxy - y^2 = 0 \quad \ldots(i)$$

Given, the lines represented by Eq.(i) makes angle α and β with X-axis.

$\therefore$ $\tan \alpha = m_1$ (say)

$\tan \beta = m_2$ (say)

From Eq.(i), $m_1 + m_2 = \frac{b}{-1} = -b$

$m_1 \cdot m_2 = -a$

Now, $\tan(\alpha + \beta) = \frac{\tan \alpha + \tan \beta}{1 - \tan \alpha \cdot \tan \beta}$

$= \frac{m_1 + m_2}{1 - m_1 \cdot m_2} = \frac{-b}{1 - (-a)} = \frac{-b}{1 + a}$

31. (a) Let ϕ be the angle between the lines represented by

$$x^2 + 2xy \sec \theta + y^2 = 0 \quad \ldots(i)$$

Here, $a = 1, b = 1, h = \sec \theta$

$\therefore$ $\tan \phi = \frac{2\sqrt{\sec^2 \theta - 1}}{1 + 1} = \frac{2\sqrt{\sec^2 \theta - 1}}{2}$

$\tan \phi = \tan \theta$

$\phi = \theta$

Hence, the angle between the lines represented by Eq. (i) is θ.

32. (a) Given, $2A$ is the angle between the lines represented by equation

$$y^2 + kxy - x^2 \tan^2 A = 0 \quad \ldots(i)$$

$\therefore$ $\tan 2A = \frac{2\sqrt{h^2 - ab}}{|a + b|} = \frac{2\sqrt{\frac{k^2}{4} + \tan^2 A}}{|1 - \tan^2 A|}$

$\frac{2 \tan A}{1 - \tan^2 A} = \frac{2\sqrt{\frac{k^2}{4} + \tan^2 A}}{|1 - \tan^2 A|}$

$\Rightarrow$ $\tan^2 A = \frac{k^2}{4} + \tan^2 A$ $\Rightarrow$ $\frac{k^2}{4} = 0$

$\Rightarrow$ $k^2 = 0$

$\therefore$ $k = 0$

33. (c) Given lines are $(lx + my)^2 - 3(mx - ly)^2 = 0 \quad \ldots(i)$

$$lx + my + x = 0 \quad \ldots(ii)$$

Eq. (i) can be written as,

$l^2x^2 + m^2y^2 + 2lm\,xy - 3(m^2x^2 + l^2y^2 - 2mlxy) = 0$

$\Rightarrow$ $(l^2 - 3m^2)x^2 + 8\,lmxy + (m^2 - 3l^2)y^2 = 0 \quad \ldots(iii)$

Let θ be the angle between the lines represented by Eq.(iii).

$\therefore$ $\tan \theta = \frac{2\sqrt{16l^2m^2 - (l^2 - 3m^2)(m^2 - 3l^2)}}{[l^2 - 3m^2 + m^2 - 3l^2]}$

$= \frac{2\sqrt{16l^2m^2 - l^2m^2 + 3l^4 + 3m^4 - 9l^2m^2}}{|-2(l^2 + m^2)|}$

$= \frac{2\sqrt{6l^2m^2 + 3l^4 + 3m^4}}{|-2(l^2 + m^2)|} = \frac{2\sqrt{3(l^2 + m^2)^2}}{2(l^2 + m^2)}$

$\tan \theta = \sqrt{3}$

$\therefore$ $\theta = 60^\circ$

One of the line represented by Eq.(iii) is

$$(l + \sqrt{3}m)x + (m - \sqrt{3}l)y = 0 \quad \ldots(iv)$$

Let ϕ be the angle between the lines (ii) and (iv).

$\therefore$ $\tan \phi = \left| \frac{-\frac{l}{m} + \frac{l + \sqrt{3}m}{m - \sqrt{3}l}}{1 + \frac{l}{m} \times \left(\frac{l + \sqrt{3}m}{m - \sqrt{3}l}\right)} \right|$

$\tan \phi = \left| \frac{-lm + \sqrt{3}l^2 + lm + \sqrt{3}m^2}{m^2 - \sqrt{3}lm + l^2 + \sqrt{3}lm} \right| = \left| \frac{\sqrt{3}(l^2 + m^2)}{(l^2 + m^2)} \right|$

$\tan \phi = \sqrt{3}$

$\therefore$ $\phi = 60^\circ$

Since, two angles of triangle are 60° each.

$\therefore$ Third angle must be 60°.

Hence, triangle is equilateral.

34. (c) Given, $ax^2 + 2hxy + by^2 = 0 \quad \ldots(i)$

and $(a + 3b)(3a + b) = 4h^2 \quad \ldots(ii)$

Let θ be the angle between the lines represented by Eq. (i).

$\therefore$ $\tan \theta = \frac{2\sqrt{h^2 - ab}}{|a + b|} = \frac{2\sqrt{\frac{(a + 3b)(3a + b)}{4} - ab}}{|a + b|}$

$= \frac{2\sqrt{\frac{3a^2 + ab + 9ab + 3b^2 - 4ab}{4}}}{|a + b|}$

$= \frac{2\sqrt{\frac{3(a^2 + b^2 + 2ab)}{4}}}{|a + b|}$

$\tan \theta = \frac{\sqrt{3}(a + b)}{|a + b|} = \pm\sqrt{3}$

$\therefore$ $\theta = 60^\circ$ or 120°

35. (d) Given, $2x^2 - 3xy + y^2 = 0 \quad \ldots(i)$

Let lines represented by Eq.(i) makes angle α and β with X-axis.

$\therefore$ $\tan \alpha = m_1$ (say), $\tan \beta = m_2$ (say)

From Eq. (i), we have

$m_1 + m_2 = \tan \alpha + \tan \beta = \frac{3}{1} = 3$

and $m_1 \cdot m_2 = \tan \alpha \cdot \tan \beta = 2$

Now, $\cot^2 \alpha + \cot^2 \beta = \frac{1}{\tan^2 \alpha} + \frac{1}{\tan^2 \beta} = \frac{\tan^2 \alpha + \tan^2 \beta}{\tan^2 \alpha \cdot \tan^2 \beta}$

$= \frac{(\tan \alpha + \tan \beta)^2 - 2 \tan \alpha \tan \beta}{\tan^2 \alpha \tan^2 \beta}$

$= \frac{(3)^2 - 2 \times 2}{(2)^2} = \frac{9 - 4}{4} = \frac{5}{4}$

36. (d) Given, $(p - q)x^2 + 2(p + q)xy + (q - p)y^2 = 0 \quad \ldots(i)$

Given, lines represented by Eq. (i) are perpendicular to each other.

$\therefore$ Coefficient of x^2 + Coefficient of $y^2 = 0$

$\Rightarrow$ $p - q + q - p = 0$

$\therefore$ p and q may have any value.

37. (c) The equation $ax^3 + bx^2y + cxy^2 + dy^3 = 0$ is a homogeneous equation of third degree, so it represents three straight lines passing through the origin.

Let one of the line be $y = mx$, then

$$ax^3 + bx^2(mx) + cx(m^2x^2) + d(mx)^3 = 0$$

$\Rightarrow$ $dm^3 + cm^2 + bm + a = 0 \quad \ldots(i)$

This is a cubic equation in m. So, it gives three values of m. Let the values of m be m_1, m_2, m_3, then

$$m_1 + m_2 + m_3 = -\frac{c}{d} \quad \ldots(ii)$$

$$m_1m_2 + m_2m_3 + m_3m_1 = \frac{b}{d} \quad \ldots(iii)$$

$$m_1 \cdot m_2 \cdot m_3 = -\frac{a}{d} \quad \ldots(iv)$$

It is given that, two of the three lines are at right angle.

$\therefore$ $m_1 \cdot m_2 = -1$

Putting $m_1 \cdot m_2 = -1$ in Eq. (iv), we get

$$m_3 = \frac{a}{d}$$

Since, m_3 is a root of Eq.(i). Therefore,

$$d\left(\frac{a}{d}\right)^3 + c\left(\frac{a}{d}\right)^2 + b\left(\frac{a}{d}\right) + a = 0$$

$$\Rightarrow \quad da^3 + ca^2d + bad^2 + ad^3 = 0$$

$$\Rightarrow \quad a^2 + ac + bd + d^2 = 0$$

38. (c) Given, $$x^2 - 2hxy - 2y^2 = 0 \quad \ldots(i)$$

Equation of bisector represented by Eq. (i) is

$$\frac{x^2 - y^2}{1+2} = \frac{xy}{-h}$$

$$\Rightarrow \quad -hx^2 - 3xy + hy^2 = 0$$

$$\Rightarrow \quad hx^2 + 3xy - hy^2 = 0 \quad \ldots(ii)$$

Since, $x - 2y = 0$ is one of the line represented by Eq. (ii).

$$\therefore \quad h(2y)^2 + 3(2y)y - hy^2 = 0$$

$$\Rightarrow \quad 4h + 6 - h = 0$$

$$\Rightarrow \quad 3h + 6 = 0$$

$$\therefore \quad h = -2$$

39. (a) Given, $$(y - mx)^2 = (x + my)^2$$

$$\Rightarrow \quad y^2 + m^2x^2 - 2mxy = x^2 + m^2y^2 + 2mxy$$

$$\Rightarrow \quad (1 - m^2)x^2 + (m^2 - 1)y^2 + 4mxy = 0 \quad \ldots(i)$$

Equation of bisector of Eq. (i) is

$$\frac{x^2 - y^2}{1 - m^2 - m^2 + 1} = \frac{xy}{2m}$$

$$\Rightarrow \quad 2mx^2 - 2my^2 = 2(1 - m^2)xy$$

$$\Rightarrow \quad 2mx^2 + 2(m^2 - 1)xy - 2my^2 = 0$$

$$\Rightarrow \quad mx^2 + (m^2 - 1)xy - my^2 = 0$$

40. (c) Given, $$ax^2 - 2hxy + by^2 = 0 \quad \ldots(i)$$

Equation of bisector of Eq. (i), $\dfrac{x^2 - y^2}{a - b} = \dfrac{xy}{-h}$

$$\Rightarrow \quad -hx^2 + hy^2 - (a - b)xy = 0$$

$$\Rightarrow \quad hx^2 - hy^2 + (a - b)xy = 0 \quad \ldots(ii)$$

Since, $y = mx$ is one of the line represented by Eq. (ii).

$$\therefore \quad hx^2 - hm^2x^2 + (a - b)mx^2 = 0$$

$$\Rightarrow \quad h - hm^2 + (a - b)m = 0$$

$$\therefore \quad h(1 - m^2) + m(a - b) = 0$$

41. (a) Given, equation of pairs of straight lines are

$$ax^2 + 2hxy + by^2 = 0 \quad \ldots(i)$$

$$a'x^2 + 2h'xy + b'y^2 = 0 \quad \ldots(ii)$$

Equation of bisector of the angle subtended by the lines represented by Eq. (i) is

$$\frac{x^2 - y^2}{a - b} = \frac{xy}{h}$$

$$hx^2 - (a - b)\,xy - hy^2 = 0 \quad \ldots(iii)$$

Equation of bisector of the angle subtended by the lines represented by Eq. (ii) is

$$\frac{x^2 - y^2}{a' - b'} = \frac{xy}{h'}$$

$$\Rightarrow \quad h'x^2 - (a' - b')xy - h'y^2 = 0 \quad \ldots(iv)$$

Eqs. (iii) and (iv) are same.

$$\therefore \quad \frac{h}{h'} = \frac{a - b}{a' - b'} = \frac{h}{h'}$$

$$\Rightarrow \quad (a - b)h' = (a' - b')h$$

42. (c) Given, $$px^2 - 2rxy + qy^2 = 0 \quad \ldots(i)$$

and $$r(1 - m^2) + m(p - q) = 0$$

$$\Rightarrow \quad r(1 - m^2) = -(p - q)m$$

$$\Rightarrow \quad \frac{1 - m^2}{m} = -\frac{(p - q)}{r} \quad \ldots(ii)$$

Equation of bisector of the angle subtended by the line represented by Eq.(i) is

$$\therefore \quad \frac{x^2 - y^2}{p - q} = \frac{xy}{-r}$$

$$\Rightarrow \quad \frac{x^2 - y^2}{xy} = \frac{p - q}{-r} = \frac{1 - m^2}{m}$$

$$\Rightarrow \quad mx^2 + (m^2 - 1)xy - my^2 = 0$$

$$\Rightarrow \quad mx^2 + m^2xy - xy - my^2 = 0$$

$$\Rightarrow \quad mx(x + my) - y(x + my) = 0$$

$$\Rightarrow \quad (x + my)(mx - y) = 0$$

$$\Rightarrow \quad x + my = 0 \Rightarrow x = -my$$

$$\text{or} \quad mx - y = 0$$

$$\Rightarrow \quad y = mx$$

43. (d) We have,

$$a(x - 1)^2 + 2h(x - 1)(y - 0) + b(y - 0)^2 = 0$$

Clearly, it represents a pair of straight lines passing through (1, 0). Since, the bisector of the angles between two lines are always perpendicular. So, the other bisector passes through (1, 0) and is perpendicular to $2x + y - 2 = 0$. Hence, its equation is

$$y - 0 = \frac{1}{2}(x - 1)$$

$$\Rightarrow \quad x - 2y - 1 = 0$$

44. (c) Given, $$x^2 + 4xy + y^2 = 0 \quad \ldots(i)$$

$$x - y = 4 \quad \ldots(ii)$$

Let θ be the angle between the lines represented by Eq. (i).

Then, $$\tan\theta = \frac{2\sqrt{4 - 1}}{2}$$

$$\tan\theta = \sqrt{3}$$

$$\therefore \quad \theta = 60^\circ$$

Eq. (i) can also be written as

$$(y + (\sqrt{3} + 2)x)(y - (\sqrt{3} - 2)x) = 0$$

One of the line represented by Eq. (i) is

$$y = (\sqrt{3} - 2)x \quad \ldots(iii)$$

Let ϕ be the angle between lines (ii) and (iii).

$$\therefore \quad \tan\phi = \frac{1 - \sqrt{3} + 2}{1 + (\sqrt{3} - 2)}$$

$$\tan\phi = \frac{3 - \sqrt{3}}{\sqrt{3} - 1} = \frac{3 - \sqrt{3}}{\sqrt{3} - 1} \times \frac{\sqrt{3} + 1}{\sqrt{3} + 1}$$

$$= \frac{1}{2}(2\sqrt{3}) = \sqrt{3}$$

$$\therefore \quad \phi = 60^\circ$$

If two out of the three angles of a triangle are 60°. So, third angle is also 60°.

Hence, triangle is equilateral.

45. (d) Given,

$$x^2(\sec^2\theta - \sin^2\theta) - 2xy\tan\theta + y^2\sin^2\theta = 0 \quad \ldots(i)$$

Let θ_1 and θ_2 be the angles subtended by the lines with X-axis.

$$\therefore \quad \tan\theta_1 = m_1 \text{ (say) and } \tan\theta_2 = m_2 \text{ (say)}$$

Now, $$\tan\theta_1 + \tan\theta_2 = m_1 + m_2 = \frac{2\tan\theta}{\sin^2\theta}$$

$$\tan\theta_1 \cdot \tan\theta_2 = m_1 \cdot m_2 = \frac{\sec^2\theta - \sin^2\theta}{\sin^2\theta}$$

Now, $\tan\theta_1 - \tan\theta_2 = m_1 - m_2$

$$= \sqrt{(m_1 + m_2)^2 - 4m_1m_2}$$

$$= \sqrt{\frac{4\tan^2\theta}{\sin^4\theta} - 4\frac{(\sec^2\theta - \sin^2\theta)}{\sin^2\theta}}$$

$$= 2\sqrt{\frac{\tan^2\theta - \tan^2\theta + \sin^4\theta}{\sin^4\theta}}$$

$\therefore$ $\tan\theta_1 - \tan\theta_2 = 2$

46. (b) Given,

$$x^2 - 2hxy - y^2 = 0 \quad \ldots(i)$$

$$x^2 - 2gxy - y^2 = 0 \quad \ldots(ii)$$

Equation of bisector of angle subtended between the lines represented by Eq. (i) is

$$\frac{x^2 - y^2}{1+1} = \frac{xy}{-h}$$

$\Rightarrow$ $-hx^2 - 2xy + hy^2 = 0$

$\Rightarrow$ $hx^2 + 2xy - hy^2 = 0 \quad \ldots(iii)$

Since, Eqs. (ii) and (iii) are same.

$\therefore$ $\frac{1}{h} = -\frac{2g}{2} = \frac{-1}{-h}$

$\Rightarrow$ $1 + hg = 0$

$\Rightarrow$ $gh = -1$

Now, $1 - gh = 1 - (-1) = 2$

47. (c) The equation $ax^3 + bx^2y + cxy^2 + dy^3 = 0$ is a homogeneous equation of third degree. So, it represent three straight lines passing through the origin.

Let one of the lines be $y = mx$.

Then, $ax^3 + bx^2(mx) + cx(m^2x^2) + d(mx)^3 = 0$

$\Rightarrow$ $dm^3 + cm^2 + bm + a = 0 \quad \ldots(i)$

This is a cubic equation in m. So, it gives three values of m. Let the values of m be m_1, m_2 and m_3. Then,

$$m_1 + m_2 + m_3 = -\frac{c}{d} \quad \ldots(ii)$$

$$m_1m_2 + m_2m_3 + m_3m_1 = \frac{b}{d} \quad \ldots(iii)$$

$$m_1 \cdot m_2 \cdot m_3 = -\frac{a}{d} \quad \ldots(iv)$$

Since, two of the three lines make complementary angles with X-axis.

$\therefore$ Let one lines make angle θ and other make angle $(90 - \theta)$.

$$m_1 = \tan\theta$$

$$m_2 = \tan(90 - \theta) = \cot\theta$$

Substituting these values in Eq. (iv), we get

$$\tan\theta \cdot \cot\theta \cdot m_3 = -\frac{a}{d}$$

$\Rightarrow$ $m_3 = -\frac{a}{d}$

Since, m_3 is a root of Eq. (i).

$\therefore$ $d\left(-\frac{a}{d}\right)^3 + c\left(-\frac{a}{d}\right)^2 + b\left(-\frac{a}{d}\right) = a = 0$

$\Rightarrow$ $-da^3 + ca^2d - bad^2 + ad^3 = 0$

$\Rightarrow$ $-ad(a^2 - ac + bd - d^2) = 0$

$\Rightarrow$ $a^2 - ac + bd - d^2 = 0$

48. (3) Equation of bisector of angle between two straight lines is

$$12x^2 - 7xy - 12y^2 = 0 \quad \ldots(i)$$

One of the line is $2y - x = 0 \quad \ldots(ii)$

Let the other lines be $y = mx$

$\therefore$ $(y - mx)(2y - x) = 0$

$$2y^2 - xy - 2mxy + mx^2 = 0$$

$$2y^2 - (2m+1)xy + mx^2 = 0 \quad \ldots(iii)$$

Equation of bisector of the angle between the lines represented by Eq. (iii) is

$$\frac{x^2 - y^2}{m - 2} = \frac{xy}{-\frac{(2m+1)}{2}}$$

$$-(2m+1)x^2 - 2(m-2)xy + (2m+1)y^2 = 0$$

$\Rightarrow$ $(2m+1)x^2 + 2(m-2)xy - (2m+1)y^2 = 0 \quad \ldots(iv)$

Eqs. (i) and (iv) are same.

$\therefore$ $\frac{12}{2m+1} = \frac{-7}{2(m-2)} = \frac{-12}{-(2m+1)}$

$$24(m-2) = -7(2m+1)$$

$\Rightarrow$ $24m - 48 = -14m - 7$

$\Rightarrow$ $38m = 41$

$\Rightarrow$ $m = \frac{41}{38}$

$\therefore$ Required equation is

$$y = \frac{41}{38}x$$

$\Rightarrow$ $41x - 38y = 0$

$\Rightarrow$ $41x - 38y = mx + ny = 0$

$\Rightarrow$ $m = 41, n = -38$

$\Rightarrow$ $m + n = 41 - 38 = 3$

49. (2) Given,

$$4x^2 - y^2 + 2y - 1 = 0$$

$\Rightarrow$ $(2x)^2 - (y^2 - 2y + 1) = 0$

$\Rightarrow$ $(2x)^2 - (y-1)^2 = 0$

$\Rightarrow$ $(2x + y - 1)(2x - y + 1) = 0$

$\Rightarrow$ $2x + y - 1 = 0$

$\Rightarrow$ $m_1 = -2$

and $2x - y + 1 = 0$

$\Rightarrow$ $m_2 = 2$

$\therefore$ Greatest slope is 2.

50. (2) Given lines,

$$x^2 - y^2 + 2y - 1 = 0$$

$$x^2 - (y-1)^2 = 0 \quad \ldots(i)$$

Equation of angle bisector of the angle between the lines represented by Eq. (i) is

$$\frac{x^2 - (y-1)^2}{1+1} = \frac{x(y-1)}{0}$$

$\Rightarrow$ $2x(y-1) = 0$

$\Rightarrow$ $x(y-1) = 0$

$\Rightarrow$ $x = 0, y - 1 = 0$

and $x + y = 3$

Now, area bounded by the lines $x = 0$, $y - 1 = 0$ and $x + y = 3$ is triangle.

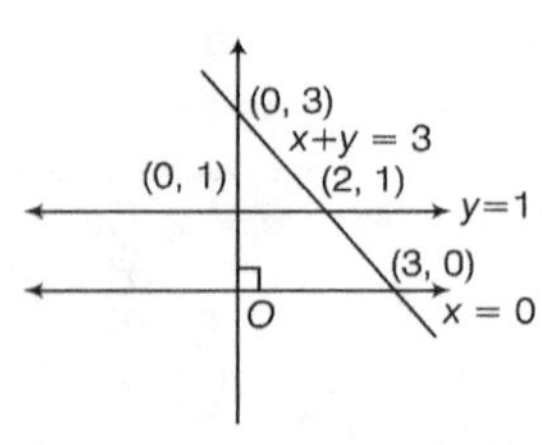

$\therefore$ Required area $= \frac{1}{2} \times 2 \times 2 = 2$ sq units.

DPP-2 General Equation of Second Degree

1. (a) We have,

$$6x^2 + 11xy - 10y^2 + x + 31y + k = 0 \quad \ldots(\text{i})$$

Eq.(i) represent pair of straight lines, then $\Delta = 0$

$$\therefore \begin{vmatrix} 6 & \frac{11}{2} & \frac{1}{2} \\ \frac{11}{2} & -10 & \frac{31}{2} \\ \frac{1}{2} & \frac{31}{2} & k \end{vmatrix} = 0$$

$$\Rightarrow 6\left\{-10k - \left(\frac{31}{2}\right)^2\right\} - \frac{11}{2}\left\{\frac{11}{2}k - \frac{31}{4}\right\} + \frac{1}{2}\left\{\frac{11}{2} \times \frac{31}{2} + \frac{10}{2}\right\} = 0$$

$$\Rightarrow 6\left\{-10k - \frac{961}{4}\right\} - \frac{11}{2}\left\{\frac{22k - 31}{4}\right\} + \frac{1}{2}\left\{\frac{341}{4} + \frac{10}{2}\right\} = 0$$

$$\Rightarrow \frac{-240k - 5766}{4} - \left(\frac{242k - 341}{8}\right) + \frac{1}{2}\left\{\frac{341 + 20}{4}\right\} = 0$$

$$\Rightarrow \left(\frac{-240k - 5766}{4}\right) - \left(\frac{242k - 341}{8}\right) + \frac{361}{8} = 0$$

$$\Rightarrow \frac{-480k - 11532}{8} - \frac{242k}{8} + \frac{341}{8} + \frac{361}{8} = 0$$

$$\Rightarrow -\frac{480k}{8} - \frac{242k}{8} + \frac{11532}{8} + \frac{341}{8} + \frac{361}{8} = 0$$

$$\Rightarrow -\frac{722k}{8} - \frac{10830}{8} = 0$$

$$\Rightarrow -\frac{722k}{8} = \frac{10830}{8}$$

$$\Rightarrow k = -\frac{10830}{722} = -15$$

2. (a) We have,

$$2x^2 - 10xy + 12y^2 + 5x - 16y - 3 = 0 \quad \ldots(\text{i})$$

From Eq.(i), we have

$$m_1 + m_2 = \frac{10}{12} = \frac{5}{6} \quad (\text{ii})$$

$$m_1 \cdot m_2 = \frac{1}{6} \quad \ldots(\text{iii})$$

$$m_1 - m_2 = \sqrt{\left(\frac{5}{6}\right)^2 - 4 \times \frac{1}{6}}$$

$$m_1 - m_2 = \sqrt{\frac{25}{36} - \frac{4}{6}} = \sqrt{\frac{25 - 24}{36}} = \frac{1}{6} \quad \ldots(\text{iv})$$

On solving Eqs. (ii) and (iv), we get

$$m_1 = \frac{1}{2}, m_2 = \frac{1}{3}$$

$\therefore$ Slope of perpendicular will be -2 and -3, respectively.

Combined equation of perpendicular

$$(y + 2x)(y + 3x) = 0$$

$$\Rightarrow y^2 + 3xy + 2xy - 16x^2 = 0$$

$$\therefore 6x^2 + 5xy + y^2 = 0$$

3. (a) We have,

$$2x^2 - xy - 6y^2 + 7x + 21y - 15 = 0 \quad \ldots(\text{i})$$

To get the equation of lines parallel to Eq. (i) and passing through origin, omitting first degree term from Eq. (i), we have

$$2x^2 - xy - 6y^2 = 0$$

4. (c) The given equation will represent a pair of perpendicular straight lines, if

$$\begin{vmatrix} 2 & 2 & 2 \\ 2 & -p & \frac{q}{2} \\ 2 & \frac{q}{2} & 1 \end{vmatrix} = 0 \quad \text{and} \quad 2 - p = 0$$

$$\Rightarrow p = 2 \text{ and } \begin{vmatrix} 2 & 2 & 2 \\ 2 & -p & \frac{q}{2} \\ 2 & \frac{q}{2} & 1 \end{vmatrix} = 0$$

$$\Rightarrow p = 2 \text{ and } \begin{vmatrix} 2 & 2 & 2 \\ 2 & -2 & \frac{q}{2} \\ 2 & \frac{q}{2} & 1 \end{vmatrix} = 0$$

$$\Rightarrow p = 2 \text{ and } \begin{vmatrix} 2 & 0 & 0 \\ 2 & -4 & \frac{q}{2} - 2 \\ 2 & \frac{q}{2} - 2 & -1 \end{vmatrix} = 0 \quad \begin{pmatrix} C_2 \to C_2 - C_1 \\ C_3 \to C_3 - C_1 \end{pmatrix}$$

$$\Rightarrow p = 2 \text{ and } 2\left\{4 - \left(\frac{q}{2} - 2\right)^2\right\} = 0$$

$$\Rightarrow p = 2 \text{ and } 4 - \left(\frac{q}{2} - 2\right)^2 = 0$$

$$\Rightarrow p = 2 \text{ and } \left(\frac{q}{2} - 2\right)^2 = 4$$

$$\Rightarrow p = 2 \text{ and } \left(\frac{q}{2} - 2\right) = \pm 2$$

$$\Rightarrow p = 2 \text{ and } q = 0 \text{ or } 8.$$

5. (d) We have,

$$xy + a^2 = a(x + y)$$

$$\Rightarrow xy + a^2 - ax - ay = 0$$

$$\Rightarrow ax - xy + ay - a^2 = 0$$

$$\Rightarrow x(a - y) - a(a - y) = 0$$

$$\Rightarrow (a - y) = 0 \text{ and } (x - a) = 0$$

$$\therefore y = a \text{ and } x = a$$

6. (c) We have,

$$4x^2 + 12xy + 9y^2 + 2gx + 2fy + c = 0 \quad \ldots(\text{i})$$

Here, $A = 4, B = 9, C = c, G = g, F = f$

and $H = 6$

Given that, lines are parallel and real.

$$\therefore \frac{A}{H} = \frac{H}{B} = \frac{G}{F} \text{ and } H^2 \geq AB$$

$$\Rightarrow \frac{4}{6} = \frac{6}{9} = \frac{g}{f} \text{ and } 36 \geq 36$$

$$\Rightarrow \frac{2}{3} = \frac{g}{f} \Rightarrow g = 2, f = 3, c \in R$$

7. (c) We have,

$$ax^2 + by^2 + cx + cy = 0 \quad \ldots(i)$$

The given equation will represent a pair of straight lines, if

$$\begin{vmatrix} a & 0 & \frac{c}{2} \\ 0 & b & \frac{c}{2} \\ \frac{c}{2} & \frac{c}{2} & 0 \end{vmatrix} = 0$$

$$\Rightarrow \quad a\left(0 - \frac{c^2}{4}\right) + \frac{c}{2}\left(0 - \frac{bc}{2}\right) = 0$$

$$\Rightarrow \quad -\frac{ac^2}{4} - \frac{bc^2}{4} = 0$$

$$\Rightarrow \quad -\frac{c^2}{4}(a + b) = 0$$

$$\therefore \quad c^2(a + b) = 0 \Rightarrow c(a + b) = 0$$

8. (a) We have, $ax^2 + (a + b)xy + by^2 + (x + y) = 0$

$$\Rightarrow \quad ax^2 + axy + bxy + by^2 + (x + y) = 0$$

$$\Rightarrow \quad ax(x + y) + by(x + y) + (x + y) = 0$$

$$\Rightarrow \quad (x + y)[ax + by + 1] = 0$$

$$\therefore \quad x + y = 0 \text{ and } ax + by + 1 = 0$$

9. (b) We have, $\lambda x^2 + 2y^2 - 5xy + 5x - 7y - 3 = 0 \quad \ldots(i)$

Since, Eq. (i) represent two straight lines

$$\therefore \quad \begin{vmatrix} \lambda & \frac{5}{2} & \frac{5}{2} \\ -\frac{5}{2} & 2 & -\frac{7}{2} \\ \frac{5}{2} & -\frac{7}{2} & 3 \end{vmatrix} = 0$$

$$\Rightarrow \quad \lambda\left(6 - \frac{49}{4}\right) + \frac{5}{2}\left(-\frac{15}{2} + \frac{35}{4}\right) + \frac{5}{2}\left(\frac{35}{4} - \frac{10}{2}\right) = 0$$

$$\Rightarrow \quad \lambda\left(\frac{24 - 49}{4}\right) + \frac{5}{2}\left(\frac{-30 + 35}{4}\right) + \frac{5}{2}\left(\frac{35 - 20}{4}\right) = 0$$

$$\Rightarrow \quad -\lambda\frac{25}{4} + \frac{25}{8} + \frac{75}{8} = 0$$

$$\Rightarrow \quad -\frac{25}{4}\lambda = -\frac{100}{8} \quad \Rightarrow \quad 25\lambda = 50$$

$$\therefore \quad \lambda = 2$$

10. (a) We have, $9x^2 + 24xy + 16y^2 + 21x + 28y + 6 = 0$

Here, $a = 9, b = 16, c = 6, h = 12, g = \frac{21}{2}$ and $f = 14$

Now, $h^2 = (12)^2 = 144$

$ab = 9 \times 16 = 144$

$\because \quad h^2 = ab$

$\therefore$ Lines are parallel.

11. (a) Let θ be the angle represented by the equation $2x^2 + 5xy + 3y^2 + 6x + 7y + 4 = 0 \quad \ldots(i)$

$$\therefore \quad \theta = \tan^{-1} m \Rightarrow \tan\theta = m$$

Now, $\tan\theta = \dfrac{2\sqrt{h^2 - ab}}{|a + b|}$

$$\Rightarrow \quad \tan\theta = \left|\frac{2\sqrt{\frac{25}{4} - 6}}{5}\right|$$

$$\Rightarrow \quad \tan\theta = \left|\frac{1}{5}\right|$$

$$\therefore \quad m = \frac{1}{5}$$

12. (b) Given equation,

$$x^2 + y^2 - 2x - 1 = 0 \quad \ldots(i)$$

$$x + y = 1 \quad \ldots(ii)$$

Making Eq. (i) homogeneous with help of Eq. (ii), we have

$$x^2 + y^2 - 2x(x + y) - (x + y)^2 = 0$$

$$\Rightarrow \quad x^2 + y^2 - 2x^2 - 2xy - x^2 - y^2 - 2xy = 0$$

$$\Rightarrow \quad -2x^2 - 4xy = 0$$

$$\Rightarrow \quad 2x^2 + 4xy = 0 \quad \ldots(iii)$$

Let θ be the angle between the lines represented by Eq. (iii).

Now, $\tan\theta = \left|\dfrac{2\sqrt{h^2 - ab}}{a + b}\right| = \left|\dfrac{2\sqrt{4 - 0}}{2}\right|$

$$\Rightarrow \quad \tan\theta = 2$$

$$\therefore \quad \theta = \tan^{-1}(2)$$

13. (c) Let $f(x, y) = 2x^2 + 3y^2 + 7xy + 8x + 14y + 8$

Then, point of intersection of the pair of lines given by $f(x, y) = 0$ is obtained by solving

$$\frac{\partial f}{\partial x} = 0 \text{ and } \frac{\partial f}{\partial y} = 0 \text{ simultaneously}$$

Now, $\dfrac{\partial f}{\partial x} = 4x + 7y + 8 = 0 \quad \ldots(i)$

and $\dfrac{\partial f}{\partial y} = 6y + 7x + 14 = 0 \quad \ldots(ii)$

On solving Eqs. (i) and (ii), we get

$$x = -2 \text{ and } y = 0$$

$\therefore$ Required point of intersection is $(-2, 0)$.

14. (a) We have,

$$x^2 + 2\sqrt{3}\,xy + 3y^2 - 3x - 3\sqrt{3}y - 4 = 0$$

Now, $d = 2\sqrt{\dfrac{g^2 - ac}{a(a + b)}} \quad \ldots(i)$

Here, $a = 1, b = 3, c = -4, h = \sqrt{3}$,

and $g = -\dfrac{3}{2}$

Putting all the values in Eq. (i), we get

$$d = 2\sqrt{\frac{\frac{9}{4} + 4}{1(1 + 3)}}$$

$$= 2\sqrt{\frac{\frac{25}{4}}{4}} = 2 \times \frac{5}{4}$$

$$\therefore \quad d = \frac{5}{2}$$

15. (c) We have,

$$x^2 + 2\sqrt{3}\,xy + 2y^2 + 4x + 4\sqrt{2}y + 1 = 0$$

Here, $a = 1, b = 2, c = 1$, and $g = 2$

$$\therefore \quad d = 2\sqrt{\frac{g^2 - ac}{a(a + b)}} = 2\sqrt{\frac{4 - 1}{3}}$$

$$d = 2$$

16. (b) We have,

$$9x^2 - 6xy + y^2 + 18x - 6y + 8 = 0$$

Here, $a = 9, b = 1, c = 8$ and $g = 9$

Now, $d = 2\sqrt{\dfrac{g^2 - ac}{a(a + b)}}$

$$= 2\sqrt{\frac{81 - 9 \times 8}{9 \times 10}} = 2\sqrt{\frac{81 - 72}{9 \times 10}}$$

$$= 2\sqrt{\frac{9}{9 \times 10}} = 2 \times \sqrt{\frac{1}{10}} = \frac{2}{\sqrt{10}}$$

17. (c) We have, $(x-2y)^2 + k(x-2y) = 0$ ...(i)

and $d = 3$

From Eq. (i), we have

$$x^2 + 4y^2 - 4xy + k(x-2y) = 0$$

$$\Rightarrow \quad x^2 - 4xy + 4y^2 + kx - 2ky = 0$$

Here, $a = 1, b = 4, c = 0, g = \dfrac{k}{2}$

Now, $$d = 2\sqrt{\frac{g^2 - ac}{a(a+b)}}$$

$$3 = 2\sqrt{\frac{\frac{k^2}{4} - 0}{5}}$$

$$3 = 2\sqrt{\frac{k^2}{4 \times 5}}$$

$$3 = \pm \frac{2k}{\sqrt{20}}$$

$$\Rightarrow \quad 3 = \pm \frac{k}{\sqrt{5}}$$

$$\therefore \quad k = 3\sqrt{5} \text{ or } -3\sqrt{5}$$

18. (a) Given, pair of lines are

$$l^2x^2 - m^2y^2 - n(lx + my) = 0 \quad \text{...(i)}$$

$$l^2x^2 - m^2y^2 + n(lx - my) = 0 \quad \text{...(ii)}$$

From Eq. (i), we have

$$(lx)^2 - (my)^2 - n(lx + my) = 0$$

$$\Rightarrow \quad (lx + my)(lx - my) - n(lx + my) = 0$$

$$\Rightarrow \quad (lx + my)(lx - my - n) = 0$$

$$\Rightarrow \quad lx + my = 0 \quad \text{...(iii)}$$

$$lx - my - n = 0 \quad \text{...(iv)}$$

and from Eq. (ii), we have

$$(lx)^2 - (my)^2 + n(lx - my) = 0$$

$$(lx + my)(lx - my) + n(lx - my) = 0$$

$$\Rightarrow \quad (lx - my)(lx + my + n) = 0$$

$$\Rightarrow \quad lx - my = 0 \quad \text{...(vi)}$$

$$\text{and} \quad lx + my + n = 0 \quad \text{...(v)}$$

Since, Eqs. (iii), (vi) and (iv), (v) are parallel.

$\therefore$ The figure bounded by these lines are parallelogram.

$$\text{Now, area of parallelogram} = \left|\frac{(0-n)(0+n)}{\begin{vmatrix} l & m \\ l & -m \end{vmatrix}}\right| = \left|\frac{-n^2}{-2lm}\right| = \frac{n^2}{2|lm|}$$

19. (b) Given,

$$\sqrt{(x-3)^2 + (y-1)^2} + \sqrt{(x+3)^2 + (y-1)^2} = 6$$

$$\Rightarrow \quad \sqrt{(x-3)^2 + (y-1)^2} = 6 - \sqrt{(x+3)^2 + (y-1)^2}$$

$$\Rightarrow \quad (x-3)^2 + (y-1)^2 = 36 + \{(x+3)^2 + (y-1)^2\} - 12\sqrt{(x+3)^2 + (y-1)^2}$$

$$\Rightarrow x^2 + 9 - 6x + y^2 + 1 - 2y = 36 + x^2 + 9 + 6x + y^2 + 1 - 2y - 12\sqrt{(x+3)^2 + (y-1)^2}$$

$$\Rightarrow \quad -12x - 36 = -12\sqrt{(x+3)^2 + (y-1)^2}$$

$$\Rightarrow \quad 12(x+3) = 12\sqrt{(x+3)^2 + (y-1)^2}$$

$$\Rightarrow \quad x + 3 = \sqrt{(x+3)^2 + (y-1)^2}$$

$$\Rightarrow \quad (x+3)^2 = (x+3)^2 + (y-1)^2$$

$$\Rightarrow \quad x^2 + 9 + 6x = x^2 + 9 + 6x + y^2 + 1 - 2y$$

$$\Rightarrow \quad (y-1)(y-1) = 0$$

$\therefore$ Pair of coincident straight lines.

20. (c) Given equations are

$$xy + 2x + 2y + 4 = 0$$

$$\Rightarrow \quad (y+2)(x+2) = 0$$

$$\Rightarrow \quad x + 2 = 0 \quad \text{...(i)}$$

$$\Rightarrow \quad y + 2 = 0 \quad \text{...(ii)}$$

$$\text{and} \quad x + y + 2 = 0 \quad \text{...(iii)}$$

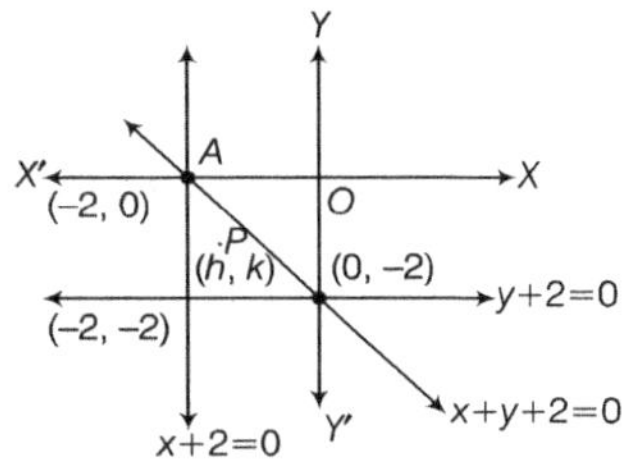

Taking mutual intersection of lines (i), (ii) and (iii), we get vertices of triangle as

$$(-2, 0), (0, -2), (-2, -2)$$

Let $P(h, k)$ be the circumcenter of ΔABC.

$$\therefore \quad PA^2 = PB^2$$

$$(h+2)^2 + k^2 = h^2 + (k+2)^2 \quad \text{...(i)}$$

$$\text{and} \quad PA^2 = PC^2$$

$$(h+2)^2 + k^2 = (h+2)^2 + (k+2)^2$$

$$\Rightarrow \quad k^2 = (k+2)^2$$

$$\Rightarrow \quad k^2 = k^2 + 4 + 4k$$

$$\Rightarrow \quad k = -1$$

Substituting $k = -1$ in Eq.(i), we get

$$(h+2)^2 + 1 = h^2 + 1$$

$$\Rightarrow \quad h^2 + 4 + 4h = h^2$$

$$\Rightarrow \quad h = -1$$

$\therefore$ Circumcenter is $(-1, -1)$.

21. (b) Given equations are

$$3x^2 + 8xy - 3y^2 = 0 \quad \text{...(i)}$$

$$\text{and} \quad 3x^2 + 8xy - 3y^2 + 2x - 4y - 1 = 0 \quad \text{...(ii)}$$

Clearly, Eq. (i) represent two straight lines passing through origin.

Let one of the line be $y = m_1x$

From Eq. (i), we have

$$3x^2 + 8m_1x^2 - 3m_1^2x^2 = 0$$

$$\Rightarrow \quad 3 + 8m_1 - 3m_1^2 = 0$$

$$\Rightarrow \quad 3m_1^2 - 8m_1 - 3 = 0$$

$$\Rightarrow \quad m_1 = \frac{8 \pm \sqrt{64 + 36}}{6}$$

$$= \frac{8 \pm \sqrt{100}}{6} = \frac{8 \pm 10}{6}$$

$$\Rightarrow \quad m_1 = 3, -\frac{1}{3}$$

$\therefore$ The lines represented by Eq. (i) are perpendicular to each other and equation of AB and BC are

$$y = 3x \text{ and } y = -\frac{1}{3}x$$

Similarly, Eq. (ii) represent two perpendicular lines.

From Eq. (ii), we have

$$3x^2 + 8xy - 3y^2 + 2x - 4y - 1 = 0$$

$$\Rightarrow \quad (x + 3y + 1)(3x - y - 1) = 0$$

$\therefore$ Equations of CD and DA are

$$x + 3y + 1 = 0$$

$$\text{and} \quad 3x - y - 1 = 0$$

22. (a) Area of quadrilateral (parallelogram)

$$= \left| \frac{(0+1)(0-1)}{\begin{vmatrix} 1 & 3 \\ 3 & -1 \end{vmatrix}} \right| = \left| \frac{-1}{-10} \right| = \frac{1}{10} \text{ sq unit}$$

23. (b) We have,

$\lambda x^2 - 10xy + 12y^2 + 5x - 16y - 3 = 0$ ...(i)

$$\Rightarrow \begin{vmatrix} \lambda & -5 & \frac{5}{2} \\ -5 & 12 & -8 \\ \frac{5}{2} & -8 & -3 \end{vmatrix} = 0$$

$\Rightarrow \lambda(-36-64) + 5(15+20) + \frac{5}{2}(40-30) = 0$

$\Rightarrow \lambda(-100) + 175 + 25 = 0$

$\Rightarrow -100\lambda = -200$

$\therefore \lambda = 2$

24. (c) Let $f(x, y) = 2x^2 - 10xy + 12y^2 + 5x - 16y - 3 = 0$

$\frac{\partial f}{\partial x} = 4x - 10y + 5$

$\frac{\partial f}{\partial y} = -10x + 24y - 16$

Now, $\frac{\partial f}{\partial x} = 0$ and $\frac{\partial f}{\partial y} = 0$ will simultaneously give the point of intersection of the lines represented by Eq. (i).

$4x - 10y + 5 = 0$ and $10x - 24y + 16 = 0$

$\Rightarrow x = -10, y = -\frac{7}{2}$

$\therefore$ Product of abscissa and ordinate, i.e.

$xy = (-10) \times \left(-\frac{7}{2}\right) = 35$

25. (d) Let θ be acute angle between the lines.

$\therefore \tan\theta = \frac{2\sqrt{h^2 - ab}}{(a+b)} = \left|\frac{2\sqrt{25-24}}{14}\right| = \left|\frac{2}{14}\right|$

$\tan\theta = \left|\frac{1}{7}\right| = 0.\overline{1428571} \in (0, 15°)$

26. (4) We have,

$ax^2 + 6xy + by^2 - 10x + 10y - 6 = 0$ ...(i)

Since, Eq. (i) represent two perpendicular lines

$$\therefore \begin{vmatrix} a & 3 & -5 \\ 3 & b & 5 \\ -5 & 5 & -6 \end{vmatrix} = 0 \text{ and } a + b = 0$$

$\Rightarrow a(-6b - 25) - 3(-18 + 25) - 5(15 + 5b) = 0$ and $a + b = 0$

$\Rightarrow -6ab - 25a - 21 - 75 - 25b = 0$

and $a + b = 0$

$\Rightarrow 6a^2 - 25a - 96 + 25a = 0$

$\Rightarrow 6a^2 - 96 = 0$

$\Rightarrow 6a^2 = 96 \Rightarrow a^2 = 16$

$\Rightarrow a = \pm 4 \Rightarrow |a| = 4$

27. (10) $x^2 - 3xy + \lambda y^2 + 3x - 5y + 2 = 0$ ...(i)

Since, equation represent pair of straight lines.

$$\therefore \begin{vmatrix} 1 & -\frac{3}{2} & \frac{3}{2} \\ -\frac{3}{2} & \lambda & -\frac{5}{2} \\ \frac{3}{2} & -\frac{5}{2} & 2 \end{vmatrix} = 0$$

$\Rightarrow 1\left(2\lambda - \frac{25}{4}\right) + \frac{3}{2}\left(-3 + \frac{15}{4}\right) + \frac{3}{2}\left(\frac{15}{4} - \frac{3\lambda}{2}\right) = 0$

$\Rightarrow \frac{8\lambda - 25}{4} + \frac{9}{8} + \frac{45 - 18\lambda}{8} = 0$

$\Rightarrow \frac{16\lambda - 50}{8} + \frac{9}{8} + \frac{45 - 18\lambda}{8} = 0$

$\Rightarrow -2\lambda + 4 = 0$

$\Rightarrow \lambda = 2$

Now, Eq. (i) becomes

$x^2 - 3xy + 2y^2 + 3x - 5y + 2 = 0$

$\Rightarrow \tan\theta = 2\left|\frac{2\sqrt{h^2 - ab}}{a+b}\right| = \left|\frac{2\sqrt{\frac{9}{4} - 2}}{3}\right| = \left|\frac{1}{3}\right|$

$\Rightarrow \sin\theta = \frac{1}{\sqrt{10}} \Rightarrow \operatorname{cosec}\theta = \sqrt{10}$

$\therefore \operatorname{cosec}^2\theta = 10$

28. (1) We have,

$x^2 + y^2 + 2gx + 2fy + 1 = 0$ ...(i)

Since, Eq. (i) represent pair of straight lines.

$$\therefore \begin{vmatrix} 1 & 0 & g \\ 0 & 1 & f \\ g & f & 1 \end{vmatrix} = 0$$

$\Rightarrow 1(1 - f^2) + g(0 - g) = 0$

$\Rightarrow (1 - f^2) - g^2 = 0$

$\Rightarrow 1 - (f^2 + g^2) = 0$

$\therefore f^2 + g^2 = 1$

29. (5) The equations of PQ and PR are given by

$y - 1 = \frac{-2 \mp \tan 45°}{1 \pm (-2)\tan 45°}(x - 2)$

$\Rightarrow y - 1 = \left(\frac{-2 \pm 1}{1 \pm 2}\right)(x - 2)$

$\Rightarrow y - 1 = -\frac{1}{3}(x - 2)$

and $y - 1 = 3(x - 2)$

$\Rightarrow x + 3y - 5 = 0$

and $3x - y - 5 = 0$

The combined equation of these two lines is

$(x + 3y - 5)(3x - y - 5) = 0$

$\Rightarrow 3x^2 - 3y^2 + 8xy - 20x - 10y + 25 = 0$ ...(i)

Given that combined equation of PQ and PR is

$3x^2 - 3y^2 + 8xy - 20x - 10y + \lambda = 0$ ...(ii)

Eqs. (i) and (ii) are same.

$\therefore \lambda = 25$

$\Rightarrow \lambda - 20 = 5$

30. (1) We have,

$\lambda^2 x^2 + (\lambda^2 - 5\lambda + 4)xy + (3\lambda - 2)y^2 - 8x + 12y - 4 = 0$ ...(i)

Eq. (i) represent circle, if

$h = 0$ and $a = b$

$\Rightarrow \lambda^2 = 3\lambda - 2$ and $\lambda^2 - 5\lambda + 4 = 0$

$\Rightarrow \lambda^2 - 3\lambda + 2 = 0$ and $\lambda^2 - 5\lambda + 4 = 0$

$\Rightarrow \lambda = \frac{3 \pm \sqrt{9-8}}{2}$ and $\lambda = \frac{5 \pm \sqrt{25-16}}{2}$

$\Rightarrow \lambda = \frac{3 \pm 1}{2}$ and $\lambda = \frac{5 \pm 3}{2}$

$\Rightarrow \lambda = 2, 1$ and $\lambda = 4, 1$

$\therefore \lambda = 1$

DPP-3 Equation of Line Joining the Origin to the Point of Intersection of a Curve and a Line

1. (a) Given, equation of curve

$$x^2 + y^2 = 4 \quad \text{and} \quad y - x = 2$$

Now, equation of straight line joining point of intersection of above curve and origin is

$$x^2 + y^2 = (y - x)^2$$

2. (c) Given, equations of curve are

$$x^2 + y^2 = 9 \qquad \text{...(i)}$$

and $$x + y = 3 \qquad \text{...(ii)}$$

Equation of line joining point of intersection of lines (i) and (ii) and origin is

$$x^2 + y^2 = (x + y)^2$$

$$\Rightarrow \quad x^2 + y^2 = x^2 + y^2 + 2xy$$

$$\Rightarrow \quad 2xy = 0$$

$$\therefore \quad xy = 0$$

3. (b) Given, equation of curves are

$$3x - 2y = 1 \qquad \text{...(i)}$$

and $$3x^2 + 5xy - 3y^2 + 2x + 3y = 0 \qquad \text{...(ii)}$$

Equation of lines joining point of intersection of Eqs.(i) and (ii) and origin is

$$3x^2 + 5xy - 3y^2 + 2x(3x - 2y) + 3y(3x - 2y) = 0$$

$$\Rightarrow \quad 3x^2 + 5xy - 3y^2 + 6x^2 - 4xy + 9xy - 6y^2 = 0$$

$$\Rightarrow \quad 9x^2 - 9y^2 + 10xy = 0$$

Now, coefficient of x^2 + coefficient of $y^2 = 9 - 9 = 0$

$\therefore$ Lines will be perpendicular to each other.

4. (b) Given, equation of curve is

$$(x - k)^2 + (y - h)^2 = c^2 \qquad \text{...(i)}$$

and equation of line

$$hx + ky = 2hk \qquad \text{...(ii)}$$

Making the equation of the curve homogeneous with the help of equation of line, we get

$$x^2 + y^2 - 2(kx + hy)\left(\frac{hx + ky}{2hk}\right) + (h^2 + k^2 - c^2)\left(\frac{hx + ky}{2hk}\right)^2 = 0$$

This is the equation of the pair of lines joining the origin to the point of intersection of the given line and the curves.

These will be at right angle, if

coefficient of x^2 + coefficient of $y^2 = 0$

$$(h^2 + k^2)(h^2 + k^2 - c^2) = 0$$

$$\Rightarrow \quad h^2 + k^2 = c^2 \qquad (\because h^2 + k^2 \neq 0)$$

5. (b) Equation of given curve is

$$x^2 + 2xy + 3y^2 + 4x + 8y - 11 = 0 \qquad \text{...(i)}$$

and the equation of the given line is

$$y - 3x = 2$$

$$\Rightarrow \quad \frac{y - 3x}{2} = 1 \qquad \text{...(ii)}$$

Making Eq. (i) homogeneous with the help of Eq. (ii), we have

$$x^2 + 2xy + 3y^2 + 4x\left(\frac{y - 3x}{2}\right) + 8y\left(\frac{y - 3x}{2}\right) - 11\left(\frac{y - 3x}{2}\right)^2 = 0$$

$$\Rightarrow \quad 7x^2 - 2xy - y^2 = 0 \qquad \text{...(iii)}$$

This is the equation of lines joining the origin to the point of intersection of Eqs. (i) and (ii).

Comparing Eq. (iii) with the equation $ax^2 + 2hxy + by^2 = 0$, we get

$$a = 7, b = -1, 2h = -2$$

If θ is the angle between the pair of lines of Eq. (iii), then

$$\tan\theta = \left|\frac{2\sqrt{h^2 - ab}}{a + b}\right| = \left|\frac{2\sqrt{1 + 7}}{7 - 1}\right|$$

$$= \frac{2\sqrt{8}}{6} = \frac{2\sqrt{2}}{3}$$

$$\therefore \quad \theta = \tan^{-1}\left(\frac{2\sqrt{2}}{3}\right)$$

6. (c) Given, equation of circle

$$x^2 + y^2 = a^2 \qquad \text{...(i)}$$

and equation of line

$$y = mx + c$$

$$\frac{y - mx}{c} = 1 \qquad \text{...(ii)}$$

Making the equation of the curve homogeneous with the help of that of line, we get

$$x^2 + y^2 - a^2\left(\frac{y - mx}{c}\right)^2 = 0$$

$$\Rightarrow \quad x^2 + y^2 - \frac{a^2}{c^2}(y^2 + m^2x^2 - 2mxy) = 0$$

$$\Rightarrow \quad \left(1 - \frac{a^2m^2}{c^2}\right)x^2 + \left(1 - \frac{a^2}{c^2}\right)y^2 + \frac{2a^2m}{c^2}xy = 0 \qquad \text{...(iii)}$$

The lines represented by Eq. (iii) will be perpendicular, if coefficient of x^2 + coefficient of $y^2 = 0$

$$\Rightarrow \quad 1 - \frac{a^2m^2}{c^2} + 1 - \frac{a^2}{c^2} = 0$$

$$\Rightarrow \quad -\frac{a^2}{c^2}(m^2 + 1) + 2 = 0$$

$$\Rightarrow \quad -a^2(m^2 + 1) = -2c^2$$

$$\therefore \quad a^2(m^2 + 1) = 2c^2$$

7. (c) Given, equation of circle is

$$x^2 + y^2 = 2 \qquad \text{...(i)}$$

and equation of line is

$$\left(\frac{y - 2\sqrt{2}x}{c}\right) = 1 \qquad \text{...(ii)}$$

Making Eq. (i) homogeneous with help of Eq. (ii), we get

$$x^2 + y^2 - 2\left(\frac{y - 2\sqrt{2}x}{c}\right)^2 = 0$$

$$\Rightarrow \quad x^2 + y^2 - \frac{2}{c^2}(y^2 + 8x^2 - 4\sqrt{2}xy) = 0$$

$$\Rightarrow \quad \left(1 - \frac{16}{c^2}\right)x^2 + \left(1 - \frac{2}{c^2}\right)y^2 + \frac{8\sqrt{2}}{c^2}xy = 0 \qquad \text{...(iii)}$$

The lines represented by Eq. (iii) will be perpendicular, if

$$1 - \frac{16}{c^2} + 1 - \frac{2}{c^2} = 0$$

$$\Rightarrow \quad 2 - \frac{18}{c^2} = 0$$

$$\Rightarrow \quad 2c^2 = 18$$

$$\Rightarrow \quad c^2 = 9$$

$$\Rightarrow \quad c^2 - 9 = 0$$

8. (d) Given, equation of curve is

$$x^2 + y^2 - 2y + \lambda = 0 \quad \dots(i)$$

and equation of line is

$$x + y = 1 \quad \dots(ii)$$

Making Eq.(i) homogeneous with the help of Eq. (ii), we get

$$x^2 + y^2 - 2y(x + y) + \lambda(x + y)^2 = 0$$

$$\Rightarrow x^2 + y^2 - 2xy - 2y^2 + \lambda(x^2 + y^2 + 2xy) = 0$$

$$\Rightarrow x(1 + \lambda)x^2 + (\lambda - 1)y^2 + (\lambda - 1)2xy = 0 \quad \dots(iii)$$

The lines represented by Eq. (iii) are perpendicular.

$$\therefore (1 + \lambda) + \lambda - 1 = 0$$

$$\Rightarrow \lambda = 0$$

9. (d) Equation of curve,

$$x^2 - y^2 + hxy + gx + fy = 0 \quad \dots(i)$$

and equation of line

$$fx - gy = \lambda$$

$$\frac{fx - gy}{\lambda} = 1 \quad \dots(ii)$$

Making Eq. (i) homogeneous with the help of Eq. (ii), we get

$$x^2 - y^2 + hxy + gx\left(\frac{fx - gy}{\lambda}\right) + fy\left(\frac{fx - gy}{\lambda}\right) = 0$$

$$\Rightarrow x^2 - y^2 + hxy + \frac{gf}{\lambda}x^2 - \frac{g^2xy}{\lambda} + \frac{f^2xy}{\lambda} - \frac{fgy^2}{\lambda} = 0$$

$$\Rightarrow \left(1 + \frac{gf}{\lambda}\right)x^2 - \left(1 + \frac{fg}{\lambda}\right)y^2 + \left(h - \frac{g^2}{\lambda} + \frac{f^2}{\lambda}\right)xy = 0 \quad \dots(iii)$$

The lines represented by Eq. (iii) are perpendicular.

$$\therefore \left(1 + \frac{gf}{\lambda}\right) - \left(1 + \frac{fg}{\lambda}\right) = 0$$

$\Rightarrow$ λ may have any value.

10. (b) Equation of curve,

$$x^2 + y^2 - ax - ay = 0 \quad \dots(i)$$

and

$$x^2 + y^2 = a^2 \quad \dots(ii)$$

Making Eq. (i) homogeneous with help of Eq. (ii), we get

$$x^2 + y^2 = a(x + y)$$

$$\Rightarrow (x^2 + y^2)^2 = a^2(x + y)^2$$

$$\Rightarrow (x^2 + y^2)^2 = (x^2 + y^2)(x + y)^2$$

$$\Rightarrow x^2 + y^2 = (x + y)^2$$

$$\Rightarrow x^2 + y^2 = x^2 + y^2 + 2xy$$

$$\Rightarrow 2xy = 0$$

$$\therefore xy = 0$$

11. (b) Given, equation of curves

$$x + ky = 1$$

and $5x^2 + 12xy - 6y^2 + 4x - 2y + 3 = 0$

Equation of lines joining point of intersection of equations, is

$$5x^2 + 12xy - 6y^2 + 4x \cdot (x + ky) - 2y(x + ky) + 3(x + ky)^2 = 0$$

Since, lines are equally inclined to X-axis.

$$\Rightarrow h = 0$$

$$\Rightarrow 12 + 4k - 2 + 6k = 0$$

$$\Rightarrow k = -1$$

12. (4) The x-coordinates of the points of intersection of the pair of lines given by $2x^2 + 2hxy + 6y^2 - 4x + 5y - 6 = 0$ and X-axis are the roots of the equation $2x^2 - 4x - 6 = 0$

or $x^2 - 2x - 3 = 0$

$$\Rightarrow (x - 3)(x + 1) = 0$$

$$\Rightarrow x = -1, 3$$

Hence, required length $= |x_1 - x_2| = 4$

13. (2) Equation of curve,

$$3x^2 + 4xy - 4x + 1 = 0 \quad \dots(i)$$

and equation of line

$$2x + y = 1 \quad \dots(ii)$$

Making Eq. (i) homogeneous with help of Eq. (ii), we get

$$3x^2 + 4xy - 4x(2x + y) + (2x + y)^2 = 0$$

$$\Rightarrow 3x^2 + 4xy - 8x^2 - 4xy + 4x^2 + y^2 + 4xy = 0$$

$$\Rightarrow -x^2 + 4xy + y^2 = 0$$

$$\Rightarrow x^2 - 4xy - y^2 = 0 \quad \dots(iii)$$

Here, $a = 1, b = -1, h = -2$

Now, $a + b = 1 - 1 = 0$

$\therefore$ Lines represented by Eq. (iii) are perpendicular to each other.

Therefore, these lines include an angle $\frac{\pi}{2}$.

But it is given that, these lines include an angle $\frac{\pi}{k}$.

$$\therefore k = 2$$

14. (1) The equation of line passing through the origin is

$$y - mx = 0 \quad \dots(i)$$

If it is at a distance d from (x_1, y_1), then

$$\left|\frac{y_1 - mx_1}{\sqrt{m^2 + 1}}\right| = d$$

$$\Rightarrow (y_1 - mx_1)^2 = d^2(1 + m^2)$$

$$\Rightarrow \left(y_1 - \frac{y}{x}x_1\right)^2 = d^2\left(1 + \frac{y^2}{x^2}\right)$$

$$\Rightarrow \left(\frac{xy_1 - yx_1}{x}\right)^2 = d^2\left(\frac{x^2 + y^2}{x^2}\right)$$

$$\Rightarrow (xy_1 - yx_1)^2 = d^2(x^2 + y^2) \quad \dots(ii)$$

But the given Eq. of lines is

$$(xy_1 - yx_1)^2 = kd^2(x^2 + y^2) \quad \dots(iii)$$

Comparing Eqs. (ii) and (iii), we get

$$k = 1$$

15. (4) Clearly, angle between the lines given by $Ax^2 + 2Hxy + By^2 = 0$ must be 60°.

$$\therefore \tan 60° = \frac{2\sqrt{H^2 - AB}}{A + B}$$

$$\Rightarrow 3(A + B)^2 = 4(H^2 - AB)$$

$$\Rightarrow (3A + B)(A + 3B) = 4H^2 \quad \dots(i)$$

But it is given that

$$(3A + B)(A + 3B) = kH^2 \quad \dots(ii)$$

$$\therefore k = 4$$

Revisal Problems (JEE Main)

1. (a) We have, $ax^2 + 2hxy + by^2 = 0$ ...(i)

Let $m_1 = m$

$\therefore$ $m_2 = m^2$

From Eq. (i), we have

$$m_1 + m_2 = \frac{-2h}{b}$$

$\Rightarrow$ $m + m^2 = \frac{-2h}{b}$ ...(ii)

and $m_1 \cdot m_2 = \frac{a}{b}$, $m \cdot m^2 = \frac{a}{b}$,

$m^3 = \frac{a}{b}$ ...(iii)

Cubing both sides of Eq. (ii), we get

$$(m + m^2)^3 = \frac{-8h^3}{b^3}$$

$\Rightarrow$ $m^3 + m^6 + 3m^3 (m + m^2) = \frac{-8h^3}{b^3}$

$\Rightarrow$ $\frac{a}{b} + \left(\frac{a}{b}\right)^2 + \frac{3a}{b}\left(\frac{-2h}{b}\right) = \frac{-8h^3}{b^3}$

$\Rightarrow$ $\frac{a}{b} + \frac{a^2}{b^2} - \frac{6ah}{b^2} = \frac{-8h^3}{b^3}$

$\Rightarrow$ $\frac{ab + a^2 - 6ah}{b^2} = \frac{-8h^3}{b^3}$

$\Rightarrow$ $b(ab + a^2 - 6ah) = -8h^3$

$\Rightarrow$ $ab^2 + a^2b - 6abh + 8h^3 = 0$

2. (b) We have,

$hxy + gx + fy + c = 0$ ...(i)

If Eq. (i) represent pair of straight line, then

$$\begin{vmatrix} 0 & \frac{h}{2} & \frac{g}{2} \\ \frac{h}{2} & 0 & \frac{f}{2} \\ \frac{g}{2} & \frac{f}{2} & c \end{vmatrix} = 0$$

$\Rightarrow$ $\frac{-h}{2}\left(\frac{hc}{2} - \frac{fg}{4}\right) + \frac{g}{2}\left(\frac{hf}{4} - 0\right) = 0$

$\Rightarrow$ $-\frac{h^2c}{4} + \frac{hfg}{8} + \frac{ghf}{8} = 0$

$\Rightarrow$ $\frac{-h^2c}{4} + \frac{hfg}{4} = 0$

$\Rightarrow$ $hfg - h^2c = 0$

$\Rightarrow$ $h(fg - hc) = 0$

$\Rightarrow$ $fg - hc = 0$ $(\because h \neq 0)$

$\therefore$ $fg = hc$

3. (c) We have,

$ax^2 + 2hxy + by^2 = 0$ ...(i)

Interchanging coefficient of x^2 and y^2 and changing sign of the term containing xy in Eq. (i), we get

$bx^2 - 2hxy + ay^2 = 0$ ...(ii)

Which is equation of pair of straight lines perpendicular to pair of lines represented by Eq. (i).

4. (a) We have,

$3x^2 + 2hxy - 3y^2 - 40x + 30y - 75 = 0$ ...(i)

If Eq. (i) represent pair of straight lines, then

$$\begin{vmatrix} 3 & h & -20 \\ h & -3 & 15 \\ -20 & 15 & -75 \end{vmatrix} = 0$$

$\Rightarrow$ $3(75 \times 3 - 15 \times 15) - h(-75h + 15 \times 20) - 20(15h - 60) = 0$

$\Rightarrow$ $3(225 - 225) - h(-75h + 300) - 20(15h - 60) = 0$

$\Rightarrow$ $75h^2 - 300h - 300h + 1200 = 0$

$\Rightarrow$ $75h^2 - 600h + 1200 = 0$

$\Rightarrow$ $h^2 - 8h + 16 = 0$

$\Rightarrow$ $h = \frac{8 \pm \sqrt{64 - 64}}{2} = \frac{8}{2} = 4$

$\therefore$ $h = 4, 4$

5. (d) Given, $ax^2 + 2hxy + by^2 = 0$ ...(i)

and $4ab = 3h^2$ ...(ii)

Let m_1 and m_2 be the slopes of the line represented by Eq. (i), we have

$\therefore$ $m_1 + m_2 = \frac{-2h}{b}$ ...(iii)

$m_1 \cdot m_2 = \frac{a}{b}$ ...(iv)

$\therefore$ $m_1 - m_2 = \sqrt{(m_1 + m_2)^2 - 4m_1 m_2} = \sqrt{\frac{4h^2}{b^2} - \frac{4a}{b}}$

$= \sqrt{\frac{4(h^2 - ab)}{b^2}} = \sqrt{\frac{4\left(h^2 - \frac{3}{4}h^2\right)}{b^2}}$ [using Eq. (iii)]

$= \sqrt{\frac{h^2}{b^2}}$

$m_1 - m_2 = \frac{h}{b}$ ...(v)

Now, $\frac{m_1 + m_2}{m_1 - m_2} = \frac{\frac{-2h}{b}}{\frac{h}{b}} = \frac{-2}{1}$

Applying componendo and dividendo, we get

$\Rightarrow$ $\frac{2m_1}{2m_2} = \frac{-1}{-3}$ $\Rightarrow$ $\frac{m_1}{m_2} = \frac{1}{3}$

$\therefore$ $m_1 : m_3 = 1 : 3$

6. (c) We know that, general equation of second degree represent pair of straight line, if

$$\begin{vmatrix} a & h & g \\ h & b & f \\ g & f & c \end{vmatrix} = 0$$

and $h^2 \geq ab$, $g^2 \geq ca$ and $f^2 \geq bc$.

Clearly, option (c) satisfy these requirements as

$$\begin{vmatrix} 4 & -2 & 0 \\ -2 & 1 & 0 \\ 0 & 0 & -4 \end{vmatrix} = 4(-4 - 0) + 2(8 - 0)$$

$= -16 + 16 = 0$

and $h^2 = (-2)^2 = 4$

$ab = 4 \times 1 = 4$

$h^2 \geq ab$

7. (b) We know that, equation of line passing through origin having slope m is

$$y = mx$$

$\therefore$ Equation of line passing through origin and having slope 3 and $-\frac{1}{3}$ are

$$y = 3x \text{ and } y = -\frac{1}{3}x$$

Now, combined equation of lines is

$$(y-3x)\left(y+\frac{1}{3}x\right)=0$$

$$\Rightarrow \quad y^2+\frac{xy}{3}-3xy-x^2=0$$

$$\Rightarrow \quad 3y^2+xy-9xy-3x^2=0$$

$$\Rightarrow \quad 3y^2-8xy-3x^2=0$$

$$\therefore \quad 3x^2+8xy-3y^2=0$$

8. (b) We have, $pq(x^2-y^2)+(p^2-q^2)xy=0$ ...(i)

$$pqx^2-pqy^2+p^2xy-q^2xy=0$$

$$pqx^2-q^2xy-pqy^2+p^2xy=0$$

$$qx(px-qy)+py(-qy+px)=0$$

$$(px-qy)(qx+py)=0$$

$\therefore$ $px-qy=0$ is one of the lines represented by Eq. (i).

9. (c) The necessary and sufficient condition that the lines represented by

$$f(x,y)=ax^2+2hxy+by^2+2gx+2fy+c=0$$

are coincident, if $\Delta=\begin{vmatrix} a & h & g\\ h & b & f\\ g & f & c\end{vmatrix}=0$

and $h^2=ab, g^2=ac, f^2=bc$.

10. (c) We have, $ax^2+4xy+y^2=0$ ...(i)

and $m_1=3m_2$ ...(ii)

From Eq. (i), we have

$$m_1+m_2=\frac{-4}{1} \quad \text{...(iii)}$$

$$m_1\cdot m_2=\frac{a}{1} \quad \text{...(iv)}$$

$$3m_2\cdot m_2=a \quad \text{[using Eq. (iii)]}$$

$$\Rightarrow \quad 3m_2^2=a$$

$$\Rightarrow \quad m_2^2=\frac{a}{3} \quad \text{...(v)}$$

From Eqs. (ii) and (iii), we have

$$3m_2+m_2=-4$$

$$4m_2=-4$$

$$m_2=-1 \quad \text{...(vi)}$$

Using Eq. (vi) in Eq. (v), we have

$$(-1)^2=\frac{a}{3} \Rightarrow a=3$$

11. (d) We have,

$$x^2+xy-12y^2=0 \quad \text{...(i)}$$

This is a homogeneous equation of second degree, it represent pair of straight lines passing through origin.

Let one of the line be,

$$y=mx$$

$$\therefore \quad x^2+mx^2-12m^2x^2=0$$

$$\Rightarrow \quad x^2(1+m-12m^2)=0$$

$$\Rightarrow \quad -12m^2+m+1=0$$

$$\Rightarrow \quad 12m^2-m-1=0$$

$$\Rightarrow \quad m=\frac{1\pm\sqrt{1+48}}{24}$$

$$m=\frac{1\pm\sqrt{49}}{24}=\frac{1\pm 7}{24}=\frac{8}{24},\frac{-6}{24}=\frac{1}{3},-\frac{1}{4}$$

$\therefore$ Lines are $y=\frac{1}{3}x \Rightarrow x-3y=0$

and $y=-\frac{1}{4}x \Rightarrow x+4y=0$

12. (a) We have,

$$2x^2+7xy+3y^2-9x-7y+k=0$$

This will represent a pair of straight lines, if

$$\begin{vmatrix} 2 & \frac{7}{2} & -\frac{9}{2}\\ \frac{7}{2} & 3 & -\frac{7}{2}\\ -\frac{9}{2} & -\frac{7}{2} & k\end{vmatrix}=0$$

$$\Rightarrow 2\left(3k-\frac{49}{4}\right)-\frac{7}{2}\left(\frac{7k}{2}-\frac{63}{4}\right)-\frac{9}{2}\left(-\frac{49}{4}+\frac{27}{2}\right)=0$$

$$\Rightarrow 2\left(\frac{12k-49}{4}\right)-\frac{7}{2}\left(\frac{14k-63}{4}\right)-\frac{9}{2}\left(\frac{-49+54}{4}\right)=0$$

$$\Rightarrow \quad \frac{12k-49}{2}-\left(\frac{98k-441}{8}\right)-\frac{45}{8}=0$$

$$\Rightarrow \quad \left(\frac{48k-196}{8}\right)-\left(\frac{98k-441}{8}\right)-\frac{45}{8}=0$$

$$\Rightarrow \quad 48k-196-98k+441-45=0$$

$$\Rightarrow \quad -50k+200=0$$

$$\therefore \quad k=4$$

13. (c) We have,

$$x^2-2cxy-7y^2=0 \quad \text{...(i)}$$

and $m_1+m_2=4m_1\cdot m_2$ ...(ii)

From Eq. (i), we have

$$m_1+m_2=\frac{2c}{-7}$$

$$m_1\cdot m_2=\frac{1}{-7}$$

Now, $\frac{m_1+m_2}{m_1\cdot m_2}=\frac{\frac{2c}{-7}}{\frac{1}{-7}} \Rightarrow \frac{m_1+m_2}{m_1\cdot m_2}=2c$

$$\Rightarrow \quad 4=2c \quad \text{[using Eq. (ii)]}$$

$$\therefore \quad c=2$$

14. (a) Given that, θ_1 be the angle between the lines represented by equation $3x^2-7xy+4y^2=0$

$$\therefore \quad \tan\theta_1=\left|\frac{2\sqrt{\frac{49}{4}-12}}{7}\right|$$

$$\tan\theta_1=\left|\frac{1}{7}\right| \quad \text{...(i)}$$

and θ_2 be the angle between the lines represented by equation $6x^2-5xy+y^2=0$

$$\therefore \quad \tan\theta_2=\left|\frac{2\sqrt{\frac{25}{4}-6}}{7}\right|$$

$$\tan\theta_2=\left|\frac{1}{7}\right| \quad \text{...(ii)}$$

From Eqs. (i) and (ii), we have

$$\tan\theta_1=\tan\theta_2$$

$$\therefore \quad \theta_1=\theta_2$$

15. (c) Equation of curves,

$$2x^2+3y^2+10x=0 \quad \text{...(i)}$$

$$3x^2+5y^2+16x=0 \quad \text{...(ii)}$$

Eq. (i) can be written as

$$x=-\frac{(2x^2+3y^2)}{10} \quad \text{...(iii)}$$

Making Eq. (ii) homogeneous with help of Eq. (iii), we get

$$3x^2 + 5y^2 - \frac{16}{10}(2x^2 + 3y^2) = 0$$

$\Rightarrow \quad 10(3x^2 + 5y^2) - 16(2x^2 + 3y^2) = 0$

$\Rightarrow \quad 30x^2 + 50y^2 - 32x^2 - 48y^2 = 0$

$\Rightarrow \quad -2x^2 + 2y^2 = 0$

$\Rightarrow \quad x^2 - y^2 = 0 \qquad \ldots\text{(iv)}$

$\because$ Coefficient of x^2 + coefficient of $y^2 = 1 - 1 = 0$

$\therefore$ Lines represented by Eq. (iv) are perpendicular.

16. (b) We have,

$$x^2 + 2\sqrt{3}xy + 3y^2 - 3x - 3\sqrt{3}y - 4 = 0 \qquad \ldots\text{(i)}$$

Here, $a = 1, h = \sqrt{3}, b = 3$

Now, $\quad h^2 = (\sqrt{3})^2 = 3$

$\quad ab = 1 \times 3 = 3$

$\Rightarrow \quad h^2 = ab$

$\therefore$ Lines represented by Eq. (i) are parallel.

17. (a) We have,

$$x^2 - 3xy + \lambda y^2 + 3x - 5y + 2 = 0 \qquad \ldots\text{(i)}$$

Let θ be the angle between the lines represented by Eq. (i).

$\therefore \quad \theta = \tan^{-1}\left(\frac{1}{3}\right) \Rightarrow \tan\theta = \frac{1}{3}$

Now, $\quad \tan\theta = \left|\frac{2\sqrt{h^2 - ab}}{a + b}\right|$

$$\tan\theta = \left|\frac{2\sqrt{\frac{9}{4} - \lambda}}{1 + \lambda}\right|$$

$$\frac{1}{3} = \left|\frac{2\frac{\sqrt{9 - 4\lambda}}{2}}{1 + \lambda}\right|$$

$$\frac{1}{3} = \left|\frac{\sqrt{9 - 4\lambda}}{1 + \lambda}\right|$$

$\Rightarrow \quad \frac{\sqrt{9 - 4\lambda}}{1 + \lambda} = \pm\frac{1}{3}$

Taking positive sign, we have

$$\frac{\sqrt{9 - 4\lambda}}{1 + \lambda} = \frac{1}{3}$$

$\Rightarrow \quad 9 - 4\lambda = \frac{(1 + \lambda)^2}{9}$

$\Rightarrow \quad 81 - 36\lambda = 1 + \lambda^2 + 2\lambda$

$\Rightarrow \quad \lambda^2 + 38\lambda - 80 = 0$

$\Rightarrow \quad \lambda^2 + 40\lambda - 2\lambda - 80 = 0$

$\Rightarrow \quad \lambda(\lambda + 40) - 2(\lambda + 40) = 0$

$\Rightarrow \quad (\lambda + 40)(\lambda - 2) = 0$

$\Rightarrow \quad \lambda = 2, -40$

$\Rightarrow \quad \lambda = 2 \qquad (\because \lambda \text{ is non-negative})$

18. (d) We have, $\quad ax^2 + 2hxy + by^2 = 0 \qquad \ldots\text{(i)}$

$$ax^2 + 2hxy + by^2 + \lambda(x^2 + y^2) = 0 \qquad \ldots\text{(ii)}$$

Equation of bisector of the angle between the lines represented by Eq. (i) is

$$\frac{x^2 + y^2}{a - b} = \frac{xy}{h}$$

$\Rightarrow \quad h(x^2 - y^2) = (a - b)xy$

$\Rightarrow \quad hx^2 - (a - b)xy - hy^2 = 0 \qquad \ldots\text{(iii)}$

and equation of bisector of the angle between the lines represented by Eq. (ii) is

$$\frac{x^2 - y^2}{a + \lambda - b - \lambda} = \frac{xy}{h}$$

$\Rightarrow \quad h(x^2 - y^2) = (a - b)xy$

$\Rightarrow \quad hx^2 - (a - b)xy - hy^2 = 0 \qquad \ldots\text{(iv)}$

Eqs. (iii) and (iv) are coincident for any value of λ because it does not depend on λ.

19. (d) Clearly, $y = x$ is one of the lines given by the equation

$$ax^2 + 2hxy + by^2 = 0$$

$\therefore \quad ax^2 + 2hx^2 + bx^2 = 0 \qquad (\text{putting } y = x)$

$\Rightarrow \quad a + b = -2h$

$\therefore \quad (a + b)^2 = 4h^2$

20. (b) We have,

$$8x^2 + 8xy + 2y^2 + 26x + 13y + 15 = 0 \qquad \ldots\text{(i)}$$

Now, distance between the lines represented by Eq. (i) is

$$d = 2\sqrt{\frac{g^2 - ac}{a(a + b)}} = 2\sqrt{\frac{169 - 15 \times 8}{8(10)}}$$

$$= \sqrt{\frac{169 - 120}{40}} = \sqrt{\frac{49}{40}} = \frac{7}{2\sqrt{5}} \text{ units}$$

21. (c) We have,

$$x^2 - 6xy + 9y^2 + 3x - 9y - 4 = 0 \qquad \ldots\text{(i)}$$

Now, $\quad d = 2\sqrt{\frac{g^2 - ac}{a(a + b)}} = 2\sqrt{\frac{\frac{9}{4} + 4}{1(1 + 9)}}$

$$= 2\sqrt{\frac{\frac{25}{4}}{10}} = \sqrt{\frac{25}{40}} = \frac{10}{2\sqrt{10}} = \sqrt{\frac{5}{2}} \text{ units}$$

22. (a) We have,

$$xy = 0 \qquad \ldots\text{(i)}$$

$$x + y = 1 \qquad \ldots\text{(ii)}$$

Equation of sides of triangle are

$$x = 0, y = 0 \text{ and } x + y = 1$$

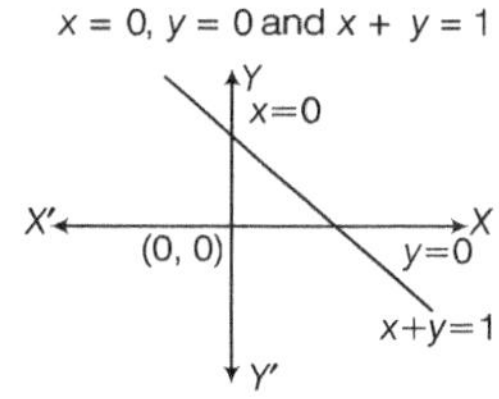

Clearly, they form a right angled triangle. So, the coordinates of the orthocenter are the coordinates of the vertex forming the right angle i.e. point of intersection of the lines $x = 0$ and $y = 0$ i.e. (0, 0).

23. (a) We have, $\quad y^2 - 9xy + 18x^2 = 0 \qquad \ldots\text{(i)}$

$$y = 9 \qquad \ldots\text{(ii)}$$

Eq. (i) is homogeneous equation of second degree, it represent pair of straight lines passing through origin.

Let one of the lines be,

$$y = mx$$

$\therefore \quad m^2x^2 - 9mx^2 + 18x^2 = 0$

$\Rightarrow \quad m^2 - 9m + 18 = 0$

$\Rightarrow \quad m = \frac{9 \pm \sqrt{81 - 72}}{2}$

$\quad m = \frac{9 \pm 3}{2} = 6, 3$

$\therefore$ Lines are $\quad y = 6x \qquad \ldots\text{(iii)}$

$\quad y = 3x \qquad \ldots\text{(iv)}$

Taking mutual intersection of sides of triangle, we get vertices of triangle as (0, 0), $\left(\frac{3}{2}, 9\right)$ and (3, 9).

$$\therefore \quad \text{Area} = \frac{1}{2}\left\|\begin{matrix} 0 & 0 & 1 \\ \frac{3}{2} & 9 & 1 \\ 3 & 9 & 1 \end{matrix}\right\|$$

$$= \frac{1}{2}\left|1\left(\frac{27}{2} - 27\right)\right|$$

$$= \frac{1}{2}\left|\frac{-27}{2}\right| = \frac{27}{4} \text{ sq units.}$$

24. (c) We have,

$$x^2 - 5x + 6 = 0$$

$$\Rightarrow \quad x^2 - 3x - 2x + 6 = 0$$

$$\Rightarrow \quad x(x-3) - 2(x-3) = 0$$

$$\Rightarrow \quad (x-3)(x-2) = 0$$

$$\Rightarrow \quad x = 2 \quad \text{...(i)}$$

$$\text{and} \quad x = 3 \quad \text{...(ii)}$$

$$y^2 - 6y + 5 = 0$$

$$\Rightarrow \quad y^2 - 5y - y + 5 = 0$$

$$\Rightarrow \quad y(y-5) - 1(y-5) = 0$$

$$\Rightarrow \quad (y-5)(y-1) = 0$$

$$\Rightarrow \quad y = 5 \quad \text{...(iii)}$$

$$\therefore \quad y = 1 \quad \text{...(iv)}$$

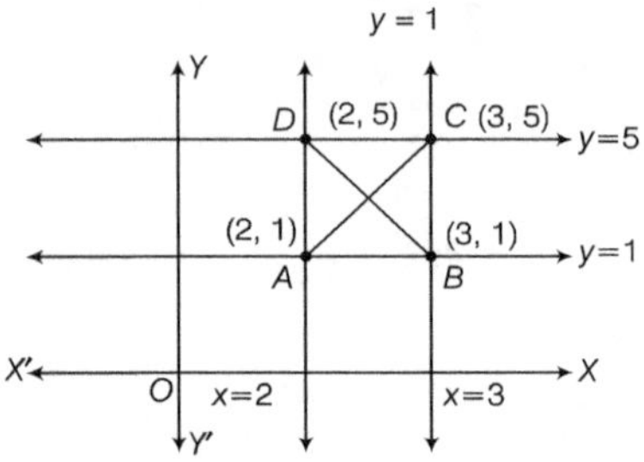

Equation of diagonal AC,

$$y - 1 = \frac{5-1}{3-2}(x-2)$$

$$y - 1 = \frac{4}{1}(x-2)$$

$$y - 1 = 4x - 8$$

$$4x - y - 7 = 0 \quad \text{...(v)}$$

Equation of diagonal BD,

$$y - 1 = \frac{5-1}{2-3}(x-3)$$

$$y - 1 = \frac{4}{-1}(x-3)$$

$$y - 1 = -4x + 12$$

$$4x + y - 13 = 0 \quad \text{...(vi)}$$

25. (b) Let the equation of lines represented by $ax^2 + 2hxy + by^2 = 0$ be $y - m_1x = 0$ and $y - m_2x = 0$ and one diagonal AC be $lx + my = 1$.

$$\therefore \quad m_1 + m_2 = \frac{-2h}{b} \text{ and } m_1m_2 = \frac{a}{b}$$

Now, on solving the equations of OA and OC with the line AC, we get the coordinates

$$A\left(\frac{1}{l+mm_1}, \frac{m_1}{l+mm_1}\right) \text{and } C\left(\frac{1}{l+mm_2}, \frac{m_2}{l+mm_2}\right)$$

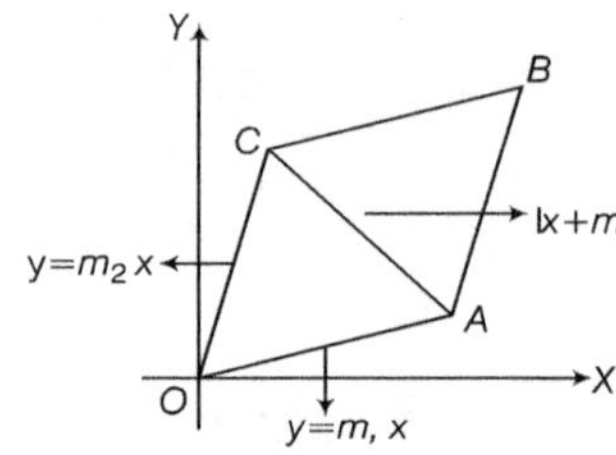

Now, find the coordinates of mid-point of AC and the equation of diagonal through this mid-point and origin.

$\therefore$ The required equation is, $x(am - hl) = (lb - mh)y$

Revisal Problems (JEE Advanced)

1. (d) We have,

$$y^3 - 4x^2y = 0$$

$$\Rightarrow \quad y(y^2 - 4x^2) = 0$$

$$\Rightarrow \quad y(y+2x)(y-2x) = 0$$

$$\Rightarrow \quad y = 0 \quad \text{...(i)}$$

$$y + 2x = 0 \quad \text{...(ii)}$$

$$y - 2x = 0 \quad \text{...(iii)}$$

Since, these lines are concurrent.

$\therefore$ They do not form triangle.

2. (d) The equation of a line passing through the origin is

$$mx - y = 0 \quad \text{...(i)}$$

If it is at a distance δ from (x_1, y_1), then

$$\left|\frac{mx_1 - y_1}{\sqrt{m^2 - 1}}\right| = \delta$$

$$\Rightarrow \quad (mx_1 - y_1)^2 = \delta^2(m^2 + 1) \quad \text{...(ii)}$$

$$\Rightarrow \quad (yx_1 - xy_1)^2 = \delta^2(x^2 + y^2) \quad \left(\because m = \frac{y}{x}\right)$$

3. (c) We have,

$$ax^2 + 2hxy + by^2 = 0 \quad \text{...(i)}$$

Shifting the origin at (x_1, y_1), the equation of the required lines are given by

$$a(x - x_1)^2 + 2h(x - x_1)(y - y_1) + b(y - y_1)^2 = 0$$

4. (c) We know that, the equation of the lines passing through the origin and perpendicular to the lines given by $2x^2 - xy - y^2 = 0$ is $-x^2 + xy + 2y^2 = 0$.

Now, shifting the origin at (1, 0) the equation of the required lines are

$$-(x-1)^2 + (x-1)(y-0) + 2(y-0)^2 = 0$$

$$\Rightarrow \quad -(x-1)^2 + (x-1)y + 2y^2 = 0$$

$$\Rightarrow \quad -x^2 - 1 + 2x + xy - y + 2y^2 = 0$$

$$\Rightarrow \quad -x^2 + 2y^2 + xy + 2x - y - 1 = 0$$

$$\Rightarrow \quad 2y^2 + xy - x^2 + 2x - y - 1 = 0$$

5. (a) Let $y = mx$ be one of the lines represented by the given equation. Then, $y = mx$ satisfies it.

$\therefore \lambda(1 - 3m^2) + m^3 - 3m = 0$

$$\Rightarrow \lambda = \frac{3m - m^3}{1 - 3m^2}$$

$$\Rightarrow \tan\phi = \frac{3\tan\theta - \tan^3\theta}{1 - 3\tan^2\theta}$$

where, $\lambda = \tan\phi$ and $m = \tan\theta$

$$\Rightarrow \tan\phi = \tan 3\theta$$

$$\Rightarrow 3\theta = \phi, \pi + \phi, 2\pi + \phi$$

$$\Rightarrow \theta = \frac{\phi}{3}, \frac{\pi}{3} + \frac{\phi}{3}, \frac{2\pi}{3} + \frac{\phi}{3}$$

Hence, the given equation represents three lines passing through the origin such that they are equally inclined with one another.

6. (a) We have,

$$x^3 + x^2y - xy^2 - y^3 = 0$$

$$\Rightarrow x^2(x + y) - y^2(x + y) = 0$$

$$\Rightarrow (x + y)(x^2 - y^2) = 0$$

$$\Rightarrow (x + y)^2(x - y) = 0$$

$$\Rightarrow x + y = 0, x + y = 0, x - y = 0$$

Clearly, two of the three lines represented by the given equation are coincident and two are perpendicular.

7. (c) We know that, the product of perpendiculars drawn from the point (x_1, y_1) to the pair of lines

$$ax^2 + 2hxy + by^2 + 2gx + 2fy + c = 0 \text{ is} \quad \ldots\text{(i)}$$

$$\left|\frac{ax_1^2 + 2hx_1y_1 + by_1^2 + 2gx_1 + 2fy_1 + c}{\sqrt{(a - b)^2 + 4h^2}}\right|$$

$\therefore$ Product of perpendicular from (0, 0) to the lines (i) is

$$= \left|\frac{a \times 0 + 2h \times 0 \times 0 + b \times 0 + 2g \times 0 + 2f \times 0 + c}{\sqrt{(a - b)^2 + 4h^2}}\right|$$

$$= \left|\frac{c}{\sqrt{(a - b)^2 + 4h^2}}\right|$$

8. (a) Let $f(x, y) = ax^2 + 2hxy + by^2 + 2gx + 2fy + c$

The point of intersection of the lines given by

$$ax^2 + 2hxy + by^2 + 2gx + 2fy + c = 0$$

is obtained by solving

$$\frac{\partial f}{\partial x} = 0$$

and $$\frac{\partial f}{\partial y} = 0$$

i.e. $$ax + hy + g = 0$$

and $$hx + by + f = 0$$

The coordinates of point of intersection are

$$P\left(\frac{hf - bg}{ab - h^2}, \frac{hg - af}{ab - h^2}\right)$$

$\therefore$ Distance between the origin and point P is given by

$$OP = \sqrt{\frac{(hf - bg)^2 + (hg - af)^2}{(ab - h^2)^2}}$$

$$= \sqrt{\frac{f^2(h^2 + a^2) + g^2(h^2 + b^2) - 2fgh(a + b)}{(ab - h^2)^2}}$$

$$= \sqrt{\frac{f^2(h^2 + a^2) + g^2(h^2 + b^2) - (a + b)(af^2 + bg^2 + ch^2 - abc)}{(ab - h^2)^2}}$$

$$(\because abc + 2fgh - af^2 - bg^2 - ch^2 = 0)$$

$$= \sqrt{\frac{ac(ab - h^2) + bc(ab - h^2) - f^2(ab - h^2) - g^2(ab - h^2)}{(ab - h^2)^2}}$$

$$= \sqrt{\frac{c(a + b) - f^2 - g^2}{(ab - h^2)}}$$

$$\therefore OP^2 = \frac{c(a + b) - f^2 - g^2}{(ab - h^2)}$$

9. (a) Let the equation of the pair of perpendicular lines be $x^2 + \lambda xy - y^2 = 0$, then

$$ay^4 + bxy^3 + cx^2y^2 + dx^3y + ex^4 = (x^4 + \lambda xy - y^2)(ex^2 + \mu xy - ay^2)$$

On equating the coefficient of like terms, we get

$$b = -a\lambda - \mu$$

$$c = -a - e + \lambda\mu$$

and $$d = \mu + e\lambda$$

Now, $$a\lambda + \mu + b = 0$$

and $$e\lambda + \mu - d = 0$$

$$\Rightarrow \frac{\lambda}{-(b + d)} = \frac{\mu}{ad + be} = \frac{1}{a - e}$$

$$\lambda = \frac{-b + d}{a - e}$$

and $$\mu = \frac{ad + be}{a - e}$$

Substituting these values in $c = -a - e + \lambda\mu$, we get

$$c + a + e = -\frac{(b + d)(ad + be)}{(a - e)^2}$$

$$\therefore (c + a + e)(a - e)^2 + (b + d)(ad + be) = 0$$

10. (c) Given,

$$ax^2 + 2hxy + by^2 + 2gx + 2fy + c = 0 \quad \ldots\text{(i)}$$

Let $$l_1x + m_1y + n_1 = 0$$

and $$l_2x + m_2y + n_2 = 0$$

are the lines represented by Eq. (i).

It is given that lines represented by Eq. (i) are equidistant from the origin.

$$\therefore \left|\frac{n_1}{\sqrt{l_1^2 + m_1^2}}\right| = \left|\frac{n_2}{\sqrt{l_2^2 + m_2^2}}\right|$$

$$\Rightarrow n_1^2(l_2^2 + m_2^2) = n_2^2(l_1^2 + m_1^2)$$

$$\Rightarrow (n_1l_2 + n_2l_1)(n_1l_2 - n_2l_1) = (n_2m_1 + n_1m_2)(n_2m_1 - n_1m_2)$$

$$\Rightarrow 4g^2(n_1l_2 - n_2l_1)^2 = 4f^2(n_2m_1 - n_1m_2)^2$$

$$\Rightarrow g^2[(n_1l_2 - n_2l_1)^2 - 4l_1l_2n_1n_2] = f^2[(n_1m_2 + n_2m_1)^2 - 4m_1m_2n_1n_2]$$

$$\Rightarrow g^2[4g^2 - 4ac] = f^2[4f^2 - 4bc]$$

$$\therefore f^4 - g^4 = c(bf^2 - ag^2)$$

11. (d) Given, equation of sides of parallelogram are

$$2x + 3y = 0 \quad \ldots\text{(i)}$$

$$2x + 3y = 5 \quad \ldots\text{(ii)}$$

$$3x - 4y = 0 \quad \ldots\text{(iii)}$$

$$3x - 4y = 3 \quad \ldots\text{(iv)}$$

The graph of these lines is given below:

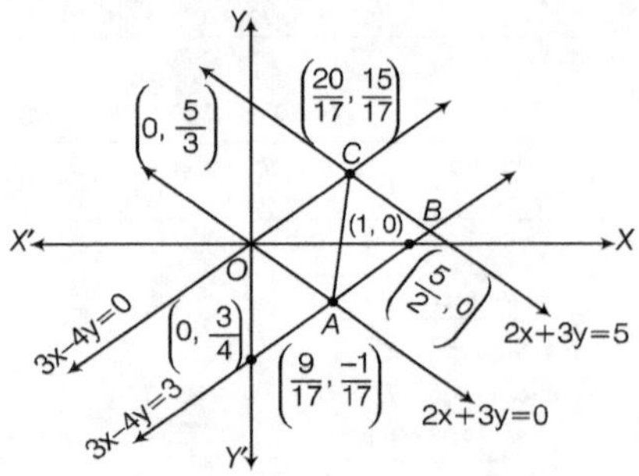

Now, equation of diagonal AC (not passing through origin) is

$$y + \frac{6}{17} = \frac{\frac{5}{17} + \frac{6}{17}}{\frac{20}{17} - \frac{9}{17}}\left(x - \frac{9}{17}\right)$$

$$y + \frac{6}{17} = \frac{\frac{21}{17}}{\frac{11}{17}}\left(x - \frac{9}{17}\right)$$

$$y + \frac{6}{17} = \frac{21}{17} \times \frac{17}{11}\left(x - \frac{9}{17}\right)$$

$$\Rightarrow \quad 11y + \frac{66}{17} = 21x - \frac{189}{17}$$

$$\Rightarrow \quad 21x - 11y - \frac{189}{17} - \frac{66}{17} = 0$$

$$\Rightarrow \quad 21x - 11y - \frac{255}{17} = 0$$

$$\therefore \quad 21x - 11y - 15 = 0$$

12. (a)

(i) Given lines are

$$3x - 4y = 11 \quad \ldots(i)$$
$$3x - 4y - 9 = 0 \quad \ldots(ii)$$
$$4x + 3y + 3 = 0 \quad \ldots(iii)$$
$$4x + 3y - 17 = 0 \quad \ldots(iv)$$

Since, lines (i), (ii) are parallel to each other and line (iii) and (iv) are parallel to each other.

$\therefore$ They enclose a parallelogram.

As lines (i) and (ii) are perpendicular to the lines (iii) and (iv).

$\therefore$ Each interior angles of parallelogram are 90°.

Now, distance between lines (i) and (ii) is

$$d_1 = \left|\frac{11 + 9}{\sqrt{16 + 9}}\right| = \left|\frac{20}{5}\right| = 4$$

and distance between parallel lines (iii) and (iv) is

$$d_2 = \left|\frac{3 + 17}{\sqrt{16 + 9}}\right| = \left|\frac{20}{5}\right| = 4$$

$$\because \quad d_1 = d_2$$

$\therefore$ The given line enclose a square.

(i) → (s)

(ii) Given lines are

$$2x + y = 1 \quad \ldots(i)$$
$$x + 2y = 1 \quad \ldots(ii)$$
$$2x + y = 3 \quad \ldots(iii)$$
$$x + 2y = 3 \quad \ldots(iv)$$

Lines (i) and (iii) are parallel to each other as their slopes are equal.

Similarly, lines (ii) and (vi) are parallel to each other.

$\therefore$ They enclose a parallelogram.

Now, distance between lines (i) and (iii) is

$$d_1 = \left|\frac{1 - 3}{\sqrt{5}}\right| = \frac{2}{\sqrt{5}}$$

and distance between lines (ii) and (iv) is

$$d_2 = \left|\frac{1 - 3}{\sqrt{5}}\right| = \frac{2}{\sqrt{5}}$$

$$\because \quad d_1 = d_2$$

$\therefore$ Enclosed figure is a rhombus.

(ii) → (r)

(iiii) Let $f(x, y) = 2x^2 - 7xy + 3y^2 + 5x + 10y - 25 = 0 \quad \ldots(i)$

$$\frac{\partial f}{\partial x} = 4x - 7y + 5 = 0 \quad \ldots(ii)$$

$$\frac{\partial f}{\partial y} = -7x + 6y + 10 = 0 \quad \ldots(iii)$$

On solving Eqs. (ii) and (iii), we have

$$x = 4, y = 3$$

$\therefore$ Point P is (4, 3).

and $\quad x + 2y = 5 \quad \ldots(iv)$

Now, point of intersection of the curves (i) and (ii) is

$$2(5 - 2y)^2 - 7(5 - 2y)y + 3y^2 + 5(5 - 2y) + 10y - 25 = 0$$

$$\Rightarrow 2(25 + 4y^2 - 20y) - 7(5y - 2y^2) + 3y^2 + 25 - 10y + 10y - 25 = 0$$

$$\Rightarrow \quad 50 + 8y^2 - 40y - 35y + 14y^2 + 3y^2 = 0$$

$$\Rightarrow \quad 25y^2 - 75y + 50 = 0$$

$$\Rightarrow \quad y^2 - 3y + 2 = 0$$

$$\Rightarrow \quad y = \frac{3 \pm \sqrt{9 - 8}}{2}$$

$$y = \frac{3 \pm 1}{2} = 2, 1$$

$\therefore$ Points A and B are (1, 2), (3, 1) respectively.

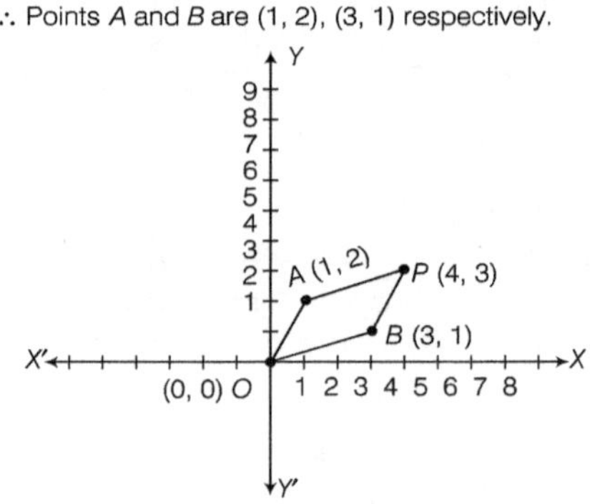

Now, $\quad OA = \sqrt{5}$
$OB = \sqrt{10}$
$AP = \sqrt{10}$
$PB = \sqrt{5}$

and $\quad$ slope $PB = \frac{2}{1}$

slope $AP = \frac{1}{3}$

slope $OA = \frac{2}{1}$

slope $OB = \frac{1}{3}$

$$\Rightarrow \quad PB \parallel OA \text{ and } AP \parallel OB$$

$$\therefore \quad OA = PB \text{ and } OB = AP$$

Hnece, $OAPB$ is a parallelogram.

(iii) → (q)

13. (b) Given, $f(\alpha, \beta) \le 0$ ...(i)

Also, $f(\alpha, \beta) = \left|\frac{\beta}{\alpha} - \frac{3}{2}\right| + (3\alpha - 2\beta)^6 + \sqrt{e\alpha + 2\beta - 2e - 6} \ge 0$ ...(ii)

From Eqs. (i) and (ii), $f(\alpha, \beta) = 0$

$$\Rightarrow \quad \frac{\beta}{\alpha} = \frac{3}{2}, 3\alpha - 2\beta = 0$$

and $e\alpha + 2\beta - 2e - 6 = 0$

On solving, $2\beta = 3\alpha$ and $e\alpha + 2\beta - 2e - 6 = 0$, we get

$$\alpha = 2, \beta = 3$$

14. (a) Given, $x = \alpha, y = \beta$ and $x + y = 8$

i.e. $x = 2, y = 3$ and $x + y = 8$

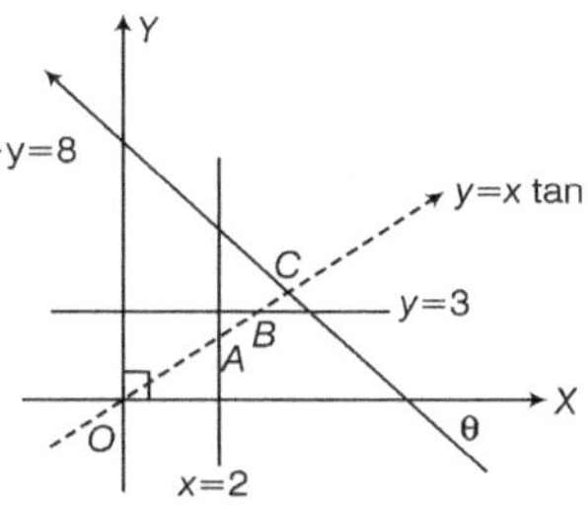

Let $OA = r_1$, $OB = r_2$ and $OC = r_3$

$\therefore \quad A(r_1 \cos\theta, r_1 \sin\theta), B(r_2 \cos\theta, r_2 \sin\theta), C(r_3 \cos\theta, r_3 \sin\theta)$

$\Rightarrow \quad r_1 \cos\theta = 2, \quad r_2 \sin\theta = 3 \quad$ and $\quad r_3(\cos\theta + \sin\theta) = 8$

Given, $r_1 r_2 r_3 = 48\sqrt{2}$

$$\Rightarrow \quad \frac{2}{\cos\theta} \cdot \frac{3}{\sin\theta} \cdot \frac{8}{(\cos\theta + \sin\theta)} = 48\sqrt{2}$$

$$\Rightarrow \quad \sqrt{2} = \sin 2\theta \cdot (\sin\theta + \cos\theta)$$

which is only possible, when $\theta = 45°$.

$\therefore$ $y = x$ is required straight line.

15. (d) Since, $OA = r_1 = \frac{2}{\cos\theta} = 2\sqrt{2}$, as $\theta = 45°$

$$OB = r_2 = \frac{3}{\sin\theta} = 3\sqrt{2}$$

$$OC = r_3 = \frac{8}{\sqrt{2}} = 4\sqrt{2}$$

$\Rightarrow \quad OA + OB + OC = 9\sqrt{2}$

Explanations for Q. Nos. (16-18)

Let $f(x, y) = ax^2 + 3xy - 2y^2 - 5x + 5y + c = 0$ ...(i)

Since, the lines represented by Eq. (i) are perpendicular

$$\therefore \quad a + b = 0$$
$$a + (-2) = 0$$
$$a = 2$$

and
$$\begin{vmatrix} 2 & \frac{3}{2} & -\frac{5}{2} \\ \frac{3}{2} & -2 & \frac{5}{2} \\ -\frac{5}{2} & \frac{5}{2} & c \end{vmatrix} = 0$$

$$\Rightarrow \quad 2\left(-2c - \frac{25}{4}\right) - \frac{3}{2}\left(\frac{3c}{2} + \frac{25}{4}\right) - \frac{5}{2}\left(\frac{15}{4} - 5\right) = 0$$

$$\Rightarrow \quad 2\left(\frac{-8c - 25}{4}\right) - \frac{3}{2}\left(\frac{6c + 25}{4}\right) - \frac{5}{2}\left(\frac{15 - 20}{4}\right) = 0$$

$$\Rightarrow \quad \frac{-16c - 50}{4} - \left(\frac{18c + 75}{8}\right) + \left(\frac{25}{8}\right) = 0$$

$$\Rightarrow \quad \frac{-32c - 100}{8} - \left(\frac{18c + 75}{8}\right) + \frac{25}{8} = 0$$

$$\Rightarrow \quad -32c - 100 - 18c - 75 + 25 = 0$$

$$\Rightarrow \quad -50c - 150 = 0$$

$$\Rightarrow \quad c = -3$$

$$\therefore \quad f(x, y) = 2x^2 + 3xy - 2y^2 - 5x + 5y - 3 = 0$$

Now, point of intersection of these lines can be obtained by simultaneously simplifying

$$\frac{\partial f}{\partial x} = 0 \quad \text{and} \quad \frac{\partial f}{\partial y} = 0$$

i.e. $4x + 3y - 5 = 0$

and $3x - 4y + 5 = 0$

Point of intersection is $\left(\frac{1}{5}, \frac{7}{5}\right)$.

Now, point of intersection of X-axis and the line given in Eq. (i) is

$$2x^2 - 5x - 3 = 0$$

$$\Rightarrow \quad x = \frac{5 \pm \sqrt{25 + 24}}{4}$$

$$x = \frac{5 \pm 7}{4} = 3, -\frac{1}{2}$$

$$\Rightarrow \quad x_1 = 3, x_2 = -\frac{1}{2}$$

Now, point of intersection of Y-axis and the lines given in Eq. (i) is

$$-2y^2 + 5y - 3 = 0$$

$$\Rightarrow \quad 2y^2 - 5y + 3 = 0$$

$$\Rightarrow \quad y = \frac{5 \pm \sqrt{25 - 24}}{4}$$

$$= \frac{5 \pm 1}{4} = \frac{6}{4}, \frac{4}{4} = \frac{3}{2}, 1$$

$$\therefore \quad y_1 = \frac{3}{2}, y_2 = 1$$

16. (b) $x_1 + x_2 + y_1 + y_2 = 3 - \frac{1}{2} + \frac{3}{2} + 1 = 5$

17. (c) Since, ΔABC is right angled triangle.

$\therefore$ Orthocenter is vertex C, i.e. $\left(\frac{1}{5}, \frac{7}{5}\right)$.

Centroid of ΔABC is $\left(\frac{9}{10}, \frac{7}{15}\right)$.

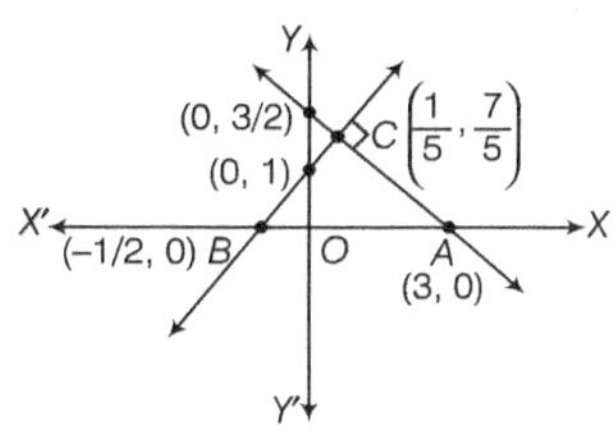

We know that, centroid divides circumcenter and orthocenter in the ratio 1 : 2.

$$\therefore \quad OG : GH = 1 : 2$$

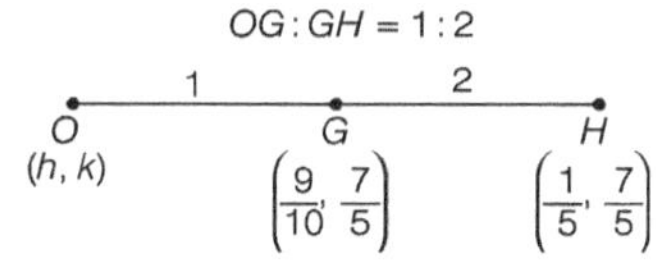

$$\frac{9}{10} = \frac{\frac{1}{5} + 2h}{3}$$

$$\frac{27}{10} = \frac{1}{5} + 2h$$

$$\Rightarrow \quad 2h = \frac{27}{10} - \frac{1}{5} = \frac{27 - 2}{10} = \frac{25}{10} = \frac{5}{2}$$

$$\frac{7}{15} = \frac{\frac{7}{5} + 2k}{3}$$

$$\frac{21}{15} = \frac{7}{5} + 2k$$

$\Rightarrow$ $$2k = \frac{21}{15} - \frac{7}{5} = \frac{21-21}{15} = 0$$

$\therefore$ $$k = 0$$

Distance between circumcenter to orthocenter is

$$= \sqrt{\left(\frac{1}{5} - \frac{5}{2}\right)^2 + \left(\frac{7}{5}\right)^2} = \sqrt{\frac{441}{400} + \frac{49}{25}}$$

$$= \sqrt{\frac{441 + 16 \times 49}{400}} = \sqrt{\frac{441 + 784}{400}}$$

$$= \sqrt{\frac{1225}{400}} = \frac{7}{4}$$

18. (d) Equation of circumcircle is

$$\left(x - \frac{5}{4}\right)^2 + y^2 = \left(\frac{7}{4}\right)^2$$

$\Rightarrow$ $$x^2 + \frac{25}{16} - \frac{5x}{2} + y^2 = \frac{49}{16}$$

$\Rightarrow$ $$\frac{16x^2 + 25 - 40x + 16y^2}{16} = \frac{49}{16}$$

$\Rightarrow$ $$16x^2 + 16y^2 - 40x + 25 = 49$$

$\Rightarrow$ $$16x^2 + 16y^2 - 40x - 24 = 0 \quad \ldots(i)$$

$\Rightarrow$ $$2x^2 + 2y^2 - 5x - 3 = 0$$

and $$x^2 + y^2 - 4y + k = 0 \quad \ldots(ii)$$

Circles (i) and (ii) are orthogonal.

$\therefore$ $$2g_1g_2 + 2f_1f_2 = c_1 + c_2$$

$\Rightarrow$ $$2\left(\frac{-5}{2}\right) \times 0 + 2 \times 0 \times (-2) = -\frac{3}{2} + k$$

$\therefore$ $$k = \frac{3}{2}$$

19. (3) The sides of the triangle are given by

$$(x^2 - y^2)(2x + 3y - 6) = 0$$

or $$x - y = 0$$
$$x + y = 0$$
$$2x + 3y - 6 = 0$$

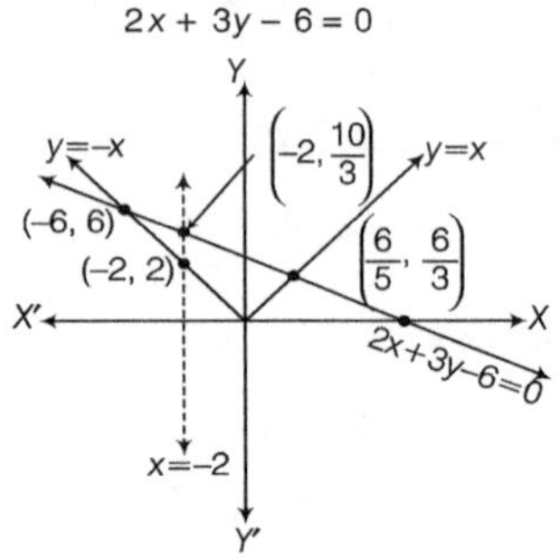

We observe that the point $(-2, a)$ moves on $x = -2$.

Thus, $$2 \le a \le \frac{10}{3}$$

$\therefore$ $$a = 3$$

20. (3) We have,

$$3x^2 + 2\lambda xy + 3y^2 + (6 - \lambda)x + (2\lambda - 6)y - 21 = 0$$

If this equation represent circle, then

$$a = b \text{ and } h = 0$$
$$3 = 3 \text{ and } \lambda = 0$$

$\therefore$ Equation of circle, $x^2 + y^2 + 2x - 2y - 7 = 0$

and $$\text{radius} = \sqrt{g^2 + f^2 - c}$$
$$= \sqrt{1+1+7} = \sqrt{9} = 3$$

21. (12) The combined equation of the straight lines joining the origin to the point of intersection of C_1 and C_2 is a homogeneous equation and is given by

$$\lambda x^2 + 3y^2 - 2\lambda xy + 3(3x^2 - 4y^2 + 8xy) = 0$$

or $$(\lambda + 9)x^2 + 2xy(12 - \lambda) - 9y^2 = 0$$

Lines given by this equation are equally inclined with X-axis.

$\therefore$ Sum of their slopes = 0

$\Rightarrow$ $$\frac{-2(12 - \lambda)}{-9} = 0 \qquad \left(\because m_1 + m_2 = \frac{-2h}{b}\right)$$

$\therefore$ $$\lambda = 12$$

22. (1) Given,

$$a_1x^2 + 2h_1xy + b_1y^2 = c_1 \quad \ldots(i)$$
$$a_2x^2 + 2h_2xy + b_2y^2 = c_2 \quad \ldots(ii)$$

The equation of a conic passing through the intersection of the given conic is

$$(a_1x^2 + 2h_1xy + b_1y^2 - c_1) + \lambda(a_2x^2 + 2h_2xy + b_2y^2 - c_2) = 0$$

This equation will represent a circle, if coefficient of x^2 = coefficient of y^2 and coefficient of $xy = 0$

$\Rightarrow$ $$(a_1 + \lambda a_2) = (b_1 + \lambda b_2)$$

and $$2h_1 + 2\lambda h_2 = 0$$

Eliminating λ from these two equations, we get

$$a_1 - b_1 = \frac{h_1}{h_2}(a_2 - b_2)$$

$\Rightarrow$ $$(a_1 - b_2)h_2 = (a_2 - b_2)h_1$$

But it is given that

$$(a_1 - b_1)h_2 = \lambda(a_2 - b_2)h_1$$

$\therefore$ $$\lambda = 1$$

23. (0)

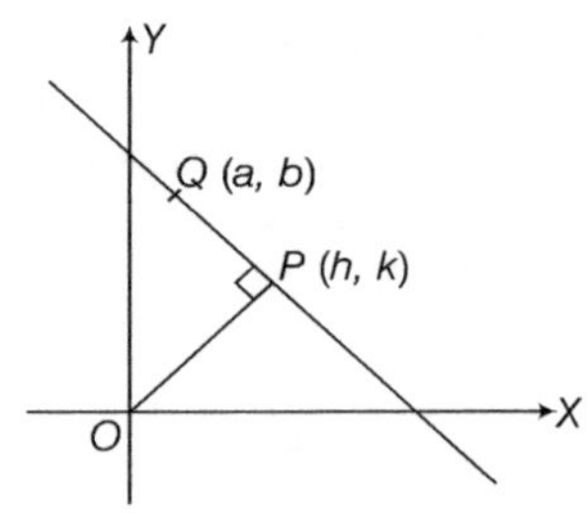

$\therefore$ Locus of P, is

$$\frac{k}{h} \cdot \left(\frac{k-b}{h-a}\right) = -1$$

$\Rightarrow$ $x(x - a) + y(y - b) = 0$, locus of P.

$\Rightarrow$ $$x^2 + y^2 - ax - by = 0$$

$\therefore$ $$\lambda = 0$$

JEE Archive

1. (c) Equation of bisectors of line $xy = 0$ are

$$y = \pm x$$

Put $y = \pm x$ in $my^2 + (1 - m^2)xy - mx^2 = 0$, we get

$$mx^2 \pm (1 - m^2)x^2 - mx^2 = 0$$

$$\Rightarrow \quad (1 - m^2)x^2 = 0$$

$$\therefore \quad m = \pm 1$$

2. (a) Given equation of pair of lines is

$$ax^2 + 2(a + b)xy + by^2 = 0$$

Here, $H = a + b$, $A = a$, $B = b$

Since, $4\theta = \pi$

$$\Rightarrow \quad \theta = \frac{\pi}{4}$$

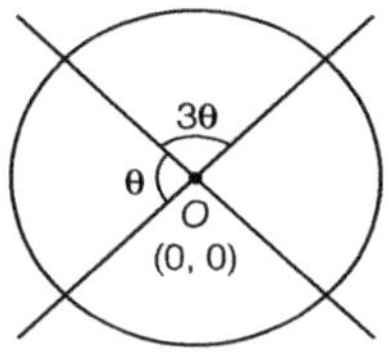

Angle between lines is given by

$$\tan \theta = \frac{2\sqrt{H^2 - AB}}{A + B}$$

$$\Rightarrow \quad \tan \frac{\pi}{4} = 1 = \frac{2\sqrt{(a + b)^2 - ab}}{a + b}$$

$$\Rightarrow \quad a^2 + b^2 + 2ab = 4(a^2 + b^2 + ab)$$

$$\Rightarrow 3a^2 + 3b^2 + 2ab = 0$$

3. (c) The given pair of lines is

$$x^2 - 2cxy - 7y^2 = 0$$

On comparing the standard equation $ax^2 + 2hxy + by^2 = 0$, we get

$$a = 1, \quad 2h = -2c, \quad b = -7$$

Now, $m_1 + m_2 = -\frac{2h}{b} = -\frac{2c}{7}$

and $m_1m_2 = \frac{a}{b} = -\frac{1}{7}$

According to the given condition,

$$m_1 + m_2 = 4m_1m_2$$

$$\therefore \quad -\frac{2c}{7} = -\frac{4}{7}$$

$$\Rightarrow \quad c = 2$$

4. (d) Since, one of the two lines is $3x + 4y = 0$. Then, $y = -\frac{3x}{4}$ will satisfy the equation

$$6x^2 - xy + 4cy^2 = 0$$

$$\therefore \quad 6x^2 - x\left(-\frac{3x}{4}\right) + 4c\left(-\frac{3x}{4}\right)^2 = 0$$

$$\Rightarrow \quad 6x^2 + \frac{3x^2}{4} + 4c \cdot \frac{9x^2}{16} = 0$$

$$\Rightarrow \quad x^2(27 + 9c) = 0$$

$$\therefore \quad c = -3$$

5. (d) The given equation is $x^2 - 2pxy - y^2 = 0$

On comparing with standard equation

$$ax^2 + 2hxy + by^2 = 0,$$

we get

$$a = 1$$
$$b = -1$$
$$h = -p$$

Equation of the bisector of angles is

$$\frac{x^2 - y^2}{a - b} = \frac{xy}{h}$$

$$\therefore \quad \frac{x^2 - y^2}{1 + 1} = \frac{xy}{-p}$$

$$\Rightarrow \quad x^2 - y^2 = -\frac{2xy}{p}$$

$$\Rightarrow \quad x^2 + \frac{2xy}{p} - y^2 = 0 \quad \ldots\text{(i)}$$

But given bisector equation of angles is

$$x^2 - 2qxy - y^2 = 0 \quad \ldots\text{(ii)}$$

On comparing Eqs. (i) and (ii), we get

$$\frac{2}{p} = -2q$$

$$\Rightarrow \quad pq = -1$$

6. (b) Let a and b be non-zero real numbers.

Therefore, the given equation

$(ax^2 + by^2 + c)(x^2 - 5xy + 6y^2) = 0$ implies either

$$x^2 - 5xy + 6y^2 = 0$$

$$\Rightarrow \quad (x - 2y)(x - 3y) = 0$$

$$\Rightarrow \quad x = 2y$$

and $x = 3y$

represent two straight lines passing through origin or

$ax^2 + by^2 + c = 0$ when $c = 0$ and a and b are of same signs

Then,

$$ax^2 + by^2 + c = 0, \; x = 0$$

and $y = 0$

which is a point specified as the origin.

When $a = b$ and c is of sign opposite to that of a, $ax^2 + by^2 + c = 0$ represents a circle.

Hence, the given equation,

$$(ax^2 + by^2 + c)(x^2 - 5xy + 6y^2) = 0$$

may represent two straight lines and a circle.

4. Circle

DPP-1 Equation of Circle

1. (c) Given equation is

$$px^2 + (2-q)xy + 3y^2 - 6qx + 30y + 6q = 0$$

$\because$ It represent an equation of circle.

$\therefore$ Coefficient of x^2 = Coefficient of y^2

$$\Rightarrow \quad p = 3$$

Also, for a circle, the coefficient of $xy = 0$

$$\Rightarrow \quad (2-q) = 0$$

$$\Rightarrow \quad q = 2$$

Hence, the values of p and q are 3 and 2, respectively.

2. (c) The equation of the sides of quadrilateral are

$$L_1 \equiv x - a = 0$$
$$L_2 \equiv y - a = 0$$
$$L_3 \equiv x - 2a = 0$$
$$L_4 \equiv y + a = 0$$

The equation of the curve passing through the vertices of the quadrilateral is

$$L_1L_3 + \lambda L_2L_4 = 0$$

$$\Rightarrow \quad (x-a)(x-2a) + \lambda(y-a)(y+a) = 0$$

$$\Rightarrow \quad x^2 - ax - 2ax + 2a^2 + \lambda(y^2 - a^2) = 0$$

$$\Rightarrow \quad x^2 - 3ax + 2a^2 + \lambda y^2 - \lambda a^2 = 0 \quad \ldots(\text{i})$$

It will represent a circle, if

Coefficient of x^2 = Coefficient of y^2

$$\Rightarrow \quad \lambda = 1$$

On substituting the value of λ in Eq. (i), we get

$$x^2 - 3ax + 2a^2 + y^2 - a^2 = 0$$

Thus, the circumcircle of the quadrilateral formed by the given line is

$$x^2 + y^2 - 3ax + a^2 = 0$$

3. (d) Let x_1 and x_2 be the x-coordinates of the points A and B respectively which are roots of equation,

$$x^2 + 2x - a^2 = 0$$

$\therefore$ Sum of roots $(x_1 + x_2) = -2$

Product of roots $(x_1x_2) = -a^2$

Again, let y_1 and y_2 be the y-coordinates of the points A and B respectively which are roots of equation,

$$y^2 + 4y - b^2 = 0$$

$\therefore$ Sum of roots $(y_1 + y_2) = -4$

Product of roots $(y_1y_2) = -b^2$

Since, AB is diameter of circle. Therefore, required equation of circle is

$$(x - x_1)(x - x_2) + (y - y_1)(y - y_2) = 0$$

$$\Rightarrow x^2 + y^2 - x(x_1 + x_2) - y(y_1 + y_2) + x_1x_2 + y_1y_2 = 0$$

$$\Rightarrow \quad x^2 + y^2 - x(-2) - y(-4) + (-a^2) + (-b^2) = 0$$

$$\Rightarrow \quad x^2 + y^2 + 2x + 4y - a^2 - b^2 = 0$$

$$\Rightarrow \quad x^2 + 2x + 1 + y^2 + 4y + 4 = a^2 + b^2 + 1 + 4 = 0$$

$$\Rightarrow \quad (x+1)^2 + (y+2)^2 = 5 + a^2 + b^2$$

$\therefore$ It is the required equation of circle.

4. (c) Let O be the origin and the centre is $C(\alpha, \beta)$, draw a perpendicular $CM \perp OX$.

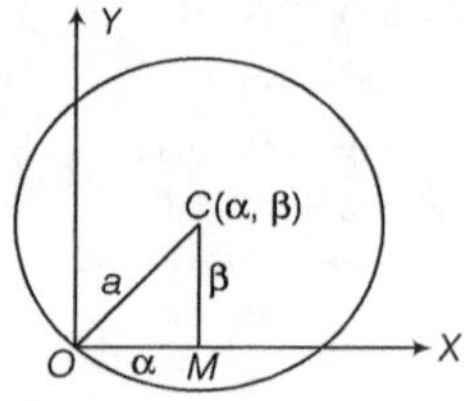

In ΔOMC,

$$OC^2 = CM^2 + OM^2$$

$$a^2 = \beta^2 + \alpha^2$$

Then, equation of the circle is

$$(x-\alpha)^2 + (y-\beta)^2 = \alpha^2 + \beta^2$$

$$\Rightarrow \quad x^2 - 2x\alpha + \alpha^2 + y^2 - 2y\beta + \beta^2 = \alpha^2 + \beta^2$$

$$\Rightarrow \quad x^2 + y^2 - 2\alpha x - 2\beta y = 0$$

5. (a) Let $x^2 + y^2 + 2gx + 2fy + c = 0 \quad \ldots(\text{i})$

be the equation of the circle passing through three points $P(4, 5)$, $Q(3, 4)$ and $R(5, 2)$. Then,

$$8g + 10f + c + 41 = 0 \quad \ldots(\text{ii})$$
$$6g + 8f + c + 25 = 0 \quad \ldots(\text{iii})$$
$$10g + 4f + c + 29 = 0 \quad \ldots(\text{iv})$$

On subtracting Eq. (iii) from Eq. (ii), we get

$$2g + 2f = -16$$

$$\Rightarrow \quad g + f = -8 \quad \ldots(\text{v})$$

On subtracting Eq. (iii) from Eq. (iv), we get

$$g - f = -1 \quad \ldots(\text{vi})$$

On adding Eqs. (v) and (vi), we get

$$2g = -9$$

$$\Rightarrow \quad g = -\frac{9}{2}$$

On substituting value of g in Eq. (v), we get

$$-\frac{9}{2} + f = -8 \quad \Rightarrow \quad f = -8 + \frac{9}{2}$$

$$\Rightarrow \quad f = \frac{-7}{2}$$

Hence, the coordinates of centre $= (-g, -f) = \left(\frac{9}{2}, \frac{7}{2}\right)$.

6. (b) Equation of circle

$$x^2 + y^2 + 2x\cos\theta + 2y\sin\theta = 8$$

$$x^2 + y^2 + 2x\cos\theta + 2y\sin\theta - 8 = 0$$

Radius $= \sqrt{g^2 + f^2 - c}$

In the given equation,

$$g = \cos\theta,\ f = \sin\theta,\ c = -8$$

Radius $= \sqrt{\cos^2\theta + \sin^2\theta + 8}$

$= \sqrt{1 + 8} \quad [\because \cos^2\theta + \sin^2\theta = 1]$

$= \sqrt{9} = 3$ units

Hence, the radius of circle is 3 units.

7. (c) Let equation of circle be

$$x^2 + y^2 = r^2 \quad \ldots \text{(i)}$$

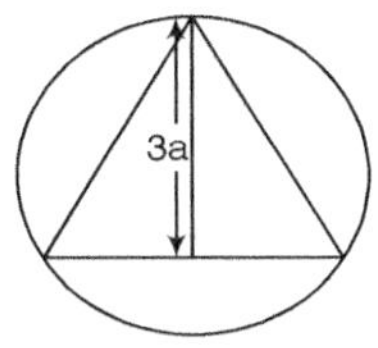

Since, in an equilateral triangle, the centroid coincides with the centre of the circle.

$\therefore$ Radius of circle $= r = \frac{2}{3}(3a) = 2a$

On putting $r = 2a$, in Eq. (i), we get

$$x^2 + y^2 = (2a)^2$$

$$\Rightarrow \quad x^2 + y^2 = 4a^2$$

8. (c) The circle is passing through the lines.

$$x = 0 \quad \ldots \text{(i)}$$

$$y = 0 \quad \ldots \text{(ii)}$$

$$\frac{x}{a} + \frac{y}{b} = 1$$

$$\Rightarrow \quad bx + ay - ab = 0 \quad \ldots \text{(iii)}$$

Let the equation of the circle be

$$x^2 + y^2 + 2gx + 2fy + c = 0$$

From the given equations of lines, the circle is passing through the point O (0, 0), A (a, 0) and B(0, b).

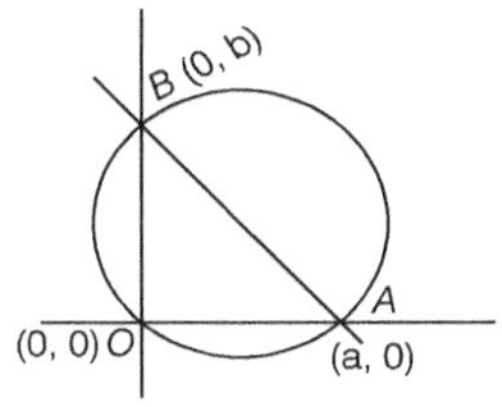

The equations of the circle passing through these three points are

$$c = 0 \quad \ldots \text{(iv)}$$

$$a^2 + 2ag = 0 \quad \ldots \text{(v)}$$

$$b^2 + 2bf = 0 \quad \ldots \text{(vi)}$$

From Eqs. (v) and (vi), we get

$$g = -\frac{a}{2} \text{ and } f = -\frac{b}{2}$$

So, the equation of the circle is

$$x^2 + y^2 + 2\left(-\frac{a}{2}\right)x + 2\left(\frac{-b}{2}\right)y + 0 = 0$$

$$\Rightarrow \quad x^2 + y^2 - ax - by = 0$$

9. (c) The equation of two diameters are $3x + y = 14$ and $2x + 5y = 18$,

where two diameters are intersecting, that point is the centre of the circle.

The intersection point of two lines is (4, 2).

If the circle is passing through origin, then equation of circle is

$$x^2 + y^2 - 2\alpha x - 2\beta y = 0$$

where, (α, β) are coordinates of centre.

So, equation of circle passing through (4, 2).

$$x^2 + y^2 - 2(4)x - 2(2)y = 0$$

$$\Rightarrow \quad x^2 + y^2 - 8x - 4y = 0$$

10. (c) The equation of the lines of quadrilateral

$$L_1 \equiv x - 1 = 0$$

$$L_2 \equiv y - 1 = 0$$

$$L_3 \equiv x - 2 = 0$$

$$L_4 \equiv y - 2 = 0$$

The equation of the curve passing through the vertices of the quadrilateral is

$$L_1L_3 + \lambda L_2L_4 = 0$$

$$\Rightarrow \quad (x - 1)(x - 2) + \lambda(y - 1)(y - 2) = 0$$

$$x^3 - 3x + 2 + \lambda(y^2 - 3y + 2) = 0$$

$$\Rightarrow \quad x^2 + \lambda y^2 - 3x - 3\lambda y + 2 + 2\lambda = 0$$

This is respectively a circle.

So, coefficient of x^2 = coefficient of y^2

$$\Rightarrow \quad \lambda = 1$$

Hence, the required equation of circle is

$$x^2 + y^2 - 3x - 3y + 4 = 0$$

11. (b) Let the equation of the required circle is

$$x^2 + y^2 + 2gx + 2fy + c = 0$$

This passes through A (1, 0) and B (0, 1).

Therefore,

$$1 + 2g + c = 0 \quad \ldots \text{(i)}$$

$$\text{and} \quad 1 + 2f + c = 0 \quad \ldots \text{(ii)}$$

On solving the Eqs. (i) and (ii), we get

$$g = -\left(\frac{c + 1}{2}\right)$$

$$\text{and} \quad f = -\left(\frac{c + 1}{2}\right)$$

Let the radius of the circle be r,

$$\therefore \quad r = \sqrt{g^2 + f^2 - c}$$

$$\Rightarrow \quad r = \sqrt{\left(\frac{c + 1}{2}\right)^2 + \left(\frac{c + 1}{2}\right)^2 - c}$$

$$\Rightarrow \quad r = \sqrt{\frac{2(c + 1)^2 - 4c}{4}}$$

$$\Rightarrow \quad r = \sqrt{\frac{c^2 + 1}{2}}$$

r is minimum, when $c = 0$.

$$\text{Then,} \quad r = \sqrt{\frac{0 + 1}{2}} = \frac{1}{\sqrt{2}}$$

$$\text{For } c = 0, \quad g = -\frac{1}{2} \text{ and } f = \frac{-1}{2}$$

$$x^2 + y^2 + 2\left(\frac{-1}{2}\right)x + 2\left(\frac{-1}{2}\right)y + 0 = 0$$

$$\Rightarrow \quad x^2 + y^2 - x - y = 0$$

12. (a) Let the coordinates of the centre be O (h, k).

The radius is 5 units and the line $3x + 4y = 7$ touches the circle at (1, 1).

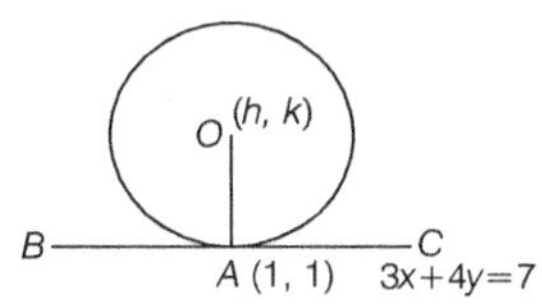

Since, the lines touches the circle, so it is the tangent of the circle.

The distance between the tangent and centre of the circle is equal to the radius.

$$\text{So,} \quad \left|\frac{3h + 4k - 7}{5}\right| = 5$$

$\Rightarrow \quad 3h + 4k - 7 = \pm 25$

$\Rightarrow \quad 3h + 4k = \pm 25 + 7$

$\Rightarrow \quad 3h + 4k = 32 \quad \ldots(i)$

or $\quad 3h + 4k = -18 \quad \ldots(ii)$

Since, OA is perpendicular to BC.

Let slope of $OA = m_1$ and slope of $BC = m_2$

Therefore, $\quad m_1 m_2 = -1$

$$\frac{k-1}{h-1} \times \left(\frac{-3}{4}\right) = -1$$

$\Rightarrow \quad 3(k-1) = 4(h-1)$

$\Rightarrow \quad 3k - 4h = -1 \quad \ldots(iii)$

From Eqs. (i) and (iii), we get

$$h = 4, k = 5$$

From Eqs. (ii) and (iii), we get

$$h = -2, k = -3$$

Hence, the coordinates of the centre are (4, 5) and (– 2, – 3).

13. (d) The centre of the circle lies at the intersection of $2x - 3y = 5$ and $3x - 4y = 7$.

So, the coordinates of centre are (1, – 1).

Area of circle $= \pi r^2 = 154$ sq units

$\Rightarrow \quad r^2 = \dfrac{154}{\pi}$

$\Rightarrow \quad r^2 = \dfrac{154 \times 7}{22}$

$\Rightarrow \quad r = 7$

Hence, the required equation of circle is

$$(x-1)^2 + (y+1)^2 = 7^2$$

$\Rightarrow \quad x^2 - 2x + 1 + y^2 + 2y + 1 = 49$

$\Rightarrow \quad x^2 + y^2 - 2x + 2y = 47$

14. (c) In the circle, AB is a diameter, where the coordinates of A are (p, q) and let the coordinates of B are (x_1, y_1).

Equation of circle in diameter form is

$$(x-p)(x-x_1) + (y-q)(y-y_1) = 0$$

$\Rightarrow x^2 - (p + x_1)x + px_1 + y^2 - (y_1 + q)y + qy_1 = 0$

$\Rightarrow x^2 - (p + x_1)x + y^2 - (y_1 + q)y + px_1 + qy_1 = 0$

Since, the circle touches X-axis.

$\therefore \quad y = 0$

$\Rightarrow \quad x^2 - (p + x_1)x + px_1 + qy_1 = 0$

Discriminate of above equation will be zero because it touches X-axis.

$\therefore \quad b^2 - 4ac = 0$

Here, $\quad b = (p + x_1)$

$a = 1$

$c = (px_1 + qy_1)$

$\Rightarrow \quad (p + x_1)^2 - 4(px_1 + qy_1) = 0$

$\Rightarrow \quad (p + x_1)^2 = 4(px_1 + qy_1)$

$p^2 + x_1^2 + 2px_1 = 4px_1 + 4qy_1$

$\Rightarrow \quad (x_1 - p)^2 = 4qy_1$

Therefore, the locus of the other end is

$$(x-p)^2 = 4qy$$

15. (c) The intersection of the two diameters is the centre of the circle.

So, the coordinates of the centre are (1, – 1).

Circumference of circle $= 2\pi r$

Here, $\quad 2\pi r = 10\pi$

$\Rightarrow \quad r = \dfrac{10\pi}{2\pi}$

$\Rightarrow \quad r = 5$ units

So, the equation of the circle is

$$(x-1)^2 + (y+1)^2 = 5^2$$

$$x^2 + 1 - 2x + y^2 + 1 + 2y = 25$$

$\Rightarrow \quad x^2 + y^2 - 2x + 2y - 23 = 0$

16. (c) The intercept on the line $y = x$, the circle $x^2 + y^2 - 2x = 0$ is AB, that is the diameter of the required circle.

Line and equation of the circle are

$y = x \quad \ldots(i)$

$x^2 + y^2 - 2x = 0 \quad \ldots(ii)$

On solving the Eqs. (i) and (ii), we get

$$x = 0, 1 \text{ and } y = 0, 1$$

These are the coordinates of the diameter.

The equation of the circle in diameter form is

$$(x - x_1)(x - x_2) + (y - y_1)(y - y_2) = 0$$

$\Rightarrow \quad (x-0)(x-1) + (y-0)(y-1) = 0$

$\Rightarrow \quad x^2 + y^2 - x - y = 0$

17. (c) Here, the given equation of circle is

$$x^2 + y^2 = a^2$$

So, radius of the circle $= OB = a$

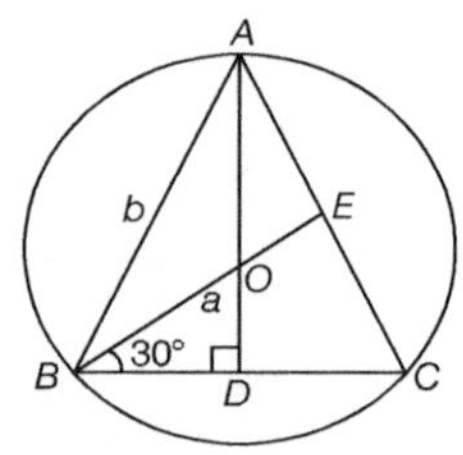

In equilateral ΔABC, let the side is b.

The median AD is the altitude of the triangle.

So, $\quad BD = DC = \dfrac{b}{2}$

In ΔBOD, $\quad \angle ODB = 90° \quad [\because AD \perp BC]$

BE is a median,so

$\angle ABE = \angle EBD = 30°$

$[\because \Delta ABC$ is equilateral$]$

In ΔBOD,

$$\cos 30° = \frac{\left(\frac{b}{2}\right)}{a}$$

$$\frac{\sqrt{3}}{2}a = \frac{b}{2}$$

$\Rightarrow \quad b = \sqrt{3}\, a$ units

18. (d) Let P be image of the origin in the line $x + y = 1$.

Since, $OA = OB$, therefore Q is the mid-point of AB.

$\therefore$ Coordinates of Q are $\left(\frac{1}{2}, \frac{1}{2}\right)$.

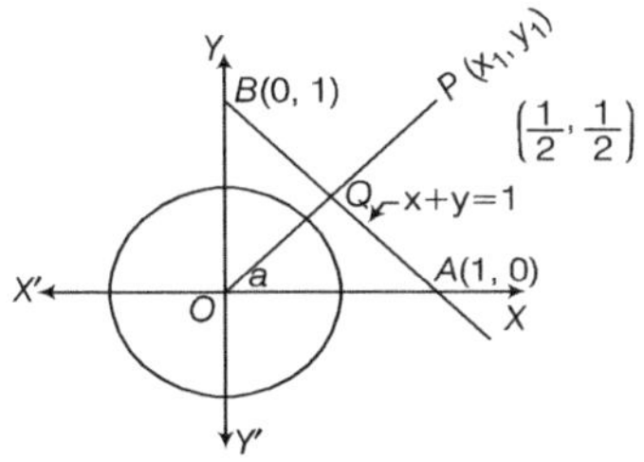

Let the coordinates of P be (x_1, y_1).

Since, Q is the mid-point of OP.

$$\therefore \quad \frac{0 + x_1}{2} = \frac{1}{2}$$

and $$\frac{0 + y_1}{2} = \frac{1}{2}$$

$$\Rightarrow \quad x_1 = 1$$

and $$y_1 = 1$$

$\therefore$ The coordinates of P are (1, 1).

19. (c) The given triangle is an equilateral triangle with side a.

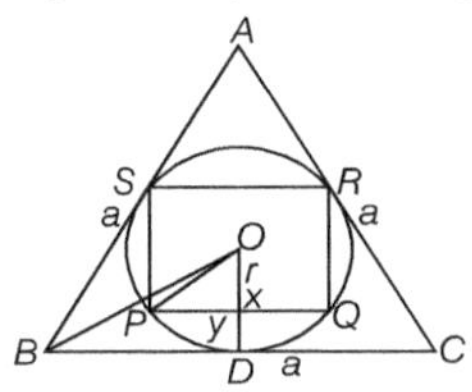

Let the radius of the circle be r,

$$OD = r$$

In ΔOBD, $\quad \angle OBD = 30°$ $\quad$ [since, ΔABC is equilateral]

Here, $$BD = \frac{a}{2}$$

Therefore, $$\tan 30° = \frac{r}{\frac{a}{2}}$$

$$\Rightarrow \quad r = \frac{a}{2\sqrt{3}}$$

$PQRS$ is a square inscribed in the circle.

Let the side $PX = y$

In ΔOPX, $\quad \angle OPX = 45°$

[since, in square every angle is of 90°]

OP is also radius of circle.

$$\therefore \quad OP = \frac{a}{2\sqrt{3}}$$

$$\cos 45° = \frac{y}{\left(\frac{a}{2\sqrt{3}}\right)}$$

$$\frac{1}{\sqrt{2}} \times \frac{a}{2\sqrt{3}} = y$$

$$\Rightarrow \quad y = \frac{a}{2\sqrt{6}}$$

Side of square $PQRS = 2 \times (PX) = 2 \times \frac{a}{2\sqrt{6}} = \frac{a}{\sqrt{6}}$ units

Area of square = Side × Side

$$= \frac{a}{\sqrt{6}} \times \frac{a}{\sqrt{6}} = \frac{a^2}{6} \text{ sq units}$$

20. (c) We have,

$$m_1 = \text{Slope of } OQ = \frac{4}{3}$$

$$m_2 = \text{Slope of } OR = \frac{-3}{4}$$

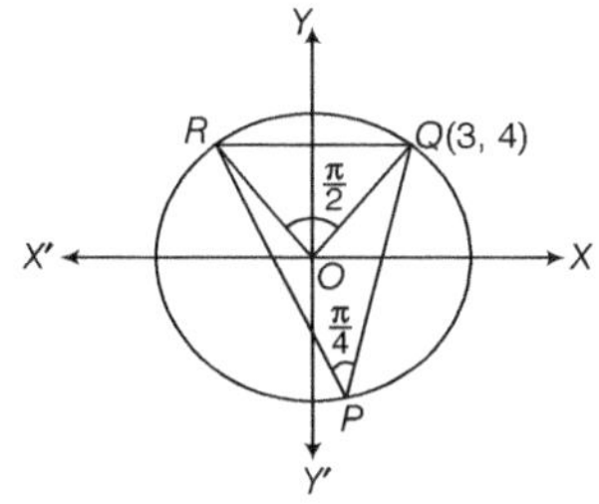

Here, $\quad OQ \perp OR$

$$\therefore \quad m_1 m_2 = -1$$

$$\Rightarrow \quad \angle QOR = \frac{\pi}{2}$$

$$\Rightarrow \quad \angle QPR = \frac{\pi}{4}$$

21. (a) The equation of the lines of square are $x^2 - 8x + 12 = 0$ and $y^2 - 14y + 45 = 0$.

So, on solving these equations, sides of the square are

$$x = 2, x = 6, y = 5 \text{ and } y = 9$$

The coordinates of both ends of the diameter are (2, 5) and (6, 9).

$\therefore$ The coordinates of the centre

$$= \left(\frac{2+6}{2}, \frac{5+9}{2}\right) = (4, 7)$$

22. (d) Let $x = a$, $x = b$, $y = c$ and $y = d$ be the sides of the square. The length of each diagonal of square is equal to the diameter of the circle.

$\therefore$ Diagonal $= 2\sqrt{1 + 4 + 93} = 14\sqrt{2}$

Let l be the length of each side of the square. Then,

$$2l^2 = (\text{Diagonal})^2$$

$$\Rightarrow \quad 2l^2 = (14\sqrt{2})^2$$

$$\Rightarrow \quad l = 14$$

Therefore, each side of the square is at a distance of 7 units from the centre (1, – 2) of the given circle.

This implies

$$a = -6, b = 8, c = -9 \text{ and } d = 5$$

Hence, the vertices of the square are (– 6, – 9), (– 6, 5), (8, – 9) and (8, 5).

23. (b) To find distance OP (i.e. radius of circle touching all the three circles internally)

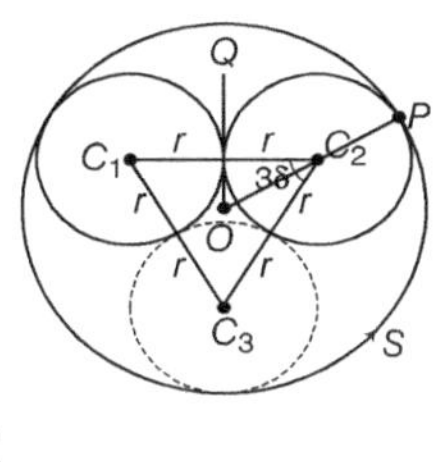

$$OP = OC_2 + r \quad \ldots(i)$$

In ΔOQC_2, $\dfrac{QC_2}{OC_2} = \cos 30°$

$$\Rightarrow \quad OC_2 = r \cdot \frac{2}{\sqrt{3}} \quad \ldots(ii)$$

$$\therefore \quad OP = r + \frac{2r}{\sqrt{3}} = \frac{(\sqrt{3} + 2)r}{\sqrt{3}}$$

24. (b) Let $\angle AOL = \theta$

$$\therefore \quad A \equiv (a \cos \theta, a \sin \theta)$$

$$M \equiv (a + a \cos \theta, a \sin \theta)$$

$$\Rightarrow \quad x = a + a \cos \theta \quad \ldots(i)$$

and $$y = a \sin \theta \quad \ldots(ii)$$

From Eqs. (i) and (ii), we get

$$(x - a)^2 + y^2 = a^2$$

$$\Rightarrow \quad x^2 + y^2 - 2ax = 0$$

$$\Rightarrow \quad x^2 + y^2 = 2ax$$

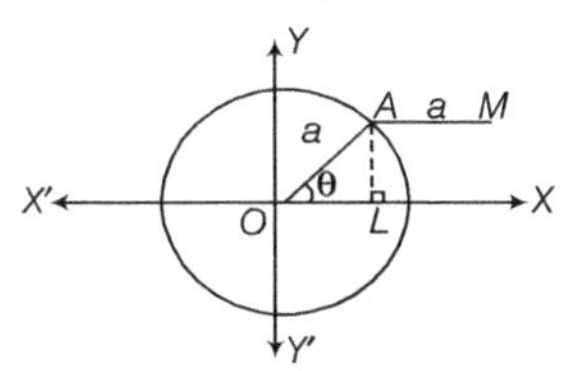

25. (c) The centre of the circle lies at the intersection of lines $3x - 4y - 7 = 0$ and $2x - 3y - 5 = 0$.

So, the coordinates of centre are $(1, -1)$.

$$\text{Area of circle} = 49\pi \qquad \text{(given)}$$

$$\pi r^2 = 49\pi \Rightarrow r^2 = 49$$

$$\Rightarrow r = 7 \text{ units}$$

Hence, the required equation is

$$(x-1)^2 + (y+1)^2 = 7^2$$

$$x^2 - 2x + 1 + y^2 + 2y + 1 = 49$$

$$\Rightarrow x^2 + y^2 - 2x + 2y - 47 = 0$$

26. (c) $\because$ Centre of the circle is mid-point of the line joining end points of diameter.

Given, equation of circle is

$$x^2 + y^2 + 2x + 4y - 3 = 0$$

That can be rewritten as

$$(x+1)^2 + (y+2)^2 = (2\sqrt{2})^2$$

Let coordinates of Q be $Q(\alpha, \beta)$.

Mid-point of $P(1, 0)$ and $Q(\alpha, \beta)$ is the centre of circle.

Coordinates of centre,

$$\frac{1+\alpha}{2} = -1 \text{ and } \frac{0+\beta}{2} = -2$$

$$\Rightarrow \alpha = -3 \text{ and } \beta = -4$$

So, required point is $(-3, -4)$.

27. (d) The equation of the circle touching Y-axis, i.e. $x = 0$ at $(0, 2)$ is

$$(x-0)^2 + (y-2)^2 + \lambda x = 0 \qquad \ldots\text{(i)}$$

It passes through $(-1, 0)$.

$$\therefore (-1-0)^2 + (0-2)^2 + (-1)\lambda = 0$$

$$\Rightarrow 1 + 4 - \lambda = 0$$

$$\Rightarrow \lambda = 5$$

Putting $\lambda = 5$ in Eq. (i), we get

$$x^2 + (y-2)^2 + 5x = 0$$

$$\Rightarrow x^2 + y^2 + 5x - 4y + 4 = 0$$

So, it passes through $(-4, 0)$.

28. (b,c) The centre and radius of the given circle are $\left(\frac{1}{2}, \frac{-3}{2}\right)$ and $\sqrt{\frac{5}{2}}$, respectively.

Let $y = mx$ be the equation of L_1

Then, p_1 = Length of the intercept on L_1

$$\Rightarrow p_1 = 2\sqrt{\left(\sqrt{\frac{5}{2}}\right)^2 - \left(\frac{m+3}{2\sqrt{m^2+1}}\right)^2}$$

$$\Rightarrow p_1 = 2\sqrt{\frac{5}{2} - \frac{(m+3)^2}{4(m^2+1)}}$$

and p_2 = Length of the intercept on L_2

$$\Rightarrow p_2 = 2\sqrt{\left(\sqrt{\frac{5}{2}}\right)^2 - (\sqrt{2})^2}$$

$$\Rightarrow p_2 = 2\sqrt{\frac{5}{2} - 2} = \sqrt{2}$$

Now,

$$p_1 = p_2$$

$$2\sqrt{\frac{5}{2} - \frac{(m+3)^2}{4(m^2+1)}} = \sqrt{2}$$

$$\Rightarrow 5 - \frac{(m+3)^2}{4(m^2+1)} = 1$$

$$\Rightarrow 7m^2 - 6m - 1 = 0$$

$$\Rightarrow (m-1)(7m+1) = 0$$

$$\Rightarrow m = 1, -\frac{1}{7}$$

So, the equation of line L_1 are

$$y - x = 0$$

$$\Rightarrow x - y = 0$$

$$7y = -x$$

$$\Rightarrow x + 7y = 0$$

29. (a,b,c,d) Given equation is

$$\frac{x - x_1}{\cos\theta} = \frac{y - y_1}{\sin\theta} = r$$

If θ is a constant and r is variable, then it represent an equation of the straight line having equation

$$(y - y_1) = \tan\theta\,(x - x_1)$$

which passes through (x_1, y_1) fixed points and having slope $\tan\theta$.

$$\frac{x - x_1}{\cos\theta} = \frac{y - y_1}{\sin\theta} = r$$

It represents the equation of circle in the polar form having equation.

$$(x - x_1)^2 + (y - y_1)^2 = r^2$$

which has coordinates of centre (x_1, y_1) and has radius r.

30. (a,b,c,d)

$x^2 + y^2 - 2x + 4y = 0$ has centre $C(1, -2)$.

Distance of $C(1, -2)$ from $3x - y = 0$

$$\Rightarrow \frac{|3+2|}{\sqrt{9+1}} = \frac{5}{\sqrt{10}} = \sqrt{\frac{5}{2}}$$

Distance of $C(1, -2)$ from $x + 3y = 0$

$$\Rightarrow \frac{|1-6|}{\sqrt{9+1}} = \sqrt{\frac{5}{2}}$$

Distance of $C(1, -2)$ from $x + 3y + 10 = 0$

$$\Rightarrow \frac{|1-6+10|}{\sqrt{1+9}} = \sqrt{\frac{5}{2}}$$

Distance of $C(1, -2)$ from $3x - y - 10 = 0$

$$\Rightarrow \frac{|3+2-10|}{\sqrt{9+1}} = \sqrt{\frac{5}{2}}$$

Since, chords equidistance from the centre are equal.

31. (c,d)

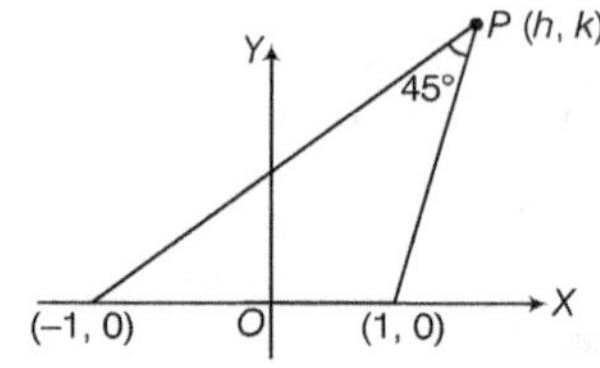

Here, $m_1 = \dfrac{k}{h-1}$ and $m_2 = \dfrac{k}{h+1}$

$$\therefore \tan 45^\circ = \left|\frac{\frac{k}{h-1} - \frac{k}{h+1}}{1 + \frac{k^2}{h^2-1}}\right|$$

$$= \left|\frac{k(h+1-h+1)}{h^2-1+k^2}\right|$$

$$\therefore h^2 + k^2 - 1 = \pm 2k$$

Locus is, $x^2 + y^2 \pm 2y - 1 = 0$

$$\therefore x^2 + (y \pm 1)^2 = (\sqrt{2})^2$$

$\Rightarrow$ Centre $(0, 1)$ or $(0, -1)$ and radius $= \sqrt{2}$.

32. (c) Consider the circle in the first quadrant as $x^2 + y^2 - 2rx - 2ry + r^2 = 0$. It passes through (h, k).

Then, $\quad r^2 - 2r(h + k) + (h^2 + k^2) = 0$

Discriminant, $\Delta = 4(h + k)^2 - 4(h^2 + k^2) \geq 0$

$$\Rightarrow \quad 8hk \geq 0$$

Since, the roots of r are real but may be equal. So, 1 or 2 such circles will exists.

33. (c) As like above we get,

$$x^2 - 2r(a + b) + (a^2 + b^2) = 0$$

We have,

$$|r_1 - r_2| = c$$

$$\Rightarrow \quad (r_1 + r_2)^2 - 4r_1r_2 = c^2$$

$$\Rightarrow \quad 8ab = c^2$$

Using, AM $\geq$ GM

$$\frac{a + b}{2} \geq \sqrt{ab}$$

$$\Rightarrow \quad a + b \geq \frac{c}{\sqrt{2}}$$

34. (b) The distance of the point (p, q) from the line $x \pm y = 0$ is $\frac{|p \pm q|}{\sqrt{2}}$.

Hence, the length of the common chord is

$$\frac{2|p \pm q|}{\sqrt{2}} = \sqrt{2}|p \pm q|$$

35. (d) Let $A\,(h, k)$ be the centre of the circle. Then, $AD \perp PQ$ and $AD = 1$ unit

$$\Rightarrow \quad \frac{2k - 3}{2h - 3\sqrt{3}} \times (-\sqrt{3}) = -1$$

and $\quad \left|\frac{\sqrt{3}\,h + k - 6}{\sqrt{3 + 1}}\right| = 1$

$$\Rightarrow \quad 2\sqrt{3}\,k - 3\sqrt{3} = 2h - 3\sqrt{3}$$

and $\quad |\sqrt{3}h + k - 6| = 2$

$$\Rightarrow \quad 2h - 2\sqrt{3}\,k = 0$$

and $\quad \sqrt{3}h + k - 6 = 2$

or $\quad \sqrt{3}h + k - 6 = -2$

$$\Rightarrow \quad h = \sqrt{3}k$$

and $\quad \sqrt{3}h + k - 6 = -2$

[$\because$ origin and (h, k) lie on the same side of $\sqrt{3}x + y - 6 = 0$]

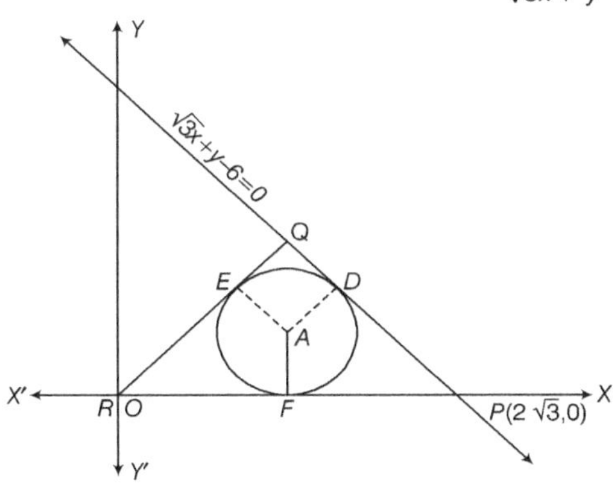

$\because \quad \sqrt{3}h + k - 6 < 0$

$$\Rightarrow \quad h = \sqrt{3}k$$

and $\quad \sqrt{3}h + k = 4$

$$\Rightarrow \quad h = \sqrt{3}, k = 1$$

The equation of circle c is

$$(x - \sqrt{3})^2 + (y - 1)^2 = 1$$

36. (a) Here, R coincides with the origin O.

$\therefore \quad RP = OP = 2\sqrt{3}$

$$\Rightarrow \quad RF = \sqrt{3}$$

Here, coordinates of F are $(\sqrt{3}, 0)$.

$\because \quad RE = RF$

$$\Rightarrow \quad RE = \sqrt{3}$$

Here, the coordinates of E are

$$(RE \cos 60°, RE \sin 60°) \text{ or } \left(\frac{\sqrt{3}}{2}, \frac{3}{2}\right).$$

37. (d) Here, RP is along X-axis.

So, its equation is $y = 0$.

The equation of QR is $y = \sqrt{3}x$.

38. (2) The intercepts made by the circle $x^2 + y^2 - 14x - 10y + 23 = 0$ on the line $7x + y - 4 = 0$ are 2, one meets at X-axis and one meets at Y-axis.

39. (1) Let the point $\left(m_i, \frac{1}{m_i}\right), i = 1, 2, 3, 4$

lie on the circle $x^2 + y^2 + 2gx + 2fy + c = 0$.

Then,

$$m_i^2 + \frac{1}{m_i^2} + 2gm_i + \frac{2f}{m_i} + c = 0, i = 1, 2, 3, 4$$

$$\Rightarrow m_i^4 + 2gm_i^3 + cm_i^2 + 2fm_i + 1 = 0, i = 1, 2, 3, 4$$

m_1, m_2, m_3 and m_4 are roots of the equation.

$$m^4 + 2gm^3 + cm^2 + 2fm + 1 = 0$$

$$\Rightarrow \quad m_1m_2m_3m_4 = \frac{1}{1} = 1$$

40. (4) The minor arc AB has circumference $\frac{2\pi r}{3}$.

[$\because$ line $3x - 4y - 8 = 0$ divides the circumference in the ratio 1 : 2]

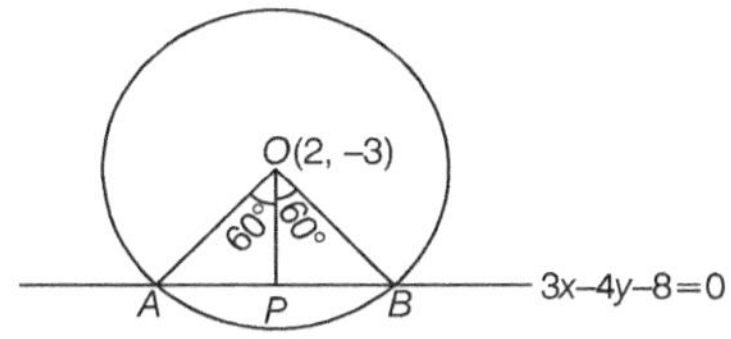

By the sector formula,

$$\frac{\theta}{360°} \times 2\pi r = \frac{2\pi r}{3}$$

$$\Rightarrow \quad \theta = 120°$$

In ΔOAP,

$$\cos 60° = \frac{2}{r}$$

$$\Rightarrow \quad r = 4$$

Hence, radius of the circle is 4 units.

DPP-2 Equation of Tangent, Equation of Normal and Length of Tangent

1. (b) The tangents are parallel to the line $4x + 3y + 5 = 0$

$\therefore$ $2 \times$ Radius = Distance between parallel lines

Here, radius $= \sqrt{g^2 + f^2 - c} = \sqrt{9 + 4 + 12} = 5$

$\because$ Tangents are parallel to $4x + 3y + 5 = 0$

So, equations of tangents are

$$4x + 3y + a_1 = 0$$

and $$4x + 3y + a_2 = 0$$

$$2 \times 5 = \left|\frac{a_1 - a_2}{\sqrt{16 + 9}}\right|$$

$\Rightarrow$ $a_1 - a_2 = 0$

$\therefore$ According to the options, the equations of tangents are

$$4x + 3y - 31 = 0$$

and $$4x + 3y + 19 = 0$$

2. (d) Given equation of the circle is

$$x^2 + y^2 = a^2 \quad \text{...(i)}$$

On differentiating Eq. (i) w.r.t. x

$$2x + 2y\frac{dy}{dx} = 0$$

$\Rightarrow$ $\frac{dy}{dx} = \frac{-2x}{2y}$

$\Rightarrow$ $\frac{dy}{dx} = \frac{-x}{y}$

$\therefore$ $\left(\frac{dy}{dx}\right)_{(a\cos\alpha,\ a\sin\alpha)} = -\cot\alpha$

3. (a) Normal of circle always passes through the centre of the circle i.e. $(20, -5)$ but it also passes through $(4, -1)$.

$\therefore$ Equation of normal,

$$y + 1 = \frac{-5 + 1}{20 - 4}(x - 4)$$

$\Rightarrow$ $4(y + 1) = -(x - 4)$

$\Rightarrow$ $4y + 4 = -x + 4$

$\Rightarrow$ $x + 4y = 0$

4. (a) When, $y = -1$

$$x^2 + 1 - 8x + 2 + 12 = 0$$

$$x^2 - 8x + 15 = 0$$

$\Rightarrow$ $x = 3, 5$

$\therefore$ $P(3, -1)$ and $Q(5, -1)$

$\Rightarrow$ Equation of normal at P and Q are

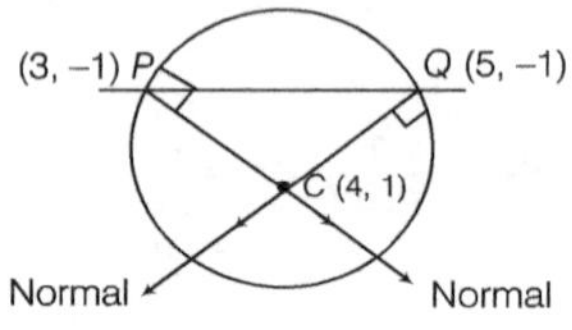

$$\frac{y - 1}{x - 4} = \frac{-1 - 1}{5 - 4}$$

$\Rightarrow$ $y - 1 = -2x + 8$

i.e. $2x + y - 9 = 0$

and $\frac{y - 1}{x - 4} = \frac{-1 - 1}{3 - 4}$

$\Rightarrow$ $y - 1 = 2x - 8$

$\Rightarrow$ $2x - y - 7 = 0$

5. (c) The given equation of line is

$$x + y = 2$$

It can be rewritten as

$$y = -x + 2$$

Slope $= -1$

If the line touches the circle, then

(length of perpendicular from the centre $(c, 0)$ = radius)

$$\left|\frac{(1)(c) - 0 - 2}{\sqrt{1 + 1}}\right| = c$$

$\Rightarrow$ $\left|\frac{c - 2}{\sqrt{2}}\right| = c$

$\Rightarrow$ $c - 2 = \pm c\sqrt{2}$

$\Rightarrow$ $c(1 \mp \sqrt{2}) = 2$

$\Rightarrow$ $c = \frac{2}{1 \mp \sqrt{2}}$

$\Rightarrow$ $c = \frac{1}{1 + \sqrt{2}}, \frac{2}{1 - \sqrt{2}}$

$\Rightarrow$ $c = 2(\sqrt{2} - 1), -2(\sqrt{2} + 1)$

Since, $(1, 0)$ lies inside the circle.

$\Rightarrow$ Distance between the point $(1, 0)$ and centre will be less than radius.

$\therefore$ $\sqrt{(c - 1)^2 + (0 - 0)^2} < c$

$\Rightarrow$ $\sqrt{(c - 1)^2} < c$

$\Rightarrow$ $|c - 1| < c$

$\Rightarrow$ $-c + 1 < c$ or $c - 1 < c$

$\Rightarrow$ $2c > 1$ $\quad -1 > 0$

$\Rightarrow$ $c > 1$

Hence, required value of $c = 2(\sqrt{2} - 1)$.

6. (b) The given equation of line is

$$3x + 4y = m$$

It can be rewritten as $y = \frac{-3x}{4} + \frac{m}{4}$

Here, slope of line, $m_1 = -\frac{3}{4}$

and $c = \frac{m}{4}$

It touches the circle $x^2 + y^2 - 10x = 0$

having centre $(-5, 0)$ and radius 5.

So, the length of the perpendicular from the centre $(-5, 0)$ = radius

$\Rightarrow$ $\left|\frac{m_1 \times (-5) - 0 + c}{\sqrt{m^2 + 1}}\right| = 5$

$\Rightarrow$ $\left|\frac{\frac{-3}{4} \times (-5) - 0 + \frac{m}{4}}{\sqrt{\left(\frac{-3}{4}\right)^2 + 1}}\right| = 5$

$\Rightarrow$ $15 + m = \pm 25$

$\Rightarrow$ $m = +25 + 15$

or $m = -25 + 15$

$\Rightarrow$ $m = 40$

or $m = -10$

7. (a) Given equation of circle is

$$x^2 + y^2 = 25$$

Here, radius of circle = 5 units

Let the equation of the line of tangent is

$$y = mx + c$$

The tangent makes an angle of 60°.

So, slope of the line of tangent,

$$m = \tan 60° = \sqrt{3}$$

Line $f = mx + c$ touches the circle

$$x^2 + y^2 = 5$$

So, the length of the perpendicular from the centre (0, 0) = Radius

$$\therefore \quad \left|\frac{m \times 0 - 0 + c}{\sqrt{m^2 + 1}}\right| = 5$$

$$\Rightarrow \quad \frac{c}{2} = \pm 5$$

$$\Rightarrow \quad c = \pm 0$$

So, the equation of tangent

$$y = \sqrt{3}x \pm 10$$

8. (b) For required circle $P(1, 8)$ and $O(3, 2)$ will be the end point of its diameter.

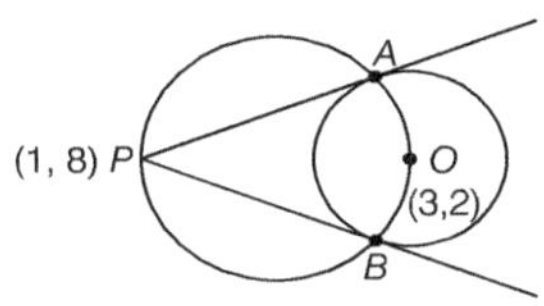

$$\therefore \quad (x - 1)(x - 3) + (y - 8)(y - 2) = 0$$

$$\Rightarrow \quad x^2 + y^2 - 4x - 10y + 19 = 0$$

9. (c) Let the equation of tangent be $y = mx + p$ ...(i)

The line $y = mx + c$ touches the circle $x^2 + y^2 = a^2$, if the length of the perpendicular from the centre (0, 0) = Radius of circle

$$\Rightarrow \quad \left|\frac{m \times 0 - 0 + p}{\sqrt{m^2 + 1}}\right| = a$$

$$\Rightarrow \quad p = \pm a\sqrt{1 + m^2}$$

On substituting the value of p in Eq. (i), we get

$$y = mx \pm a\sqrt{1 + m^2}$$

10. (b) The line $x + 2y = 5$ is the tangent to the circle $x^2 + y^2 = 5$.

The normal is perpendicular to the tangent.

Let the slope of normal be m_1.

The equation of line $x + 2y = 5$ can be rewritten as

$$y = \frac{-x}{2} + \frac{5}{2}$$

Here, slope of line, $m_1 = -\frac{1}{2}$

If lines are perpendicular to each other, then

$$m_1 m_2 = -1$$

$$\Rightarrow \quad m_2 = 2$$

Equation of the normal,

$$(y - 0) = 2(x - 0)$$

$$\Rightarrow \quad y = 2x$$

$$\therefore \quad 2x - y = 0$$

11. (c) $y = x + c$ intersect in two coincident points,

when $c = \pm r\sqrt{1 + m^2}$

$= \pm \sqrt{2}$

12. (b) The length of the chord is $2b$ and the equation of line

$$y = mx + c$$

The radius of circle is a and centre is (0, 0).

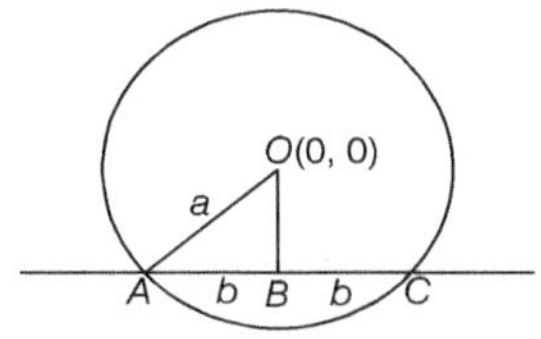

In ΔOAB,

$$OB = \sqrt{a^2 - b^2}$$

[using Pythagoras theorem]

The equation of line OB is given by the distance between OB.

$$\frac{c}{\sqrt{m^2 + 1}} = \sqrt{a^2 - b^2}$$

$$\Rightarrow \quad c^2 = (a^2 - b^2)(m^2 + 1)$$

13. (b) Here, distance $PC = \sqrt{64 + 36} = 10$

$$CQ = r = 5$$

The point nearest to P is $Q(6, 4)$.

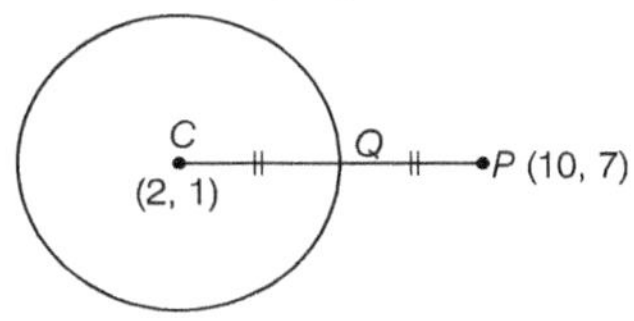

$\therefore$ Tangent at $Q(6, 4)$ is

$$x(6) + y(4) - 2(x + 6) - (y + 4) - 20 = 0$$

$$\Rightarrow \quad 6x + 4y - 2x - 12 - y - 4 - 20 = 0$$

$$\Rightarrow \quad 4x + 3y - 36 = 0$$

14. (d) The radius of the given circle is r.

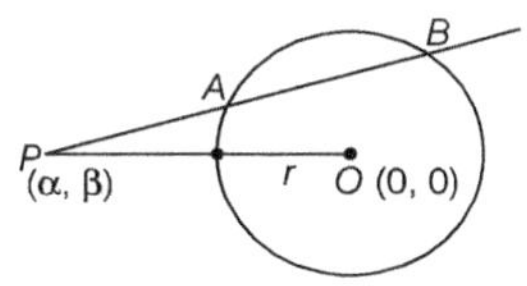

It is given that, $PA \cdot PB = PO^2$

$$PA \cdot PB = \alpha^2 - \beta^2$$

15. (a) Let (p, q) be the point on

$$x^2 + y^2 + 2gx + 2fy + \alpha = 0$$

Then, the length of the tangent from (p, q) to $x^2 + y^2 + 2gx + 2fy + \beta = 0$ is given by

$$l = \sqrt{p^2 + q^2 + 2gp + 2fq + \beta}$$

Since, (p, q) lies on $x^2 + y^2 + 2gx + 2fy + \alpha = 0$

$$\therefore \quad p^2 + q^2 + 2gx + 2fq = -\alpha$$

Hence, $l = \sqrt{\beta - \alpha}$

16. (c) Here, the length of the chord $AB = 2\sqrt{3}$ units

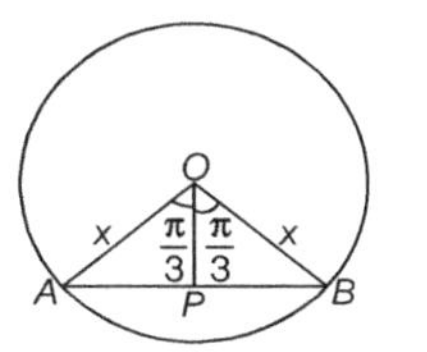

The perpendicular drawn from the centre to the circle divides the chord into two equal parts.

So, $AP = PB = \sqrt{3}$

In ΔOAP, $\angle OAP = 30^\circ$

$$\cos 30^\circ = \frac{x}{\sqrt{3}}$$

$$\Rightarrow \quad \frac{\sqrt{3}}{2} = \frac{x}{\sqrt{3}} \Rightarrow x = 2$$

Area of circle $= \pi r^2 = \pi\,(2)^2$ sq units $= 4\pi$ sq units

17. (c) Let $P(h, k)$ be the mid-point of the chord subtending the angle $\frac{2\pi}{3}$ at the centre.

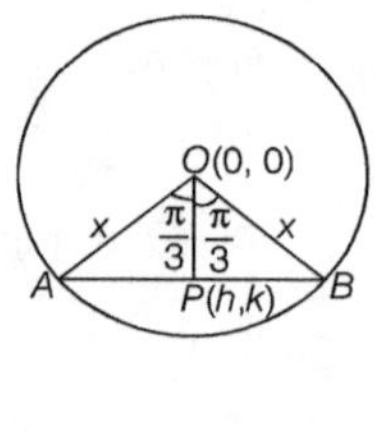

In ΔOPA, we have

$$\cos\frac{\pi}{3} = \frac{OP}{OA}$$

$$\Rightarrow \quad 2OP = OA$$

$$\Rightarrow \quad 4OP^2 = OA^2$$

$$\Rightarrow \quad 4(h^2 + k^2) = 9$$

$\therefore$ Locus of (h, k) is $x^2 + y^2 = \frac{9}{4}$

18. (c) The line $5x - 2y + 6 = 0$ cuts Y-axis at $Q\,(0, 3)$. Clearly, PQ is the length of the tangent drawn from Q on the circle $x^2 + y^2 + 6x + 6y - 2 = 0$

$\therefore \quad PQ = \sqrt{0 + 9 + 6 \times 0 + 6 \times 3 - 2} = 5$ units

19. (a) Clearly, $y = mx - b\sqrt{1 + m^2}$ is a tangent to $x^2 + y^2 = b^2$. It will touch the circle $(x - a)^2 + y^2 = b^2$,

$$\left|\frac{ma - 0 - b\sqrt{1 + m^2}}{\sqrt{m^2 + 1}}\right| = b$$

$$\Rightarrow \quad |ma - b\sqrt{1 + m^2}| = b\sqrt{1 + m^2}$$

$$\Rightarrow \quad ma - b\sqrt{1 + m^2} = \pm\, b\sqrt{1 + m^2}$$

$$\Rightarrow \quad ma = 2b\sqrt{1 + m^2}$$

or $\quad ma = 0$

$$\Rightarrow \quad m^2a^2 = 4b^2 + 4b^2m^2$$

$$\Rightarrow \quad m = \pm\frac{2b}{\sqrt{a^2 - 4b^2}}$$

$$\Rightarrow \quad m = \frac{2b}{\sqrt{a^2 - 4b^2}}$$

20. (d) Equation of tangent to the circle at (1, 7) is

$$x - \left(\frac{y + 7}{2}\right) + 6 = 0$$

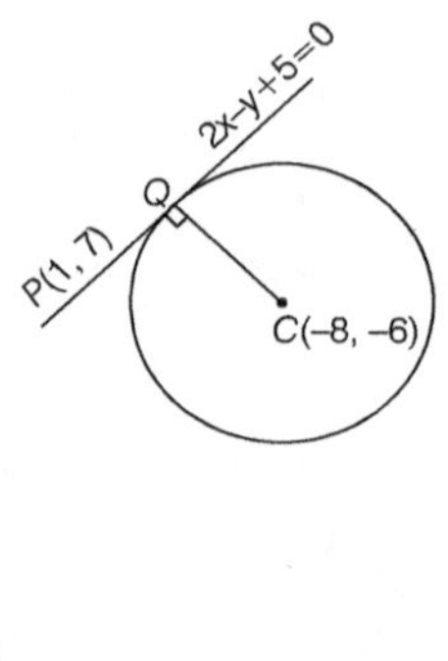

$$\Rightarrow \quad 2x - y + 5 = 0$$

Centre $= (-8, -6)$

Equation of $CQ = x + 2y + k = 0$

$$-8 - 12 + k = 0$$

$$\Rightarrow \quad k = 20$$

$$PQ \equiv 4x - 2y + 10 = 0$$

$$CQ \equiv x + 2y + 20 = 0$$

$$\Rightarrow \quad 5x + 30 = 0$$

$$\Rightarrow \quad x = -6$$

$$\Rightarrow \quad -6 + 2y + 20 = 0$$

$$\Rightarrow \quad y = -7$$

$\therefore$ Coordinates of Q are $(-6, -7)$.

21. (b) Let $Q = \frac{\pi}{2k}$

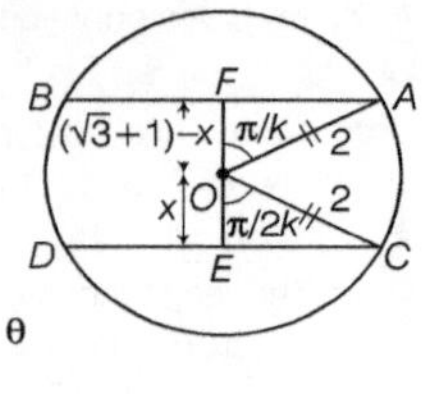

$$\therefore \quad \cos\theta = \frac{x}{2}$$

and $\quad \cos 2\theta = \frac{(\sqrt{3} + 1) - x}{2}$

$$\Rightarrow \quad \cos 2\theta = \frac{\sqrt{3} + 1}{2} - \cos\theta$$

$$\Rightarrow \quad \cos 2\theta + \cos\theta = \frac{\sqrt{3}}{2} + \frac{1}{2}$$

$$\therefore \quad \theta = 30^\circ = \frac{\pi}{2k}$$

$$\Rightarrow \quad \frac{\pi}{2k} = \frac{\pi}{6} \Rightarrow k = 3$$

22. (c) Let the centre be $(3, k)$ and radius r.

Clearly, equation of circle is

$$(x - 3)^2 + (y - k)^2 = r^2 \quad \ldots\text{(i)}$$

$\because$ Circle passes through (3, 0), therefore we have

$$(3 - 3)^2 + (0 - k)^2 = r^2$$

$$\Rightarrow \quad k^2 = r^2$$

$$\Rightarrow \quad k = -r$$

[$\because$ centre lies in the IV quadrant]

Now, Eq. (i) becomes

$$(x - 3)^2 + (y + r)^2 = r^2$$

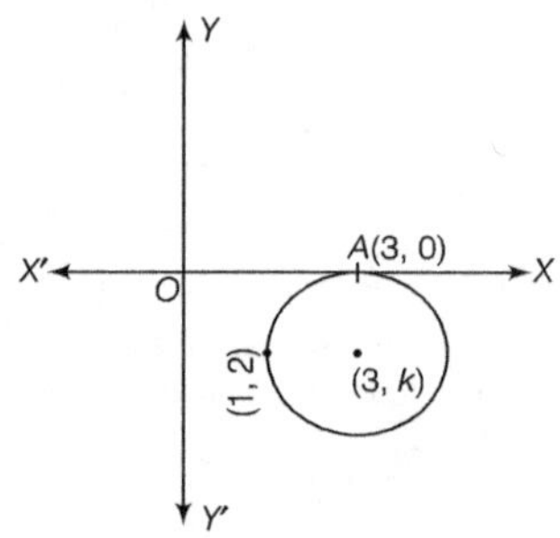

$\because$ Circle also passes through $(1, -2)$ therefore, we have

$$4 + (r - 2)^2 = r^2$$

$$\Rightarrow \quad 4 + r^2 + 4 - 4r = r^2$$

$$\Rightarrow \quad r = 2$$

$$\therefore \quad k = -2$$

Now, equation of circle is

$$(x - 3)^2 + (y + 2)^2 = 4$$

$$\Rightarrow \quad x^2 + 9 - 6x + y^2 + 4y + 4 = 4$$

$$\Rightarrow \quad x^2 + y^2 - 6x + 4y + 9 = 0$$

Clearly, it passes through $(5, -2)$.

23. (a) Here, $C(h, k)$ lies on $x - 2y = 4$

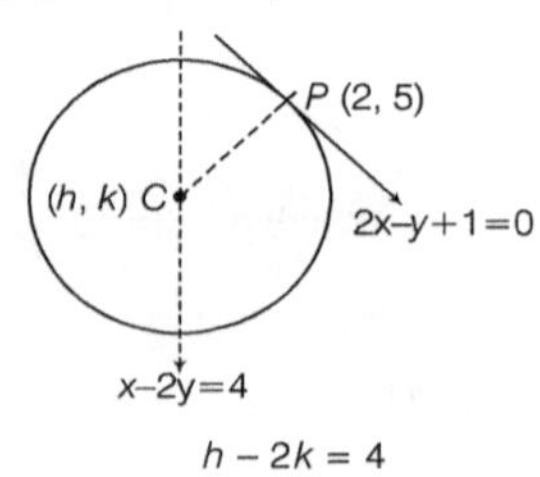

$$\Rightarrow \quad h - 2k = 4 \quad \ldots\text{(i)}$$

Also, CP is perpendicular to $2x - y + 1 = 0$

$$\Rightarrow \quad \frac{k-5}{h-2} \times 2 = -1$$

$$\Rightarrow \quad 2k - 10 = -h + 2$$

or $$h + 2k = 12 \quad \text{...(ii)}$$

From Eqs.(i) and (ii), we get

$$h = 8 \text{ and } k = 2$$

$$\therefore \quad \text{Radius} = \sqrt{(8-2)^2 + (2-5)^2}$$
$$= \sqrt{36 + 9} = \sqrt{45}$$
$$= 3\sqrt{5} \text{ units}$$

24. (a) Since, the tangents are perpendicular.
So, locus of perpendicular tangents of circle $x^2 + y^2 = 169$ is a director circle having equation $x^2 + y^2 = 338$.

25. (c) Equation of circle C is

$$(x+3)^2 + (y-5)^2 = 9 + 25 - 30 = 4$$
$$\Rightarrow \quad (x+3)^2 + (y-5)^2 = 2^2$$

Centre = (3, – 5)

If L_1 is a diameter, then

$$2(3) + 3(-5) + p - 3 = 0$$
$$\Rightarrow \quad p = 12$$

$\therefore$ L_1 is $\quad 2x + 3y + 9 = 0$

and L_2 is $\quad 2x + 3y + 15 = 0$

Distance of centre of circle from L_2

$$= \left|\frac{2(3) + 3(-5) + 15}{\sqrt{2^2 + 3^2}}\right| \quad \text{(radius of circle)}$$
$$= \frac{6}{\sqrt{13}} < 2$$

$\therefore$ L_2 is a chord of circle C.

Statement II is incorrect.

26. (c) The equation of the circle passing through the points (-3, 5) and (5, – 1) is

$$(x+3)(x-5) + (y-5)(y+1) = 0$$
$$\Rightarrow \quad x^2 - 2x - 15 + y^2 - 4y - 5 = 0$$
$$\Rightarrow \quad (x-1)^2 + (y-2)^2 = (5)^2$$

Here $a = 1, \; b = 2$ and $r = 5$

Then, $a + b + r = 1 + 2 + 5 = 8$

27. (a)

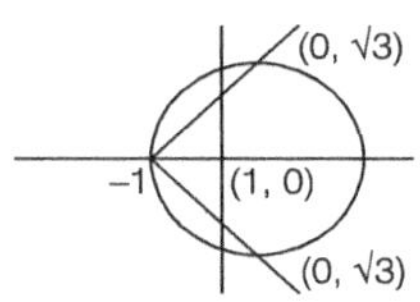

From the figure,

There are three points of intersection.

Hence, correct answer is 3.

Alternate Method

Director's circle, is

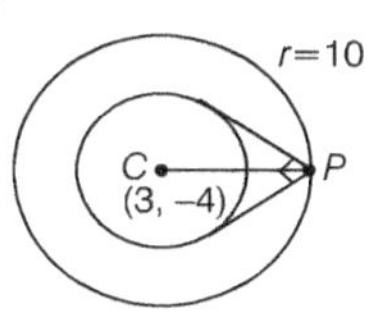

$$(x-3)^2 + (y+4)^2 = 2(10)^2 \quad \text{...(i)}$$

(since, tangent drawn from any point on director's circle subtends 90°)

Here, $P(13, 6)$ lie on (i)

as $(13-3)^2 + (6+4)^2 = 2(10)^2$

$\therefore$ Angle between the tangents drawn from the point $P(13, 6)$ is 90°.

28. (c) Let the mid-point of the chord AB be $C(h, k)$.

In ΔCOB,

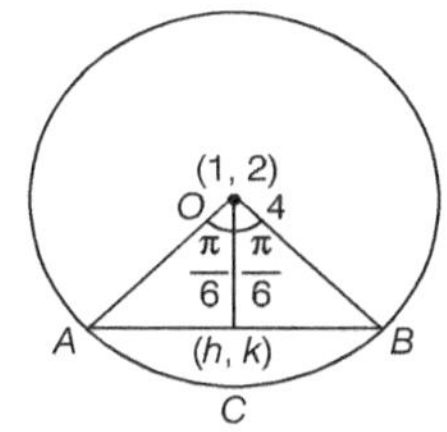

$$\cos 30° = \frac{\sqrt{(h-1)^2 + (k-2)^2}}{4}$$
$$(h-1)^2 + (k-2)^2 = 16 \times \frac{3}{4}$$
$$h^2 + 1 - 2h + k^2 + 4 - 4k = 12$$
$$h^2 + k^2 - 2h - 4k - 7 = 0$$
$$x^2 + y^2 - 2x - 4y - 7 = 0$$

So, the correct option is (c).

29. (d) Since, $S_1 < 0$

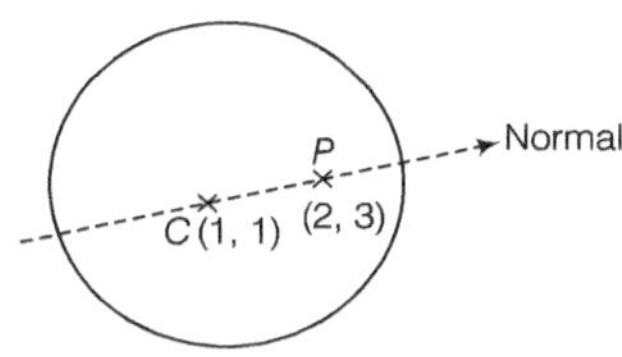

i.e. $\quad 4 + 9 - 4 - 6 - 23 < 0$

$\therefore$ P lies inside the circle and one normal is possible.

30. (b) From figure, $A \cap B \cap C$ is 1.

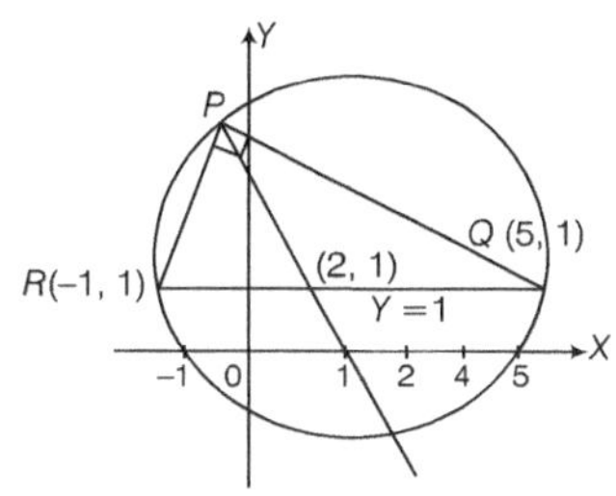

31. (c) When $y = 1$

$$x^2 - 4x - 5 = 0$$
$$\Rightarrow \quad (x-5)(x+1) = 0$$
$$\therefore \quad x = -1, 5$$

Thus, $(x+1)^2 + (y-1)^2 + (x-5)^2 + (y-1)^2 = QR^2 = 36$

32. (c) Equation of director's circle is

$$(x-2)^2 + (y-1)^2 = (3\sqrt{2})^2 = 18$$
$$\text{Area} = \pi(r_1^2 - r_2^2)$$
$$= \pi(18 - 9)$$
$$= 9\pi$$

33. (a) QR is chord of contact $x + 1 = 0$

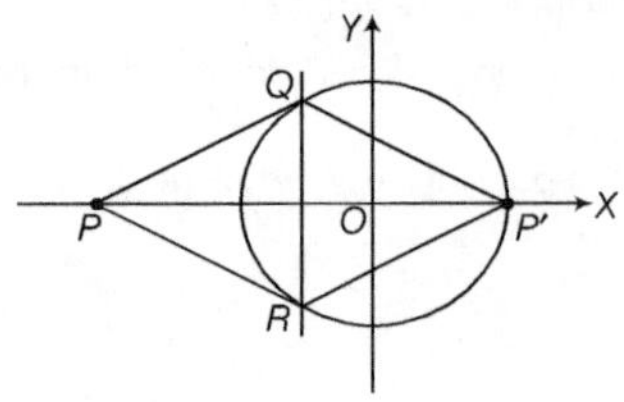

Circumcircle of PQR is

$$S + \lambda L = 0$$

$\Rightarrow \quad x^2 + y^2 - 4 + \lambda(x + 1) = 0$

passes through $P(-4, 0)$.

$\therefore \quad \lambda = 4$

$\therefore$ Circumcentre $(-2, 0)$

34. (d) $P'(2, 0)$ such that

$PQ\,P'R$ is a rhombus.

$\Delta PQP' : \Delta P'QR = 1:1$

35. (a) Area of rhombus $PQP'R = 6\sqrt{3}$ sq units

36. (5) $y = 2x$ and $y = 3x$ are tangents.

$$\Rightarrow \quad \tan\theta = \frac{3-2}{1+3(2)} = \frac{1}{7} \quad \ldots(i)$$

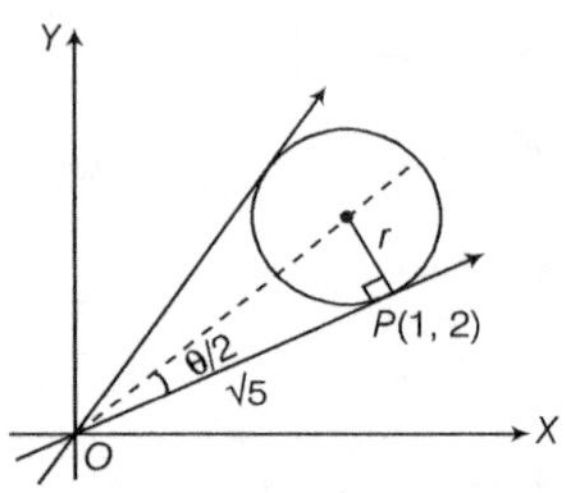

$$\therefore \quad \tan\frac{\theta}{2} = \frac{r}{\sqrt{5}}$$

$$\Rightarrow \quad r = \sqrt{5}\tan\frac{\theta}{2} \quad \ldots(ii)$$

From Eqs. (i) and (ii), we get

$$\tan\theta = \frac{2\tan\frac{\theta}{2}}{1-\tan^2\frac{\theta}{2}}$$

$$\Rightarrow \quad \frac{1}{7} = \frac{2\cdot\frac{r}{\sqrt{5}}}{1-\frac{r^2}{5}}$$

$\Rightarrow \quad r^2 + 14\sqrt{5}r - 5 = 0$

$\therefore \quad r = \sqrt{250} - \sqrt{245}$

Thus, $\quad \alpha - \beta = 5$

37. (1)

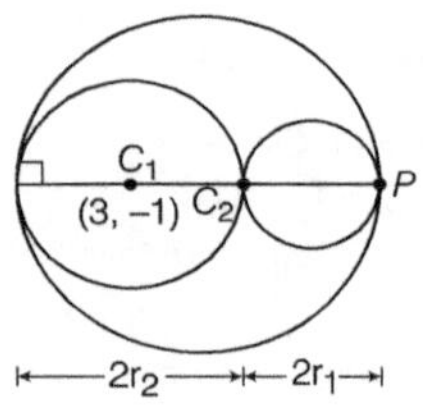

From figure, $\quad 2(r_1 + r_2) = 6$

$\Rightarrow \quad r_1 + r_2 = 3 \quad \ldots(i)$

Also, $\quad 2r_2 - 2\sqrt{2} = 3$

$$\Rightarrow \quad r_2 = \frac{3+2\sqrt{2}}{2}$$

$$\therefore \quad r_1 = \frac{3-2\sqrt{2}}{2}$$

$$\Rightarrow \quad (r_1 + r_2)^2 - (r_2 - r_1)^2 = 4r_1r_2 = 4\cdot\left(\frac{1}{4}\right) = 1$$

38. (8) Equation of circle having $(-3, 5)$ and $(5, -1)$ as end points of diameter, is

$$(x + 3)(x - 5) + (y - 5)(y + 1) = 0$$

$\Rightarrow \quad x^2 - 2x + y^2 - 4y - 20 = 0$

$\Rightarrow \quad (x-1)^2 + (y-2)^2 = 25 = 5^2$

$\therefore \quad a + b + r = 8$

39. (3) The graph of $|y| = x + 1$ is shown as

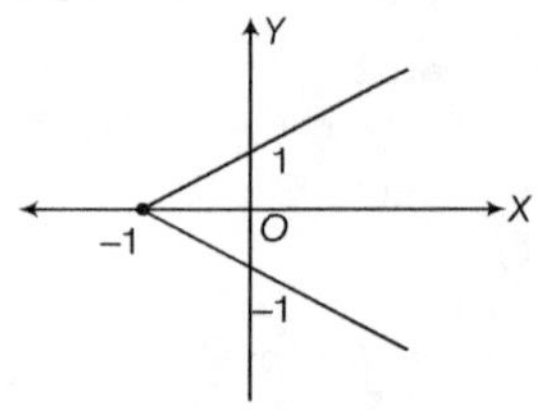

and equation of circle $(x-1)^2 + y^2 = 4$, is

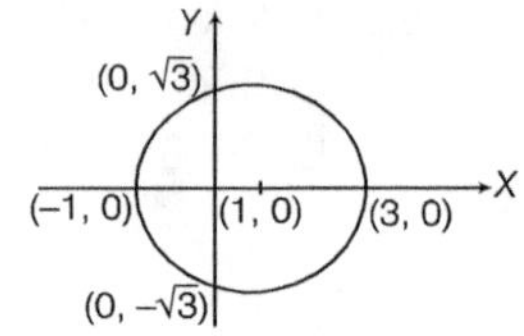

$\therefore$ Number of solutions is point of intersection of (i) and (ii).

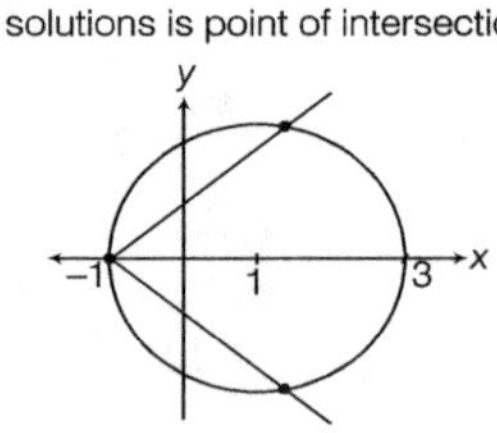

Hence, 3 solutions.

40. (2) As, $\quad x^2 = y^2$

$\Rightarrow \quad y = \pm x$

and $\quad (x - A)^2 + y^2 = 1$

has centre on X-axis.

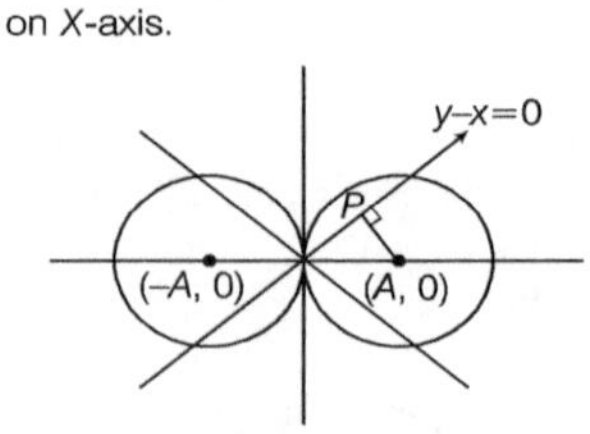

Thus, $\quad AP < r$

$$\Rightarrow \quad \frac{|A|}{\sqrt{2}} < 1$$

$\Rightarrow \quad |A| < \sqrt{2}$

i.e. $\quad -\sqrt{2} < A < \sqrt{2}$

$\Rightarrow A = \{-1, 0, 1\}$ but $A \neq 0$ as only two solutions.

Thus, A can take only 2 values.

DPP-3 **Angle between Circles, Radical Axis and Length of Tangent**

1. (b) Equation of two circles $x^2 + y^2 - 5x - 3 = 0$ ...(i)

and $3x^2 + 3y^2 + 2x + 4y - 6 = 0$

$\Rightarrow \quad x^2 + y^2 + \frac{2}{3}x + \frac{4}{3}y - 2 = 0$ ...(ii)

Since, length of tangent is equal

$$\sqrt{S_1} = \sqrt{S_2}$$

$\Rightarrow \quad S_1 = S_2$

$\Rightarrow x_1^2 + y_1^2 - 5x_1 - 3 = x_1^2 + y_1^2 + \frac{2}{3}x_1 + \frac{4}{3}y_1 - 2$

$\Rightarrow \quad -5x_1 - 3 - \frac{2}{3}x_1 - \frac{4}{3}y_1 + 2 = 0$

$\Rightarrow \quad 17x_1 + 4y_1 + 3 = 0$

$\therefore$ Locus of point $17x + 4y + 3 = 0$

2. (d) The equation of the circle having centre $(2, -2)$ is

$$S_1 \equiv (x-2)^2 + (y+2)^2 = 0$$

$\Rightarrow \quad S_1 \equiv x^2 + y^2 - 4x + 4y + 8 = 0$ ...(i)

$S_2 \equiv x^2 + y^2 - 2x + 3y + \frac{1}{3} = 0$ ...(ii)

The radical axis of two circles, $S_1 = 0$ and $S_2 = 0$ is given by $S_1 - S_2 = 0$

$$(-4x + 4y + 8) - \left(-2x + 3y + \frac{1}{3}\right) = 0$$

$\Rightarrow \quad -2x + y + \frac{23}{3} = 0$

$\Rightarrow \quad -6x + 3y + 23 = 0$

$\Rightarrow \quad 6x - 3y - 23 = 0$

3. (c) The given equations of the circles are

$$(x-p)^2 + y^2 = p^2$$

$\Rightarrow \quad x^2 + y^2 - 2px = 0$ ...(i)

$$x^2 + (y-q)^2 = q^2$$

$\Rightarrow \quad x^2 + y^2 - 2qx = 0$ ...(ii)

and $x^2 + y^2 = p^2 + q^2$ ...(iii)

From Eqs. (i) and (iii), we get

$$2px = p^2 + q^2$$

$\Rightarrow \quad x = \frac{p^2 + q^2}{2p}$

From Eqs. (ii) and (iii), we get

$$2qy = p^2 + q^2$$

$\Rightarrow \quad y = \frac{p^2 + q^2}{2q}$

$\therefore$ Coordinates of centre $\left(\frac{p^2+q^2}{2p}, \frac{p^2+q^2}{2q}\right)$.

4. (b) Let $S_1 \equiv x^2 + y^2 - 4 = 0$

$S_2 \equiv x^2 + y^2 - 8x + 12 = 0$

The equation of common tangent of the two circles is

$$S_1 - S_2 = 0$$

$$x^2 + y^2 - 4 - (x^2 + y^2 - 8x + 12) = 0$$

$\Rightarrow \quad -4 + 8x - 12 = 0$

$\Rightarrow \quad 8x = 16$

$\Rightarrow \quad x = 2$

5. (b) The radical axis of two circles is always perpendicular to the line joining the centres of the circles.

6. (c) Since, tangents from $P(\alpha, \beta)$ to both the circles are equal. Then, radical axis of two circles is given by

$$S_1 - S_2 = 0$$

$\Rightarrow \quad (x^2 + y^2 - 4x - 5) - (x^2 + y^2 + 6x - 2y + 6) = 0$

$\Rightarrow \quad -10x + 2y - 11 = 0$

It passes through the point $P(\alpha, \beta)$.

So, $-10\alpha + 2\beta - 11 = 0$

$\Rightarrow \quad 10\alpha - 2\beta + 11 = 0$

7. (c) Both the circles have same radius

i.e. $C_1P = C_2P = r$

where, C_1 and C_2 are their centres.

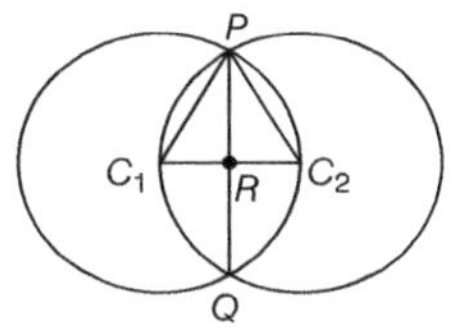

PQ is the common chord of both the circles and $C_1C_2 = r$.

In ΔC_1PR,

$$C_1P = r, C_1R = \frac{C_1C_2}{2} = \frac{r}{2}$$

Using Pythagoras theorem in ΔC_1PR,

$$C_1P^2 = PR^2 + C_1R^2$$

$\Rightarrow \quad PR = \sqrt{r^2 - \left(\frac{r}{2}\right)^2}$

$\Rightarrow \quad PR = \frac{\sqrt{3}r}{2}$

$\therefore \quad PQ = 2PR = 2 \times \frac{\sqrt{3}r}{2} = \sqrt{3}r$

8. (c) The given equations of the circle are

$x^2 + y^2 = a^2$ ...(i)

$(x-c)^2 + y^2 = a^2$ ...(ii)

$x^2 + (y-b)^2 = a^2$ ...(iii)

From Eqs. (i) and (ii) of radical axis, we get

$$c^2 - 2xc = 0$$

$\Rightarrow \quad x = \frac{c}{2}$

From Eqs. (iii) and (i) of radical axis, we get

$$b^2 - 2by = 0$$

$\Rightarrow \quad y = \frac{b}{2}$

So, the coordinates of centre $\left(\frac{c}{2}, \frac{b}{2}\right)$.

9. (b) The given equations of the circles are

$$S_1 \equiv x^2 + y^2 + x - y + 2 = 0$$

and $S_2 \equiv x^2 + y^2 - \frac{4}{3}x - 4 = 0$

The radical axis of two circles $S_1 = 0$ and $S_2 = 0$ is given by

$$S_1 - S_2 = 0$$

$\Rightarrow \quad (x - y + 2) - \left(\frac{4}{3}x - 4\right) = 0$

$\Rightarrow \quad x + \frac{4}{3}x - y + 2 + 4 = 0$

$\Rightarrow \quad 7x - 3y + 18 = 0$

10. (c) The required point is the radical centre of the three given circles.

The radical axes of these three circles in pairs are

$$3x - 24 = 0, 16y + 120 = 0$$

and $$-3x + 16y + 80 = 0$$

Solving any two of these three, we get

$$x = 8, y = -\frac{15}{2}$$

∴ The required point is $\left(8, \frac{-15}{2}\right)$.

11. (c) The equation of the common chord of the given circles is

$$5ax + (c - d)y + (a + 1) = 0 \quad \text{...(i)}$$

(using $S_1 - S_2 = 0$)

But the equation of the common chord is given as

$$5x + by - a = 0 \quad \text{...(ii)}$$

On comparing Eqs. (i) and (ii), we get

$$\frac{5a}{5} = \frac{c-d}{b} = \frac{a+1}{-a}$$

$$\Rightarrow \frac{a}{1} = \frac{c-d}{b} = -\frac{a+1}{a}$$

$$\Rightarrow a = \frac{-(a+1)}{a}$$

$$\Rightarrow a^2 + a + 1 = 0$$

Clearly, $a^2 + a + 1 = 0$ has no real roots. So, there is no real value of a for which $5x + by - a = 0$ passes through P and Q.

12. (c) The point of intersection of two circles $x^2 + y^2 + 5x - 8y + 1 = 0$ and $x^2 + y^2 - 3x + 7y - 25 = 0$ will be the common chord of these circles.

The common chord of circle is given by

$$S_1 - S_2 = 0$$

$$\Rightarrow 8x - 15y + 26 = 0$$

The centre of the given circle $x^2 + y^2 - 2x = 0$ is (1, 0).

The distance of $8x - 5y + 26 = 0$ from point (1, 0) is

$$\left|\frac{8 + 26}{\sqrt{64 + 225}}\right| = \frac{34}{17} = 2 \text{ units}$$

13. (a) Given circles will intersect orthogonally, if

$$2(1 \times 0 + k \times k) = 6 + k$$

(using, $2(g_1 g_2 + f_1 f_2) = c_1 + c_2$)

$$\Rightarrow 2k^2 - k - 6 = 0$$

$$\Rightarrow (2k + 3)(k - 3) = 0$$

$$\therefore k = 2, -\frac{3}{2}$$

14. (c) If two circles touch each other orthogonally, then

$$2(g_1 g_2 + f_1 f_2) = c_1 + c_2$$

Here, (g_1, f_1) is (3, 4)

and (g_2, f_2) is (7, 8)

Then, $2(21 + 32) = c_1 + c_2$

$$\Rightarrow c_1 + c_2 = 106 \quad \text{...(i)}$$

Since, radius of both circles are equal.

So, $r_1 = \sqrt{g_1^2 + f_1^2 - c_1}$

and $r_2 = \sqrt{g_2^2 + f_2^2 - c_2}$

$$r_1 = r_2$$

$$\Rightarrow r_1^2 = r_2^2$$

$$\Rightarrow -c_1 + c_2 = 88 \quad \text{...(ii)}$$

On solving Eqs. (i) and (ii), we get

$$c_1 = 9 \text{ and } c_2 = 97$$

Hence, $r_1 = \sqrt{9 + 16 - 9} \Rightarrow r_1 = 4$ units

15. (a) Since, circle touches the X-axis and also touches circle with the centre at (0, 3) and radius 2. Then,

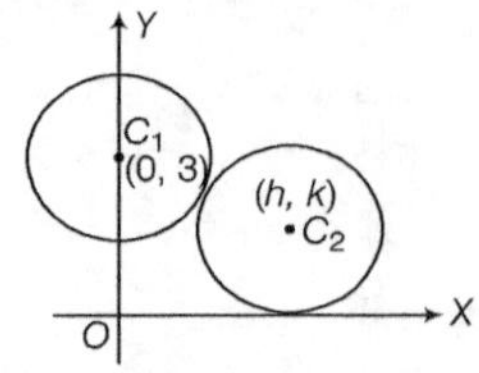

$$C_1 C_2 = r_1 + r_2$$

$$\therefore h^2 + (k - 3)^2 = (|k| + 2)^2$$

$$\Rightarrow h^2 + k^2 + 9 - 6k = k^2 + 4 + 4|k|$$

∴ Locus of centre of circle is

$$x^2 = -5 + 6y + 4|y|$$

$$\Rightarrow x^2 = 10y - 5 \quad (\because y > 0)$$

This equation represents a parabola. Thus, locus of the centre of the circle is a parabola.

16. (c) Let the circle be $x^2 + y^2 + 2gx + 2fy + c = 0$. ...(i)

It cuts the circle $x^2 + y^2 = 4$ orthogonally.

Then, $2(g_1 g_2 + f_1 f_2) = c_1 + c_2$

$$\Rightarrow 2(g \cdot 0 + f \cdot 0) = c - 4 \quad \Rightarrow c = 4$$

and it passes through (1, 2).

$$\therefore 1 + 4 + 2g + 4f + 4 = 0$$

$$2g + 4f + 9 = 0$$

So, locus of $(-g, -f)$ is

$$-2x - 4y + 9 = 0$$

$$\Rightarrow 2x + 4y - 9 = 0$$

17. (a) Let the circle be $x^2 + y^2 + 2gx + 2fy + c = 0$ ...(i)

and given circle is $x^2 + y^2 = p^2$ ...(ii)

Both circles cut orthogonally, then

$$2g_1 g_2 + 2f_1 f_2 = c_1 + c_2$$

$$\Rightarrow 2 \cdot (g \cdot 0 + f \cdot 0) = c - p^2$$

$$\Rightarrow c = p^2$$

So, $c = p^2$ and it passes through (a, b).

$$\therefore a^2 + b^2 + 2ga + 2fb + p^2 = 0$$

So, locus of $(-g, -f)$ is

$$a^2 + b^2 - 2ax - 2by + p^2 = 0$$

$$\Rightarrow 2ax + 2by - (a^2 + b^2 + p^2) = 0$$

18. (a) Clearly, the centre of the circles is the point of intersection of the line $x - 2y = 4$ and normal at P i.e. a line passing through P and perpendicular to the tangent at P.

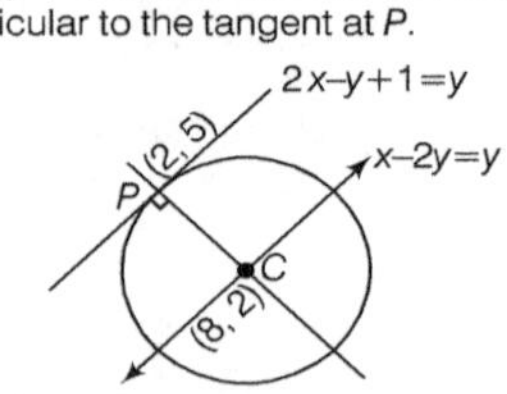

The equation of CP is

$$(y - 5) = -\frac{1}{2}(x - 2)$$

$$\Rightarrow x + 2y = 12 \quad \text{...(i)}$$

Solving this equation with $x - 2y = 4$.

We obtain the coordinates of centre C (8, 2).

$$\text{Radius} = CP = \sqrt{(8-2)^2 + (2-5)^2}$$

$$= \sqrt{(36 + 9)} = \sqrt{45} = 3\sqrt{5} \text{ units}$$

19. (d) Let the equation of the circle be

$$x^2 + y^2 + 2gx + 2fy + c = 0 \quad \ldots(i)$$

It cuts the circles

$$x^2 + y^2 + 4x - 6y + 9 = 0 \quad \ldots(ii)$$

and $\quad x^2 + y^2 - 5x + 4y - 2 = 0$ orthogonally. $\ldots$(iii)

By using $2(g_1 g_2 + f_1 f_2) = c_1 + c_2$ in Eqs. (i) and (ii), we get

$$g \cdot 2 + f \cdot (-3) = c + 9$$

$$\Rightarrow \quad 2g - 3f - c = 9 \quad \ldots(iv)$$

Again, in Eqs. (i) and (iii), we get

$$g \cdot \left(\frac{-5}{2}\right) + f \cdot (2) = c - 2$$

$$\Rightarrow \quad -\frac{5g}{2} + 2f - c = -2 \quad \ldots(v)$$

From Eqs. (iv) and (v), we get

$$\Rightarrow \quad 9g - 10f = 22$$

$$\Rightarrow \quad 9g - 10f - 22 = 0$$

On taking locus of $(-g, -f)$,

$$-9x + 10y - 22 = 0$$

$$\Rightarrow \quad 9x - 10y + 22 = 0$$

20. (d) The coordinates of the centre and radius of given circle are $(-2, 3)$ and $\sqrt{4 + 9 - 9\sin^2\alpha - 13\cos^2\alpha} = 2\sin\alpha$.

Let the coordinates of P be (h, k).

Clearly, CP bisects $\angle TPT' = 2\alpha$

$\therefore \quad \angle CPT = \angle CPT' = \alpha$

Now, in ΔCPT, we have

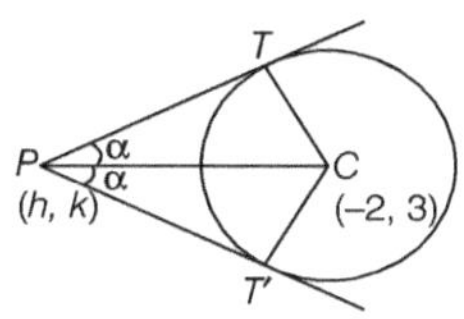

$$\sin\alpha = \frac{CT}{CP}$$

$$\Rightarrow \quad \sin\alpha = \frac{2\sin\alpha}{\sqrt{(h-2)^2 + (k-3)^2}}$$

$$\Rightarrow \quad (h+2)^2 + (k-3)^2 = 4$$

$$\Rightarrow h^2 + k^2 + 4h - 6k + 9 = 0$$

Hence, the locus of (h, k) is

$$x^2 + y^2 + 4x - 6y + 9 = 0$$

21. (b) The equations of the circles are

$$(x-1)^2 + y^2 = 10 \quad \ldots(i)$$

$$x^2 + (y-2)^2 = 5 \quad \ldots(ii)$$

The centres of these circles are C_1 (1, 0) and C_2 (0, 2) respectively. Let r_1 and r_2 be their radii.

Then, $\quad r_1 = \sqrt{10}$ and $r_2 = \sqrt{5}$

$$C_1 C_2 = \sqrt{(1-0)^2 + (0-2)^2} = \sqrt{5}$$

Suppose circles (i) and (ii) intersect at P. Then,

$$\cos\angle C_1 P C_2 = \frac{C_1P^2 + C_2P^2 - C_1C_2^2}{2C_1 \cdot P \cdot C_1P}$$

$$= \frac{10 + 5 - 5}{2 \times \sqrt{50}}$$

$$= \frac{10}{2 \times 5\sqrt{2}} = \frac{1}{\sqrt{2}}$$

$$\therefore \quad \angle C_1 P C_2 = 45° = \frac{\pi}{4}$$

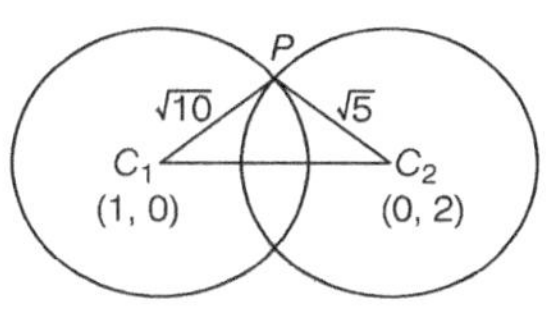

22. (a) The given equations of the circles are

$$S_1 : x^2 + y^2 + 2x - 4 = 0$$

$$S_2 : x^2 + y^2 - y + 1 = 0$$

Tangent from point $P(0, 5)$ on S_1 and S_2 are equal.

$$[\sqrt{S_1}(0,5)] = [\sqrt{S_2}(0,5)]$$

$$S_1(0,5) = S_2(0,5)$$

So, Statement I is correct.

Statement II is correct explanation for Statement I because radical axis of two circles is the locus of a point which moves in such a way that the length of the tangents drawn from it to the two circles are equal.

23. (b) The equation of the common chord of the given circles is

$$2x + 2y - 14 = 0$$

$$\Rightarrow \quad x + y - 7 = 0 \quad \ldots(i)$$

Since, $x^2 + y^2 - 6x - 4y + 9 = 0$ bisects the circumference of the circle

$$x^2 + y^2 - 8x - 6y + 23 = 0$$

Therefore, Eq. (i) is the diameter of the circle

$$x^2 + y^2 - 8x - 6y + 23 = 0$$

Then, centre of the circle C_1 lies on the circumference of C_2 because common chord is diameter of the circle.

24. (a) Statement I is correct and Statement II is correct explanation of Statement I.

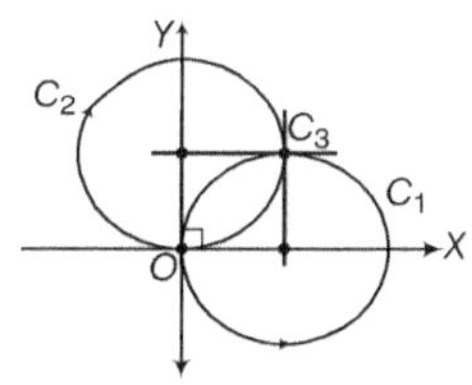

Thus, every member of C_1 intersects any member of C_2 at right angle at the point other than origin.

25. (c) $r^2 = a^2 + b^2 + 1 = (a+1)^2 + b^2 + 4$

$$= (a^2) + (b+1)^2 + 4$$

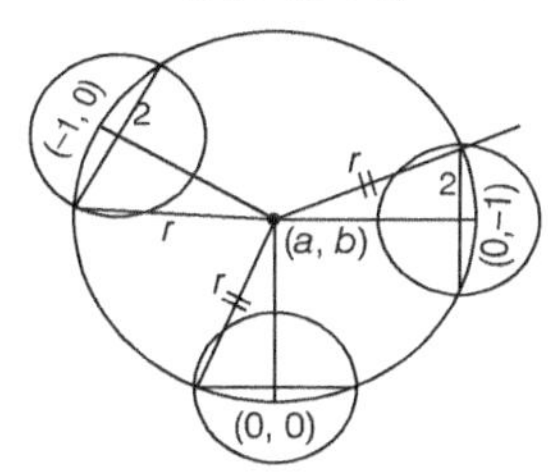

$$\therefore \quad a^2 + b^2 + 1 = a^2 + 2a + 1 + b^2 + 4$$

$$\Rightarrow \quad a = -2$$

and $\quad a^2 + (b+1)^2 + 4 = (a+1)^2 + b^2 + 4$

$$\Rightarrow \quad 2a = 2b \quad \Rightarrow \quad b = -2$$

$$\therefore \quad r^2 = 9 \quad \Rightarrow \quad r = 3 \text{ units}$$

26. (d) $S_1 - S_2 = 0 \Rightarrow x = 1$

$S_2 - S_3 = 0 \Rightarrow y = 1$

$\therefore$ Radical centre (1, 1)

Radius $L_T = \sqrt{S_1} = 1$

$\therefore$ Equation of circle is $(x-1)^2 + (y-1)^2 = 1$.

$\Rightarrow$ Radius = 1 and $a = 1$, $b = 1 \Rightarrow a + b + r = 3$

27. (c) Family of circles touches the line $x = 1$ at (1, 0) is

$$(x-1)^2 + (y-0)^2 + \lambda(x-1) = 0$$

Passing through (3, 2)

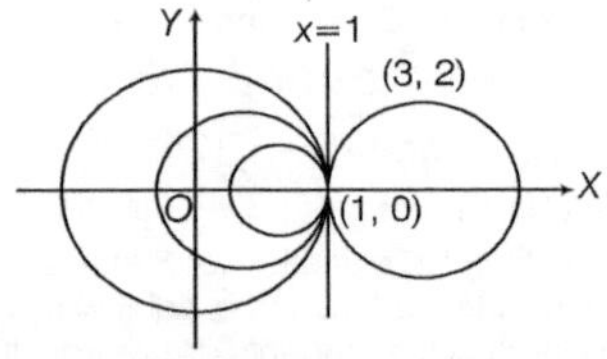

$\Rightarrow \quad 4 + 4 + 2\lambda = 0 \quad \Rightarrow \lambda = -4$

$\therefore \quad x^2 + y^2 - 6x + 5 = 0$

Hence, radius $= \sqrt{9-5} = 2$ units

28. (a, b, d) Since, (a, 0) is a point on the diameter along X-axis.

$\therefore \quad -2 \le a \le 2 \Rightarrow 0 \le a^2 \le 4$

Let $f(x) = x^2 - 4x - a^2$. Clearly, discriminant of $f(x) = 0$ is positive. So, it has real and distinct roots.

Now, $\quad f(0)\,f(-1) = (-a^2)(5-a^2) < 0 \quad (\because 0 \le a^2 \le 4)$

$\Rightarrow \quad f(x) = 0$ has root in $(-1, 0]$

Also, $\quad f(2)\,f(5) = (-4-a^2)(5-a^2) < 0$

$\Rightarrow \quad f(x) = 0$ has a root in $[2, 5]$

29. (a,c) Let tangents intersect at (α, β), then the line joining the two points at which tangents are drawn is

(i) common chord of C_1 and C_2 is

$$C_1 - C_2 = 0$$

$\Rightarrow \quad 2x + 4y - 18 = 0$

or $\quad x + 2y - 9 = 0$

(ii) chord of contact when tangents are drawn from (α, β) to C_1

$\Rightarrow \quad 2x + 4y - 18 = 0$

and $\quad \alpha x + \beta y - 25 = 0$

must be identical on comparing, we get

$$\frac{\alpha}{2} = \frac{\beta}{4} = \frac{25}{18}$$

$\therefore$ Tangents intersect at $\left(\frac{25}{9}, \frac{50}{9}\right)$.

DPP-4 Location of Two Circles, Number of Circles, Tangents and Family of Circles

1. (c) The given equations of the circles are

$$x^2 + y^2 = 4 \quad \ldots(i)$$

$$x^2 + y^2 - 8x + 12 = 0 \quad \ldots(ii)$$

Centres and radii of these circles are

$$C_1(0, 0), C_2(4, 0)$$

and $\quad r_1 = 2, r_2 = \sqrt{16-12} = 2$

Now, $\quad C_1C_2 = \sqrt{(4-0)^2 + (0-0)^2} = 4$

and $\quad r_1 + r_2 = 2 + 2 = 4$

$\because \quad C_1C_2 = r_1 + r_2$

It means that two circles touch each other externally. So, number of common tangents are 3.

2. (d) The given equations of the circles are

$$x^2 + y^2 - 4x - 2y - 4 = 0 \quad \ldots(i)$$

$$x^2 + y^2 - 12x - 8y - 36 = 0 \quad \ldots(ii)$$

The centres of the two circles are $C_1(2, 1), C_2(6, 4)$ and radius are

$$r_1 = \sqrt{4+1+4} = 3$$

$$r_2 = \sqrt{36+16+36} = \sqrt{88} = 9.38$$

$$C_1C_2 = \sqrt{(4)^2 + (3)^2} = 5$$

$$r_1 + r_2 = 12.38$$

$$|C_1C_2| \ne r_1 + r_2$$

Therefore, two circles not touch but they intersect each other.

3. (a) Given equations of the circles are

$$x^2 + y^2 = r^2 \quad \ldots(i)$$

and $\quad x^2 + y^2 - 10x + 16 = 0 \quad \ldots(ii)$

The centres and radii of given circles are

$$C_1(0, 0), C_2(5, 0) \text{ and } r_1 = r$$

$$r_2 = \sqrt{25-16} = 3$$

$$C_1C_2 = 5$$

For intersection of two circles

$$r_2 - r_1 < C_1C_2 < r_1 + r_2$$

$$3 - r < 5 < r + 3$$

$\Rightarrow \quad r < 8$ and $r > 2$

$\therefore \quad 2 < r < 8$

4. (b) The given equations of the circles are

$$x^2 + y^2 = 4 \quad \ldots(i)$$

$$x^2 + y^2 - 6x - 8y - 24 = 0 \quad \ldots(ii)$$

Centres and radii of these circles are

$$C_1(0, 0), C_2 = (3, 4) \text{ and } r_1 = 2$$

$$r_2 = \sqrt{9+16+24} = 7$$

$$C_1C_2 = \sqrt{9+16} = 5$$

$$r_2 - r_1 = 5$$

$$|C_1C_2| = |r_1 - r_2|$$

$\therefore$ There is one common tangent.

5. (d) The given equations of the circles are

$$x^2 + y^2 + 2hx - (r_1^2 - h^2) = 0 \quad \ldots(i)$$

$$x^2 + y^2 + 2kx - (r_2^2 - k^2) = 0 \quad \ldots(ii)$$

The centres and radii of the circles are

$$C_1(-h, 0) \text{ and } (-k, 0)$$

$$R_1 = \sqrt{h^2 + r_1^2 - h^2} = r_1$$

$$R_2 = \sqrt{k^2 - r_2^2 - k^2} = r_2$$

Since, they touch externally

$$|C_1C_2| = R_1 + R_2$$

$$\sqrt{h^2 + k^2} = r_1 + r_2$$

$\Rightarrow \quad h^2 + k^2 = (r_1 + r_2)^2$

$\therefore \quad h^2 + k^2 = r_1^2 + r_2^2 + 2r_1r_2$

6. (d) $S_1 : x^2 + y^2 + 4x + d = 0$ having $C_1(-2, 0)$ and $r_1 = \sqrt{4-d}$.

Also, $S_2 : x^2 + y^2 + 4fy + d = 0$ having $C_2(0, -2f)$ and $r_2 = \sqrt{4f^2 - d}$.

Since, circles touch each other.

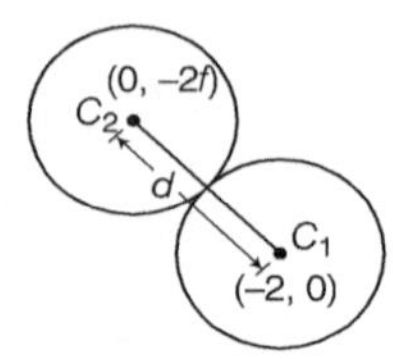

$\therefore \quad d = r_1 + r_2$

$\Rightarrow \quad \sqrt{4 + 4f^2} = \sqrt{4-d} + \sqrt{4f^2 - d}$

or $\quad 4 + 4f^2 = 4 - d + 4f^2 - d + 2\sqrt{(4-d)(4f^2-d)}$

$$\Rightarrow \quad 2(d) = 2\sqrt{16f^2 - 4d - 4df^2 + d^2}$$

or $$(d)^2 = 4f^2(4-d) + d^2 - 4d$$

$$\Rightarrow \quad 4d = 4f^2(4-d)$$

$$\Rightarrow \quad f^2 = \frac{d}{4-d}$$

or $$f = \pm\sqrt{\frac{d}{4-d}}$$

7. (d) The given equations of the circles are

$$x^2 + y^2 - 2x - 4y + 1 = 0 \quad \ldots(i)$$

$$x^2 + y^2 - 12x - 16y + 91 = 0 \quad \ldots(ii)$$

The centres and radii of the circles are

$$C_1(1, 2), C_2(6, 8) \text{ and } r_1 = \sqrt{1 + 4 - 1} = 2$$

$$r_2 = \sqrt{36 + 64 - 91} = 3$$

$$|C_1 C_2| = \sqrt{(6-1)^2 + (8-2)^2}$$

$$= \sqrt{25 + 36} = 7.81$$

$$r_1 + r_2 = 2 + 3 = 5$$

$$|C_1 C_2| > r_1 + r_2$$

$\therefore$ There are two direct and two transverse common tangents. So, there are total 4 tangents.

8. (b) The given equations of the circles are

$$x^2 + y^2 - 2x - 4y = 0 \quad \ldots(i)$$

$$x^2 + y^2 - 8y - 4 = 0 \quad \ldots(ii)$$

The centres and radii are

$$C_1(1, 2) \quad \text{and} \quad C_2(0, 4)$$

$$r_1 = \sqrt{1 + 4}, \quad r_2 = \sqrt{16 + 4} = \sqrt{20} = 2\sqrt{5}$$

$$= \sqrt{5}$$

$$C_1C_2 = \sqrt{1 + 4} = \sqrt{5}$$

Here, $$|C_1C_2| = |r_1 - r_2|$$

$$\sqrt{5} = \sqrt{5}$$

Hence, two circles touch each other internally.

9. (a) The given equations of the circles are

$$x^2 + y^2 + 2x - 8y + 8 = 0 \quad \ldots(i)$$

$$x^2 + y^2 + 10x - 2y + 22 = 0 \quad \ldots(ii)$$

The centres and radii are

$$C_1(-1, 4) \quad \text{and} \quad C_2(-5, 1)$$

$$r_1 = \sqrt{1 + 16 - 8} = 3$$

$$r_2 = \sqrt{25 + 1 - 22} = 2$$

$$|C_1C_2| = \sqrt{16 + 9} = 5$$

$$|C_1C_2| = r_1 + r_2$$

$$5 = 5$$

So, the two circles touch each other externally. The equation of the common tangent is obtained by subtracting the two equations

$$4x + 3y + 7 = 0$$

If two circles touch each other externally, then the point of contact divides the line segment joining their centers in the ratio of their radii, i.e.

$$r_1 : r_2$$

$$3 : 2$$

Hence, the point of contact $= \left(\frac{-15-2}{3+2}, \frac{3+8}{3+2}\right)$

$$= \left(\frac{-17}{5}, \frac{11}{5}\right)$$

10. (a) If a circle S_1 bisects the circumference of another circle S_2, then it will be the diameter of circle S_2. Then, their radial axis passes through the centre of S_1.

11. (b) The given equation of the circle is $x^2 + y^2 = a^2$

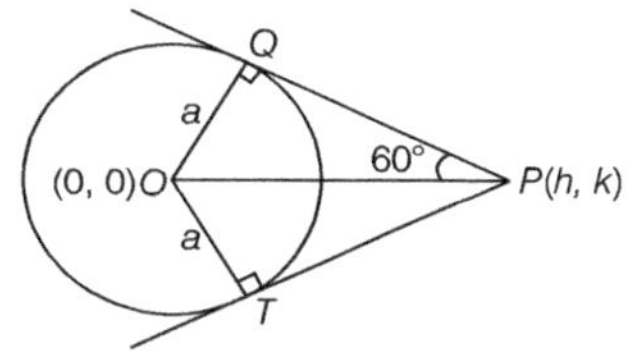

where, O (0, 0) is centre of the circle and a is the radius of the circle. Let the coordinates of point P be (h, k)

$$OP = \sqrt{(h-0)^2 + (k-0)^2} = \sqrt{h^2 + k^2}$$

In ΔPOQ, $$\frac{a}{\sqrt{h^2 + k^2}} = \sin 60°$$

$$\Rightarrow \quad \frac{a}{\sqrt{h^2 + k^2}} = \frac{\sqrt{3}}{2} \Rightarrow \frac{a^2}{h^2 + k^2} = \frac{3}{4}$$

$$\Rightarrow \quad 3h^2 + 3k^2 = 4a^2$$

On taking locus of (h, k), $3x^2 + 3y^2 = 4a^2$

12. (b) Here, the centre is (1, 2) and radius $r = 5$

Let the coordinates of C are (p, q).

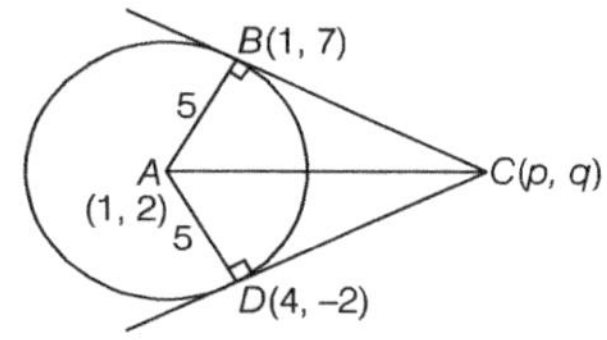

Here, A (1, 2), B (1, 7), C (p, q) and D (4, – 2)

Distance between A and C,

$$AC = \sqrt{(p-1)^2 + (q-2)^2} = \sqrt{p^2 + q^2 - 2p - 4q + 5} \ldots(i)$$

In ΔABC, AC is hypotenuse using Pythagoras theorem,

$$AC = \sqrt{(AB)^2 + (BC)^2} = \sqrt{25 + \{(p-1)^2 + (q-7)^2\}}$$

$$= \sqrt{p^2 + q^2 - 2p - 14q + 75} \quad \ldots(ii)$$

From Eqs. (i) and (ii), we get

$$q = 7 \text{ and } m_{BC} \cdot m_{AB} = -1$$

$$\Rightarrow \quad P = 16$$

$\therefore$ Area of $ABCD = 2$ (Area ΔABC)

$$= 2 \times \frac{1}{2} \times 5 \times 15 = 75 \text{ sq units}$$

13. (d) The equation of the common chord of the circle is given by

$$S_1 - S_2 = 0$$

$$\Rightarrow \quad 2y + 1 = 0$$

$$y = -\frac{1}{2}$$

The equation of a circle passing through the intersection of the given circles is

$$(x^2 + y^2 + 3x + 2y + 1) + \lambda(x^2 + y^2 + 3x + 4y + 2) = 0$$

$$\Rightarrow \quad x^2 + y^2 + x\left(\frac{3 + 3\lambda}{\lambda + 1}\right) + 2y\left(\frac{1 + 2\lambda}{\lambda + 1}\right) + \frac{1 + 2\lambda}{\lambda + 1} = 0 \quad \ldots(i)$$

Since, $2y + 1 = 0$ is a diameter of this circle. Therefore, its centre is $\left(\frac{-3}{2}, \frac{-2\lambda + 1}{\lambda + 1}\right)$ lies on it.

Now, $$-2\left(\frac{2\lambda + 1}{\lambda + 1}\right) + 1 = 0$$

$$\Rightarrow \quad -4\lambda - 2 + \lambda + 1 = 0 \Rightarrow \lambda = \frac{-1}{3}$$

On putting $\lambda = -\frac{1}{3}$ in Eq. (i), the equation of the required circle is

$$2x^2 + 2y^2 + 6x + 2y + 1 = 0$$

$$\Rightarrow \quad 2(x^2 + y^2) + 6x + 2y + 1 = 0$$

14. (a) The equation of the common chord is

$$S_1 - S_2 = 0$$

$$(x^2 + y^2 + 2gx + 2fy + c) - (x^2 + y^2 + 2ax + 2by + d) = 0$$

$$2gx + 2fy + c - 2ax - 2by - d = 0$$

It passes through $(-a, -b)$, i.e. the centre of the second circle.

$$\therefore \quad 2g(-a) + 2f(-b) + c - 2a(-a) - 2b(-b) - d = 0$$

$$\Rightarrow \quad -2ag - 2fb + c + 2a^2 + 2b^2 - d = 0$$

$$\Rightarrow \quad 2a(-g + a) + 2b(b - f) + (c - d) = 0$$

$$\Rightarrow \quad 2a(g - a) + 2b(f - b) = (c - d)$$

15. (c) Here, $m_{PQ} = -\frac{4}{3}$.

$\therefore$ Coordinates of P, are

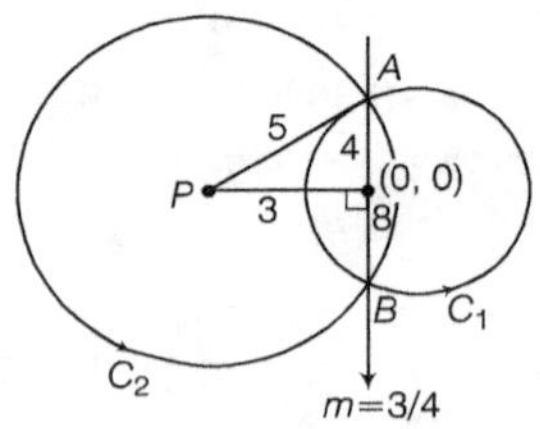

$$(0 \pm 3\cos\theta, 0 \pm 3\sin\theta) \quad \ldots(i)$$

Here, $\tan\theta = -\frac{4}{3} \Rightarrow \sin\theta = \frac{4}{5}$

and $\cos\theta = -\frac{3}{5}$.

Thus, centre coordinates are $\left(\mp\frac{9}{5}, \pm\frac{12}{5}\right)$.

16. (b) Clearly, OP is a chord of the circle passing through O (0, 0) and P(1, 0).

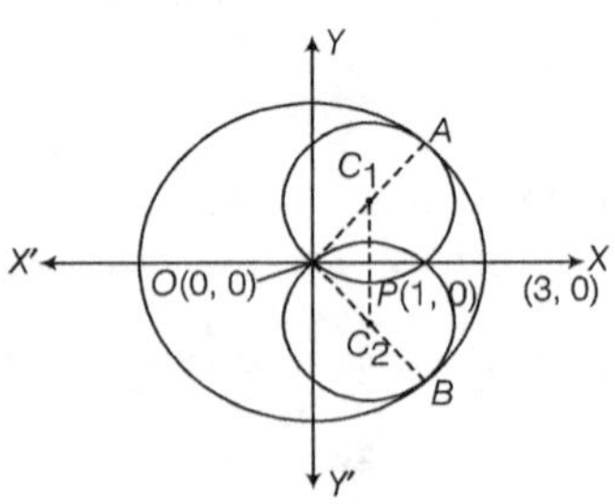

So, its entire lies on the perpendicular bisector of OP,

i.e. $x = \frac{1}{2}$

Let the radius of the circle passing through O and Q be r. Then,

$$OA = 2r \Rightarrow 3 = 2r$$

$$\Rightarrow \quad r = \frac{3}{2} \Rightarrow OC_1 = \frac{3}{2}$$

In ΔOPC_1, we have

$$OC_1^2 = OP^2 + PC_1^2$$

$$\Rightarrow \quad \frac{9}{4} = \frac{1}{4} + PC_1^2$$

$$\Rightarrow \quad PC_1 = \pm\sqrt{2}$$

$\therefore$ The coordinates of the centre $\left(\frac{1}{2}, \sqrt{2}\right)$ or $\left(\frac{1}{2}, -\sqrt{2}\right)$.

17. (d) Let C_1 (h, k) be the centre. Since, circle touches Y-axis.

$$\therefore \quad \text{Radius, } r_1 = h$$

The centre of the circle

$$x^2 + y^2 - 6x - 6y + 14 = 0 \text{ is } C_2\ (3, 3).$$

and radius $r_2 = \sqrt{9 + 9 - 14} = 2$

Since, desired circle touches the given circle externally,

$$C_1C_2 = r_1 + r_2$$

$$\sqrt{(h-3)^2 + (k-3)^2} = h + 2$$

$$\Rightarrow \quad (h-3)^2 + (k-3)^2 = (h+2)^2$$

$$\Rightarrow \quad h^2 + 9 - 6h + k^2 - 6k + 9 = h^2 + 4h + 4$$

$$\Rightarrow \quad k^2 - 10h - 6k + 14 = 0$$

$\therefore$ The locus is $y^2 - 10x - 6y + 14 = 0$

18. (d) Let (h, k) be the centre of the circle and r be the radius. Then,

$$\sqrt{(h-0)^2 + (k-1)^2} = r + 1 \text{ and } (k) = r$$

$$\Rightarrow \quad h^2 + (k-1)^2 = [(k) + 1]^2$$

$$\Rightarrow \quad h^2 = 4k, \text{ when } k \geq 0 \text{ or } h = 0, \text{ when } k < 0$$

Hence, the locus of (h, k) is

$$x^2 = 4y, \text{ when } y \geq 0$$

or $\quad x = 0$, when $y < 0$

or $\quad \{x^2 = 4y : y \geq 0\} \cup \{(0, y) : y < 0\}$

19. (b) Let $CD = \alpha$. Then, $AB = 2\alpha$

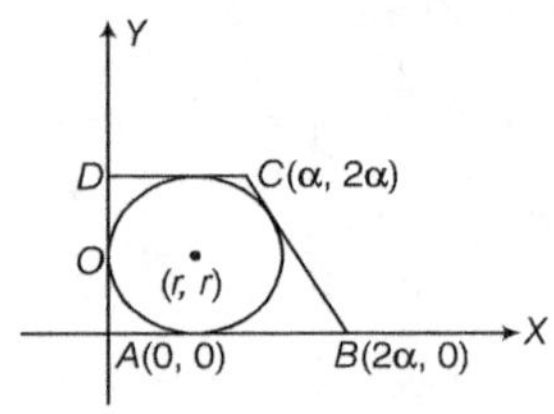

The equation of BC is

$$y = -\frac{2r}{\alpha}(x - 2\alpha)$$

or $\quad 2rx + \alpha y - 4r\alpha = 0$

It touches the circle $(x - r)^2 + (y - r)^2 = r^2$

$$\therefore \quad \left|\frac{2r^2 + \alpha r - 4\alpha r}{\sqrt{4r^2 + \alpha^2}}\right| = r$$

$$\Rightarrow \quad (2r^2 - 3\alpha r)^2 = r^2(\alpha^2 + 4r^2)$$

$$\Rightarrow \quad (2r - 3\alpha)^2 = \alpha^2 + 4r^2$$

$$\Rightarrow \quad -12r\alpha + 9\alpha^2 = \alpha^2 \quad \Rightarrow \alpha = \frac{3r}{2}$$

Now, area of quadrilateral $ABCD = 18$

$$\Rightarrow \quad \frac{1}{2} \times (AB + CD) \times 2r = 18$$

$$\Rightarrow \quad 3\alpha r = 18 \quad \Rightarrow \alpha r = 6$$

$$\Rightarrow \quad 3r^2 = 12 \quad \left[\because \alpha = \frac{3r}{2}\right]$$

$$\therefore \quad r = 2 \text{ units}$$

20. (c) Solving $y = 0$

and $x^2 + y^2 - 2x - 8 = 0$

$$\Rightarrow \quad x^2 - 2x - 8 = 0$$

i.e. $\quad x = 4, -2$

$\therefore$ $A(4, 0)$ and $B(-2, 0)$

$\Rightarrow$ Distance between A and B

= 6 units

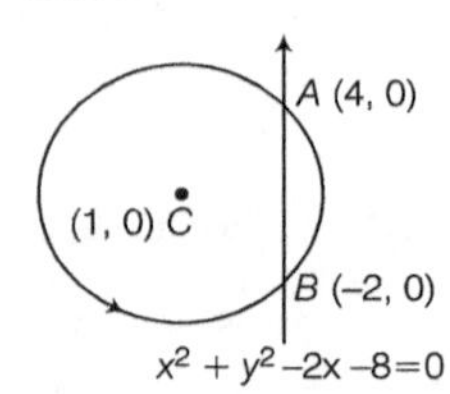

21. (c) Now, tangent at $A(4, 0)$, is $3x - \lambda y - 12 = 0$ and tangent at $B(-2, 0)$ is $-3x - \lambda y - 6 = 0$.

On solving, $x = 1,\ y = -\dfrac{9}{\lambda}$

It lies on $x + 2y + 5 = 0$

$\Rightarrow \quad 1 - \dfrac{18}{\lambda} + 5 = 0$

$\therefore$ Required equation of circle, is

$$x^2 + y^2 - 2x - 6y - 8 = 0$$

having radius, $R = \sqrt{1 + 9 + 8} = 3\sqrt{2}$

$\therefore \quad \dfrac{a/2}{R} = \cos 30^\circ$

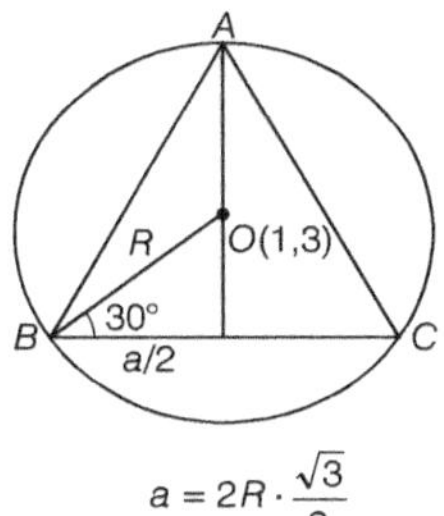

$\Rightarrow \quad a = 2R \cdot \dfrac{\sqrt{3}}{2}$

$\Rightarrow \quad a = \sqrt{3}R = 3\sqrt{6}$

Thus, area of equilateral $\Delta = \dfrac{\sqrt{3}}{4}(3\sqrt{6})^2$

$= \dfrac{27\sqrt{3}}{2}$ sq units

22. (d) As, $x^2 + y^2 - 2x - 6y - 8 = 0$ and $x^2 + y^2 - 10x + 2y + C = 0$ is orthogonal.

$\Rightarrow \quad 2[(-1)(-5) + (-3)(1)] = C - 8$

$\Rightarrow \quad 2[2] = C - 8$

$\therefore \quad C = 12$

23. (a) Let A, B, C and D be the complex numbers $\sqrt{2}, -\sqrt{2}, \sqrt{2}i$ and $-\sqrt{2}\,i$, respectively.

$$\Rightarrow \frac{PA^2 + PB^2 + PC^2 + PD^2}{QA^2 + QB^2 + QC^2 + QD^2}$$

$$= \frac{|z_1 - \sqrt{2}|^2 + |z_1 + \sqrt{2}|^2 + |z_1 + \sqrt{2}\,i|^2 + |z_1 - \sqrt{2}\,i|^2}{|z_2 + \sqrt{2}|^2 + |z_2 - \sqrt{2}|^2 + |z_2 - \sqrt{2}\,i|^2 + |z_2 + \sqrt{2}\,i|^2}$$

$$= \frac{|z_1|^2 + 2}{|z_2|^2 + 2} = \frac{3}{4} = 0.75$$

24. (c) Let C be the centre of the required circle. Now, draw a line parallel to L at a distance of r_1 (radius of C_1) from it.

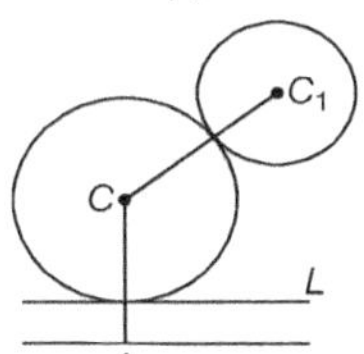

Now, $Cp_1 = AC \Rightarrow C$ lies on parabola.

25. (c) $\because AG = \sqrt{3}$

$\therefore \quad AT_1 = T_1G = \dfrac{1}{\sqrt{2}}$

[as A is the focus, T_1 is the vertex and BD is the direction of parabola]

Also, T_2T_3 is latusrectum.

$\therefore \quad T_2T_3 = 4 \times \dfrac{1}{\sqrt{2}}$

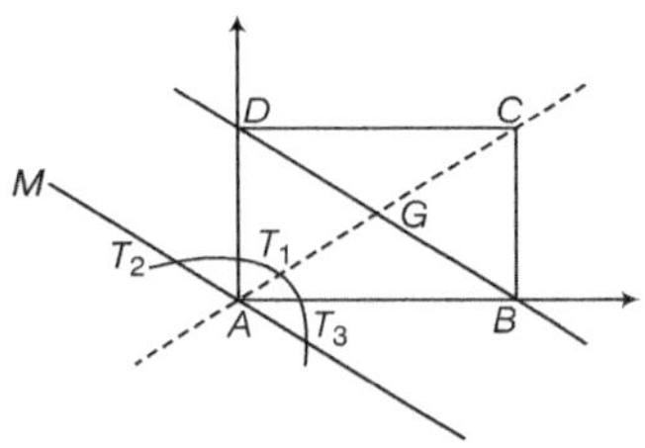

Here, area of $\Delta T_1T_2T_3 = \dfrac{1}{2} \times \dfrac{1}{\sqrt{2}} \times \dfrac{4}{\sqrt{2}}$

$= 1$ sq unit

26. (d) Let $A\ (h, k)$ be the centre of the circle.

Then, $AD \perp PQ$ and $AD = 1$

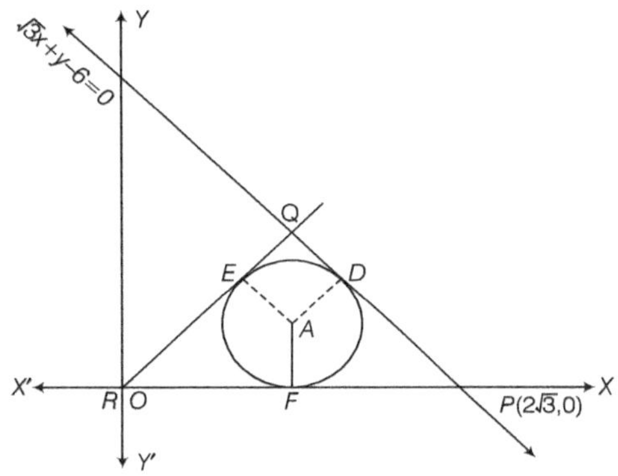

$$\frac{2k - 3}{2h - 3\sqrt{3}} \times (-\sqrt{3}) = -1$$

and $\left|\dfrac{\sqrt{3}\,h + k - 6}{\sqrt{3 + 1}}\right| = 1$

$\Rightarrow \quad 2\sqrt{3}k - 3\sqrt{3} = 2h - 3\sqrt{3}$

and $\quad |\sqrt{3}h + k - 6| = 2$

$\Rightarrow \quad 2h - 2\sqrt{3}k = 0$

and $\quad \sqrt{3}h + k - 6 = 2$

or $\quad \sqrt{3}h + k - 6 = -2$

$\Rightarrow \quad h = \sqrt{3}k$ and $\sqrt{3}h + k - 6 = -2$

[$\because$ origin of (h, k) lie on the same side of $\sqrt{3}x + y - 6 = 0$]

$\therefore \quad \sqrt{3}h + k - 6 < 0$

$\Rightarrow \quad h = \sqrt{3}k$ and $\sqrt{3}h + k = 4 \Rightarrow k = \sqrt{3},\ k = 1$

Hence, the equation of circle C is $(x - \sqrt{3})^2 + (y - 1)^2 = 1$

27. (a) Since, R coincides with the origin O.

$\therefore \quad RP = OP = 2\sqrt{3}$

$\Rightarrow \quad RF = \sqrt{3}$

Here, coordinates of F are $(\sqrt{3}, 0)$

$\because \quad RE = RF \Rightarrow RE = \sqrt{3}$

Hence, coordinates of E are

$(RE \cos 60^\circ, RE \sin 60^\circ)$ or $\left(\dfrac{\sqrt{3}}{2}, \dfrac{3}{2}\right)$.

28. (d) Clearly, RP is along X-axis.

So, its equation is $y = 0$.

$\therefore$ The equation of QR is $y = \sqrt{3}x$

29. (a) The equation tangent at $P(\sqrt{3}, 1)$ to the circle $x^2 + y^2 = 4$ is $\sqrt{3}x + y = 4$. Its slope is $-\sqrt{3}$.

So, slope of a line perpendicular to PT is $\frac{1}{\sqrt{3}}$.

The equation of tangents of slope $\frac{1}{\sqrt{3}}$ to the circle $(x-3)^2 + (y-0)^2 = 1^2$ are

$$y - 0 = \frac{1}{\sqrt{3}}(x-3) \pm \sqrt{1 + \left(\frac{1}{\sqrt{3}}\right)^2}$$

or $\quad \sqrt{3}y = x - 3 \pm 2$

or $\quad x - \sqrt{3}y - 1 = 0$

and $\quad x - \sqrt{3}y - 5 = 0$

or $\quad x - \sqrt{3}y = 1$

and $\quad x - \sqrt{3}y = 5$

30. (d) The centres of two circles are at $O(0, 0)$ and $C(3, 0)$ and their radii are 2 and 1 unit, respectively.

$$OC = \sqrt{(3-0)^2 + (0-0)^2} = 3$$

i.e. distance between centres is equal to the $r_1 + r_2 = 2 + 1 = 3$.
So, two circles touch each other externally and one of the common tangent is given by

$$S_1 - S_2 = 0$$

$\Rightarrow \quad 6x - 12 = 0$

$\Rightarrow \quad x = 2$

The point P of intersection of direct common tangent divides OC externally in the ratio of their radius, i.e. 2 : 1.
So, coordinates of P are (6, 0). The equation of a line through (6, 0) is

$$y - 0 = m(x-6)$$

$\Rightarrow \quad mx - y - 6m = 0$, where m is the slope of the line.

It touches the circle $x^2 + y^2 = 4$, then

$$\left|\frac{0 - 0 - 6m}{\sqrt{m^2+1}}\right| = 2$$

$\Rightarrow \quad m^2 + 1 = 9m^2$

$\Rightarrow \quad m = \pm \frac{1}{2\sqrt{2}}$

Therefore, the equations of the common tangent are

$$\pm \frac{x}{2\sqrt{2}} - y \mp \frac{3}{\sqrt{2}} = 0$$

or $\quad \pm x - 2\sqrt{2} \mp 6 = 0$

31. (a)

(i) Two intersecting circles have a common tangent and a common normal.

$\therefore$ **(i) → (p, q)**

(ii) Two mutually external circles have a common tangent and a common normal.

$\therefore$ **(ii) → (p, q)**

(iii) If two circles, one strictly inside the other, then they have a common normal but don't have a common tangent.

$\therefore$ **(iii) → (q, r)**

(iv) Two branches of a hyperbola have a common normal but don't have a common tangent.

$\therefore$ **(iv) → (q, r)**

32. (8) In ΔCPC_2, $\quad (r+1)^2 = \alpha^2 + 3^2$

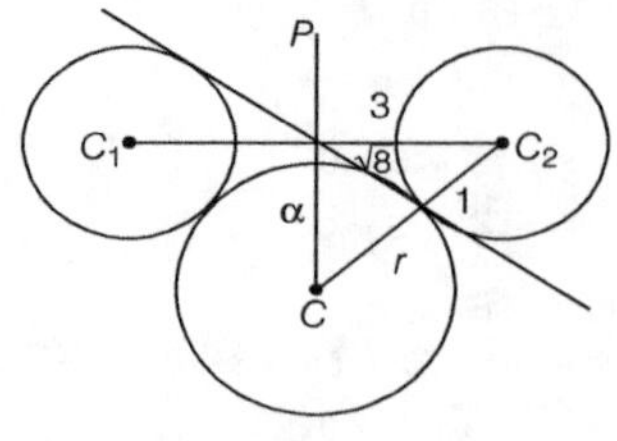

$\Rightarrow \quad r^2 + 2r + 1 = \alpha^2 + 9 \quad$ ...(i)

Also, In ΔCQP,

$\Rightarrow \quad \alpha^2 = r^2 + (\sqrt{8})^2 \quad$ ...(ii)

From Eqs. (i) and (ii), $\quad 2r = 16$

$r = 8$ units

33. (5) The given equation of the circles are

$$x^2 + y^2 - 20x + 64 = 0$$

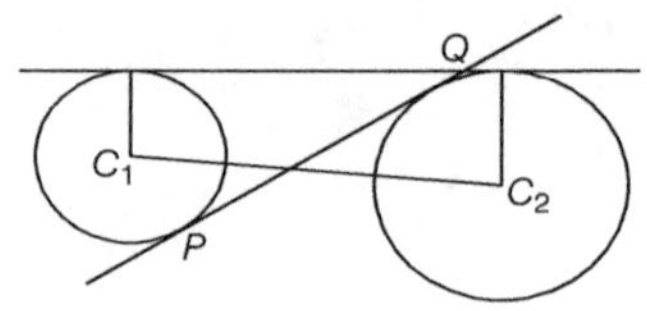

Here, $\quad C_1(10, 0)$ and $r_1 = 6$

and $\quad x^2 + y^2 + 30x + 144 = 0$

Here, $\quad C_2(-15, 0)$ and $r_2 = 9$

$$|C_1 - C_2| = \sqrt{(10+15)^2 + (0-0)} = 25$$

$$r_1 + r_2 = 15$$

$$|C_1C_2| = r_1 + r_2$$

So, there are two direct and two transverse common tangents.
The length of the transverse common tangent is less than the direct tangent.
The length is given by

$$k = \sqrt{(C_1C_2)^2 - (r_1 + r_2)^2}$$

$$= \sqrt{(25)^2 - (9+6)^2} = \sqrt{400}$$

$$k = 20$$

So, value of $\frac{k}{4} = \frac{20}{4} = 5$

34. (0) The straight line $y = mx$ is outside the circle $x^2 + y^2 - 20y + 90 = 0$, that means it is not intersecting the circle.
By putting $y = mx$ in the equation of circle.

$$x^2 - (mx)^2 - 20(mx) + 90 = 0$$

$\Rightarrow \quad x^2 - m^2x^2 - 20mx + 90 = 0$

$\Rightarrow \quad x^2(1 - m^2) - 20mx + 90 = 0$

If there is no intersection point, then discriminant < 0

$$b^2 - 4ac < 0$$

$\Rightarrow \quad 400m^2 - 4(1 - m^2)90 < 0$

$\Rightarrow \quad m^2 - 9 < 0$

$\Rightarrow \quad m = -3, +3$

$\therefore$ Sum of values of $m = 0$.

35. (3) Centre of smallest circle is mid-point of $C_1(0, 1)$ and $C_2(4, 9)$.

i.e. $\quad C(2, 5) = (h, k)$

$\therefore \quad |h - k| = 3$

DPP-5 $S' = T, S \cdot S' = T^2, T = 0$ and Pole, Polar

1. (a) Equation of pair of tangent is

$$SS_1 = T^2$$

$$\Rightarrow \quad \left(x^2 + y^2 - \frac{1}{4}\right)\left(1 - \frac{1}{4}\right) = \left(x \times 0 + y \cdot 1 - \frac{1}{4}\right)^2$$

$$\Rightarrow \quad \left(x^2 + y^2 - \frac{1}{4}\right)\left(\frac{3}{4}\right) = \left(y - \frac{1}{4}\right)^2$$

2. (c) Let pole be (x_1, y_1)

$\therefore$ Equation of polar, is

$$2x(x_1) + 2y(y_1) - \frac{3}{2}(x + x_1) + \frac{5}{2}(y + y_1) - 7 = 0$$

$$\Rightarrow \quad 2xx_1 + 2yy_1 - \frac{3}{2}x - \frac{3}{2}x_1 + \frac{5}{2}y + \frac{5}{2}y_1 - 7 = 0$$

$$\Rightarrow \quad x\left(2x_1 - \frac{3}{2}\right) + y\left(2y_1 + \frac{5}{2}\right) - \frac{3}{2}x_1 + \frac{5}{2}y_1 - 7 = 0$$

is same as, $\quad 9x + y - 28 = 0$

$$\Rightarrow \quad \frac{2x_1 - 3/2}{9} = \frac{2y_1 + 5/2}{1} = \frac{(-3/2)x_1 + (5/2)y_1 - 7}{-28}$$

On solving $\quad x_1 = 3, y_1 = -1$

3. (a) Let pole be (h, k).

Then, equation of polar is $xh + yk = 5$ same as

$$x + 2y = 1$$

$$\therefore \quad \frac{h}{1} = \frac{k}{2} = \frac{5}{1}$$

$$\Rightarrow \quad h = 5, k = 10$$

Thus, pole is (5, 10).

4. (d) Since, $y = -q$ intersect $x^2 + y^2 = px + qy$ at two real and distinct points.

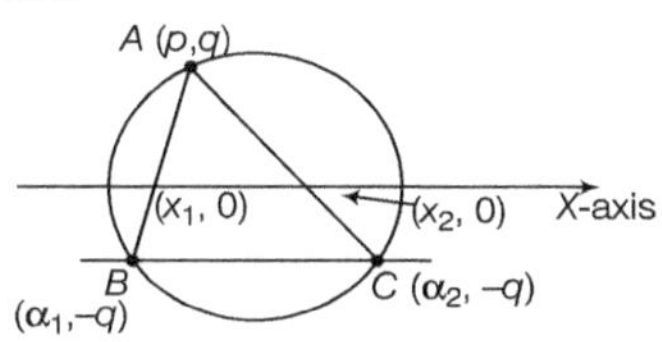

$$\Rightarrow \quad x^2 + q^2 = px - q^2$$

$$\Rightarrow \quad x^2 - px + 2q^2 = 0$$

Here, $\quad D > 0 \quad \Rightarrow \quad p^2 - 8q^2 > 0$

or $\quad p^2 > 8q^2$

5. (b) Here, radius of smaller circle is $AC = \sqrt{1 + 9 - 6} = 2$ units

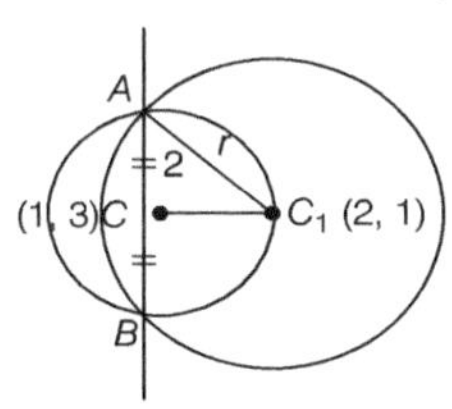

$$\therefore \quad r^2 = AC^2 + C_1C^2$$

$$r^2 = 4 + [(2 - 1)^2 + (1 - 3)^2]$$

$$\therefore \quad r^2 = 9 \quad \Rightarrow r = 3 \text{ units}$$

6. (b) Let $P(x_1, y_1)$ be a point on the circle $x^2 + y^2 = a^2$. Then,

$$x_1^2 + y_1^2 = a^2 \quad \ldots\text{(i)}$$

The equation of the chord of contact of tangents drawn from (x_1, y_1) to the $x^2 + y^2 = b^2$ is

$$xx_1 + yy_1 = b^2 \quad \ldots\text{(ii)}$$

This touches the circle

$$x^2 + y^2 = c^2$$

$\therefore$ Length of the perpendicular from (0, 0) to line (ii) is equal to the radius of $x^2 + y^2 = c^2$.

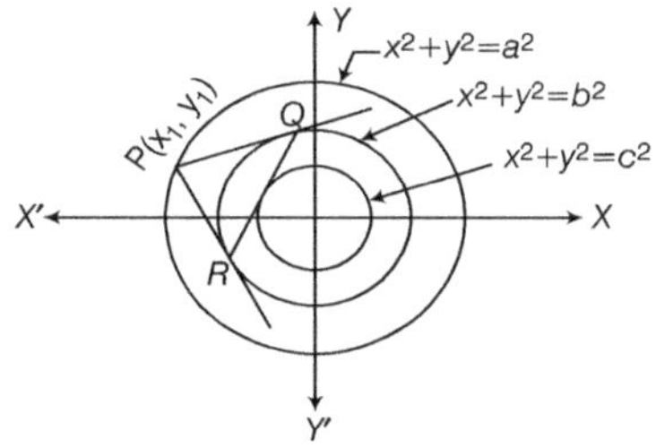

$$\Rightarrow \quad \left|\frac{0 \times x_1 + 0 \times y_1 - b^2}{\sqrt{x_1^2 + y_1^2}}\right| = c$$

$$\Rightarrow \quad \frac{b^2}{\sqrt{a^2}} = c$$

$$\Rightarrow \quad b^2 = ac$$

$\therefore$ a, b, c are in GP.

7. (a) Locus of mid-point (h, k) is $S_1 = T$

$$\Rightarrow \quad h^2 + k^2 - 9 = xh + yk - 9$$

or $\quad xh + yk - (h^2 + k^2) = 0 \quad \ldots\text{(i)}$

Also, let the point on $4x - 5y = 0$ be $\left(\lambda, \frac{4\lambda - 20}{5}\right)$

$\therefore$ Chord of contact of tangents is

$$x\lambda + y \cdot \left(\frac{4\lambda - 20}{5}\right) - 9 = 0 \quad \ldots\text{(ii)}$$

Since, Eqs.(i) and (ii) is same

$$\therefore \quad \frac{h}{\lambda} = \frac{k}{\frac{4\lambda - 20}{5}} = \frac{-(h^2 + k^2)}{-9}$$

$$(4\lambda - 20)h = 5\lambda k$$

or $\quad \lambda = \dfrac{20h}{4h - 5k} \quad \ldots\text{(ii)}$

and $\quad \dfrac{h}{\lambda} = \dfrac{h^2 + k^2}{9}$

$$\frac{9h \cdot (4h - 5k)}{20h} = h^2 + k^2 \quad \text{[using Eq. (iii)]}$$

or locus is

$$9(4x - 5y) = 20(x^2 + y^2)$$

or $\quad 20(x^2 + y^2) - 36x + 45y = 0$

8. (b) Let (x_1, y_1) be the pole of the given straight line w.r.t. $x^2 + y^2 = c^2$. Then, the equation of the polar, i.e. the straight line is

$$xx_1 + yy_1 = c^2 \quad \ldots\text{(i)}$$

It is given that, (x_1, y_1) lies on

$$x^2 + y^2 = 9c^2$$

$$\therefore \quad x_1^2 + y_1^2 = 9c^2$$

Now, length of the perpendicular from the centre $O(0, 0)$ of the circle $x^2 + y^2 = \frac{c^2}{9}$ to the line (i) is equal to the radius of the circle.

Hence, the straight line touches the circle $x^2 + y^2 = \frac{c^2}{9}$.

9. (d) From the figure, ΔOAB is a right angled triangle right angle at B.

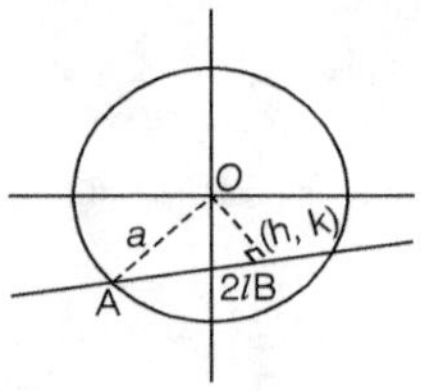

Using Pythagoras theorem,

$$h^2 + k^2 + l^2 = a^2$$

$$\Rightarrow \quad h^2 + k^2 = a^2 - l^2$$

$\therefore$ Locus of (h, k) is $\quad x^2 + y^2 = a^2 - l^2$

10. (a) The given equation of the circle is

$$x^2 + y^2 - 2x + 4y + 3 = 0$$

having centre $(1, -2)$ and radius $= \sqrt{2}$

Equation of any line through $(6, -5)$ is

$$y + 5 = m(x - 6)$$

$$\Rightarrow \quad mx - y - 6m - 5 = 0$$

If it touches the circle, then length of perpendicular from centre $(1, -2)$ to the circle is equal to the radius $\sqrt{2}$.

$$\therefore \quad \frac{m + 2 - 6m - 5}{\sqrt{m^2 + 1}} = \pm\sqrt{2}$$

$$\frac{-5m - 3}{\sqrt{m^2 + 1}} = \pm\sqrt{2}$$

$$\Rightarrow \quad (-5m - 3)^2 = 2(m^2 + 1)$$

$$\Rightarrow \quad 25m^2 + 9 + 30m = 2m^2 + 2$$

$$23m^2 + 23m + 7m + 7 = 0$$

$$\Rightarrow \quad 23m(m + 1) + 7(m + 1) = 0$$

$$\Rightarrow \quad m = \frac{-7}{25}, -1$$

So, equations of tangents are, if $m = -\frac{7}{23}$

$$y + 5 = -\frac{7}{23}(x - 6)$$

$$\Rightarrow \quad 23y + 115 = -7x + 42$$

$$\Rightarrow \quad 7x + 23y + 73 = 0 \qquad \ldots(i)$$

If $m = -1$, $\quad y + 5 = -1(x - 6)$

$$\Rightarrow \quad y + 5 = -x + 6 \quad \Rightarrow \quad x + y - 1 = 0$$

So, pair of tangent is

$$(7x + 23y + 73)(x + y - 1) = 0$$

$$7x^2 + 23y^2 + 30xy + 66x + 50y - 73 = 0$$

11. (b) The given equation of the circle is

$$x^2 + y^2 - 6x - 8x + 21 = 0$$

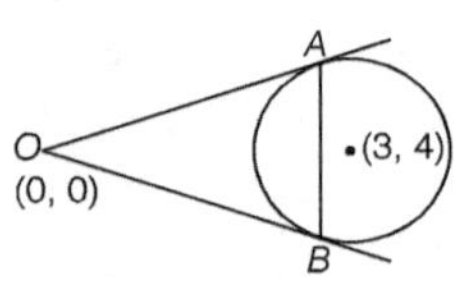

Here, $\quad r = \sqrt{9 + 16 - 21} = 2$

The length of $AB = \dfrac{2r\sqrt{S_1}}{\sqrt{S_1 + r^2}}$

$$\sqrt{S_1} = \text{Length of the tangent}$$

$$= \sqrt{x(0) + y(0) - 6(0) - 8(0) + 21} = \sqrt{21}$$

$$\therefore \quad \text{Length of } AB = \frac{2 \times 2 \times \sqrt{21}}{\sqrt{21 + 4}} = \frac{4\sqrt{21}}{5}$$

12. (c) If the lines $l_1x + m_1y + n_1 = 0$ and $l_2x + m_2y + n_2 = 0$ are conjugate lines with respect to the circle $x^2 + y^2 = a^2$, then

$$a^2(l_1l_2 + m_1m_2) = n_1n_2$$

13. (c) Equation of chord is $2x - y + 3 = 0$

$\therefore$ Equation of diameter is $x + 2y = \lambda$ passing through $(-1, 2)$

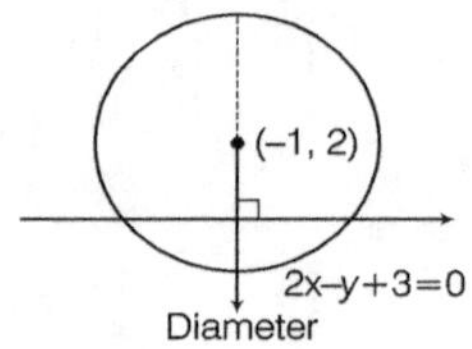

$$\Rightarrow \quad -1 + 4 = \lambda$$

$$\text{or} \quad x + 2y = 3$$

14. (b) The equation of chord of the circle is

$$x^2 + y^2 - 8x + 6y + 9 = 0$$

Here, $g = -4$ and $f = 3$

The equation of the chord of contact is

$$xx_1 + yy_1 + g(x + x_1) + f(y + y_1) + c = 0$$

Here, $\quad (x_1, y_1) = (8, 1)$

$$\Rightarrow \quad 8x + y + (-4)(x + 8) + 3(y + 1) + 9 = 0$$

$$\Rightarrow \quad 4x + 4y = 20$$

$$\Rightarrow \quad x + y = 5$$

15. (c) Let the pole be (x_1, y_1). Then, the equation of the polar is

$$xx_1 + yy_1 = a^2 \qquad \ldots(i)$$

Since, Eq.(i) and $lx + my + n = 0$ represents the same straight line.

$$\therefore \quad \frac{x_1}{l} = \frac{y_1}{m} = -\frac{a^2}{n}$$

$$\Rightarrow \quad x_1 = \frac{-a^2l}{n},\ y_1 = -\frac{a^2m}{n}$$

Hence, the pole is $\left(\dfrac{-a^2l}{n}, -\dfrac{a^2m}{n}\right)$.

16. (b) Equation of chord bisected at $P(h, k)$ is

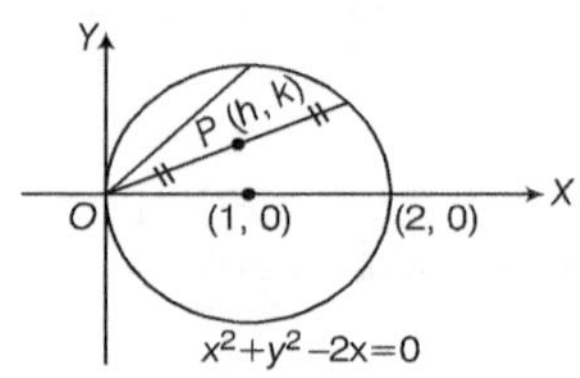

$$S_1 = T$$

$$\Rightarrow \quad h^2 + k^2 - 2h = xh + yk - (x + h) \text{ passing through } (0, 0)$$

$$\Rightarrow \quad h^2 + k^2 - 2h = -h$$

$\therefore$ Locus is $\quad x^2 + y^2 = x$

17. (b) Here, $A(x_1, y_1)$ and $B(x_2, y_2)$ are conjugate points w.r.t. $x^2 + y^2 = r^2$.

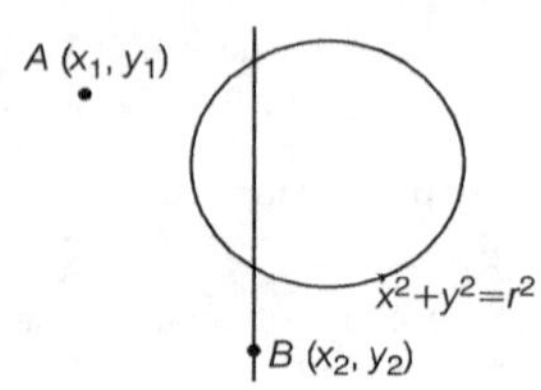

$\therefore$ Chord of contact for $A(x_1, y_1)$ is $T = 0$

$\Rightarrow xx_1 + yy_1 = r^2$ passes through $B(x_2, y_2)$

$$\Rightarrow \quad x_1x_2 + y_1y_2 = r^2 \quad \ldots(i)$$

Also, $$l_1^2 = x_1^2 + y_1^2 - r^2$$

and $$l_2^2 = x_2^2 + y_2^2 - r^2 \quad \ldots(ii)$$

$$\therefore \quad AB^2 = (x_1 - x_2)^2 + (y_1 - y_2)^2$$
$$= x_1^2 + y_1^2 + x_2^2 + y_2^2 - 2(x_1x_2 + y_1y_2)$$
$$= l_1^2 + r^2 + l_1^2 + r^2 - 2r^2 = l_1^2 + l_2^2$$

18. (b) $\because$ $$\tan\alpha = \frac{a}{\sqrt{S_1}}$$

$$\therefore \quad \alpha = \tan^{-1}\left(\frac{a}{\sqrt{S_1}}\right)$$

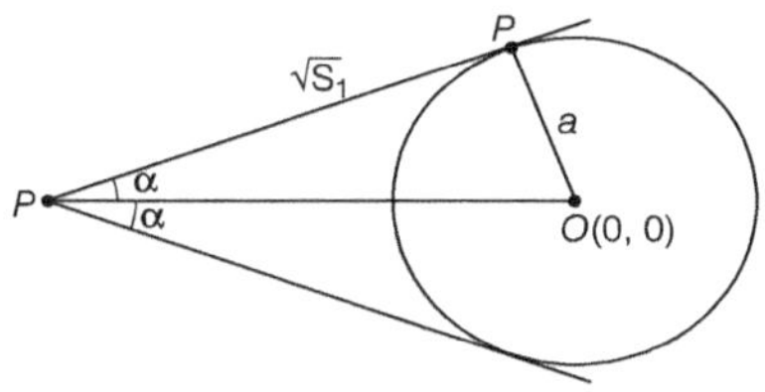

Thus, angle between tangents $= 2\alpha = 2\tan^{-1}\left(\frac{a}{\sqrt{S_1}}\right)$

19. (a) Equation of AB is $S_1 = T$

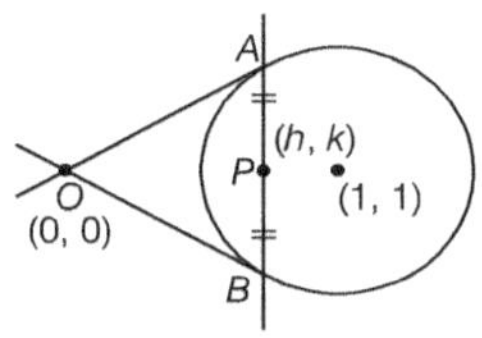

$$\therefore \quad h^2 + k^2 - 2h - 2k = xh + yk - (x + h) - (y + k)$$

or $$h^2 + k^2 - h - k = xh + yk - x - y$$

or $$1 = \frac{x(h-1) + y(k-1)}{h^2 + k^2 - h - k} \quad \ldots(i)$$

Now, product of pair of tangents OA and OB is

$$x^2 + y^2 - 2x\cdot\left[\frac{x(h-1) + y(k-1)}{h^2 + k^2 - h - k}\right] - 2y\left[\frac{x(h-1) + y(k-1)}{h^2 + k^2 - h - k}\right] = 0$$

Since, OA and OB are equally inclined to axes.

$\therefore m_1 + m_2 = 0 \Rightarrow -2(k-1) - 2(h-1) = 0$

or $$h - 1 + k - 1 = 0$$

or $$x + y = 2 \quad \text{(locus)}$$

20. (c) Radius of circle are r_1, r_2 and line $y = x + 1$

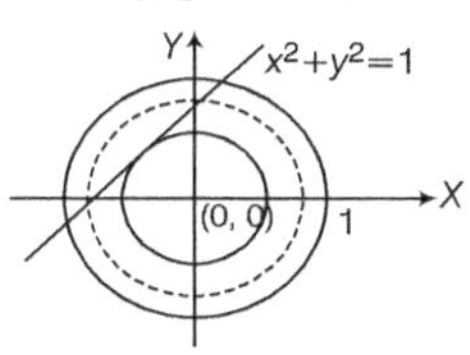

Perpendicular from (0, 0) on line $y = x + 1 \Rightarrow \frac{1}{\sqrt{2}}$

Now, $r_1 > \frac{1}{\sqrt{2}}$ but $r_1 = 1 - 2d$

Hence, $1 - 2d > \frac{1}{\sqrt{2}}$; $\frac{\sqrt{2}-1}{\sqrt{2}} > 2d$

$$\therefore \quad d < \frac{\sqrt{2}-1}{2\sqrt{2}} \Rightarrow d \in \left(0, \frac{\sqrt{2}-1}{2\sqrt{2}}\right)$$

21. (a) Let from $P(x_1, y_1)$ to the circle $x^2 + y^2 = a^2$ is a chord of contact of tangents.

Since, it passes through $(k, 0)$, therefore

$$x_1 \cdot k + y_1 \cdot 0 = a^2$$

$$\Rightarrow \quad x_1k = a^2$$

$\therefore$ Locus of $P(x_1, y_1)$, polar is

$$kx = a^2$$

22. (b) Let (h, k) be the mid-point of a chord of the circle $x^2 + y^2 = 16$.

Then, equation of the chord is

$$hx + ky - 16 = h^2 + k^2 - 16$$

$$\Rightarrow \quad hx + ky = h^2 + k^2$$

If this is parallel to $4x - 2y + 5 = 0$.

Then, $$-\frac{h}{k} = 2 \quad \Rightarrow \quad -h = 2k$$

$$\Rightarrow \quad h + 2k = 0$$

Hence, the locus of (h, k) is $x + 2y = 0$.

Explanations for Q. Nos. (23-25)

Tangent, $y - 2 = m(x - 4)$

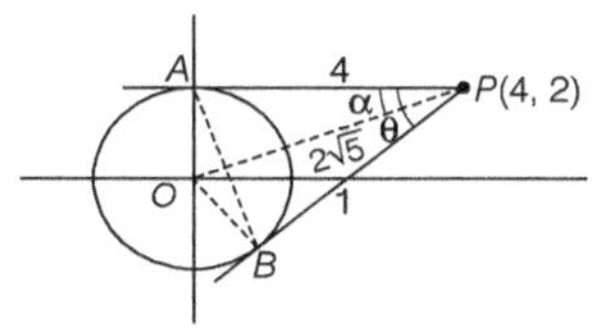

$$\Rightarrow \quad mx - y + (2 - 4m) = 0$$

$$\therefore \quad p = \frac{|2 - 4m|}{\sqrt{m^2 + 1}} = 2$$

$$(1 - 2m)^2 = 1 + m^2$$

$$\Rightarrow \quad m = 0, \frac{4}{3}$$

$$\therefore \quad \theta \in (45^\circ, 60^\circ)$$

Hence, equation of tangent is $y = 2$ and (with infinite intercept on X-axis)

or $$y - 2 = \frac{4}{3}(x - 4)$$

$$\Rightarrow \quad 3y - 6 = 4x - 16$$

$$\Rightarrow \quad 4x - 3y - 10 = 0$$

x-intercept $= \frac{10}{4}$

Variable line with mid-point (h, k) is

$\frac{x}{2h} + \frac{y}{2k} = 1$. It touches the circle $x^2 + y^2 = 4$.

$$\therefore \quad \left|\frac{-1}{\sqrt{\frac{1}{4h^2} + \frac{1}{4k^2}}}\right| = 2 \quad \text{or} \quad \frac{1}{4x^2} + \frac{1}{4y^2} = \frac{1}{4}$$

or $$x^{-2} + y^{-2} = 1$$

23. (d) **24.** (b) **25.** (a)

26. (b) $y = f(x) = \frac{x^2}{4} - ax + (a^2 + a - 2)$

For zeroes to be on either side of origin.

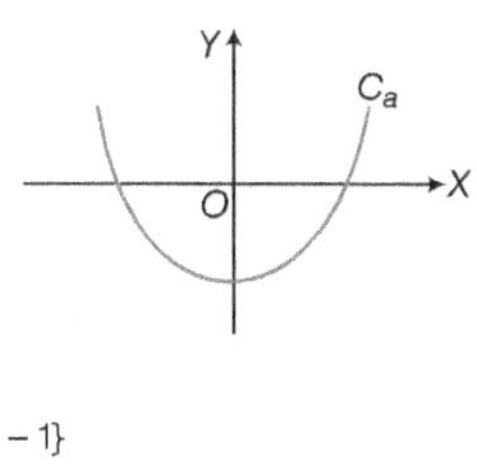

$$f(0) < 0$$

$$a^2 + a - 2 < 0$$

$$\Rightarrow \quad (a + 2)(a - 1) < 0$$

$$\Rightarrow \quad -2 < a < 1$$

$\Rightarrow$ 2 integers, i.e. $\{0, -1\}$

27. (a) Vertex of C_a is $(2a, a-2)$

Hence, $h = 2a$ and $k = a - 2$

$\therefore$ $h = 2(k+2)$ or locus is

$$x = 2y + 4$$

28. (b) Let $y = mx + c$ be a common tangent to

$$y = \frac{x^2}{4} - 3x + 10, \text{ for } a = 3 \quad \text{...(i)}$$

and $$y = 2 - \frac{x^2}{4} \quad \text{...(ii)}$$

where, $m = m_1$ or m_2 and $c = c_1$ or c_2

Solving $y = mx + c$ with Eq. (i)

$$mx + c = \frac{x^2}{4} - 3x + 10$$

or $$\frac{x^2}{4} - (m+3)x + (10 - c) = 0$$

$D = 0$, gives

$$(m+3)^2 = 10 - c$$

$\Rightarrow$ $$c = 10 - (m+3)^2 \quad \text{...(iii)}$$

Similarly, $$mx + c = 2 - \frac{x^2}{4}$$

$\Rightarrow$ $$\frac{x^2}{4} + mx + (c-2) = 0$$

$D = 0$, gives

$$m^2 = c - 2 \Rightarrow c = 2 + m^2 \quad \text{...(iv)}$$

From Eqs. (iii) and (iv),

$$10 - (m+3)^2 = 2 + m^2$$

$\Rightarrow$ $$2m^2 + 6m + 1 = 0$$

$\Rightarrow$ $$m_1 + m_2 = -3$$

29. (0) $P = a^2 + b^2 - 2(t^2 - 3t + 1)a - 2(t^2 + 2t)b + t$

$= -(2a + 2b)t^2 + (6a - 4b + 1)t + (a^2 + b^2 - 2a)$

This power is independent of the parameter t.

If and only if, $2a + 2b = 0 \Rightarrow a = -b$.

and $$6a - 4b + 1 = 0$$

$\therefore$ $$a = -\frac{1}{10} \text{ and } b = \frac{1}{10}$$

$\Rightarrow$ $$a + b = 0$$

30. (1) Locus will be a line with slope $\left(-\frac{1}{2}\right)$ and passing through the centre (2, 1) of the circle $y - 1 = -\frac{1}{2}(x-2)$

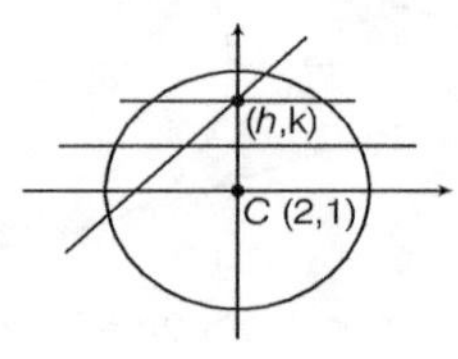

$\Rightarrow$ $$x + 2y - 4 = 0$$

$\therefore$ $$|a + b + c| = 1$$

Revisal Problems (JEE Main)

1. (c) Solving the equations of lines in pairs, we obtain that the vertices of ΔABC are $A(0, 6)$, $B(-2\sqrt{3}, 0)$ and $C(2\sqrt{3}, 0)$.

Clearly, $$AB = BC = CA$$

So, ΔABC is an equilateral triangle.

Therefore, centroid of the ΔABC coincides with the circumcentre. Coordinates of the circumcentre are $O'(0, 2)$ and radius $= O'A = 4$.

Hence, the equation of the circumcircle is

$$(x-0)^2 + (y-2)^2 = 4^2$$

$\Rightarrow$ $$x^2 + y^2 - 4y = 12$$

2. (a) The radius of a circle with centre $P(a, b)$ passing through the centre of the circle

$$x^2 + y^2 - 2gx + f^2 = 0, \text{ i.e. } Q(g, 0).$$

The radius of the circle will be distance between $P(a, b)$ and $Q(g, a)$.

$\therefore$ $$PQ = \sqrt{(a-g)^2 + (b-0)^2} = \sqrt{(a-g)^2 + b^2}$$

3. (a)

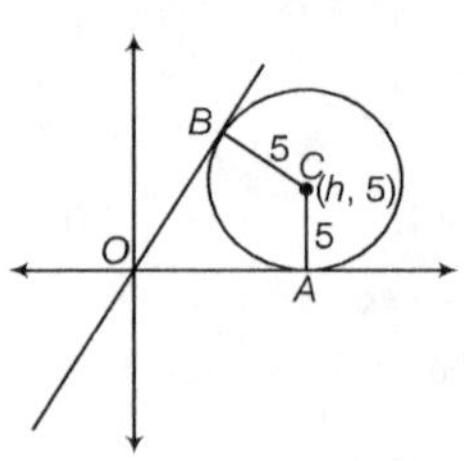

The given equation of the line is

$$4y = 3x$$

$\Rightarrow$ $$4y - 3x = 0$$

$\Rightarrow$ $$3x - 4y = 0$$

The perpendicular distance of the line from the centre (h, k) will be equal to radius.

$\therefore$ $$\left|\frac{3h - 4 \times 5}{5}\right| = 5 \Rightarrow \left|\frac{3h - 20}{5}\right| = 5$$

$\Rightarrow$ $$3h - 20 = \pm 25$$

$$h = 15 \text{ or } h = \frac{-5}{3}$$

The required equation of the circle is

$$(x - 15)^2 + (y - 5)^2 = 5^2$$

$\Rightarrow$ $$x^2 - 30x + 225 + y^2 - 10y + 25 = 25$$

$\therefore$ $$x^2 + y^2 - 30x - 10y + 225 = 0$$

4. (b) Area of rectangle $= 10 \times 8 = 80$ sq units

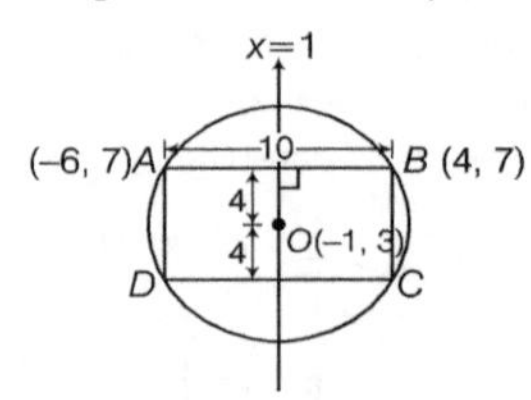

5. (c) The given equation of the circle is

$$x^2 + y^2 - 2x + 2y - 2 = 0$$

$\Rightarrow$ $$x^2 - 2x + 1 + y^2 + 2y + 1 - 2 - 2 = 0$$

$\Rightarrow$ $$(x-1)^2 + (y+1)^2 = 4$$

$\Rightarrow$ $$(x-1)^2 + (y+1)^2 = 2^2$$

The parametric coordinates are

$$x = a + r\cos\theta = 1 + 2\cos\alpha \quad (\because a = 1 \text{ and } r = 2)$$

and $$y = b + r\sin\theta = -1 + 2\sin\theta$$

$\therefore$ Parametric coordinates are $(1 + 2\cos\alpha, -1 + 2\sin\alpha)$.

6. (c) Intersecting points of the line $y = 2x$ and $x^2 + y^2 = 10x$ are

$\Rightarrow \quad x^2 + 4x^2 = 10x$

$\Rightarrow \quad 5x^2 - 10x = 0$

$\Rightarrow \quad 5x(x-2) = 0$

$\therefore \quad x = 0, 2$

Put $x = 0$ in $y = 2x$,

$\Rightarrow \quad y = 0$

and $x = 2$ in $y = 2x$,

$\Rightarrow \quad y = 4$

Hence, intersecting points are (0, 0), (2, 4).

Now, equation of the circle with (0, 0) and (2, 4) as ends of the diameter is

$$(x-0)(x-2) + (y-0)(y-4) = 0$$

$\Rightarrow \quad x^2 + y^2 - 2x - 4y = 0$

7. (b) The given equation of lines are

$$x\sin\alpha - y\cos\alpha = a \quad \text{...(i)}$$

$$x\cos\alpha + y\sin\alpha = a \quad \text{...(ii)}$$

On squaring and adding both the equations, we get

$$x^2 + y^2 = 2a^2$$

8. (d) If the straight line $y = mx$ is outside the circle $x^2 + y^2 - 20y + 90 = 0$, then there will be no point of intersection.

Putting $y = mx$ in equation of circle,

$$x^2 + m^2x^2 - 20(mx) + 90 = 0$$

$\Rightarrow \quad x^2(1 + m^2) - 20mx + 90 = 0$

If there is no point of intersection, then $D < 0$

$\Rightarrow \quad b^2 - 4ac < 0$

$\Rightarrow \quad 400m^2 - 4(1 + m^2)(90) < 0$

$\Rightarrow \quad 400m^2 < 4(1 + m^2)(90)$

$\Rightarrow \quad 10m^2 < 9 + 9m^2$

$\Rightarrow \quad m^2 < 9$

$\Rightarrow \quad m < \pm 3$

$\therefore \quad |m| < 3$

9. (a) The point (1, 2) lies inside the circle

$$x^2 + y^2 - 4x - 2y + c = 0$$

Here, centre = (2, 1), radius = $\sqrt{4 + 1 - c} = \sqrt{5 - c}$

If the point lies inside the circle, then distance of the point (1, 2) from centre (2, 1) should be less than radius.

Now, Distance = $\sqrt{(2-1)^2 + (1-2)^2} = \sqrt{2}$

$\therefore \quad \sqrt{2} < \sqrt{5 - c}$

$\Rightarrow \quad 2 < 5 - c$

$\Rightarrow \quad c < 5 - 2$

$\therefore \quad c < 3$

10. (c) Let the required circle be

$$x^2 + y^2 + 2gx + 2fy + c = 0 \quad \text{...(i)}$$

This passes through the origin.

Therefore, $c = 0$

The centre $(-g, -f)$ of circle (i) lies on the line $x + y = 4$.

$\therefore \quad -g - f = 4$

$\Rightarrow \quad g = -f - 4 \quad \text{...(ii)}$

Since, Eq. (i) cuts the circle $x^2 + y^2 - 4x + 2y + 4 = 0$ orthogonally.

$\therefore \quad 2(-2g + f) = c + 4$

$\Rightarrow \quad -2g + f = 2 \quad \text{...(iii)}$

On solving Eqs. (ii) and (iii), we get

$$g = -2 \text{ and } f = -2$$

Hence, the equation of circle is

$$x^2 + y^2 - 4x - 4y = 0$$

11. (b) The given equations of the circle are

$$S_1 \equiv x^2 + y^2 + 5x - 8y + 1 = 0$$

$$S_2 \equiv x^2 + y^2 - 3x + 7y - 25 = 0$$

The equation of the common chord is

$$S_1 - S_2 = 0$$

$$(5x - 8y + 1) - (-3x + 7y - 25) = 0$$

$\Rightarrow \quad 8x - 15y + 26 = 0$

The another equation of circle

$$x^2 + y^2 = 2x$$

$\Rightarrow \quad x^2 + y^2 - 2x = 0$

having centre (1, 0) and radius 1.

The distance of $8x - 15y + 26 = 0$ from the (1, 0) is

$$\left|\frac{8 + 26}{\sqrt{64 + 225}}\right| = \frac{34}{17} = 2 \text{ units}$$

12. (c) The centres of the circles are

$$C_1(-2, -3) \text{ and } C_2(-3, -2)$$

and corresponding radii are

$$r_1 = \sqrt{4 + 9 - 4} = 3$$

and $\quad r_2 = \sqrt{9 + 4 - 4} = 3$

The equation of the common chord is

$$S_1 - S_2 = 0$$

$$-2x + 2y = 0$$

$\Rightarrow \quad x - y = 0$

$$C_1M = \left|\frac{-2 + 3}{\sqrt{1 + 1}}\right| = \frac{1}{\sqrt{2}}$$

$$C_1P = r_1 = 3$$

$$PM = \sqrt{(3)^2 - \left(\frac{1}{\sqrt{2}}\right)^2}$$

$$= \sqrt{9 - \frac{1}{2}} = \sqrt{\frac{17}{2}}$$

$$PM = 2 \times \sqrt{\frac{17}{2}} = \sqrt{34} \text{ units}$$

13. (a) The required circle is

$$(x^2 + y^2 - 6) + \lambda(x^2 + y^2 - 6x + 8) = 0$$

[using $S_1 + \lambda S_2 = 0$]

It passes through (1, 1).

$\therefore \quad (1 + 1 - 6) + \lambda(1 + 1 - 6 + 8) = 0$

$\Rightarrow \quad -4 + 4\lambda = 0$

$\Rightarrow \quad \lambda = 1$

So, the required equation is

$$2x^2 + 2y^2 - 6x + 2 = 0$$

$\Rightarrow \quad x^2 + y^2 - 3x + 1 = 0$

14. (b) The common chord of the given circle is

$$6x + 14y + c + d = 0 \quad \text{...(i)}$$

Since, $x^2 + y^2 + 4x + 22y + c = 0$ bisect the circumference of the circle $x^2 + y^2 - 2x + 8y - d = 0$.

So, the common chord (i) passes through the centre of the second circle, i.e. (1, −4).

$\therefore \quad 6 - 56 + c + d = 0 \Rightarrow c + d = 50$

15. (b)The centre of the given circle is (4, 3) and touches the line

$$x + y = 1 \quad \ldots\text{(i)}$$

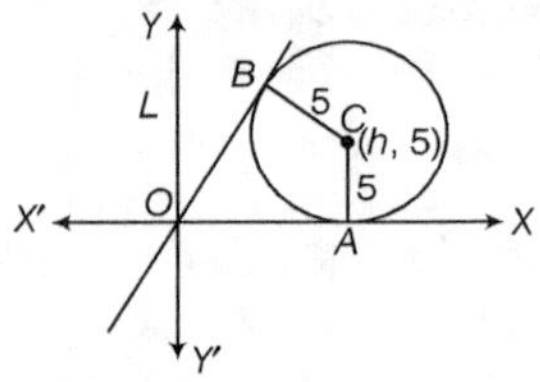

The equation of tangent is

$$(y - 3) = m(x - 4)$$

$$\Rightarrow \quad y - 3 = 1(x - 4) \quad [\because m = 1]$$

$$\Rightarrow \quad y = x - 1 \quad \ldots\text{(ii)}$$

On solving Eqs. (i) and (ii), we get

$$x = 1, y = 0$$

$\therefore$ Point of contant is (1, 0).

16. (a) Let x_1, x_2 and y_1, y_2 be the roots of the equations $x^2 + 2ax - b^2 = 0$ and $x^2 + 2pyx - q^2 = 0$ respectively.

Then, $\quad x_1 + x_2 = -2a, x_1x_2 = -b^2$

and $\quad y_1 + y_2 = -2p, y_1y_2 = -q^2$

The equation of the circle with $P(x_1, y_1)$ and $Q(x_2, y_2)$ as the end point of the diameter, is

$$(x - x_1)(x - x_2) + (y - y_1)(y - y_2) = 0$$

$$\Rightarrow x^2 + y^2 - x(x_1 + x_2) - y(y_1 + y_2) + x_1x_2 + y_1y_2 = 0$$

$$\Rightarrow \quad x^2 + y^2 + 2ax + 2py - b^2 - q^2 = 0$$

$$\text{Radius} = \sqrt{a^2 + p^2 - (-b^2 - q^2)} = \sqrt{a^2 + p^2 + b^2 + q^2}$$

17. (a) $\frac{k}{1} = \tan 30°$

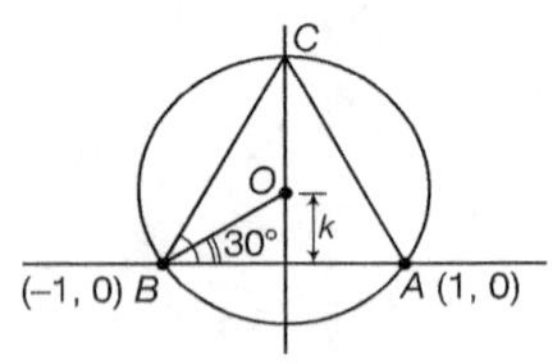

$$k = \frac{1}{\sqrt{3}} \quad \ldots\text{(i)}$$

$\therefore$ Centre $\left(0, \frac{1}{\sqrt{3}}\right)$ and radius $= \frac{2}{\sqrt{3}}$

Thus, equation of circle is

$$(x)^2 + \left(y - \frac{1}{\sqrt{3}}\right)^2 = \frac{4}{3}$$

18. (a) On comparing

$$ax^2 + 2hxy + by^2 + 2gx + 2fy + c = 0$$

with $\quad x^2 + y^2 + 2gx + 2fy + c = 0$

We find that the given equation will represent a circle, if

$$a = b \neq 0 \quad \text{and} \quad h = 0$$

19. (b) $x = 3 + 2\cos 2t$

and $\quad y = 4 + 2\sin 2t$

$\therefore \quad (x + 3)^2 + (y - 4)^2 = 2^2 \quad (\cos^2 2t + \sin^{-2} 2t)$

Thus, required locus is

$$(x + 3)^2 + (y - 4)^2 = 4$$

20. (c) Here, $|OP| = \frac{|1|}{\sqrt{m^2 + 1}}$

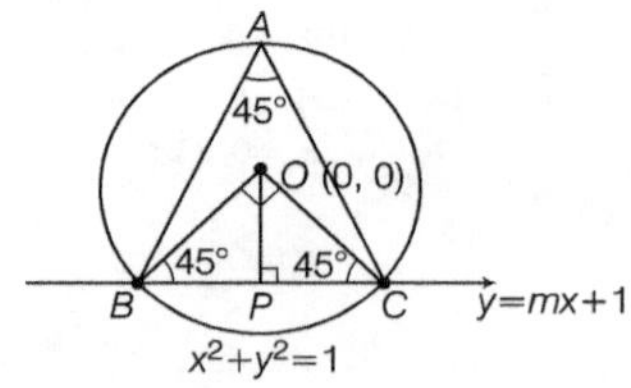

and in ΔOPC, $\quad \sin 45° = \frac{OP}{OC}$

$$\Rightarrow \quad \frac{1}{\sqrt{2}} = \frac{1}{\sqrt{m^2 + 1}}$$

or $\quad m^2 + 1 = 2$

$$m = \pm 1$$

21. (a) Let $P(h, k)$ be any point of a circle of radius 3 units and having its centre on $x^2 + y^2 = 25$.

Then, $\quad 5 - 3 \leq OP \leq 5 + 3$

$$\Rightarrow \quad 2 \leq OP \leq 8$$

$$\Rightarrow \quad 4 \leq OP^2 \leq 64$$

$$\Rightarrow \quad 4 \leq h^2 + k^2 \leq 64$$

Hence, the locus of $P(h, k)$ is

$$4 \leq x^2 + y^2 \leq 64$$

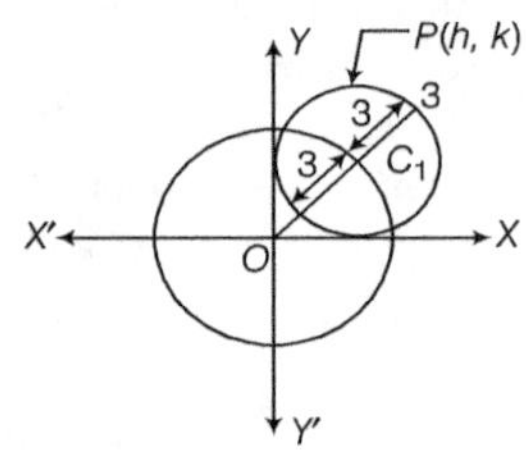

22. (a) Here, $2\sqrt{g^2 - c} = 5$

and $\quad g = -\frac{9}{2}$

$$\therefore \quad g^2 - c = \frac{25}{4}$$

or $\quad c = \frac{81 - 25}{4} = 14$

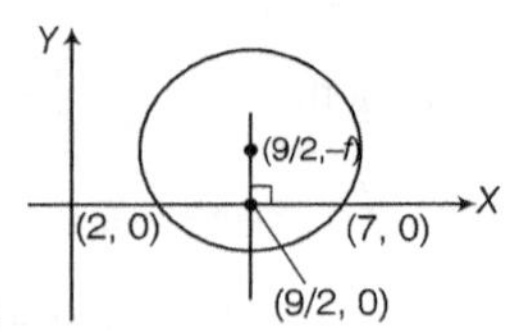

$\therefore$ Required equation of circle, is

$$x^2 + y^2 - 9x + 2fy + 14 = 0$$

23. (b) The centre and radius of the given circle are $\left(\frac{1}{2}, -\frac{3}{2}\right)$ and $\sqrt{\frac{5}{2}}$, respectively.

Let $y = mx$ be the equation of L_1. Then,

p_1 = Length of the intercept on L_1

$$= 2\sqrt{\left(\sqrt{\frac{5}{2}}\right)^2 - \left(\frac{m + 3}{2\sqrt{m^2 + 1}}\right)^2} = 2\sqrt{\frac{5}{2} - \frac{(m + 3)^2}{4(m^2 + 1)}}$$

and $\quad p_2$ = Length of the intercept on L_1,

$$= 2\sqrt{\left(\sqrt{\frac{5}{2}}\right)^2 - (\sqrt{2})^2} = 2\sqrt{\frac{5}{2} - 2} = \sqrt{2}$$

Now, $p_1 = p_2$

$\Rightarrow \quad 2\sqrt{\frac{5}{2} - \frac{(m+3)^2}{4(m^2+1)}} = \sqrt{2}$

$\Rightarrow \quad 5 - \frac{(m+3)^2}{4(m^2+1)} = 1$

$\Rightarrow \quad 7m^2 - 6m - 1 = 0$

$\Rightarrow \quad (m-1)(7m+1) = 0$

$\therefore \quad m = 1, -\frac{1}{7}$

So, the equations of L_1 are

$y = x$ and $7y = -x$

$\Rightarrow \quad x + 7y = 0$ and $x - y = 0$

24. (a) In figure, $\tan\theta = \frac{RS}{PR}$

and $\tan\left(\frac{\pi}{2} - \theta\right) = \frac{PQ}{PR}$

$\Rightarrow \quad 1 = \frac{RS}{PR} \times \frac{PQ}{PR}$

$\Rightarrow \quad PR = \sqrt{PQ \times RS}$

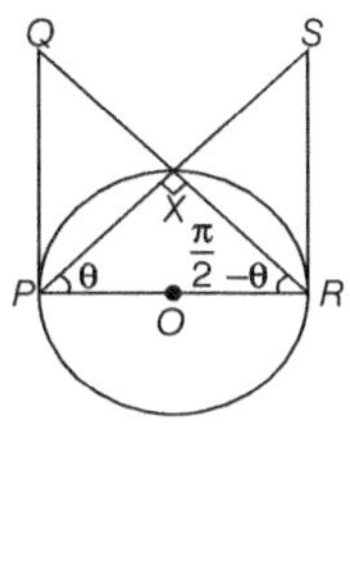

25. (b) $BC = 2a + 2$

Also, $\frac{1}{a} = \tan 30°$

$\therefore \quad a = \sqrt{3}$

Side $BC = 2(\sqrt{3} + 1)$

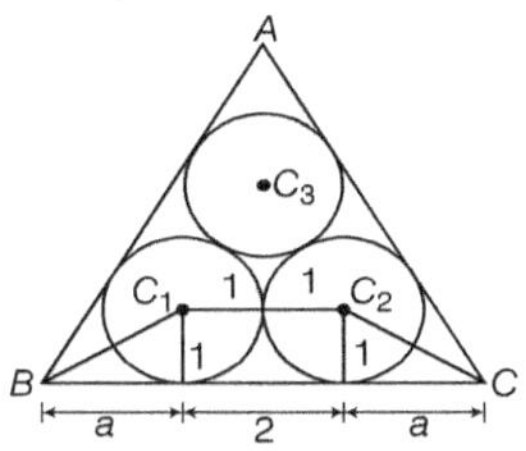

Area of equilateral $\Delta = \frac{\sqrt{3}}{4}(BC)^2 = \frac{\sqrt{3}}{4} \cdot 4(3 + 1 + 2\sqrt{3})$

$= \sqrt{3}(4 + 2\sqrt{3}) = (6 + 4\sqrt{3})$ sq units

26. (d) The family of circles having (h, k) as the centre and touching X-axis is

$$(x-h)^2 + (y-k)^2 = k^2 \quad \ldots(i)$$

It passes through $(-1, 1)$.

$\therefore \quad (h+1)^2 + (1-k)^2 = k^2$

$\Rightarrow \quad h^2 + 2h + 2 - 2k = 0$

$\Rightarrow \quad 4 - 4(2 - 2k) \geq 0 \quad \Rightarrow k \geq \frac{1}{2}$

[$\because h$ is real, so disc > 0]

27. (b) When, $a = b$ and c is of sign opposite of that of a.

$\therefore \quad ax^2 + by^2 + c = 0$ is circle

and $x^2 - 5xy + 6y^2 = 0$ is a pair of straight line.

28. (c)

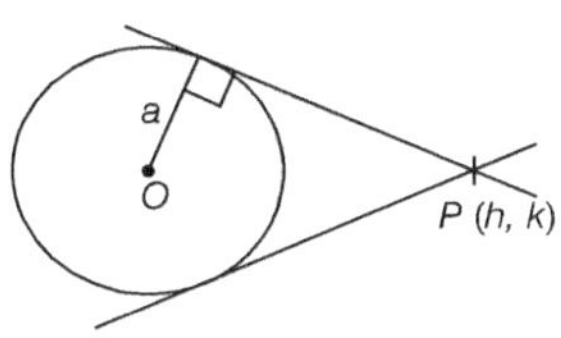

$m_1 = \tan\alpha, \quad m_2 = \tan\beta$

Equation of tangent, is

$y = mx \pm a\sqrt{1 + m^2}$, passes through (h, k).

$\Rightarrow \quad (k - mh)^2 = a^2(1 + m^2)$

$\Rightarrow \quad k^2 - 2khm + m^2h^2 = a^2 + a^2m^2$

$\Rightarrow \quad m^2(a^2 - h^2) + (2kh)m + (a^2 - k^2) = 0$

Given, $\cot\alpha + \cot\beta = 0$

$\Rightarrow \quad \frac{1}{m_1} + \frac{1}{m_2} = 0$

or $\quad m_1 + m_2 = 0$

$\Rightarrow \quad 2kh = 0$

or locus is $xy = 0$

29. (b) Let $P(x_1, y_1)$ be a point on the circle

$$x^2 + y^2 + 2gx + 2fy + c = 0$$

Then, the length of the tangents drawn from $P(x_1, y_1)$ to the circle $x^2 + y^2 + 2gx + 2fy + c\sin^2\alpha + (g^2 + f^2)\cos^2\alpha = 0$ is given by

$PQ = PR$

$\Rightarrow \quad PQ = \sqrt{x_1^2 + y_1^2 + 2gx_1 + 2fy_1 + c\sin^2\alpha + (g^2 + f^2)\cos^2\alpha}$

$= \sqrt{-c + c\sin^2\alpha + (g^2 + f^2)\cos^2\alpha}$

$= \sqrt{(g^2 + f^2 - c)}\cos\alpha$

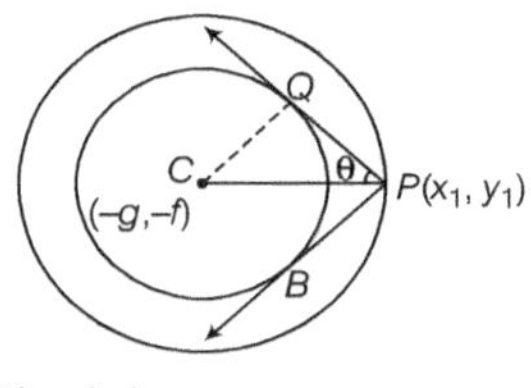

The radius of the circle

$x^2 + y^2 + 2gx + 2fy + c\sin^2\alpha + (g^2 + f^2)\cos^2\alpha = 0$

is $\quad CQ = CR$

$\Rightarrow \quad CQ = \sqrt{[g^2 + f^2 - c\sin^2\alpha - (g^2 + f^2)]\cos^2\alpha}$

$\Rightarrow \quad CQ = (\sqrt{g^2 + f^2 - c})\sin\alpha$

In ΔCPQ, we have

$\tan\theta = \frac{CQ}{PQ} = \frac{\sqrt{(g^2 + f^2 - c)}\sin\alpha}{\sqrt{(g^2 + f^2 - c)}\cos\alpha}$

$\tan\theta = \tan\alpha \quad \Rightarrow \quad \theta = \alpha$

Hence, $\angle QPR = 2\alpha$

30. (c) The given equation of the line is

$x\cos\alpha + y\sin\alpha = p$

$\Rightarrow \quad x\cos\alpha + y\sin\alpha - p = 0 \quad \ldots(i)$

The given equation of the circle is

$x^2 + y^2 - 2q\,x\cos\alpha - 2q\,y\sin\alpha = 0$

having centre $(q\cos\alpha, q\sin\alpha)$ and

radius $= \sqrt{q^2\cos^2\alpha + q^2\sin^2\alpha} = q$

The length of the perpendicular from the centre will be equal to the radius.

$\therefore \quad \left|\frac{q\cos^2\alpha + q\sin^2\alpha - p}{1}\right| = q$

$\Rightarrow \quad q - p = \pm q$

$\Rightarrow \quad q - p = q$

or $\quad q - p = -q$

$\Rightarrow \quad p = 0$

or $\quad p = 2q$

31. (b) Let the coordinates of A and B be $(a, 0)$ and $(0, b)$, respectively.

Since, the ΔOAB is a right angled triangle, then the centre of the circumcircle is the mid-point of AB, i.e. $\left(\frac{a}{2}, \frac{b}{2}\right)$ and the radius is equal to $\frac{1}{2}AB = \frac{1}{2}\sqrt{a^2 + b^2}$.

The equation of the circumcircle is

$$x^2 + y^2 - ax - by = 0$$

The equation of the tangent at the origin to this circle is

$$ax + by = 0$$

$\therefore$ $$m = \frac{a^2}{\sqrt{a^2 + b^2}}$$

and $$n = \frac{b^2}{\sqrt{a^2 + b^2}}$$

$\Rightarrow$ $$m + n = \sqrt{a^2 + b^2} = \text{diameter of circle}$$

32. (a) Equation of tangent is $x \cdot (1) + y(\sqrt{3}) = 4$

$\therefore$ $A(4, 0)$

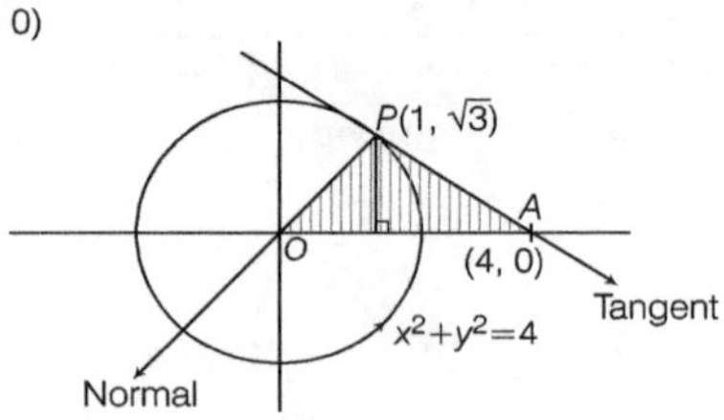

Thus, area of $\Delta = \frac{1}{2} \times 4 \times \sqrt{3} = 2\sqrt{3}$ sq units

33. (d) **Case I** : When circle lies below X-axis.

Locus of centre is $(0, y)$, when $y \le 0$

Case II : When circle lies above X-axis.

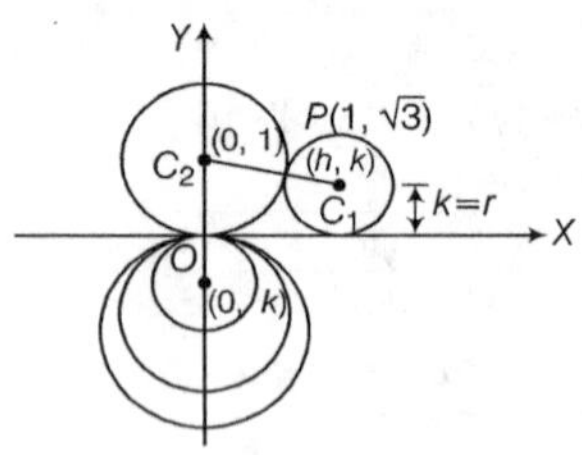

$$C_1C_2 = r_1 + r_2$$

$\Rightarrow$ $$\sqrt{h^2 + (k-1)^2} = k + 1$$

or $$h^2 + (k-1)^2 = (k+1)^2$$

$\Rightarrow$ $$h^2 = 4k \quad \text{or} \quad x^2 = 4y$$

34. (c) Here, $x^2 + y^2 = 16$

and $(x - h)^2 + (y - k)^2 = (h - 1)^2 + (k - 2)^2$

or $x^2 + y^2 - 2hx - 2ky + 2h + 4k - 5 = 0$

are orthogonal.

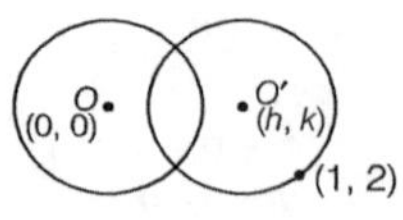

$\Rightarrow$ $2(g_1g_2 + f_1f_2) = c_1 + c_2$

$\therefore$ $2(0 + 0) = 2h + 4k - 5 - 16$

or $2x + 4y - 21 = 0$

35. (c) Here,

$$PT_1^2 = x_1^2 + y_1^2 - a^2$$
$$PT_2^2 = x_1^2 + y_1^2 - b^2$$
$$PT_3^2 = x_1^2 + y_1^2 - c^2$$

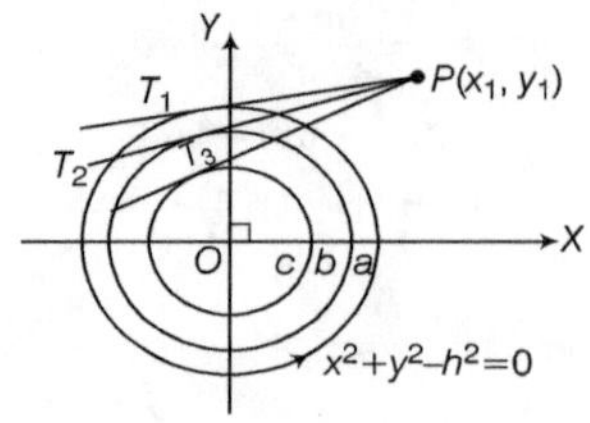

Since, PT_1^2, PT_2^2, PT_3^2 are in an AP.

$\Rightarrow$ $$2\,PT_2^2 = PT_1^2 + PT_3^2$$

or $$2(x_1^2 + y_1^2 - b^2) = (x_1^2 + y_1^2 - a^2) + (x_1^2 + y_1^2 - c^2)$$

or $$2b^2 = a^2 + c^2$$

$\therefore$ a^2, b^2, c^2 are in an AP.

36. (a) Radical axis of the circle is given by

$$S_1 - S_2 = 0$$

$$x^2 + y^2 - \frac{7}{3}x + \frac{8}{3}y + \frac{11}{3} - (x^2 + y^2 - 3x - 4y + 5) = 0$$

$\Rightarrow$ $$-\frac{7}{3}x + 3x + \frac{8}{3}y + 4y + \frac{11}{3} - 5 = 0$$

$\Rightarrow$ $$2x + 20y = 4$$

$\Rightarrow$ $$x + 10y = 2$$

37. (b) The equation of the radical axis

$$S_1 - S_2 = 0$$

$$(x^2 + y^2 - 16) - [(x-4)^2 + (y-5)^2 - 25] = 0$$

$\Rightarrow$ $$(x^2 + y^2 - 16) - (x^2 + y^2 - 8x - 10y + 16) = 0$$

$\Rightarrow$ $$8x + 10y = 32$$

$\Rightarrow$ $$4x + 5y = 16$$

38. (b) Let the equation of circle through (a, b) be

$$x^2 + y^2 + 2gx + 2fy + c = 0 \quad \text{...(i)}$$

where, $$a^2 + b^2 + 2ag + 2fb + c = 0 \quad \text{...(ii)}$$

Since, the circle $x^2 + y^2 = 4$ cuts the circle (i) orthogonally.

$\therefore$ $$2g \times 0 + 2f \times 0 = c - 4$$

$\Rightarrow$ $$c = 4$$

On substituting the value of c in Eq. (ii), we get

$$a^2 + b^2 + 2ag + 2fb + 4 = 0$$

Hence, the locus of $(-g, -f)$ is

$$a^2 + b^2 - 2ax - 2by + 4 = 0$$

$\Rightarrow$ $$2ax + 2by - (a^2 + b^2 + 4) = 0$$

39. (a) $S + \lambda L = 0$

$\Rightarrow$ $$(x^2 + y^2 - a^2) + \lambda\left(x \cdot \frac{\sqrt{3}}{2} + y \cdot \frac{1}{2} + 1\right) = 0$$

or $$x^2 + y^2 + \frac{\sqrt{3}}{2}\lambda x + \frac{\lambda}{2}y + (\lambda - a^2) = 0$$

Having centre $\left(-\frac{\sqrt{3}}{4}\lambda, -\frac{\lambda}{4}\right)$ lying on $x \cdot \frac{\sqrt{3}}{2} + \frac{y}{2} + 1 = 0$

$\Rightarrow$ $$\frac{\sqrt{3}}{2}\left(-\frac{\sqrt{3}}{4}\lambda\right) - \frac{\lambda}{8} + 1 = 0$$

$\Rightarrow$ $$-3\lambda - \lambda + 8 = 0$$

or $$\lambda = 2$$

40. (c) The equation of the common chord of the circles $(x-a)^2+y^2=a^2$ and $x^2+(y+b)^2=b^2$ is

$$ax+by=0 \quad \text{...(i)}$$

The equation of the required circle is

$$\{(x-a)^2+y^2-a^2\}+\lambda(ax+by)=0$$

$$\Rightarrow \quad x^2+y^2+x(a\lambda-2a)+\lambda by=0$$

Since, Eq. (i) is a diameter of circle (ii).

$$\therefore \quad a\left\{\frac{-(a\lambda-2a)}{2}\right\}+b\left(\frac{-\lambda b}{2}\right)=0 \Rightarrow \lambda=\frac{2a^2}{a^2+b^2}$$

Putting the value of λ in Eq. (ii), the equation of the required circle is

$$(a^2+b^2)(x^2+y^2)=2ab(bx-ay)$$

41. (a) The equation of the required circle is

$$x^2+y^2-a^2+\lambda(x\cos\alpha+y\sin\alpha-p)=0$$

The centre of this circle is $\left(\frac{-\lambda}{2}\cos\alpha, \frac{-\lambda}{2}\sin\alpha\right)$

which lies on $x\cos\alpha+y\sin\alpha=p$

$$\therefore \quad \lambda=-2p$$

Putting the value of λ in the equation of the required circle is

$$x^2+y^2-a^2-2px\cos\alpha-2py\sin\alpha+2p^2=0$$

$$\Rightarrow \quad x^2+y^2-a^2-2p(x\cos\alpha+y\sin\alpha-p)=0$$

42. (a) The equation of the family of co-axial circle having $x=\frac{a}{2}$ as radial axis with respect to $x^2+y^2=a^2$ is

$$x^2+y^2-a^2+\lambda\left(x-\frac{a}{2}\right)=0 \quad \text{...(i)}$$

It is passes through $(2a, 0)$, then

$$4a^2-a^2+\lambda\left(2a-\frac{a}{2}\right)=0$$

$$\Rightarrow \quad \lambda=-2a$$

On putting $\lambda=-2a$ in Eq. (i), we get

$$x^2+y^2-2ax=0$$

43. (a) The given equations of circle are

$$x^2+y^2+2ax+c=0 \quad \text{...(i)}$$

and $$x^2+y^2+2by+c=0 \quad \text{...(ii)}$$

The equation of the common chord is

$$2ax-2by=0$$

$$\Rightarrow \quad ax-by=0$$

The equation of a circle passing through the intersection of the given circle is

$$(x^2+y^2+2ac+c)+\lambda(x^2+y^2+2by+c)=0$$

$$\Rightarrow \quad x^2(1+\lambda)+y^2(1+\lambda)+2ax+2by\lambda+c(1+\lambda)=0$$

$$\Rightarrow \quad x^2+y^2+\frac{2ax}{1+\lambda}+\frac{2by\lambda}{1+\lambda}+c=0 \quad \text{...(i)}$$

Since, $ax-by=0$ is a diameter of the circle. Therefore, its centre $\left(\frac{-a}{1+\lambda}, \frac{-b\lambda}{1+\lambda}\right)$ lies on it.

$$\therefore \quad a\left(\frac{-a}{1+\lambda}\right)-b\left(\frac{-b\lambda}{1+\lambda}\right)=0$$

$$\Rightarrow \quad -a^2+b^2\lambda=0 \quad \Rightarrow \quad \lambda=\frac{a^2}{b^2}$$

On putting $\lambda=\frac{a^2}{b^2}$ in Eq. (i), we get

$$x^2+y^2+\frac{2ax}{\left(1+\frac{a^2}{b^2}\right)}+\frac{2by\left(\frac{a^2}{b^2}\right)}{\left(1+\frac{a^2}{b^2}\right)}+c=0$$

$$\Rightarrow \quad x^2+y^2+\frac{2ab^2x}{a^2+b^2}+\frac{2b(a^2)y}{a^2+b^2}y+c=0$$

$$\Rightarrow \quad x^2+y^2+\frac{2ab^2}{a^2+b^2}x+\frac{2a^2b}{a^2+b^2}y+c=0$$

44. (c) Given circle, $S_1\equiv x^2+y^2-4=0$ ...(i)

bisects the circumference of the circle.

$$S_2\equiv x^2+y^2-2x+6y+a=0 \quad \text{...(ii)}$$

The common chord of the given circle is

$$S_1-S_2=0$$

$$\Rightarrow \quad -4+2x-6y-a=0$$

$$\Rightarrow \quad (2x-6y)+(-4-a)=0$$

It passes through the centre of the second circle $(1,-3)$.

$$\therefore \quad 2+18-4-a=0$$

$$\Rightarrow \quad a=16$$

45. (b) Let (h, k) be the coordinates of the centre of the circle of which the given chord is the diameter. Then, (h, k) is the mid-point of the chord and so its equation is

$$S'=T$$

$$\Rightarrow \quad h^2+k^2-2ah=hx+ky-a(x+h)$$

$$\Rightarrow \quad x(h-a)+ky=h^2+k^2-ah$$

If it passes through $(0, 0)$, then

$$h^2+k^2-ah=0$$

Hence, the locus of (h, k) is $x^2+y^2-ax=0$.

46. (d) $S: x^2+y^2-2x-7$

$$\therefore \quad S_1: 1+25-2-7 \text{ at } (1, 5)$$

$$\Rightarrow \quad S_1>0$$

Thus, $(1, 5)$ lies outside and 2 circles can be drawn from point $P(1, 5)$ as its centre, touching the circle

$$x^2+y^2-2x=7$$

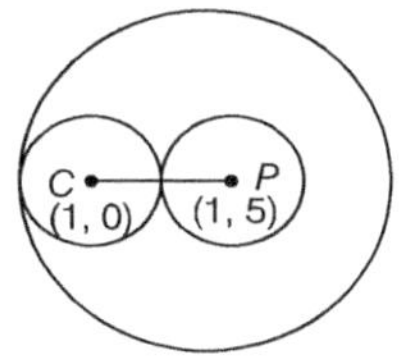

47. (a) **Statement I** The given circles

$$S_1\equiv x^2+y^2-2x+4y+4=0$$

$$S_2\equiv x^2+y^2+4x-2y+c=0$$

The equation of the common chord is

$$6x-6y+c-4=0$$

It passes through the centre $(1,-2)$, if it is longest, then

$$6+12+c-4=0$$

$$\Rightarrow \quad c=-14$$

Statement II If the circles intersect, then the common chord is longest, it is the diameter of the smaller circle.

48. (d) Locus of P is the polar of $A(6, 8)$ w.r.t. the circle $S=0$, i.e.

$$x^2+y^2-6x-8y+5=0$$

$$\therefore \quad xx_1+yy_1+g(x+x_1)+f(y+y_1)+c=0$$

$$6x+8y-3(x+6)-4(y+8)+5=0$$

$$\Rightarrow \quad 3x+4y-45=0$$

Also, $A(6, 8)$ lies outside the circle.

as $\quad S_1: 48+64-36-64+5>0$

$\therefore$ Statement I is incorrect and Statement II is correct.

49. (d) At $O(0, 0), 2x + y - 3$

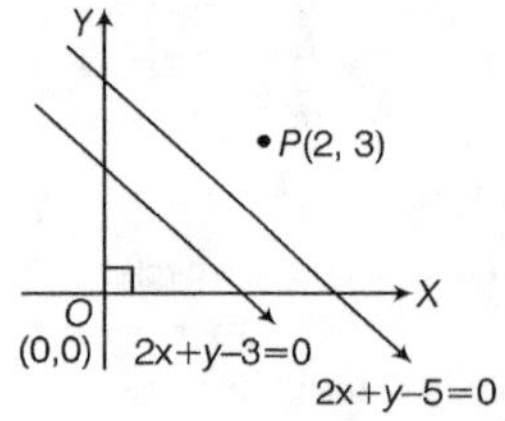

$$\Rightarrow \quad 0 + 0 - 3 < 0$$

At $P(2, 3) \Rightarrow 4 + 3 - 3 > 0$

$\therefore$ O and P lies on opposite side of $2x + y - 3 = 0$.

At $O(0, 0)$, $\quad 2x + y - 5$

$\Rightarrow \quad -5 < 0$

at $P(2, 3)$

$\Rightarrow \quad 4 + 3 - 5 > 0$

$\therefore$ O and P lies on opposite side. Thus, point lies on the same side of the line and $P(2, 3)$ does not lie between straight lines.

50. (a) $\dfrac{AK}{OA} = \cos\theta = \dfrac{AB}{AK}$

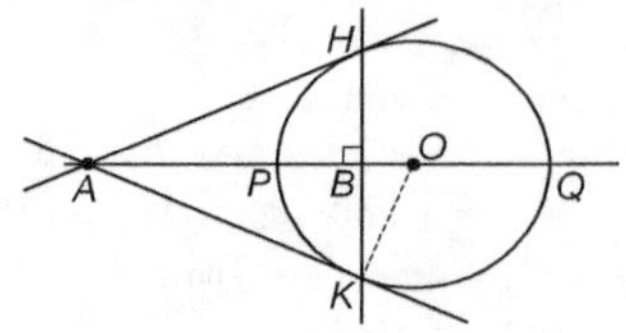

$\Rightarrow \quad (AK)^2 = (OA)(AB) = (AP)(AQ) \quad [\because (AK)^2 = AP \cdot AQ]$

Also, $\quad OA = \dfrac{AP + PQ}{2}$

$[AQ - AO = r = AO - AP \Rightarrow 2AO = AQ + QP]$

$\Rightarrow \quad (AP)(AQ) = AB \cdot \left[\dfrac{AP + AQ}{2}\right]$

$\Rightarrow AB = \dfrac{2(AP)(AQ)}{AP + AQ}$

Thus, Statement I is correct and the Statement II is correct explanation of Statement I.

Revisal Problems (JEE Advanced)

1. (d) Given, the vertices of a triangle are $A(0, 6)$, $B(5, 5)$ and $C(-1, 1)$.

Let the equation of the circumcircle of ΔABC be

$$x^2 + y^2 + 2gx + 2fy + c = 0 \quad \ldots\text{(i)}$$

It passes through $A(0, 6)$, $B(5, 5)$ and $C(-1, 1)$.

$\therefore \quad 36 + 12f + c = 0 \quad \ldots$(ii)

$50 + 10g + 10f + c = 0 \quad \ldots$(iii)

and $\quad 2 - 2g + 2f + c = 0 \quad \ldots$(iv)

On solving Eqs. (ii), (iii) and (iv), we get

$g = -2, f = -3$ and $c = 0$

On substituting the values of g, f and c in Eq. (i), we get

$x^2 + y^2 - 4x - 6y = 0$

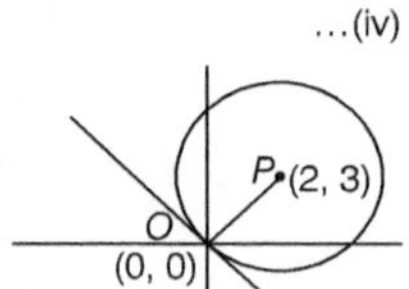

Slope of $OP = \dfrac{3 - 0}{2 - 0} = \dfrac{3}{2}$

Hence, the slope of tangent $= \dfrac{-2}{3}$

So, the required equation of tangent is $y = \dfrac{-2}{3}x$.

2. (d) Equation of the line is

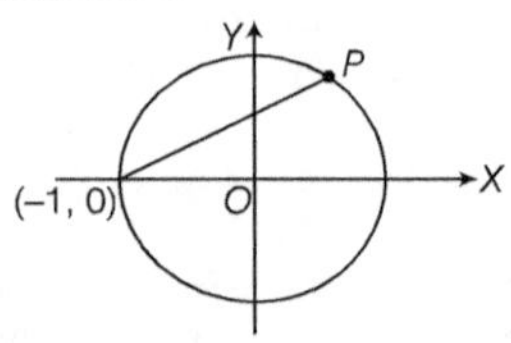

$$(y - 0) = m(x + 1) \quad \ldots\text{(i)}$$

On solving it with $x^2 + y^2 = 1$, we get

$$x^2 + m^2(x + 1)^2 = 1$$

$(m^2 + 1)x^2 + 2m^2x + (m^2 - 1) = 0, \ m \in Q$.

$\therefore \quad x_1 x_2 = \dfrac{m^2 - 1}{m^2 + 1}$

But $x_1 = -1 \Rightarrow x_2 = \dfrac{1 - m^2}{m^2 + 1} = x$ coordinate of P.

Since, $m \in Q$. Hence, x will be rational.

If x is rational, then y is also rational from Eq. (i).

3. (c) Solving $y = 7x + 5$ and the circle

$$x^2 + y^2 = 1$$

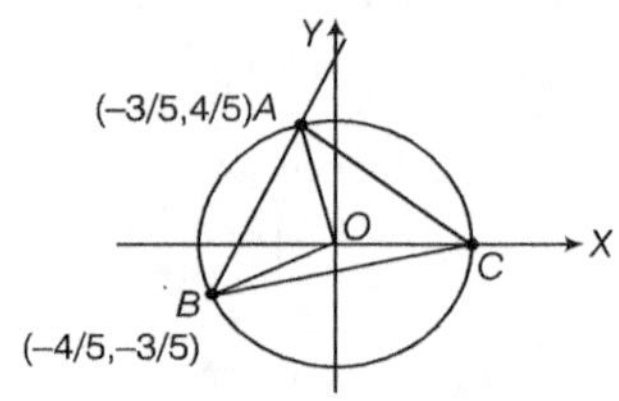

$A\left(-\dfrac{3}{5}, \dfrac{4}{5}\right)$ and $B\left(-\dfrac{4}{5}, -\dfrac{3}{5}\right)$.

$m_{OA} = -\dfrac{4}{3}, \ m_{OB} = \dfrac{3}{4}$

Hence, $\quad \angle AOB = 90°$

$\Rightarrow \quad \angle ACB = \dfrac{\pi}{4} = \tan^{-1}(1)$

4. (d) Let $O(0, 0)$, $P(6, 8)$ and $Q(12, 6)$.

As shown in the figure, the shortest route consists of tangent OT, minor arc TR and tangent RQ.

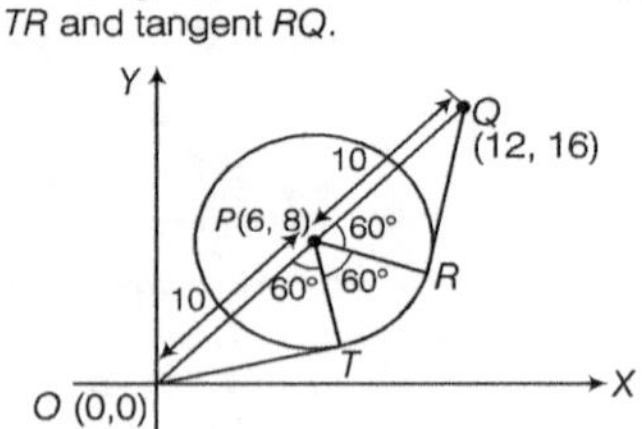

Since, $OP = 10$, $PT = 5$ and $\angle OTP = 90°$, it follows that $\angle OPT = 60°$ and $OT = 5\sqrt{3}$.

By similar reasoning,

$$\angle QPR = 60° \text{ and } QR = 5\sqrt{3}$$

Because O, P and Q are collinear.

$\angle RPT = 60°$, so arc TR is of length $5\dfrac{\pi}{3}$.

Hence, the length of the shortest route is $\left(10\sqrt{3} + \dfrac{5\pi}{3}\right)$.

5. (b)

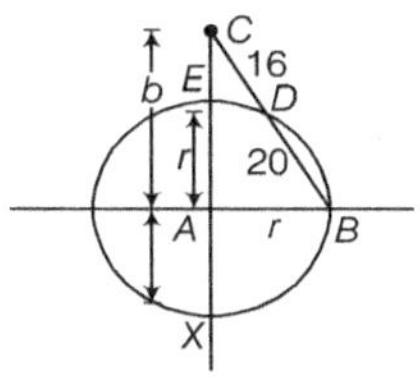

From figure, $b^2 + r^2 = (36)^2$ ...(i)

Also, $CD \cdot CB = CE \cdot CX$

$\Rightarrow \quad 16 \cdot (36) = (b - r)(b + r)$

$\therefore \quad b^2 - r^2 = 16 \cdot (36)$ ...(ii)

From Eqs. (i) and (ii), we get

$$2b^2 = 36(36 + 16) = 36 \cdot (52)$$

$\Rightarrow \quad b = 6\sqrt{26}$ units

6. (c) The pair of tangents are given by $SS_1 = T^2$

$\therefore \quad (x^2 + y^2 + 4x + 6y + 9)(9) = (2x + 3y + 9)^2$

$$9(x^2 + y^2) = (2x + 3y)^2$$

7. (a) Let equation of the line be $y = x + c$.

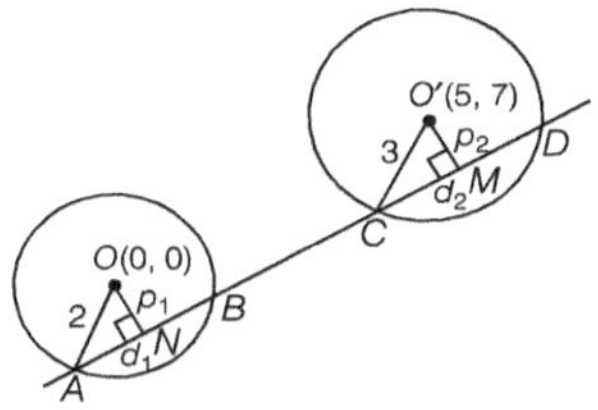

$$y - x = c \quad \text{...(i)}$$

Perpendicular from (0, 0) on Eq. (i) is $\left|\frac{-c}{\sqrt{2}}\right| = \frac{c}{\sqrt{2}}$.

In ΔAON, $\sqrt{2^2 - \left(\frac{c}{\sqrt{2}}\right)^2} = AN$

and in ΔCPM, $\sqrt{3^2 - 2 - \frac{c}{\sqrt{2}}} = CM$

Perpendicular from (5, 7) on line $y - x = c = \frac{2-c}{\sqrt{2}}$.

Given, $AN = CM = 4 - \frac{c^2}{2} = 9 - \frac{(2-c)^2}{2}$

$\Rightarrow \quad c = -\frac{3}{2}$

$\therefore$ Equation of line $y = x - \frac{3}{2}$ or $2x - 2y - 3 = 0$

8. (d)

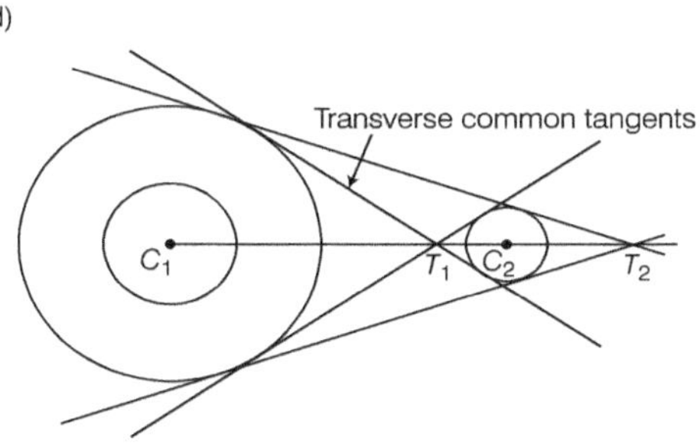

When circles neither intersect nor touch each other, then four tangents can be drawn to two circles. Two of them are direct common tangents and the other two are transverse common tangents.

9. (a) Let the equation of the variable line is

$$xx_1 + yy_1 - (x_1^2 + y_1^2) = 0$$

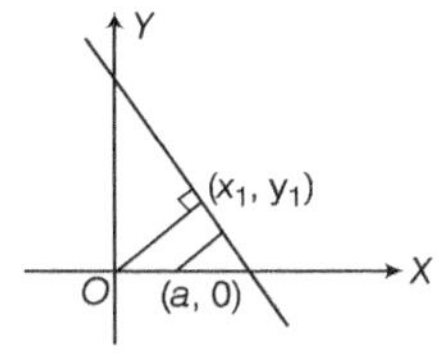

$$P_1P_2 = \left|\frac{x_1^2 + y_1^2}{\sqrt{x_1^2 + y_1^2}}\right| \left|\frac{ax_1 - (x_1^2 + y_1^2)}{\sqrt{x_1^2 + y_1^2}}\right| = k^2$$

i.e. $|x_1^2 + y_1^2 - ax_1| = k^2$

Locus is $x^2 + y^2 - ax \pm k^2 = 0$

$\therefore \quad r^2 = \frac{a^2}{4} - (\pm k^2)$

$\Rightarrow \quad r^2 = \frac{a^2}{4} - k^2$ [not possible]

Thus, $r^2 = \frac{a^2}{4} + k^2$

10. (a) Given, the equation of circle

$$x^2 + y^2 = 6x - 8y$$

$\Rightarrow \quad x^2 + y^2 - 6x + 8y = 0$

$\therefore$ Centre of circle is (3, − 4)

and equation of line is $3x + 4y - 25 = 0$.

Hence, distance of centre from the line $= \left|\frac{9 - 16 - 25}{\sqrt{9 + 16}}\right| = \frac{32}{5}$ units

Now, Shortest distance = Distance of centre from the line − Radius

$$= \frac{32}{5} - 5 = \frac{7}{5} \text{ units}$$

11. (c) Let the pole of first be (x_1, y_1).

$\therefore$ Equation of polar is, $xx_1 + yy_1 - a^2 = 0$

is same as $p_1x + q_1y + r_1 = 0$.

$\Rightarrow \quad \frac{x_1}{p_1} = \frac{y_1}{q_1} = \frac{a^2}{-r_1}$

$\Rightarrow \quad x_1 = \frac{-a^2p_1}{r_1}$ and $y_1 = \frac{-a^2q_1}{r_1}$

Since, (x_1, y_1) lie on $p_2x + q_2y + r_2 = 0$

$\Rightarrow \quad p_2\left(\frac{-a^2p_1}{r_1}\right) + q_2\left(\frac{-a^2q_1}{r_1}\right) + r_2 = 0$

$\therefore \quad a^2(p_1p_2 + q_1q_2) = r_1r_2$

12. (a) Equation of circle is $x^2 + y^2 = 1$ and

equation of line passing through $(x_1, -1)$ and $(x_2, -1)$ is

$$y + 1 = \frac{2}{x_2 - x_1}(x - x_1)$$

$\Rightarrow \quad (x_2 - x_1)y + x_2 - x_1 = 2x - 2x_1$

$\Rightarrow \quad 2x - (x_2 - x_1)y - (x_1 + x_2) = 0$ ...(i)

Since, the above line is tangent to circle.

$\therefore$ Distance of line from centre of circle = Radius

$\Rightarrow \quad \left|\frac{-(x_1 + x_2)}{\sqrt{4 + (x_2 - x_1)^2}}\right| = 1$

$\Rightarrow \quad (x_1 + x_2)^2 = 4 + (x_2 - x_1)^2$

$\Rightarrow \quad x_1^2 + x_2^2 + 2x_1x_2 = 4 + x_2^2 + x_1^2 - 2x_1x_2$

$\Rightarrow \quad 4x_1x_2 = 4$

$\therefore \quad x_1x_2 = 1$

13. (b) In ΔADO,

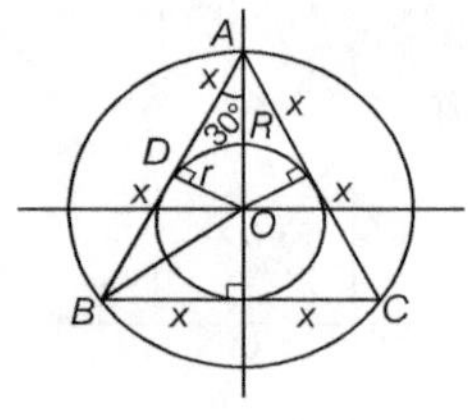

$$\sin 30° = \frac{r}{R}$$

$$\Rightarrow \quad \frac{1}{2} = \frac{r}{R}$$

or $$R = 2r$$

14. (c) Equation of chord of contact of the pair of tangent from (0, 0) and (g, f) are

$$gx + fy + c = 0 \quad \ldots(i)$$

and $$gx + fy + \frac{g^2 + f^2 + c}{2} = 0 \quad \ldots(ii)$$

Distance between them, is

$$\Rightarrow \quad \left| \frac{\frac{g^2 + f^2 + c}{2} - c}{\sqrt{g^2 + f^2}} \right| = \frac{g^2 + f^2 - c}{2\sqrt{g^2 + f^2}}$$

15. (a) Slope of the given line $= -\frac{5}{2}$

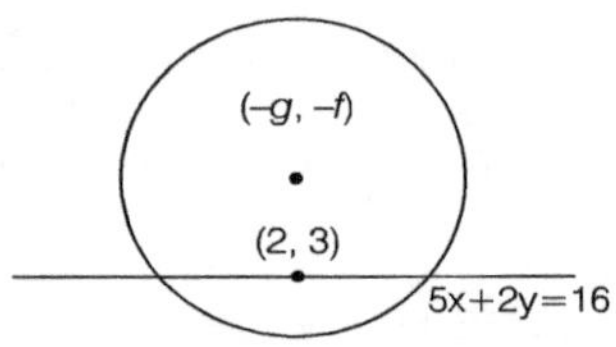

$$\Rightarrow \quad -\frac{5}{2}\left(\frac{3 + f}{2 + g}\right) = -1$$

$$\Rightarrow \quad 15 + 5f = 4 + 2g$$

$\therefore$ Locus is, $2x - 5y + 11 = 0$

16. (d) If $OA \cdot OC = OB \cdot OD$ [power of a point]

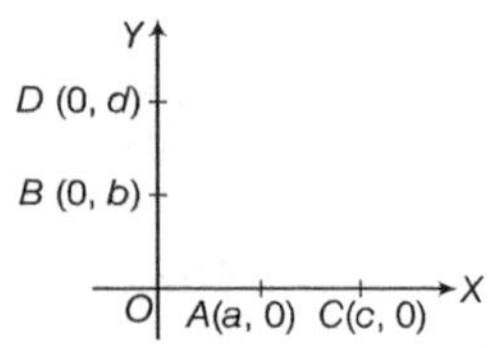

$$\Rightarrow \quad ac = bd$$

17. (b) In ΔOPC,

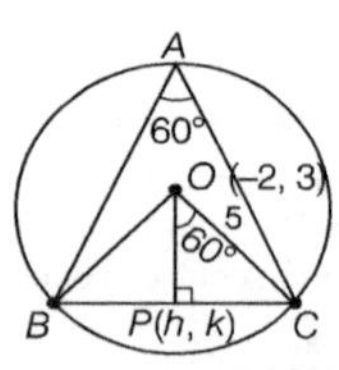

$$\frac{OP}{5} = \cos 60° \Rightarrow OP = 2.5$$

$$\Rightarrow \quad \sqrt{(h + 2)^2 + (k - 3)^2} = 2.5$$

or $$(x + 2)^2 + (y - 3)^2 = 6.25$$

18. (d) Let $(x_1, 0)$ be a mid-point of the chords drawn from the point (p, q). Then, the equation of the chord bisected at $(x_1, 0)$ is

$$x_1^2 - px_1 = xx_1 + 0y - p\left(\frac{x + x_1}{2}\right) - q\left(\frac{y + 0}{2}\right)$$

[using $S' = T$]

$$\Rightarrow \quad 2x_1^2 - 2px_1 = 2xx_1 - p(x + x_1) - qy$$

This passes through (p, q).

$$\therefore \quad 2x_1^2 - 2px_1 = 2px_1 - p(p + x_1) - q^2$$

$$\Rightarrow \quad 2x_1^2 - 3px_1 + (p^2 + q) = 0$$

Now, x_1 is real.

$$\therefore \quad 9p^2 - 8(p^2 + q^2) > 0$$

$$\Rightarrow \quad p^2 - 8q^2 > 0$$

$$\Rightarrow \quad p^2 > 8q^2$$

19. (d) The particle which moves clockwise is moving three times as fast as the particle moving anti-clockwise. This tells us that in the time that the clockwise particle travels $\frac{3}{4}$th of the way around the circle, the anti-clockwise particle will travel (1/4)th, of the way around the circle and so the II particle will not meet $P(0, 1)$. Using the same logic, they will meet at $Q(-1, 0)$, when they meet the 2nd time.

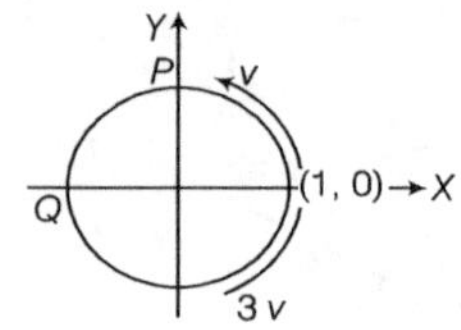

20. (d) Given, $x^2 + y^2 + 2x - 1 = 0$ has centre $(-1, 0)$ and $r = \sqrt{2}$ line $x - y + c = 0$.

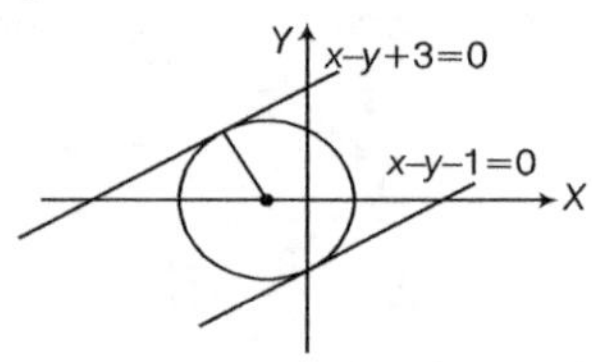

$$\Rightarrow \quad \frac{|c - 1|}{\sqrt{2}} = \sqrt{2} \Rightarrow c - 1 = \pm 2$$

$$\therefore \quad c = 3 \text{ or } -1$$

21. (c) Let the equation of circle be

$$x^2 + y^2 + 2gx + 2fy + c = 0$$

Since, $\left(a, \frac{1}{a}\right), \left(b, \frac{1}{b}\right), \left(c, \frac{1}{c}\right)$

and $\left(d, \frac{1}{d}\right)$ lie on the circle of radius 4.

$$\therefore \quad a^2 + \frac{1}{a^2} + 2ga + \frac{2f}{a} + c = 0$$

$$\Rightarrow \quad a^4 + 2ga^3 + ca^2 + 2fa + 1 = 0$$

$$\Rightarrow \quad abcd = 1$$

22. (d) The centre of the given circle is $(1, -2)$.

Hence, no vertex of the square can have its x-coordinate equal to 1 and its y-coordinate equal to -2.

23. (a) The two circles are

$$x^2 + y^2 - 2x + 2y + 1 = 0$$

and $$x^2 + y^2 - 10x - 6y + 9 = 0$$

$\therefore$ Slope of tangent at $A(1, 0)$ for

$$x^2 + y^2 - 2x + 2y + 1 = 0 \Rightarrow m_1 = 0$$

and slope of tangent for $x^2 + y^2 - 10x - 6y + 9 = 0$ is

$$m_2 = -\frac{4}{3}$$

Thus, $\tan\theta = \left|\frac{-4/3 - 0}{1+0}\right|$

or $\tan\theta = \frac{4}{3}$

$\Rightarrow$ $\theta = \tan^{-1}\left(\frac{4}{3}\right) = \cot^{-1}\left(\frac{3}{4}\right)$

24. (c) The line joining (4, 3) and (2, 1) is also along a diameter. So, the centre of the circle is the intersection of the diameter

$$2x - y = 2$$

and $y - 3 = \frac{3-1}{4-2}(x-4)$

On solving these two equations, the coordinates of centre of the circle are (1, 0).

Also, the radius of circle

= Distance between (1, 0) and (2, 1) = $\sqrt{2}$ units

$\therefore$ Equation of circle is

$$(x-1)^2 + y^2 = 2$$

$\Rightarrow$ $x^2 + y^2 - 2x - 1 = 0$

25. (a) Circles with centres (2, 0) and (−2, 0) each with radius 4 units is shown below.

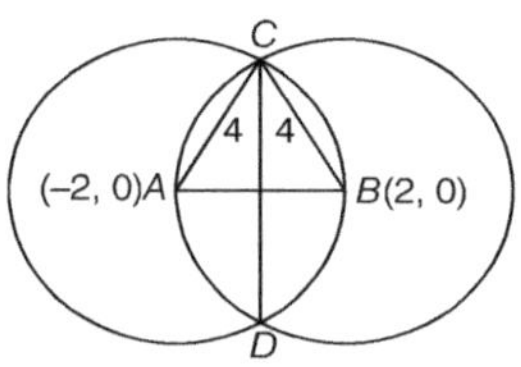

$\Rightarrow$ Y-axis is their common chord.

$\therefore$ ΔABC is equilateral.

Hence, area of $ADBC = \frac{2\sqrt{3}}{4}(4)^2 = 8\sqrt{3}$ sq units

So, the correct option is (a).

26. (d) Coordinates of $C(-a, a\sqrt{3})$ on $x^2 + y^2 = 4a^2$.

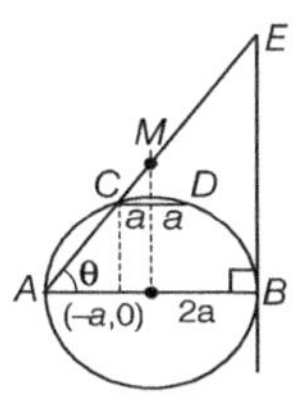

$\therefore$ $\tan\theta = \left|\frac{a\sqrt{3}}{a}\right| = \sqrt{3}$

$\Rightarrow$ $\theta = 60°$

Now, in ΔABE, $\cos 60° = \frac{4a}{AE}$

$\Rightarrow$ $AE = 8a = 2(AB)$

27. (c) Equation of line PQ, is $(y-k) = \frac{-h}{k}(x-h)$

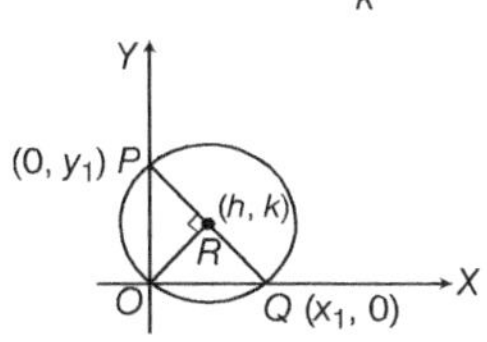

$\Rightarrow$ $hx + ky = h^2 + k^2$

Also, $2a = \sqrt{x_1^2 + y_1^2}$

$\Rightarrow$ $x_1^2 + y_1^2 = 4a^2$

$\therefore$ Points $Q\left(\frac{h^2+k^2}{h}, 0\right)$ and $P\left(0, \frac{h^2+k^2}{k}\right)$

$\Rightarrow$ $\left(\frac{h^2+k^2}{h}\right)^2 + \left(\frac{h^2+k^2}{k}\right)^2 = 4a^2$

or $(x^2+y^2)^2 \cdot \left(\frac{1}{x^2} + \frac{1}{y^2}\right) = 4a^2$

28. (a) Radius, $r = 3k$ and P is centroid.

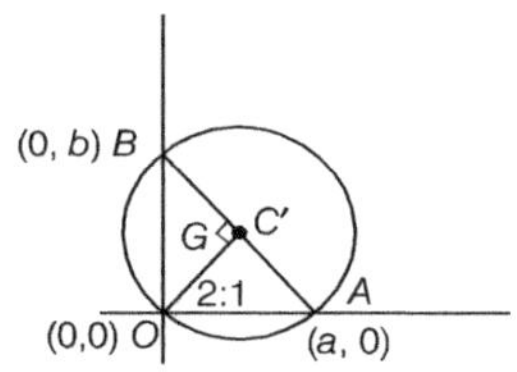

$$g^2 + f^2 = (3k)^2 \quad \text{...(i)}$$

Because $C = 0$

$\therefore$ $3h_1 = a,\ 3k_1 = b$

Also, $2g = a$ and $2f = b$

$\therefore$ $g = \frac{3h_1}{2}$ and $f = \frac{3k_1}{2}$ ...(ii)

From Eqs. (i) and (ii), we get

$$\frac{9h_1^2}{4} + \frac{9k_1^2}{4} = 9k^2$$

or locus is $x^2 + y^2 = 4k^2$.

29. (d) $\because c = f^2$

and $(g-3)^2 + (f-3)^2 = [\sqrt{9+9-14} + \sqrt{g^2+f^2-c}]^2$

$\Rightarrow$ $g^2 + f^2 - 6g - 6f + 18 = (2+g)^2$

$\Rightarrow$ $f^2 - 10g - 6f + 14 = 0$

Locus of centre (g, f) is

$$y^2 - 10x - 6y + 14 = 0$$

30. (c)

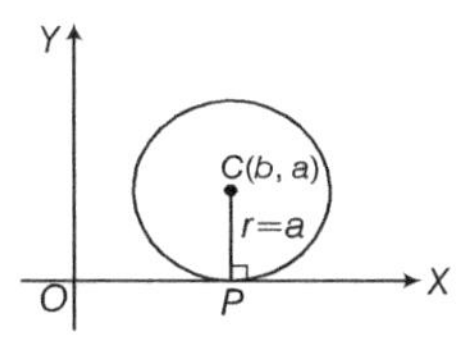

$\therefore$ Equation of circle, is

$$(x-b)^2 + (y-a)^2 = a^2$$

$\Rightarrow$ $x^2 + y^2 - 2bx - 2ay + b^2 = 0$

31. (a) Let $C(\cos\theta, \sin\theta)$;

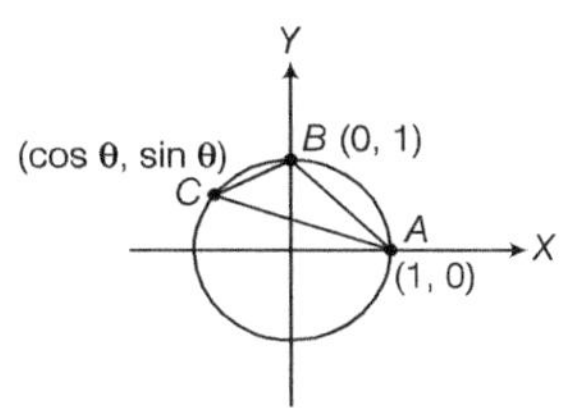

$H(h, k)$ be the orthocentre of the ΔABC.

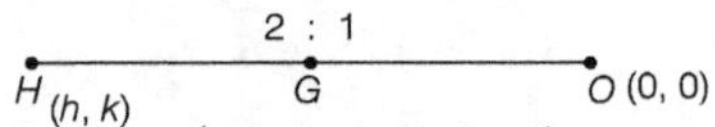

$$\left(\frac{\cos\theta+1}{3}, \frac{\sin\theta+1}{3}\right)$$

$\Rightarrow$ $h = 1 + \cos\theta$

and $k = 1 + \sin\theta$

$\therefore$ $(h-1)^2 + (k-1)^2 = 1$

or $(x-1)^2 + (y-1)^2 = 1$

$\Rightarrow$ $x^2 + y^2 - 2x - 2y + 1 = 0$

32. (c) In ΔCDB,

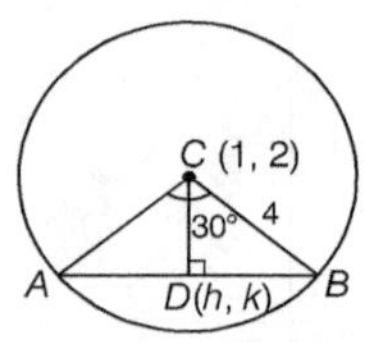

$$\frac{CD}{4} = \cos 30°$$

$\Rightarrow$ $\sqrt{(h-1)^2 + (k-2)^2} = 4\cdot\frac{\sqrt{3}}{2}$

or locus is, $(x-1)^2 + (y-2)^2 = 12$

or $x^2 + y^2 - 2x - 4y - 7 = 0$

33. (d) Since, AC is the diameter, so $\angle ADC = 90°$

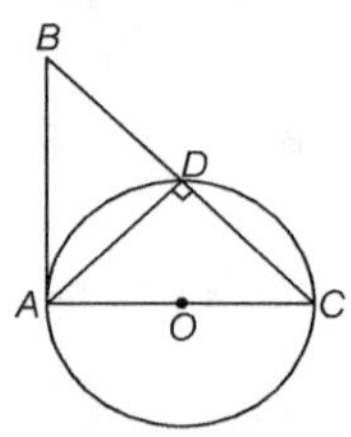

Now, in ΔABC, ΔDAC

$\Rightarrow$ $\angle A = \angle D = 90°$

$\angle C = \angle C =$ Common

Hence, by AA similarity, $\Delta ABC \sim \Delta DAC$.

So, $\frac{AB}{AD} = \frac{BC}{AC} = \frac{AC}{CD}$

$AC = \frac{AD \times BC}{AB}$

$\Rightarrow$ $AC = \frac{AD\cdot\sqrt{AB^2 + AC^2}}{AB}$

$\Rightarrow$ $AC \cdot AB = AD\sqrt{AB^2 + AC^2}$

$\Rightarrow$ $AC^2 \cdot AB^2 = AD^2 \cdot AB^2 + AD^2 \cdot AC^2$

$\Rightarrow AC^2\,[AB^2 - AD^2] = AD^2 AB^2$

$\therefore$ $AC = \frac{AD\cdot AB}{\sqrt{AB^2 - AD^2}}$

34. (c) Given, equation of line is

$x - 2y + 10 = 0$

$\Rightarrow$ $x = 2y - 10$ ...(i)

and equation of circle

$x^2 + y^2 = 100$

$\Rightarrow$ $(2y-10)^2 + y^2 = 100$ [using Eq. (i)]

$\Rightarrow$ $4y^2 + 100 - 40y + y^2 = 100$

$\Rightarrow$ $5y^2 - 40y = 0$

$\Rightarrow$ $5y(y-8) = 0$

$\therefore$ $y = 0, 8$

So, $x = -10, 6$

$\therefore$ The intersecting points are $(-10, 0)$ and $(6, 8)$.

According to the question, the point B lies in Ist quadrant.

So, coordinates of B are $(6, 8)$.

Now, slope of line $x - 2y + 10 = 0$ is

$$m_1 = -\frac{-1}{-2} = \frac{1}{2}$$

Hence, slope of required line is

$$m_2 = -2$$

So, the equation of line

$y - 8 = -2(x-6)$

$\Rightarrow$ $y - 8 = -2x + 12$

$\therefore$ $y = -2x + 20$

So, y intercept $= 20$

35. (a) Let $A(a, b)$ and $G(h, k)$.

Now, A, G, O are collinear.

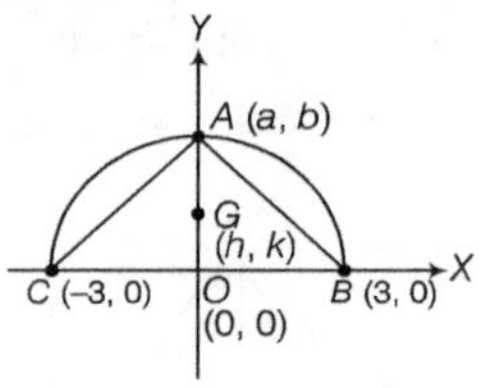

$\Rightarrow$ $h = \frac{a}{3}$

$\Rightarrow$ $a = 3h$

and similarly $b = 3k$.

Now, (a, b) lies on the circle $x^2 + y^2 = 9$

$\therefore$ $a^2 + b^2 = 9$

or $(3h)^2 + (3k)^2 = 9$

$\Rightarrow$ $x^2 + y^2 = 1$

36. (a,b) Line $y = 1 + c(x+3)$ is tangent to circle $x^2 + y^2 = 1$

$\therefore$ $p = r$

$\Rightarrow$ $\left|\frac{1+3c}{\sqrt{1+c^2}}\right| = 1$

$\Rightarrow$ $(1+3c)^2 = 1 + c^2$

or $9c^2 + 6c + 1 = c^2 + 1$

$\Rightarrow$ $8c^2 + 6c = 0$

$\therefore$ $c = 0, -3/4$

37. (b,d) Note that, Δ is right angled at (0,6). Centre of the circle is (2,3). Slope of the line joining origin to the centre is 3/2. Take parametric equation of a line through (2,3) with

$$\tan\theta = -2/3 \text{ as } \frac{x-2}{\cos\theta} = \frac{y-3}{\sin\theta} = \pm r$$

where, $r = \sqrt{13}$

$\therefore$ $x = 2 \pm r\cos\theta$

and $y = 3 \pm r\sin\theta$

Also, $x = 2 \pm \sqrt{13}\cdot\frac{3}{\sqrt{13}}$

and $y = 3 \pm \sqrt{13}\cdot\frac{2}{\sqrt{13}}$

or $(x = 5, y = 1)$ and $(x = -1, y = 5)$

Required points are (5,1) and (–1,5).

38. (a,b,d)

$$S_1: x^2 + y^2 = 4$$

and $\quad S_2: x^2 + y^2 - 2x - 4y + 4 = 0$

Centre (0,0), radius = 2 and centre (1,2), radius = 1

(a) d = Distance between centres = $\sqrt{5}$ units

$$r_1 + r_2 = 3$$

$\Rightarrow \quad |r_1 - r_2| = 1$

$\therefore \quad |r_1 - r_2| < d < r_1 + r_2$

$\Rightarrow$ These are intersecting.

Hence, number of tangents = 2

(b) $P(h, k)$ power of point P is same w.r.t. these two circles.

$\therefore \quad \sqrt{h^2 + k^2 - 4} = \sqrt{h^2 + k^2 - 2h - 4k + 4}$

$\Rightarrow \quad -4 = -2h - 4k + 4$

or $\quad x + 2y - 4 = 0$

(c) y-intercept of S_1 is $2\sqrt{4} = 4$

y-intercept of S_2 is $2\sqrt{4-4} = 0$

$\therefore$ Sum of y-intercept = 4

(d) $2(0 + 0) = -4 + 4$

i.e. $\quad 2(g_1g_2 + f_1f_2) = C_1 + C_2$

$\therefore$ Circles are orthogonal.

39. (a,c) $(d^2 - b)l^2 + 2dl + 1 = bm^2$

$\Rightarrow \quad d^2l^2 + 2dl + 1 = b(l^2 + m^2)$

$\Rightarrow \quad \left|\dfrac{dl + 1}{\sqrt{l^2 + m^2}}\right| = (\sqrt{b})^2$

$\Rightarrow \quad$ Centre $(d, 0)$ and radius = $\sqrt{b}$

$\Rightarrow \quad (x - d)^2 + (y - 0)^2 = b$

40. (b,c) The equation of circle

$$x^2 + y^2 - 8x - 16y + 60 = 0 \qquad \ldots(i)$$

Equation of chord of contact from $(-2, 0)$ is

$$-2x - 4(x - 2) - 8y + 60 = 0$$

$$3x + 4y - 34 = 0 \qquad \ldots(ii)$$

$$x^2 + \left(\frac{34 - 3x}{4}\right)^2 - 8x - 16\left(\frac{34 - 3x}{4}\right) + 60 = 0$$

$\Rightarrow 16x^2 + 1156 - 204x + 9x^2 - 128x - 2176 + 192x + 960 = 0$

$\Rightarrow \quad 5x^2 - 28x - 12 = 0$

$\Rightarrow \quad (x - 6)(5x + 2) = 0$

$\therefore \quad x = 6, -\dfrac{2}{5}$

So, the points are $(6, 4), \left(-\dfrac{2}{5}, \dfrac{44}{5}\right)$.

41. (a,c,d)

Case I If lines form a Δ, then $n = 4$

i.e. 3 excircles and 1 incircle.

Case II If lines are concurrent or all 3 parallel, then $n = 0$

Case III If two are parallel and third cuts, then $n = 2$

42. (b,c,d)

Radical centre is $(1, -1)$.

Radius of the circle orthogonally to S_1, S_2, S_3, is 1 and has the equation

$$x^2 + y^2 - 2x + 2y + 1 = 0$$

43. (a) (i) The straight line $y = kx, \forall K \in I$ touches or passes outside the circle

$$x^2 + y^2 - 20y + 90 = 0$$

Then, the perpendicular distance from the centre (0, 10) should be less than or equal to radius (i.e. $\sqrt{100 - 90} = \sqrt{10}$)

$\therefore \quad \left|\dfrac{-10}{\sqrt{k^2 + 1}}\right| \ge 10 \quad \Rightarrow \quad \dfrac{100}{k^2 + 1} \ge 10$

$\Rightarrow \quad k^2 \le 9$

(i) → (p,q,r)

(ii) If the circles intersect each other orthogonally, then

$$2(g_1g_2 + f_1f_2) = c_1 + c_2 \qquad \ldots(i)$$

The given equations of the circles are

$$S_1 \equiv x^2 + y^2 + px + py - 7 = 0$$

$$S_2 \equiv x^2 + y^2 - 10x + 2py + 1 = 0$$

On putting values in Eq.(i), we get

$$2\left[\left(\frac{p}{2}\right)(-5) + \left(\frac{p}{2}\right)(p)\right] = -7 + 1$$

$\Rightarrow \quad -5p + p^2 + 6 = 0$

$\Rightarrow \quad (p - 3)(p - 2) = 0$

$\therefore \quad p = 3, 2$

(ii) → (q,r)

(iii) Given, equations of the circles are

$$x^2 + y^2 + 2\lambda x + 4 = 0 \qquad \ldots(i)$$

and $\quad x^2 + y^2 - 4\lambda y + 8 = 0 \qquad \ldots(ii)$

represent real circles, then their radii should be greater equal to zero.

Case I For Eq. (i), $\quad r_1 = \sqrt{\lambda^2 - 4} \ge 0$

$\Rightarrow \quad \lambda^2 - 4 \ge 0 \Rightarrow \lambda^2 \ge 4$

Case II For Eq. (ii),

$$r_2 = \sqrt{(2\lambda)^2 - 8} \ge 0$$

$\Rightarrow \quad 4\lambda^2 - 8 \ge 0 \quad \Rightarrow \lambda^2 \ge 2$

(iii) → (q,r,s)

(iv) According to the graph, the possible abscissae are 1, 5.

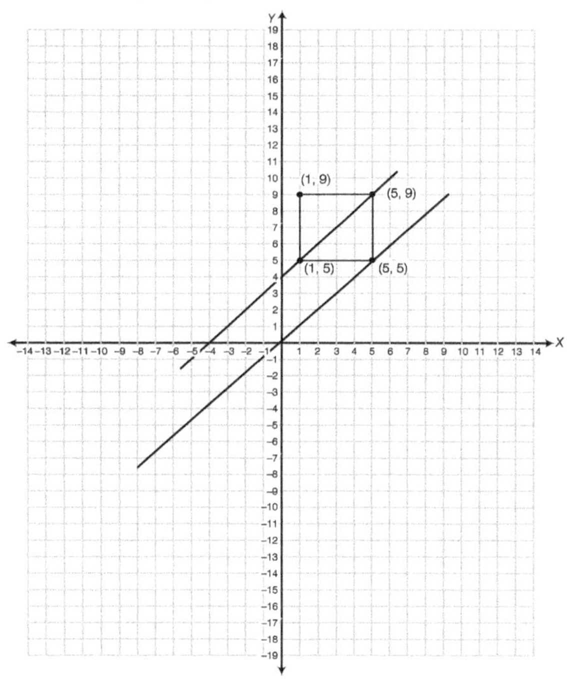

(iv) → (p,s)

44. (b) (i) Two intersecting circles have a common tangent and a common normal.

(i) → (p,q)

(ii) Two circles touching each other have a common tangent and common normal.

(ii) → (p,q)

(iii) Two non-concentric circles are strictly inside the other have only a common normal.

(iii) → (q)

(iv) Two concentric circles of different radii do not have a radical axis but have a common normal.

(iv) → (q,s)

45. (b)

Explanations for Q. Nos. (46-47)

$C_1 : (x-1)^2 + (y+1)^2 = 1$

and $C_2 : (x-1)^2 + (y+1)^2 = \sqrt{2}$

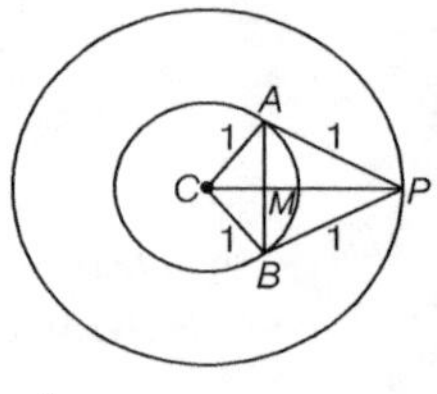

C_2 is director circle of C_1.
Since, ΔPAB is right angled triangle and $\angle APB = 90°$. So, orthocentre of ΔPAB is P and circumcentre is mid-point of chord AB.
Hence, locus of M is

$$(x-1)^2 + (y+1)^2 = \left(\frac{1}{\sqrt{2}}\right)^2.$$

46. (b) **47.** (d)

48. (b) As lines are perpendicular

$\therefore \quad a - 2 = 0$

$\Rightarrow \quad a = 2$ [coefficient of x^2 + coefficient of $y^2 = 0$]

Using $\Delta = 0 \Rightarrow c = -3$

$[\Delta = abc + 2fgh - af^2 - bg^2 - ch^2]$

Hence, two lines are

$$x + 2y - 3 = 0 \text{ and } 2x - y + 1 = 0$$

x – intercepts, $x_1 = 3, x_2 = \frac{-1}{2}$

y – intercepts, $y_1 = \frac{3}{2}, y_2 = 1$

$\Rightarrow x_1 + x_2 + y_1 + y_2 = 5$

49. (c)

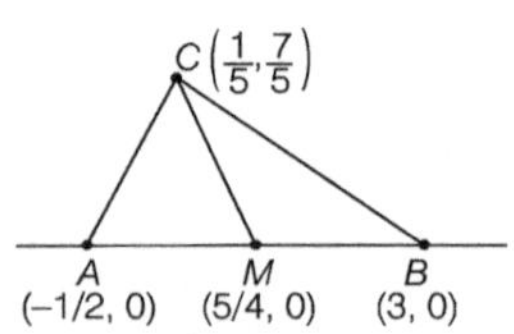

$$(CM)^2 = \left(\frac{5}{4} - \frac{1}{5}\right)^2 + \frac{49}{25}$$

$$= \frac{49}{16}$$

$\therefore \quad CM = \frac{7}{4}$ units

50. (d) Circumcircle of ΔABC is $\left(x + \frac{1}{2}\right)(x-3) + y^2 = 0$

$\Rightarrow \quad (2x+1)(x-3) + 2y^2 = 0$

$\Rightarrow \quad x^2 + y^2 - \frac{5}{2}x - \frac{3}{2} = 0$

and $x^2 + y^2 - 4y + k = 0$ are orthogonal.

$\therefore \quad k = \frac{3}{2}$

51. (d) Locus of S is a part of circle with OP as diameter passing inside the circle C.

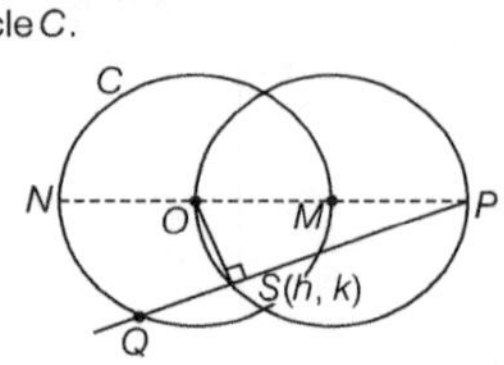

52. (d) $(PR) \cdot (PQ) = PT^2$

$= (PN)(PM) = (d - r)(d + r) = d^2 - r^2$

$= (PS - SR)(PS + SQ) = PS^2 - SQ^2$

$= PS^2 - (SQ)(SR)$

$\therefore \quad (PQ)(PR) \neq (PS)^2$

53. (a) Using Ptolemy's theorem,

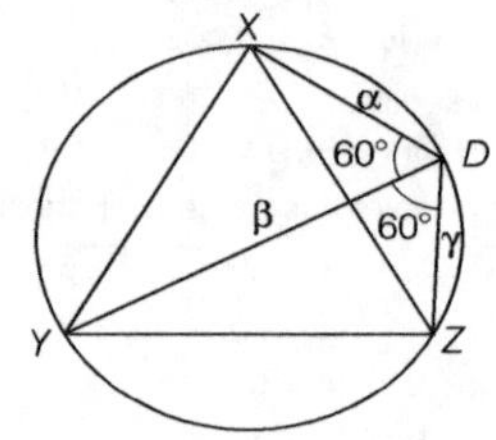

$(YD)(XZ) = (XY)(ZD) + (YZ)(XD)$

$= XZ(ZD + XD), \quad (\because XY = YZ = ZX)$

$\beta = \gamma + \alpha$

54. (5) Let the circle $x^2 + y^2 + 2gx + 2fy + c = 0$ as centre lie on $2x - 2y + 9 = 0$

$\therefore \quad -2g + 2f + 9 = 0 \quad \ldots(i)$

as it cuts $x^2 + y^2 = 4$, orthogonally.

$\therefore \quad 2g(0) + 2f(0) = c - 4 \Rightarrow c = 4$

Equation of circle

$$x^2 + y^2 + (2f + 9)x + 2fy + 4 = 0$$

or $\quad (x^2 + y^2 + 9x + 4) + 2f(x + y) = 0$

It passes through point of intersection of

$$x^2 + y^2 + 9x + 4 = 0$$

and $\quad x + y = 0$

$\therefore$ Points are $\left(-\frac{1}{2}, \frac{1}{2}\right)$ and $(-4, 4)$.

Hence, $a = \frac{-1}{2}, c = -4, b = \frac{1}{2}, d = 4$

Thus, $\quad 2b + d = 5$

55. (4) First equation is

$(x-6)^2 + (y-3)^2 = 49 \quad \ldots(i)$

$(x-2)^2 + (y-6)^2 = 40 + k \quad \ldots(ii)$

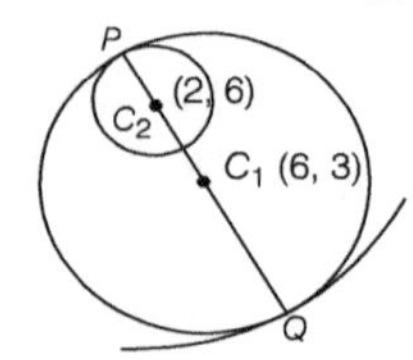

C_1: centre (6, 3) and $r_1 = 7$
C_2: centre (2, 6); $r_2 = \sqrt{40 + k}$
The two circles will intersect is

$C_2P \geq 7 - 5$

$\Rightarrow \quad \sqrt{40 + k} \geq 2$

$\Rightarrow \quad k \geq -36 \ldots(iii)$

and $\quad C_2Q \leq 7 + 5$

$\Rightarrow \quad 40 + k \leq 12^2$

$\Rightarrow \quad k \leq 104 \quad \ldots(iv)$

$\therefore \quad -36 \leq k \leq 104$

$\Rightarrow \quad a = -36$

$b = 104$

Thus, $\quad |3a + b| = 4$

56. (5) Given equations of circle C_1 is

$$x^2 + y^2 - 20x + 64 = 0$$

and C_2

$$x^2 + y^2 + 30x + 144 = 0$$

Centre and radius of C_1 are (10, 0) and 6 and C_2 are (– 15, 0) and 9, respectively.

PQ will be indirect tangent for circles C_1 and C_2.

So, length of PQ, i.e. $l = \sqrt{d^2 - (r_1 + r_2)^2}$

where, d is distance between two circles and r_1 and r_2 are their radii

Now, $l = \sqrt{625 - (9+6)^2} = \sqrt{625 - 225} = \sqrt{400} = 20$

So, the value of $(l - 15)$ is 5.

57. (7) $\quad CC_1 = CC_2 = CC_3 = r + 2$

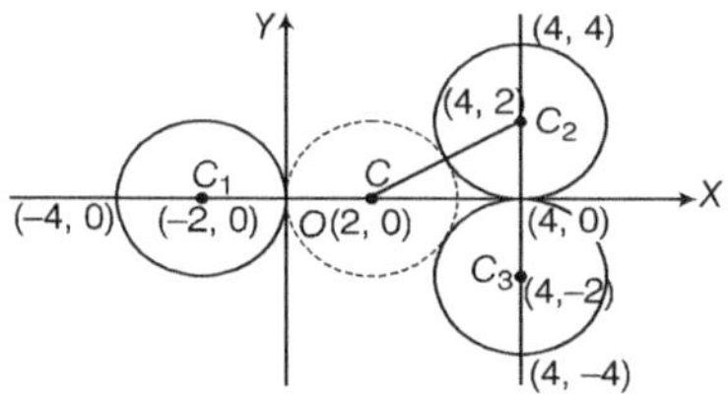

$\Rightarrow \quad (r+2)^2 = (r-4)^2 + 4 = (r-4)^2 + 4$

$\Rightarrow \quad r^2 + 4r + 4 = r^2 - 8r + 20$

$\Rightarrow \quad 12r = 16 \quad$ or $\quad r = \frac{4}{3} = \frac{p}{q}$

$\Rightarrow \quad p + q = 7$

58. (2) The intersection point of tangents lie on $x^2 + y^2 = 2$ (director's circle)

$\therefore$ Maximum value of square of ordinate of the point from where tangents are drawn is $(\sqrt{2})^2 = 2$

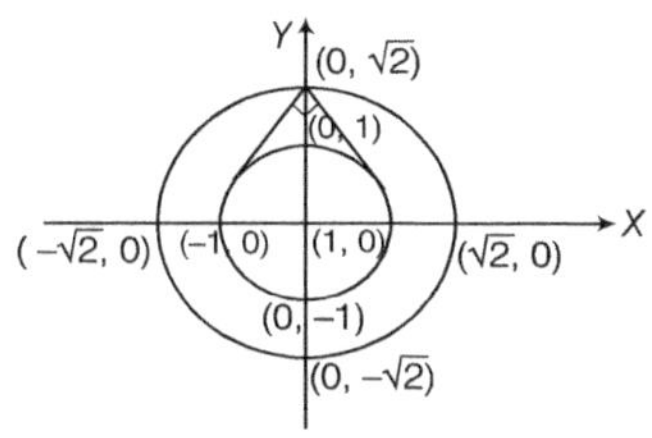

59. (3) Let $\quad \sum_{i=1}^{6} x_i = \alpha$ and $\sum_{i=1}^{6} y_i = \beta$

Let 0 be the orthocentre of the triangle made by $(x_1, y_1), (x_2, y_2)$ and (x_3, y_3).

$\Rightarrow$ 0 is

$$(x_1 + x_2 + x_3, y_1 + y_2 + y_3) = (\alpha_1, \beta_1)$$

Similarly, let G be the centroid of the triangle made by other three points,

$\Rightarrow \quad G\left(\frac{x_4 + x_5 + x_6}{3}, \frac{y_4 + y_5 + y_6}{3}\right)$

$\Rightarrow \quad G\left(\frac{\alpha - \alpha_1}{3}, \frac{\beta - \beta_1}{3}\right)$

The point dividing OG in the ratio 3:1 is $\left(\frac{\alpha}{4}, \frac{\beta}{4}\right) = (2, 1)$

$\Rightarrow \quad h + k = 2 + 1 = 3$

60. (9) $f(x, y) = x^2 + y^2 + 2gx + 2fy + c = 0$

Now, $\quad f(0, \lambda) = \lambda^2 + 2f\lambda + c = 0$

and its roots are (1, 1).

$\Rightarrow \quad f = -1, c = 1$

$f(\lambda, 0) = \lambda^2 + 2g\lambda + c = 0$ and its roots are $\left(\frac{1}{5}, 5\right)$

$\therefore \quad \frac{1}{5} + 5 = -2g$

$\Rightarrow \quad g = \frac{-13}{5}$

and $\quad \frac{1}{5} \times 5 = c$

Thus, $\quad g = \frac{-13}{3},$

$f = -1, c = 1$

$\Rightarrow \quad r = \sqrt{\frac{169}{25} + 1 - 1} = \frac{13}{5}$

$\therefore \quad 5r - 4 = 9$

JEE Main & AIEEE Archive

1. (d) Let the coordinates of the centre of T be $(0, k)$.

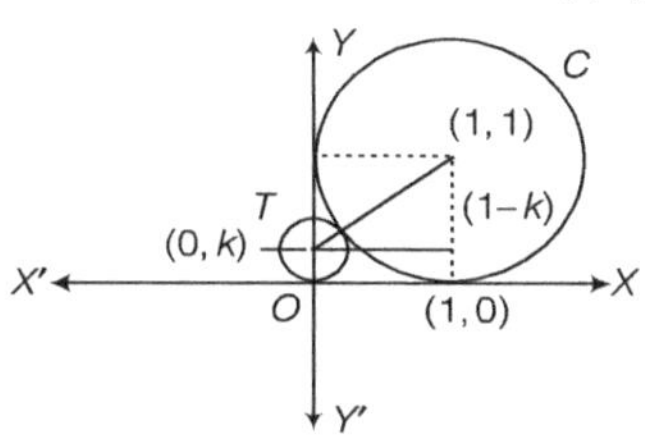

Distance between their centres

$$k + 1 = \sqrt{1 + (k-1)^2} \Rightarrow k + 1 = \sqrt{1 + k^2 + 1 - 2k}$$

$\Rightarrow \quad k + 1 = \sqrt{k^2 + 2 - 2k}$

$\Rightarrow \quad k^2 + 1 + 2k = k^2 + 2 - 2k \quad \Rightarrow \quad k = \frac{1}{4}$

So, the radius of circle T is k, i.e. $\frac{1}{4}$ units.

2. (c) Let the equation of the circle be $(x-3)^2 + (y-0)^2 + \lambda y = 0$

As it passes through $(1, -2)$.

$\therefore \quad (1-3)^2 + (-2)^2 + \lambda(-2) = 0$

$\Rightarrow \quad 4 + 4 - 2\lambda = 0 \Rightarrow \lambda = 4$

$\therefore$ Equation of circle is $(x-3)^2 + y^2 + 4y = 0$.

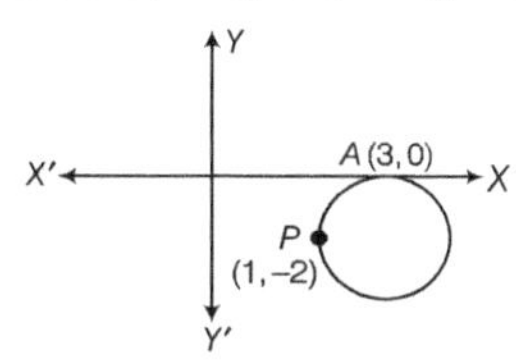

By hit and trial method, we see that point $(5, -2)$ satisfies equation of circle.

3. (a) **Given**

(i) A circle which touches X-axis at the point (1, 0).

(ii) The circle also passes through the point (2, 3).

To find The length of the diameter of the circle.

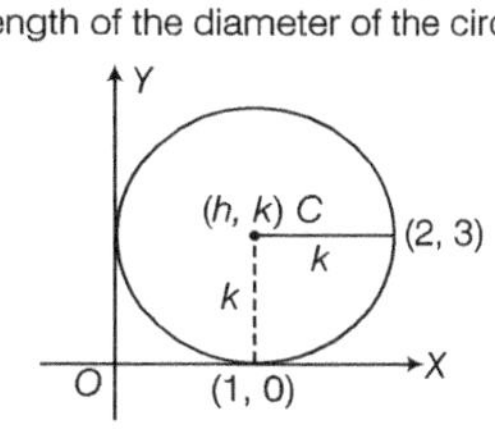

Let us assume that the coordinates of the centre of the circle are $C(h, k)$ and its radius is r.

Now, since the circle touches X-axis at (1, 0), hence its radius should be equal to ordinate of centre.

$\Rightarrow \quad r = k$

Hence, the equation of the circle is

$$(x - h)^2 + (y - k)^2 = k^2$$

Also, given that the circle passes through points (1, 0) and (2, 3). Hence, substituting them in the equation of the circle, we get

$$(1 - h)^2 + (0 - k)^2 = k^2 \quad \ldots(i)$$

$$(2 - h)^2 + (3 - k)^2 = k^2 \quad \ldots(ii)$$

From Eq. (i), we get $\quad h = 1$

On substituting value of h in Eq. (ii), we get

$$(2 - 1)^2 + (3 - k)^2 = k^2 \Rightarrow k = \frac{5}{3}$$

$\therefore$ The diameter of the circle is $2k = \dfrac{10}{3}$ units

4. (a) $x^2 + y^2 - ax = 0$ and $x^2 + y^2 = c^2$ touch each other.

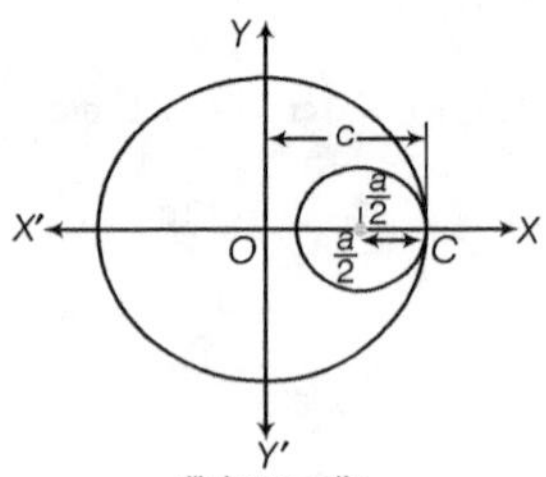

(i) Internally

(i) If circles touch internally,

$$\left|c - \frac{a}{2}\right| = \frac{a}{2} \Rightarrow c - \frac{a}{2} = \frac{a}{2}$$

$\Rightarrow \quad c = a, c > 0$

$\therefore \quad |a| = c$

(ii) If circles touch externally,

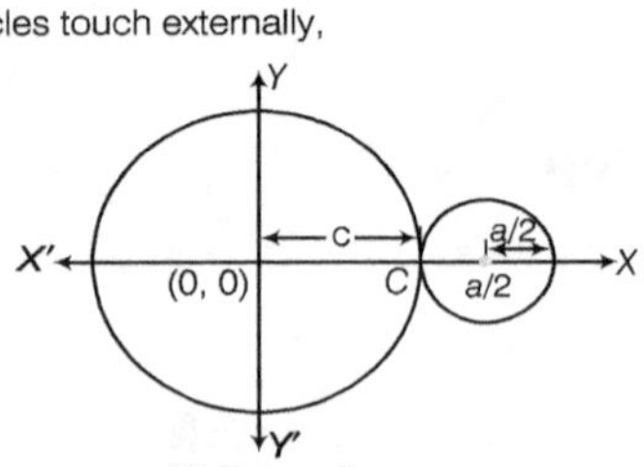

(ii) Externally

$$\left|c + \frac{a}{2}\right| = \frac{a}{2} \Rightarrow c + \frac{a}{2} = \frac{a}{2}$$

$\therefore$ $c = 0$, i.e. not possible as $c > 0$.

Hence, the circles should touch internally and $|a| = c$.

5. (c) Circle whose diametric end points are (1, 0) and (0, 1) will be of smallest radius.

$\Rightarrow \quad (x - 1)(x - 0) + (y - 0)(y - 1) = 0$

$\Rightarrow \quad x^2 + y^2 - x - y = 0$

6. (b) Since, the coordinates of the centre of the circle are (2, 4).

Also, $\quad r^2 = 4 + 16 + 5 = 25$

The line will intersect the circle at two distinct points, if the distance of (2, 4) from $3x - 4y = m$ is less than radius of the circle.

$\therefore \quad \dfrac{|6 - 16 - m|}{5} < 5 \Rightarrow -25 < 10 + m < 25$

$\Rightarrow \quad -35 < m < 15$

7. (d) Given equation can be rewritten as

$$(x + 1)^2 + (y + 2)^2 = (2\sqrt{2})^2$$

Let required point be $Q(\alpha, \beta)$.

Then, mid-point of $P(1, 0)$ and $Q(\alpha, \beta)$ is the centre of the circle.

i.e. $\quad \dfrac{\alpha + 1}{2} = -1 \quad \Rightarrow \alpha = -3$

and $\quad \dfrac{\beta + 0}{2} = -2 \quad \Rightarrow \beta = -4$

So, required point is $(-3, -4)$.

8. (b) Equation of circle which touches X-axis and coordinates of centre are (h, k), is

$$(x - h)^2 + (y - k)^2 = k^2$$

Since it is passing through $(-1, 1)$, then

$$(-1 - h)^2 + (1 - k)^2 = k^2$$

$\Rightarrow \quad h^2 + 2h - 2k + 2 = 0$

For real circles, $\quad D \geq 0$

$\Rightarrow \quad (2)^2 - 4(-2k + 2) \geq 0$

$\therefore \quad k \geq \dfrac{1}{2}$

9. (c) The given equations of diameters are

$$3x - 4y - 7 = 0 \quad \ldots(i)$$

and $\quad 2x - 3y - 5 = 0 \quad \ldots(ii)$

On solving Eqs. (i) and (ii), we get

$$x = 1 \text{ and } y = -1$$

So, the intersection of two diameters is the centre of circle, is $(-1, 1)$.

Let r be the radius of circle, then Area of circle = 49π sq units

$\Rightarrow \quad \pi r^2 = 49\pi \Rightarrow r = 7$ units

$\therefore$ Equation of required circle is

$$(x - 1)^2 + (y + 1)^2 = 49$$

$\Rightarrow \quad x^2 - 2x + 1 + y^2 + 2y + 1 = 49$

$\Rightarrow \quad x^2 + y^2 - 2x + 2y - 47 = 0$

10. (c) Let the coordinates of a point P be (h, k) which is mid-point of the chord AB.

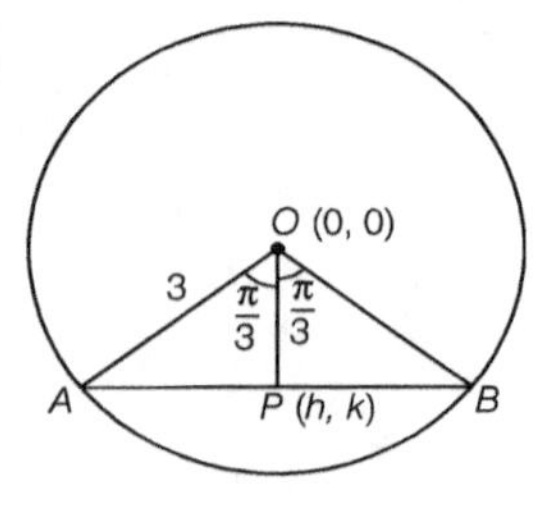

Now, $OP = \sqrt{(h - 0)^2 + (k - 0)^2}$

$= \sqrt{h^2 + k^2}$

In ΔAOP, $\quad \cos\dfrac{\pi}{3} = \dfrac{OP}{OA}$

$\Rightarrow \quad \dfrac{1}{2} = \dfrac{\sqrt{h^2 + k^2}}{3}$

$\Rightarrow \quad h^2 + k^2 = \dfrac{9}{4}$

Hence, the required locus is $x^2 + y^2 = \dfrac{9}{4}$

11. (c) Let equation of circles be

$$S_1 \equiv x^2 + y^2 + 2ax + cy + a = 0$$

and $\quad S_2 \equiv x^2 + y^2 - 3ax + dy - 1 = 0$

Chord through intersection points P and Q of the given circles is $S_1 - S_2 = 0$.

$\therefore$ $(x^2 + y^2 + 2ax + cy + a) - (x^2 + y^2 - 3ax + dy - 1) = 0$

$\Rightarrow \quad 5ax + (c - d)y + a + 1 = 0$

On comparing it with $5x + by - a = 0$, we get

$$\frac{5a}{5} = \frac{c - d}{b} = \frac{a + 1}{-a}$$

$\Rightarrow \quad a(-a) = a + 1$

$\Rightarrow \quad a^2 + a + 1 = 0$

which gives no real value of a.

Hence, the line will passes through P and Q for no value of a.

12. (a) Let the equation of circle be $x^2 + y^2 + 2gx + 2fy + c = 0$. It cuts the circle $x^2 + y^2 = p^2$ orthogonally.

$\because \quad 2g_1g_2 + 2f_1f_2 = c_1 + c_2$

$\therefore \quad 2g(0) + 2f(0) = c - p^2 \Rightarrow c = p^2$

Also, it passes through (a, b).

$\therefore \quad a^2 + b^2 + 2ga + 2fb + p^2 = 0$

So, locus of $(-g, -f)$ is $a^2 + b^2 - 2ax - 2by + p^2 = 0$

$\Rightarrow \quad 2ax + 2by - (a^2 + b^2 + p^2) = 0$

13. (b) Let the equation of circle be

$$x^2 + y^2 + 2gx + 2fy + c = 0$$

It cuts the circle $x^2 + y^2 = 4$ orthogonally, if

$$2g_1g_2 + 2f_1f_2 = c_1 + c_2$$

$\therefore \quad 2g \cdot 0 + 2f \cdot 0 = c - 4 \Rightarrow c = 4$

Equation of circle is

$$x^2 + y^2 + 2gx + 2fy + 4 = 0$$

Since, it passes through the point (a, b).

$\therefore \quad a^2 + b^2 + 2ag + 2fb + 4 = 0$

Locus of centre $(-g, -f)$ will be

$$a^2 + b^2 - 2xa - 2yb + 4 = 0$$

$\Rightarrow \quad 2ax + 2by - (a^2 + b^2 + 4) = 0$

Alternate Method

Let the centre of required circle be $(-g, -f)$. This circle cuts the circle $x^2 + y^2 = 4$ orthogonally. The centre and radius of circle $x^2 + y^2 = 4$ are $(0, 0)$ and 2, respectively.

$\therefore \quad g^2 + f^2 = 4 + (a + g)^2 + (b + f)^2$

$\Rightarrow \quad g^2 + f^2 = 4 + a^2 + g^2 + 2ag + 2bf + b^2 + f^2$

$\Rightarrow \quad 4 + a^2 + b^2 + 2ag + 2bf = 0$

Hence, the locus of centre is

$$2ax + 2by - (a^2 + b^2 + 4) = 0$$

14. (a) Since, the coordinates of one end of a diameter of a circle A are (p, q) and let the coordinates of other end B be (x_1, y_1).

Equation of circle in diameter form is

$$(x - p)(x - x_1) + (y - q)(y - y_1) = 0$$

$\Rightarrow \quad x^2 - (p + x_1)x + px_1 + y^2 - (q + y_1)y + qy_1 = 0$

$\Rightarrow \quad x^2 - (p + x_1)x + y^2 - (y_1 + q)y + px_1 + qy_1 = 0$

Since, this circle touches X-axis.

i.e., $\quad y = 0$

$\therefore \quad x^2 - (p + x_1)x + px_1 + qy_1 = 0$

Also, the discriminant of above equation will be

$$(p + x_1)^2 = 4(px_1 + qy_1)$$

$\Rightarrow \quad p^2 + x_1^2 + 2px_1 = 4px_1 + 4qy_1$

$\Rightarrow \quad x_1^2 - 2px_1 + p^2 = 4qy_1$

Hence, locus of point B is $(x - p)^2 = 4qy$.

15. (a) Given lines $2x + 3y + 1 = 0$ and $3x - y - 4 = 0$ are the diameters of circle.

The intersection of two lines is the centre of circle $(1, -1)$.

Circumference of circle $= 10\pi$ (given)

$\Rightarrow \quad 2\pi r = 10\pi \Rightarrow r = 5$

$\therefore$ Equation of circle having centre $(1, -1)$ and radius 5 is

$$(x - 1)^2 + (y + 1)^2 = 5^2$$

$\Rightarrow \quad x^2 + 1 - 2x + y^2 + 2y + 1 = 25$

$\Rightarrow \quad x^2 + y^2 - 2x + 2y - 23 = 0$

16. (a) Given, equation of line is $y = x$. ...(i)

And equation of circle is $x^2 + y^2 - 2x = 0$ ...(ii)

From Eqs. (i) and (ii), we get

$$x^2 + x^2 - 2x = 0 \Rightarrow 2x(x - 1) = 0$$

$\Rightarrow \quad x = 0, \quad x = 1$

On putting the values of x in Eq. (i) respectively, we get

$$y = 0, \quad y = 1$$

Let coordinates of A be $(0, 0)$ and coordinates of B be $(1, 1)$.

$\therefore$ Equation of circle when AB as a diameter, is

$$(x - 0)(x - 1) + (y - 0)(y - 1) = 0$$

$\Rightarrow \quad x^2 - x + y^2 - y = 0$

$\Rightarrow \quad x^2 + y^2 - x - y = 0$

17. (a) The equation of first circle is $(x - 1)^2 + (y - 3)^2 = r^2$

whose centre is $C_1(1, 3)$ and radius $r_1 = r$

and equation of second circle

$$x^2 + y^2 - 8x + 2y + 8 = 0$$

whose centre is $C_2(4, -1)$ and radius

$$r_2 = \sqrt{4^2 + 1^2 - 8} = \sqrt{17 - 8} = \sqrt{9} = 3$$

Two circles intersect at two distinct points, then

$$r_1 - r_2 < C_1C_2 < r_1 + r_2$$

$\Rightarrow \quad r - 3 < \sqrt{(4 - 1)^2 + (-1 - 3)^2} < r + 3$

$\Rightarrow \quad r - 3 < \sqrt{9 + 16} < r + 3 \quad \Rightarrow \quad r - 3 < 5 < r + 3$

$\Rightarrow \quad r - 3 < 5$ and $5 < r + 3 \quad \Rightarrow \quad r < 8$ and $2 < r$

$\therefore \quad 2 < r < 8$

18. (c) Given equation of diameters of a circle are

$$2x - 3y = 5 \quad \text{...(i)}$$

and $\quad 3x - 4y = 7$ ...(ii)

The intersection of Eqs. (i) and (ii), we get the coordinates of centre $(1, -1)$.

Given $\quad$ area of circle $= \pi r^2$

$\Rightarrow \quad 154 = \frac{22}{7} r^2$

$\therefore \quad r = 7$

Equation of the circle whose centre $(1, -1)$ and radius 7 is

$$(x - 1)^2 + (y + 1)^2 = 7^2$$

$\Rightarrow \quad x^2 - 2x + 1 + y^2 + 2y + 1 = 49$

$\Rightarrow \quad x^2 + y^2 - 2x + 2y = 47$

19. (b) Given equation of circle

$$x^2 + y^2 - 4x - 2y - 20 = 0$$

whose centre is $C(2, 1)$ and radius is 5.

At point $(10, 7)$,

$$S_1 = 10^2 + 7^2 - 4 \times 10 - 2 \times 7 - 20$$
$$= 75 > 0$$

So, P lies outside the circle.

Now, $\quad PC = \sqrt{(2 - 10)^2 + (1 - 7)^2}$

$$= \sqrt{8^2 + 6^2}$$
$$= \sqrt{10^2} = 10$$

$\therefore$ Greatest distance between circle and the point is $P = 10 + 5 = 15$ units.

JEE Advanced & IIT JEE Archive

1. (b,c)

(i) The general equation of a circle is

$$x^2 + y^2 + 2gx + 2fy + c = 0$$

where, centre and radius are given by $(-g, -f)$ and $\sqrt{g^2 + f^2 - c}$, respectively.

(ii) If the two circles $x^2 + y^2 + 2g_1x + 2f_1y + c_1 = 0$ and $x^2 + y^2 + 2g_2x + 2f_2y + c_2 = 0$ are orthogonal, then $2g_1g_2 + 2f_1f_2 = c_1 + c_2$.

Let circle be $x^2 + y^2 + 2gx + 2fy + c = 0$

It passes through (0, 1).

$$\therefore \quad 1 + 2f + c = 0 \quad \ldots(i)$$

Orthogonal with $x^2 + y^2 - 2x - 15 = 0$

$$2g(-1) = c - 15$$

$$\Rightarrow \quad c = 15 - 2g \quad \ldots(ii)$$

Orthogonal with $x^2 + y^2 - 1 = 0$

$$c = 1 \quad \ldots(iii)$$

$$\Rightarrow \quad g = 7 \text{ and } f = -1$$

Centre is $(-g, -f) \equiv (-7, 1)$

Radius $= \sqrt{g^2 + f^2 - c} = \sqrt{49 + 1 - 1} = 7$

2. (a,c) Here, the length of intercept is on Y-axis.

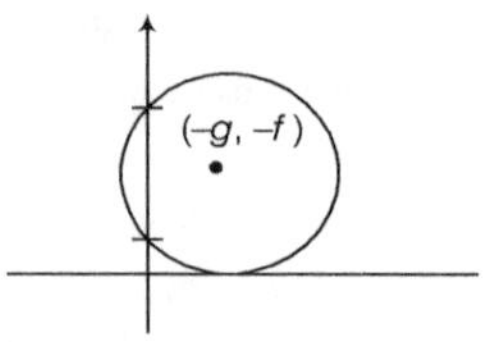

$$\therefore \quad 2 = \sqrt{f^2 - c}$$

and if circle touches X-axis

$\Rightarrow g^2 = c$

Here, $x^2 + y^2 + 2gx + 2fy + c = 0$

Passes through (3, 0)

$$\Rightarrow \quad 9 + 6g + c = 0 \quad \ldots(i)$$

$$g^2 = c \quad \ldots(ii)$$

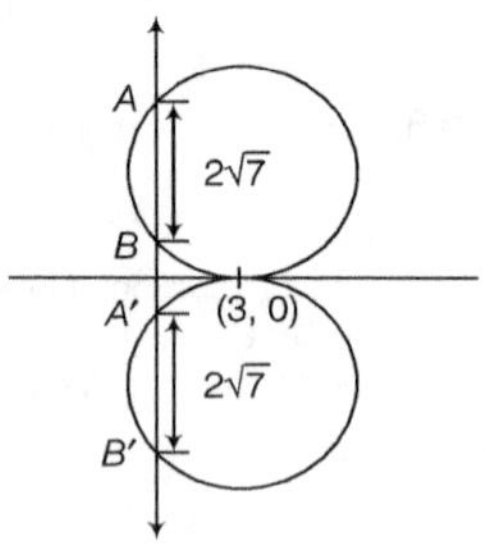

and $\quad 2\sqrt{f^2 - c} = 2\sqrt{7}$

$$f^2 - c = 7 \quad \ldots(iii)$$

From Eqs. (i) and (ii), we get

$$g^2 + 6g + 9 = 0$$

$$\Rightarrow \quad (g + 3)^2 = 0$$

$$g = -3 \text{ and } c = 9$$

$$\therefore \quad f^2 = 16, \ f = \pm 4$$

Hence, $\quad x^2 + y^2 - 6x \pm 8y + 9 = 0$

3. (d) Here, equation of common tangent be

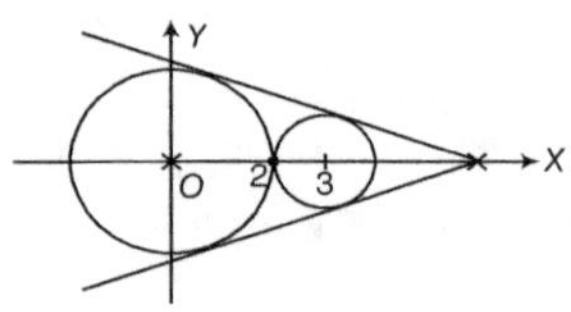

$$y = mx \pm 2\sqrt{1 + m^2}$$

which is also the tangent to

$$(x - 3)^2 + y^2 = 1$$

$$\Rightarrow \quad \frac{|3m - 0 + 2\sqrt{1 + m^2}|}{\sqrt{m^2 + 1}} = 1$$

$$\Rightarrow \quad 3m + 2\sqrt{1 + m^2} = \pm\sqrt{1 + m^2}$$

$$\Rightarrow \quad 3m = -3\sqrt{1 + m^2}$$

or $\quad m = -\sqrt{1 + m^2}$

$$\Rightarrow \quad m^2 = 1 + m^2 \text{ or } 9m^2 = 1 + m^2$$

$$\Rightarrow \quad m \in \phi \text{ or } m = \pm \frac{1}{2\sqrt{2}}$$

$$\therefore \quad y = \pm \frac{1}{2\sqrt{2}} x \pm 2\sqrt{1 + \frac{1}{8}}$$

$$\Rightarrow \quad y = \pm \frac{x}{2\sqrt{2}} \pm \frac{6}{2\sqrt{2}} \Rightarrow 2\sqrt{2}y = \pm(x + 6)$$

4. (a) Here, tangent to $x^2 + y^2 = 4$ at $(\sqrt{3}, 1)$ is $\sqrt{3}x + y = 4 \quad \ldots(i)$

As, L is perpendicular to $\sqrt{3}x + y = 4$

$\Rightarrow x - \sqrt{3}y = \lambda$ which is tangent to

$$(x - 3)^2 + y^2 = 1 \Rightarrow \frac{|3 - 0 - \lambda|}{\sqrt{1 + 3}} = 1$$

$$\Rightarrow \quad |3 - \lambda| = 2 \Rightarrow 3 - \lambda = 2, -2$$

$$\therefore \quad \lambda = 1, 5$$

$$\Rightarrow \quad L: x - \sqrt{3}y = 1, \ x - \sqrt{3}y = 5$$

5. (a) If $S: ax^2 + 2hxy + by^2 + 2gx + 2fy + C$

then equation of chord bisected at $P(x_1, y_1)$ is $T = S_1$

or $\quad axx_1 + h(xy_1 + yx_1) + byy_1 + g(x + x_1) + f(y + y_1) + C$

$$= ax_1^2 + 2hx_1y_1 + by_1^2 + 2gx_1 + 2fy_1 + C$$

Description of Situation As equation of chord of contact is $T = 0$

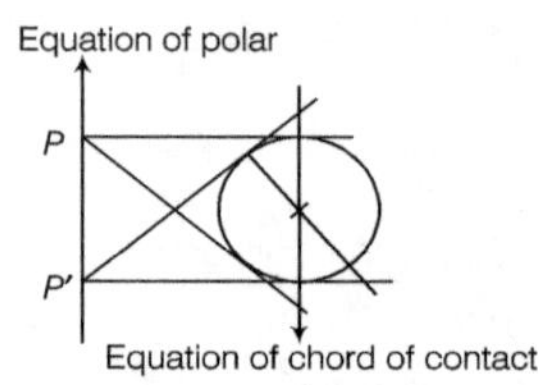

Here, equation of chord of contact w.r.t. P is

$$x\lambda + y\left(\frac{4\lambda - 20}{5}\right) = 9$$

$$5\lambda x + (4\lambda - 20)y = 45 \quad \ldots(i)$$

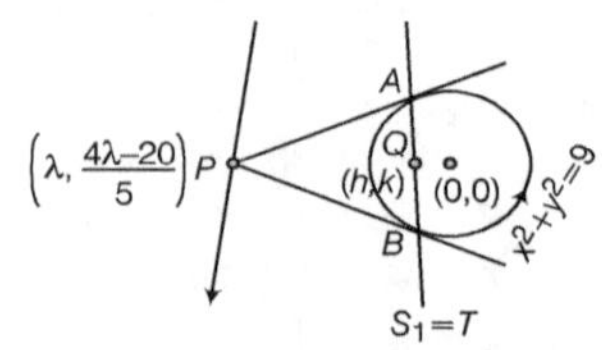

and equation of chord bisected at the point $Q(h, k)$ is

$$xh + yk - 9 = h^2 + k^2 - 9$$

$$\Rightarrow \quad xh + ky = h^2 + k^2 \quad \ldots(ii)$$

From Eqs. (i) and (ii), we get

$$\frac{5\lambda}{h} = \frac{4\lambda - 20}{k} = \frac{45}{h^2 + k^2}$$

$$\therefore \quad \lambda = \frac{20h}{4h - 5k} \text{ and } \lambda = \frac{9h}{h^2 + k^2}$$

$$\Rightarrow \quad \frac{20h}{4h - 5k} = \frac{9h}{h^2 + k^2}$$

or $\quad 20(h^2 + k^2) = 9(4h - 5k)$

or $\quad 20(x^2 + y^2) = 36x - 45y$

6. (d) Equation of circle passing through a point (x_1, y_1) and touching the straight line L, is given by

$$(x - x_1)^2 + (y - y_1)^2 + \lambda L = 0$$

$\therefore$ Equation of circle passing through (0, 2) and touching $x = 0$ is $\quad (x-0)^2 + (y-2)^2 + \lambda x = 0 \quad \ldots(i)$

Also, it passes through $(-1, 0)$

$\Rightarrow \quad 1 + 4 - \lambda = 0$

$\therefore \quad \lambda = 5$

$\therefore$ Eq. (i) becomes, $x^2 + y^2 - 4y + 4 + 5x = 0$

$\Rightarrow \quad x^2 + y^2 + 5x - 4y + 4 = 0$

For x-intercept, put $y = 0$

$\Rightarrow \quad x^2 + 5x + 4 = 0$

$(x+1)(x+4) = 0$

$\therefore \quad x = -1, -4$

7. (3) Let $\theta = \dfrac{\pi}{2k}$,

$$\cos\theta = \frac{x}{2}$$

$\Rightarrow \quad \cos 2\theta = \dfrac{\sqrt{3}+1-x}{2}$

$\Rightarrow \quad 2\cos^2\theta - 1 = \dfrac{\sqrt{3}+1-x}{2}$

$\Rightarrow \quad 2\left(\dfrac{x^2}{4}\right) - 1 = \dfrac{\sqrt{3}+1-x}{2}$

$\Rightarrow \quad x^2 + x - 3 - \sqrt{3} = 0$

$\Rightarrow \quad x = \dfrac{-1 \pm \sqrt{1+12+4\sqrt{3}}}{2} = \dfrac{-1 \pm \sqrt{13+4\sqrt{3}}}{2}$

$= \dfrac{-1+2\sqrt{3}+1}{2} = \sqrt{3}$

$\therefore \quad \cos\theta = \dfrac{\sqrt{3}}{2} \quad \Rightarrow \quad \theta = \dfrac{\pi}{6}$

Hence, required angle $= \dfrac{\pi}{k} = 2\theta = \dfrac{\pi}{3} \quad \Rightarrow \quad k = 3$

8. (8) $(r+1)^2 = \alpha^2 + 9$

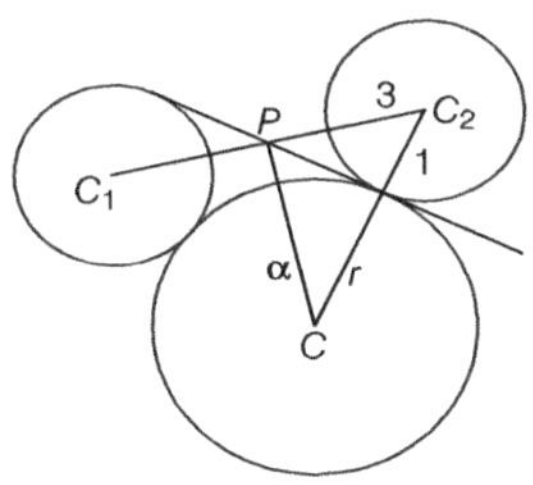

$$r^2 + 8 = \alpha^2$$

$\Rightarrow \quad r^2 + 2r + 1 = r^2 + 8 + 9$

$\Rightarrow \quad 2r = 16$

$\Rightarrow \quad r = 8$ units

9. (b) For required circle, $P(1, 8)$ and $O(3, 2)$ will be the end point of its diameter.

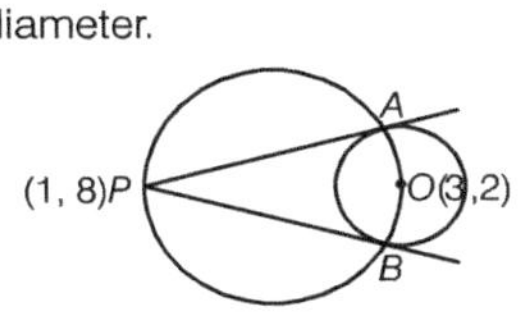

$\therefore \quad (x-1)(x-3) + (y-8)(y-2) = 0$

$\Rightarrow \quad x^2 + y^2 - 4x - 10y + 19 = 0$

10. (b) For the points of intersection of the two given curves

$$C_1 : y^2 = 4x$$

and $\quad C_2 : x^2 + y^2 - 6x + 1 = 0$

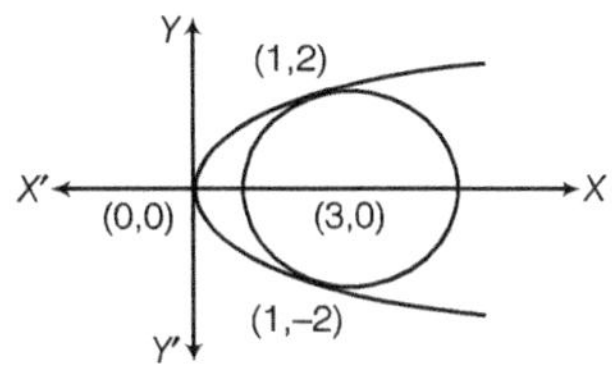

We have, $\quad x^2 + 4x - 6x + 1 = 0 \Rightarrow x^2 - 2x + 1 = 0$

$\Rightarrow \quad (x-1)^2 = 0 \Rightarrow x = 1, 1$ (equal real roots)

$\Rightarrow \quad y = 2, -2$

Thus, the given curves touch each other at exactly two points (1, 2) and (1, – 2).

11. (d) Let centre of circle C be (h, k).

Then, $\quad \left|\dfrac{\sqrt{3}h + k - 6}{\sqrt{3+1}}\right| = 1 \quad \Rightarrow \quad \sqrt{3}h + k - 6 = 2, -2$

$\Rightarrow \quad \sqrt{3}h + k = 4 \quad \ldots(i)$

(Rejecting '2' because origin and centre of C are on the same side of PQ)

The point $(\sqrt{3}, 1)$ satisfies Eq. (i).

$\therefore$ Equation of circle C is $(x - \sqrt{3})^2 + (y-1)^2 = 1$.

12. (a) Slope of line joining centre of circle to point D is

$$\tan\theta = \frac{\frac{3}{2} - 1}{\frac{3\sqrt{3}}{2} - \sqrt{3}} = \frac{1}{\sqrt{3}}$$

It makes an angle 30° with X-axis.

$\therefore$ Points E and F will make angle 150° and –90° with X-axis.

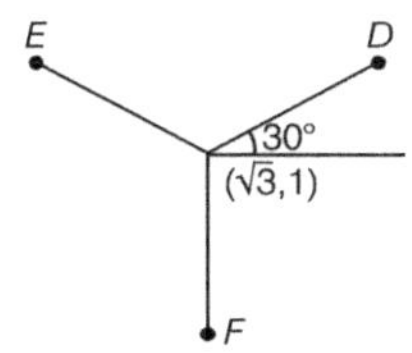

$\therefore$ E and F are given by

$$\frac{x-\sqrt{3}}{\cos 150^\circ} = \frac{y-1}{\sin 150^\circ} = 1$$

and $\quad \dfrac{x-\sqrt{3}}{\cos(-90^\circ)} = \dfrac{y-1}{\sin(-90^\circ)} = 1$

$\therefore \quad E = \left(\dfrac{\sqrt{3}}{2}, \dfrac{3}{2}\right)$ and $F = (\sqrt{3}, 0)$

13. (d) Clearly, points E and F satisfy the equations given in option (d).

14. (c) Equation of given circle C is

$$(x-3)^2 + (y+5)^2 = 9 + 25 - 30$$

i.e. $\quad (x-3)^2 + (y+5)^2 = 2^2$

Centre = (3, – 5)

If L_1 is diameter, then $\quad 2(3) + 3(-5) + p - 3 = 0$

$\Rightarrow \quad p = 12$

$\therefore \quad L_1$ is $\quad 2x + 3y + 9 = 0$

L_2 is $\quad 2x + 3y + 15 = 0$

Distance of centre of circle from $L_2 = \left|\dfrac{2(3) + 3(-5) + 15}{\sqrt{2^2 + 3^2}}\right|$

$= \dfrac{6}{\sqrt{13}} < 2$ (radius of circle)

$\therefore L_2$ is a chord of circle C.

Statement II is false.

15. (b) $18 = \frac{1}{2}(3\alpha)(2r) \Rightarrow \alpha r = 6$

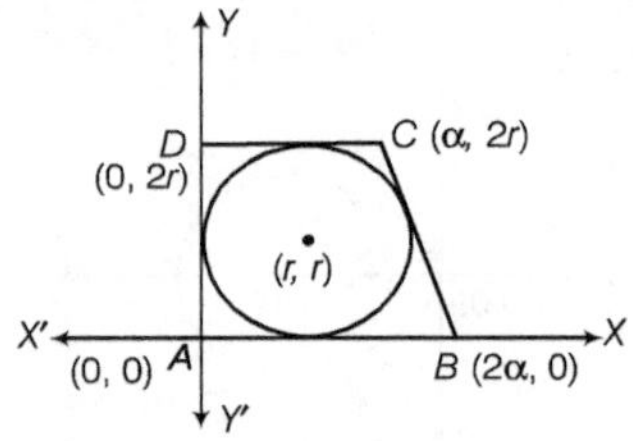

Line, $y = -\frac{2r}{\alpha}(x - 2\alpha)$ is tangent to circle

$$(x - r)^2 + (y - r)^2 = r^2$$

$$2\alpha = 3r \text{ and } \alpha r = 6$$

$$r = 2$$

Alternate Method

$$\frac{1}{2}(x + 2x) \times 2r = 18$$

$$xr = 6 \quad \text{... (i)}$$

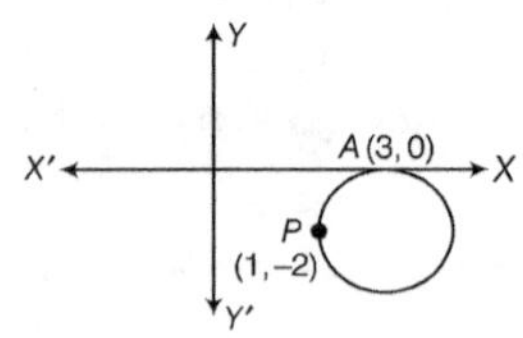

In ΔAOB, $\tan\theta = \frac{x - r}{r}$

and in ΔDOC,

$\tan(90° - \theta) = \frac{2x - r}{r}$

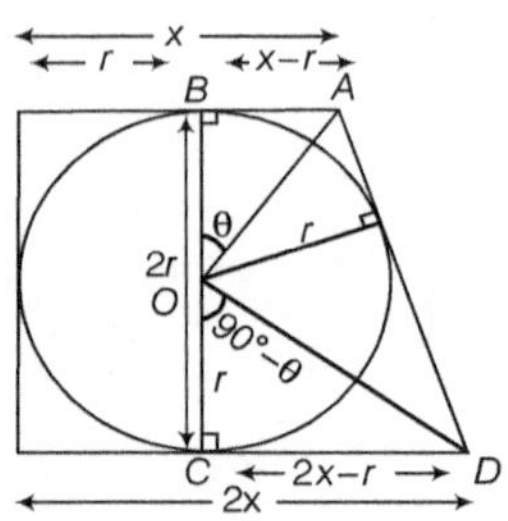

$\therefore \quad \frac{x - r}{r} = \frac{r}{2x - r}$

$\Rightarrow \quad x(2x - 3r) = 0$

$\Rightarrow \quad x = \frac{3r}{2}$ ... (ii)

From Eqs. (i) and (ii), we get, $r = 2$

16. (a) Since, the tangents are perpendicular.

So, locus of perpendicular tangents to circle $x^2 + y^2 = 169$ is a director circle having equation $x^2 + y^2 = 338$

17. (b)

(i) When two circles are intersecting, they have a common normal and common tangent.

(ii) Two mutually external circles have a common normal and common tangent.

(iii) When one circle lies inside of other, then they have a common normal but no common tangent.

(iv) Two branches of a hyperbola have a common normal but no common tangent.

(i) → (p,q), (ii) → (p,q), (iii) → (q,r), (iv) → (q,r)

18. (a) Let the equation of circles are

$$C_1 : (x - 1)^2 + (y - 1)^2 = (1)^2$$

and $$C_2 : (x - 1)^2 + (y - 1)^2 = (\sqrt{2})^2$$

$\therefore$ Coordinates of $P(1 + \cos\theta, 1 + \sin\theta)$

and $Q(1 + \sqrt{2}\cos\theta, 1 + \sqrt{2}\sin\theta)$

$\therefore \; PA^2 + PB^2 + PC^2 + PD^2$

$$= \{(1 + \cos\theta)^2 + (1 + \sin\theta)^2\} + \{(\cos\theta - 1)^2 + (1 + \sin\theta)^2\}$$
$$+ \{(\cos\theta - 1)^2 + (\sin\theta - 1)^2\} + \{(1 + \cos\theta)^2 + (\sin\theta - 1)^2\} = 12$$

Similarly, $QA^2 + QB^2 + QC^2 + QD^2 = 16$

$$\therefore \quad \frac{\Sigma PA^2}{\Sigma QA^2} = \frac{12}{16} = 0.75$$

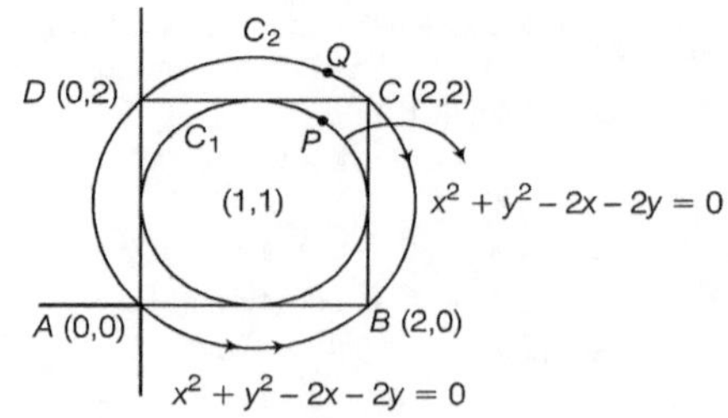

19. (c) Let C be the centre of the required circle.

Now, draw a line parallel to L at a distance of r_1 (radius of C_1) from it.

Now, $\quad CC_1 = AC$

$\Rightarrow C$ lies on a parabola.

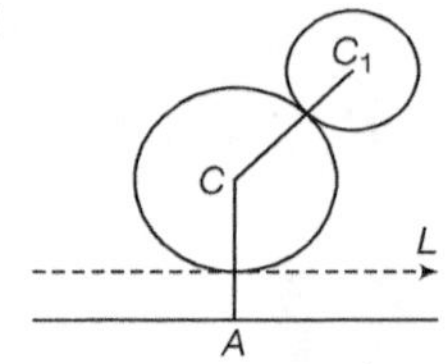

20. (c) Since, $\quad AG = \sqrt{2}$

$\therefore \quad AT_1 = T_1G = \frac{1}{\sqrt{2}}$

As A is the focus, T_1 is the vertex and BD is the directrix of parabola.

Also, T_2T_3 is latusrectum.

$\therefore \quad T_2T_3 = 4 \cdot \frac{1}{\sqrt{2}}$

Hence, area of $\Delta T_1T_2T_3$

$$= \frac{1}{2} \times \frac{1}{\sqrt{2}} \times \frac{4}{\sqrt{2}}$$

$$= 1 \text{ sq unit}$$

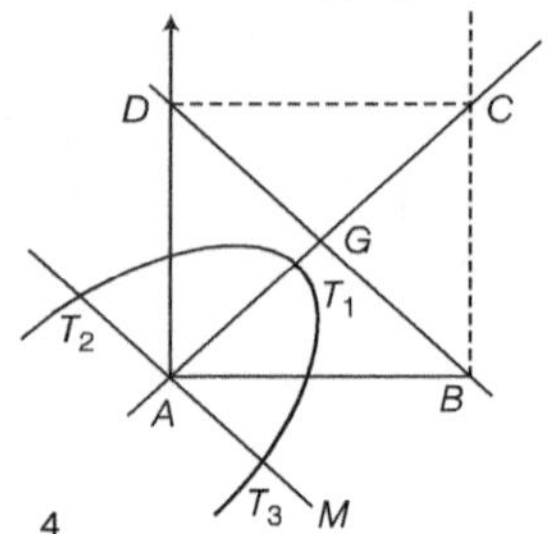

21. (a) Let the locus of centre of circle be (h, k) touching $(y - 1)^2 + x^2 = 1$ and X-axis shown as

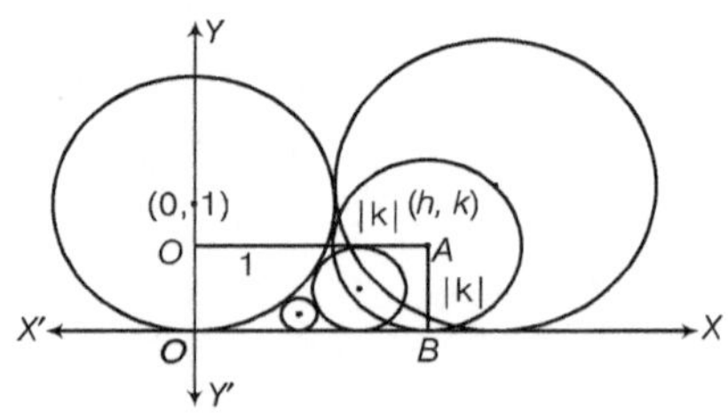

Clearly, from figure,

distance between O and A is always $1 + |k|$,

i.e. $\quad \sqrt{(h - 0)^2 + (k - 1)^2} = 1 + |k|$

$\Rightarrow \quad h^2 + k^2 - 2k + 1 = 1 + k^2 + 2|k|$

$\Rightarrow \quad h^2 = 2|k| + 2k$

$\Rightarrow \quad x^2 = 2|y| + 2y$

where, $\quad |y| = \begin{cases} y, & y \geq 0 \\ -y, & y < 0 \end{cases}$

$\therefore \quad x^2 = 2y + 2y, \; y \geq 0$

and $\quad x^2 = -2y + 2y, \; y < 0$

$\Rightarrow \quad x^2 = 4y, \text{ when } y \geq 0$

and $\quad x^2 = 0, \text{ when } y < 0$

$\therefore \quad \{(x, y) : x^2 = 4y, \text{when } y \geq 0\} \cup \{(0, y) : y < 0\}$

Rough Work

Rough Work

Rough Work

Rough Work

Last 3 Years' Questions
JEE Main & Advanced
(2016-2018)

JEE Main

2016

1. Consider $f(x) = \tan^{-1}\left(\sqrt{\frac{1+\sin x}{1-\sin x}}\right)$, $x \in \left(0, \frac{\pi}{2}\right)$. A normal to $y = f(x)$ at $x = \frac{\pi}{6}$ also passes through the point **[Straight Line]**

(a) $(0, 0)$
(b) $\left(0, \frac{2\pi}{3}\right)$
(c) $\left(\frac{\pi}{6}, 0\right)$
(d) $\left(\frac{\pi}{4}, 0\right)$

2. If one of the diameters of the circle, given by the equation, $x^2 + y^2 - 4x + 6y - 12 = 0$, is a chord of a circle S, whose centre is at $(-3, 2)$, then the radius of S is **[Circle]**

(a) $5\sqrt{2}$
(b) $5\sqrt{3}$
(c) 5
(d) 10

2017

3. Let k be an integer such that the triangle with vertices $(k, -3k)$, $(5, k)$ and $(-k, 2)$ has area 28 sq units. Then, the orthocentre of this triangle is at the point **[Point]**

(a) $\left(2, -\frac{1}{2}\right)$
(b) $\left(1, \frac{3}{4}\right)$
(c) $\left(1, -\frac{3}{4}\right)$
(d) $\left(2, \frac{1}{2}\right)$

2018

4. A straight line through a fixed point $(2, 3)$ intersects the coordinate axes at distinct points P and Q. If O is the origin and the rectangle $OPRQ$ is completed, then the locus of R is **[Straight Line]**

(a) $3x + 2y = 6$
(b) $2x + 3y = xy$
(c) $3x + 2y = xy$
(d) $3x + 2y = 6xy$

5. Let the orthocentre and centroid of a triangle be $A(-3, 5)$ and $B(3, 3)$, respectively. If C is the circumcentre of this triangle, then the radius of the circle having line segment AC as diameter, is **[Circle]**

(a) $\sqrt{10}$
(b) $2\sqrt{10}$
(c) $3\sqrt{\frac{5}{2}}$
(d) $\frac{3\sqrt{5}}{2}$

6. If the tangent at $(1, 7)$ to the curve $x^2 = y - 6$ touches the circle $x^2 + y^2 + 16x + 12y + c = 0$, then the value of c is **[Circle]**

(a) 195
(b) 185
(c) 85
(d) 95

JEE Advanced

2016

1. Consider a pyramid $OPQRS$ located in the first octant ($x \geq 0,\ y \geq 0,\ z \geq 0$) with O as origin, and OP and OR along the X-axis and the Y-axis, respectively. The base $OPQR$ of the pyramid is a square with

$OP = 3$. The point S is directly above the mid-point T of diagonal OQ such that $TS = 3$. Then, **[Straight Line, Type 2]**

(a) the acute angle between OQ and OS is $\frac{\pi}{3}$

(b) the equation of the plane containing the ΔOQS is $x - y = 0$

(c) the length of the perpendicular from P to the plane containing the ΔOQS is $\frac{3}{\sqrt{2}}$

(d) the perpendicular distance from O to the straight line containing RS is $\sqrt{\frac{15}{2}}$

2. Let RS be the diameter of the circle $x^2 + y^2 = 1$, where S is the point $(1, 0)$. Let P be a variable point (other than R and S) on the circle and tangents to the circle at S and P meet at the point Q. The normal to the circle at P

intersects a line drawn through Q parallel to RS at point E. Then, the locus of E passes through the point(s) **[Circle, Type 2]**

(a) $\left(\frac{1}{3}, \frac{1}{\sqrt{3}}\right)$ (b) $\left(\frac{1}{4}, \frac{1}{2}\right)$ (c) $\left(\frac{1}{3}, -\frac{1}{\sqrt{3}}\right)$ (d) $\left(\frac{1}{4}, -\frac{1}{2}\right)$

3. Let $a, \lambda, \mu \in R$. Consider the system of linear equations $ax + 2y = \lambda$ and $3x - 2y = \mu$. Which of the following statement(s) is/are correct? **[Straight Line, Type 2]**

(a) If $a = -3$, then the system has infinitely many solutions for all values of λ and μ

(b) If $a \neq -3$, then the system has a unique solution for all values of λ and μ

(c) If $\lambda + \mu = 0$, then the system has infinitely many solutions for $a = -3$

(d) If $\lambda + \mu \neq 0$, then the system has no solution for $a = -3$

2017

4. For how many values of p, the circle $x^2 + y^2 + 2x + 4y - p = 0$ and the coordinate axes have exactly three common points? **[Circle, Type 3]**

2018

Paragraph (Q. 5 and 6)

Let S be the circle in the XY-plane defined by the equation $x^2 + y^2 = 4$.

(*There are two questions based on Paragraph "X", the question given below is one of them*)

[Circle]

5. Let E_1E_2 and F_1F_2 be the chords of S passing through the point P_0 (1, 1) and parallel to the X-axis and the Y-axis, respectively. Let G_1G_2 be the chord of S passing through P_0 and having slope -1. Let the tangents to S at E_1 and E_2 meet at E_3, then tangents to S at F_1 and F_2 meet at F_3, and the tangents to S at G_1 and G_2 meet at G_3. Then, the points E_3, F_3 and G_3 lie on the curve

(a) $x + y = 4$ (b) $(x-4)^2 + (y-4)^2 = 16$

(c) $(x-4)(y-4) = 4$ (d) $xy = 4$

6. Let P be a point on the circle S with both coordinates being positive. Let the tangent to S at P intersect the coordinate axes at the points M and N. Then, the mid-point of the line segment MN must lie on the curve

(a) $(x + y)^2 = 3xy$ (b) $x^{2/3} + y^{2/3} = 2^{4/3}$

(c) $x^2 + y^2 = 2xy$ (d) $x^2 + y^2 = x^2y^2$

Note ***Type-1*** *Only one correct option,* ***Type-2*** *More then one correct option,* ***Type-3*** *Single digit Answer,* ***Type-4*** *Numerical value based*

7. Let T be the line passing through the points $P(-2, 7)$ and $Q(2, -5)$. Let F_1 be the set of all pairs of circles (S_1, S_2) such that T is tangent to S_1 at P and tangent to S_2 at Q, and also such that S_1 and S_2 touch each other at a point, say M. Let E_1 be the set representing the locus of M as the pair (S_1, S_2) varies in F_1. Let the set of all straight line segments joining a pair of distinct points of E_1 and passing through the point $R(1, 1)$ be F_2. Let E_2 be the set of the mid-points of the line segments in the set F_2. Then, which of the following statement(s) is (are) TRUE?
[Circle, Type 2]

(a) The point $(-2, 7)$ lies in E_1

(b) The point $\left(\frac{4}{5}, \frac{7}{5}\right)$ does NOT lie in E_2

(c) The point $\left(\frac{1}{2}, 1\right)$ lies in E_2

(d) The point $\left(0, \frac{3}{2}\right)$ does NOT lie in E_1

8. Let s, t, r be non-zero complex numbers and L be the set of solutions $z = x + iy \, (x, y \in R, i = \sqrt{-1})$ of the equation $sz + t\bar{z} + r = 0$, where $\bar{z} = x - iy$. Then, which of the following statement(s) is (are) TRUE?
[Circle, Type 2]

(a) If L has exactly one element, then $|s| \neq |t|$

(b) If $|s| = |t|$, then L has infinitely many elements

(c) The number of elements in $L \cap \{z : |z - 1 + i| = 5\}$ is at most 2

(d) If L has more than one element, then L has infinitely many elements

Answer *with* Explanations

JEE Main

1. *(b)* We have,

$$f(x) = \tan^{-1}\sqrt{\frac{1+\sin x}{1-\sin x}}, x \in \left(0, \frac{\pi}{2}\right)$$

$$\Rightarrow f(x) = \tan^{-1}\sqrt{\frac{\left(\cos\frac{x}{2} + \sin\frac{x}{2}\right)^2}{\left(\cos\frac{x}{2} - \sin\frac{x}{2}\right)^2}}$$

$$= \tan^{-1}\left(\frac{\cos\frac{x}{2} + \sin\frac{x}{2}}{\cos\frac{x}{2} - \sin\frac{x}{2}}\right)$$

$$\left(\because \cos\frac{x}{2} > \sin\frac{x}{2} \text{ for } 0 < \frac{x}{2} < \frac{\pi}{4}\right)$$

$$= \tan^{-1}\left(\frac{1 + \tan\frac{x}{2}}{1 - \tan\frac{x}{2}}\right)$$

$$= \tan^{-1}\left[\tan\left(\frac{\pi}{4} + \frac{x}{2}\right)\right] = \frac{\pi}{4} + \frac{x}{2}$$

$$\Rightarrow f'(x) = \frac{1}{2} \Rightarrow f'\left(\frac{\pi}{6}\right) = \frac{1}{2}$$

Now, equation of normal at $x = \frac{\pi}{6}$ is given by

$$\left(y - f\left(\frac{\pi}{6}\right)\right) = -2\left(x - \frac{\pi}{6}\right)$$

$$\Rightarrow \left(y - \frac{\pi}{3}\right) = -2\left(x - \frac{\pi}{6}\right)$$

$$\left[\because f\left(\frac{\pi}{6}\right) = \frac{\pi}{4} + \frac{\pi}{12} = \frac{4\pi}{12} = \frac{\pi}{3}\right]$$

which passes through $\left(0, \frac{2\pi}{3}\right)$.

2. *(b)* Given equation of a circle is $x^2 + y^2 - 4x + 6y - 12 = 0$, whose centre is $(2, -3)$ and radius

$$= \sqrt{2^2 + (-3)^2 + 12} = \sqrt{4 + 9 + 12} = 5$$

Now, according to given information, we have the following figure.

$$x^2 + y^2 - 4x + 6y - 12 = 0$$

Clearly, $AO \perp BC$, as O is mid-point of the chord.

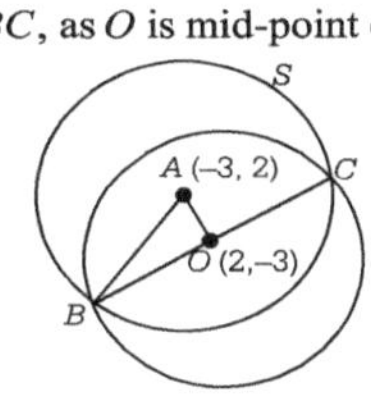

Now, in ΔAOB, we have

$$OA = \sqrt{(-3-2)^2 + (2+3)^2}$$

$$= \sqrt{25 + 25} = \sqrt{50} = 5\sqrt{2}$$

and $\quad OB = 5$

$$\therefore \quad AB = \sqrt{OA^2 + OB^2}$$

$$= \sqrt{50 + 25}$$

$$= \sqrt{75} = 5\sqrt{3}$$

3. *(d)* Given, vertices of triangle are $(k, -3k)$, $(5, k)$ and $(-k, 2)$.

$$\therefore \quad \frac{1}{2}\begin{vmatrix} k & -3k & 1 \\ 5 & k & 1 \\ -k & 2 & 1 \end{vmatrix} = \pm 28$$

$$\Rightarrow \quad \begin{vmatrix} k & -3k & 1 \\ 5 & k & 1 \\ -k & 2 & 1 \end{vmatrix} = \pm 56$$

$$\Rightarrow k(k-2) + 3k(5+k) + 1(10+k^2) = \pm 56$$

$\Rightarrow \quad k^2 - 2k + 15k + 3k^2 + 10 + k^2 = \pm 56$

$\Rightarrow \quad 5k^2 + 13k + 10 = \pm 56$

$\Rightarrow \quad 5k^2 + 13k - 66 = 0$

or $\quad 5k^2 + 13k - 46 = 0$

$\Rightarrow \quad k = 2 \qquad [\because k \in I]$

Thus, the coordinates of vertices of triangle are $A(2, -6)$, $B(5, 2)$ and $C(-2, 2)$.

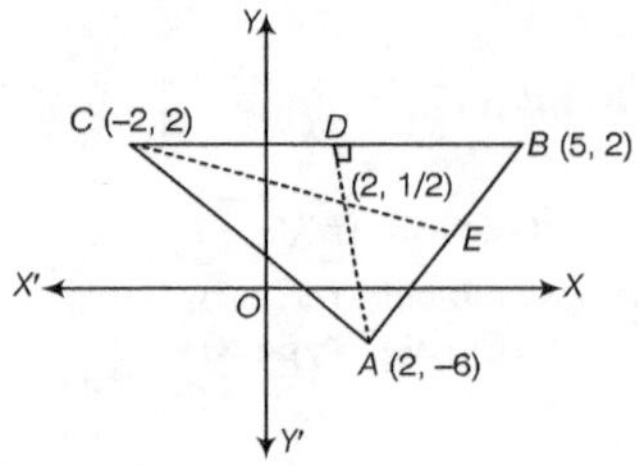

Now, equation of altitude from vertex A is

$$y - (-6) = \frac{-1}{\left(\frac{2-2}{-2-5}\right)}(x-2) \Rightarrow x = 2 \qquad \text{...(i)}$$

Equation of altitude from vertex C is

$$y - 2 = \frac{-1}{\left[\frac{2-(-6)}{5-2}\right]}[x - (-2)]$$

$$\Rightarrow \quad 3x + 8y - 10 = 0 \qquad \text{...(ii)}$$

On solving Eqs. (i) and (ii), we get $x = 2$ and $y = \frac{1}{2}$.

$\therefore$ Orthocentre $= \left(2, \frac{1}{2}\right)$

4. *(c)*

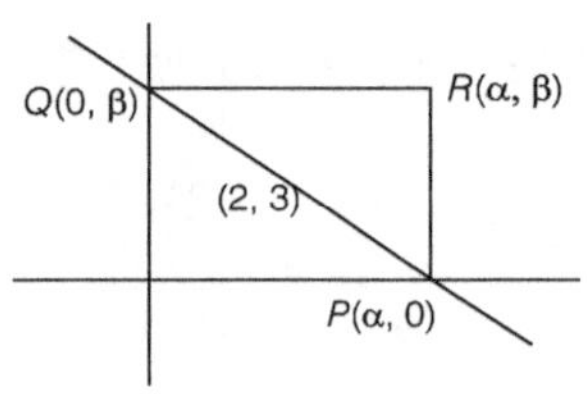

Equation of line PQ is $\frac{x}{\alpha} + \frac{y}{\beta} = 1$

Since this line is passes through fixed point (2, 3).

$\therefore \quad \frac{2}{\alpha} + \frac{3}{\beta} = 1$

$\therefore$ Locus of R is $2\beta + 3\alpha = \alpha\beta$

i.e. $\quad 2y + 3x = xy$

$\Rightarrow \quad 3x + 2y = xy$

5. *(c)* **Key idea** *Otrhocentre, centroid and circumcentre are collinear and centroid divide orthocentre and circumcentre in* $2:1$ *ratio.*

We have orthocentre and centroid of a triangle be $A(-3, 5)$ and $B(3, 3)$ respectively and C circumcentre.

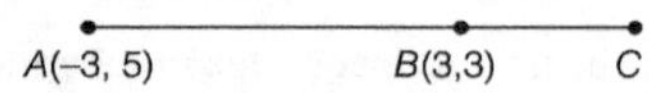

Clearly, $AB = \sqrt{(3+3)^2 + (3-5)^2} = \sqrt{36+4} = 2\sqrt{10}$

We know that, $AB : BC = 2:1 \quad \Rightarrow \quad BC = \sqrt{10}$

Now, $\quad AC = AB + BC = 2\sqrt{10} + \sqrt{10} = 3\sqrt{10}$

Since, AC is a diameter of circle.

$\therefore \quad r = \frac{AC}{2}$

$\Rightarrow \quad r = \frac{3\sqrt{10}}{2} = 3\sqrt{\frac{5}{2}}$

6. *(d)* **Key idea** *Equation of tangent to the curve* $x^2 = 4ay$ *at* (x_1, y_1) *is* $xx_1 = 4a\left(\frac{y + y_1}{2}\right)$

Tangent to the curve $x^2 = y - 6$ at (1, 7) is

$$x = \frac{y+7}{2} - 6$$

$$\Rightarrow \quad 2x - y + 5 = 0 \qquad \text{...(i)}$$

Equation of circle is $x^2 + y^2 + 16x + 12y + c = 0$

Centre $(-8, -6)$

$$r = \sqrt{8^2 + 6^2 - c} = \sqrt{100 - c}$$

Since, line $2x - y + 5 = 0$ also touches the circle.

$\therefore \quad \sqrt{100 - c} = \left|\frac{2(-8) - (-6) + 5}{\sqrt{2^2 + 1^2}}\right|$

$\Rightarrow \quad \sqrt{100 - c} = \left|\frac{-16 + 6 + 5}{\sqrt{5}}\right|$

$\Rightarrow \quad \sqrt{100 - c} = |-\sqrt{5}|$

$\Rightarrow \quad 100 - c = 5 \Rightarrow c = 95$

JEE Advanced

1. *(b, c, d)* Given, square base

$OP = OR = 3$

$\therefore \quad P\ (3,0,0),\ R = (0,3,0)$

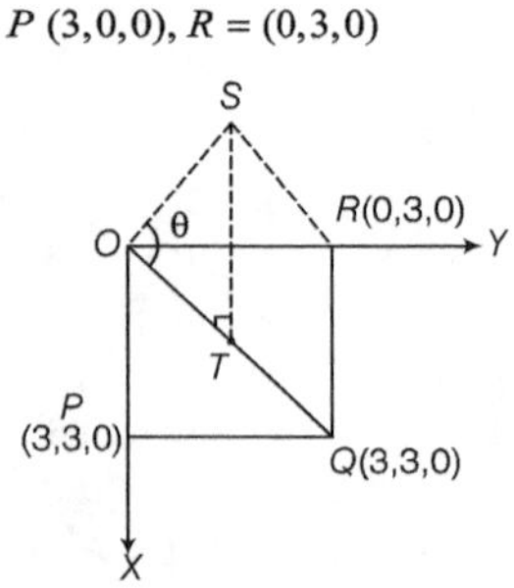

Also, mid-point of OQ is $T\left(\frac{3}{2}, \frac{3}{2}, 0\right)$.

Since, S is directly above the mid-point T of diagonal OQ and $ST = 3$.

i.e. $\quad S\left(\frac{3}{2}, \frac{3}{2}, 3\right)$

Here, DR's of OQ (3, 3, 0) and DR's of $OS\left(\frac{3}{2}, \frac{3}{2}, 3\right)$.

$$\therefore \quad \cos\theta = \frac{\frac{9}{2} + \frac{9}{2}}{\sqrt{9+9+0}\sqrt{\frac{9}{4} + \frac{9}{4} + 9}}$$

$$= \frac{9}{\sqrt{18} \cdot \sqrt{\frac{27}{2}}} = \frac{1}{\sqrt{3}}$$

∴ Option (a) is incorrect.

Now, equation of the plane containing the ΔOQS is

$$\begin{vmatrix} x & y & z \\ 3 & 3 & 0 \\ 3/2 & 3/2 & 3 \end{vmatrix} = 0 \Rightarrow \begin{vmatrix} x & y & z \\ 1 & 1 & 0 \\ 1 & 1 & 2 \end{vmatrix} = 0$$

$$\Rightarrow \quad x(2-0) - y(2-0) + z(1-1) = 0$$

$$\Rightarrow \quad 2x - 2y = 0 \text{ or } x - y = 0$$

∴ Option (b) is correct.

Now, length of the perpendicular from $P(3, 0, 0)$ to the plane containing ΔOQS is

$$\frac{|3-0|}{\sqrt{1+1}} = \frac{3}{\sqrt{2}}$$

∴ Option (c) is correct.

Here, equation of RS is

$$\frac{x-0}{3/2} = \frac{y-3}{-3/2} = \frac{z-0}{3} = \lambda$$

$$\Rightarrow \quad x = \frac{3}{2}\lambda,\ y = -\frac{3}{2}\lambda + 3,\ z = 3\lambda$$

To find the distance from $O(0, 0, 0)$ to RS.

Let M be the foot of perpendicular.

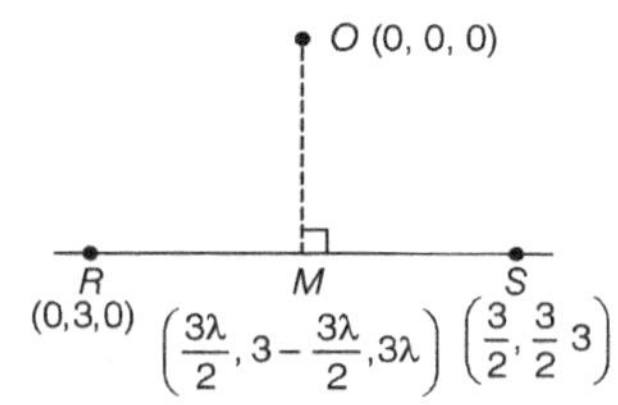

$$\because \quad \overline{OM} \perp \overline{RS} \Rightarrow \overline{OM} \cdot \overline{RS} = 0$$

$$\Rightarrow \quad \frac{9\lambda}{4} - \frac{3}{2}\left(3 - \frac{3\lambda}{2}\right) + 3(3\lambda) = 0$$

$$\Rightarrow \quad \lambda = \frac{1}{3}$$

$$\therefore \quad M\left(\frac{1}{2}, \frac{5}{2}, 1\right)$$

$$\Rightarrow \quad OM = \sqrt{\frac{1}{4} + \frac{25}{4} + 1} = \sqrt{\frac{30}{4}} = \sqrt{\frac{15}{2}}$$

∴ Option (d) is correct.

2. *(a, c)* Given, RS is the diameter of $x^2 + y^2 = 1$.

Here, equation of the tangent at $P(\cos\theta, \sin\theta)$ is $x\cos\theta + y\sin\theta = 1$.*

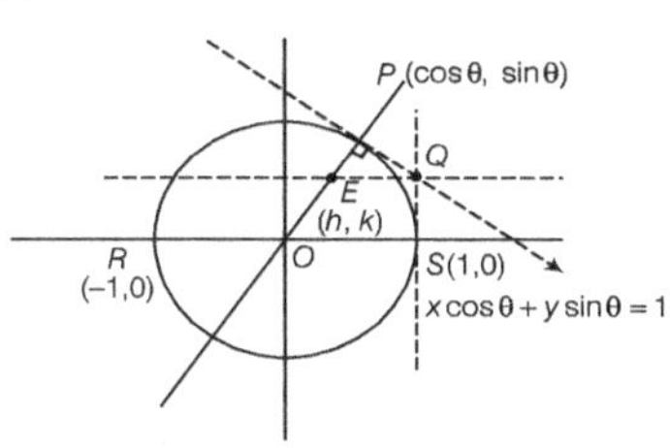

Intersecting with $x = 1$,

$$y = \frac{1-\cos\theta}{\sin\theta}$$

$$\therefore \quad Q\left(1, \frac{1-\cos\theta}{\sin\theta}\right)$$

∴ Equation of the line through Q parallel to RS is

$$y = \frac{1-\cos\theta}{\sin\theta} = \frac{2\sin^2\frac{\theta}{2}}{2\sin\frac{\theta}{2}\cos\frac{\theta}{2}} = \tan\frac{\theta}{2} \quad \ldots\text{(i)}$$

Normal at $\quad P: y = \frac{\sin\theta}{\cos\theta} \cdot x$

$$\Rightarrow \quad y = x\tan\theta \quad \ldots\text{(ii)}$$

Let their point of intersection be (h, k).

Then, $\quad k = \tan\frac{\theta}{2}$

and $\quad k = h\tan\theta$

$$\therefore \quad k = h\left(\frac{2\tan\frac{\theta}{2}}{1-\tan^2\frac{\theta}{2}}\right)$$

$$\Rightarrow \quad k = \frac{2h \cdot k}{1-k^2}$$

$$\Rightarrow \quad k(1-k^2) = 2hk$$

∴ Locus for point $E: 2x = (1 - y^2)$ ...(iii)

When $x = \frac{1}{3}$,

then $\quad 1 - y^2 = \frac{2}{3}$

$$\Rightarrow \quad y^2 = 1 - \frac{2}{3}$$

$$\Rightarrow \quad y = \pm\frac{1}{\sqrt{3}}$$

$\therefore \left(\frac{1}{3}, \pm\frac{1}{\sqrt{3}}\right)$ satisfy $2x = 1 - y^2$.

When $x = \frac{1}{4}$, then

$$1 - y^2 = \frac{2}{4}$$

$$\Rightarrow \quad y^2 = 1 - \frac{1}{2}$$

$$\Rightarrow \quad y = \pm\frac{1}{\sqrt{2}}$$

$\therefore \left(\frac{1}{4}, \pm\frac{1}{2}\right)$ does not satisfy $1 - y^2 = 2x$.

3. *(b, c, d)* Here, $ax + 2y = \lambda$

and $\quad 3x - 2y = \mu$

For $a = -3$, above equations will be parallel or coincident, i.e. parallel for $\lambda + \mu \neq 0$ and coincident, if $\lambda + \mu = 0$ and if $a \neq -3$, equations are intersecting, i.e. unique solution.

4. *(2)* The circle and coordinate axes can have 3 common points, if it passes through origin. $[p = 0]$

If circle is cutting one axis and touching other axis.

Only possibility is of touching X-axis and cutting Y-axis. $[p = -1]$

5. *(a)*

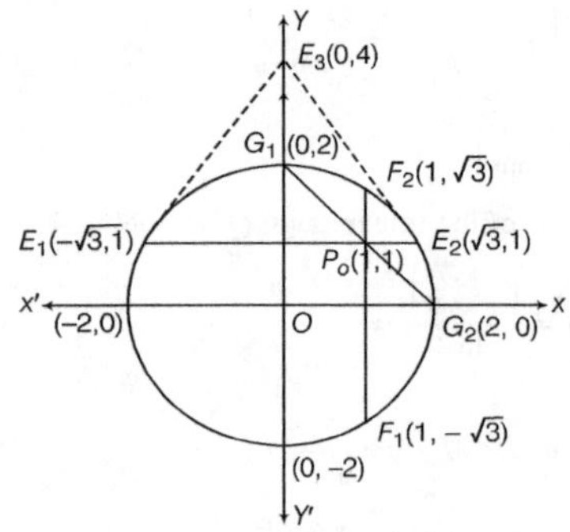

Equation of tangent at $E_1(-\sqrt{3}, 1)$ is $-\sqrt{3}x + y = 4$ and at $E_2(\sqrt{3}, 1)$ is $\sqrt{3}x + y = 4$

Intersection point of tangent at E_1 and E_2 is (0, 4)

$\therefore$ Coordinates of E_3 is (0, 4)

Similarly, equation of tangent at $F_1(1, -\sqrt{3})$ and $F_2(1, \sqrt{3})$ are $x - \sqrt{3}y = 4$ and $x + \sqrt{3}y = 4$, respectively and intersection point is (4, 0), i.e., $F_3(4, 0)$ and equation of tangent at $G_1(0, 2)$ and $G_2(2, 0)$ are $2y = 4$ and $2x = 4$, respectively and intersection point is (2, 2) i.e., $G_3(2, 2)$.

Point $E_3(0, 4)$, $F_3(4, 0)$ and $G_3(2, 2)$ satisfies the line $x + y = 4$.

6. *(d)* We have,

$$x^2 + y^2 = 4$$

Let $P(2\cos\theta, 2\sin\theta)$ be a point on a circle.

$\therefore$ Tangent at P is

$$2\cos\theta x + 2\sin\theta y = 4$$

$$\Rightarrow \quad x\cos\theta + y\sin\theta = 2$$

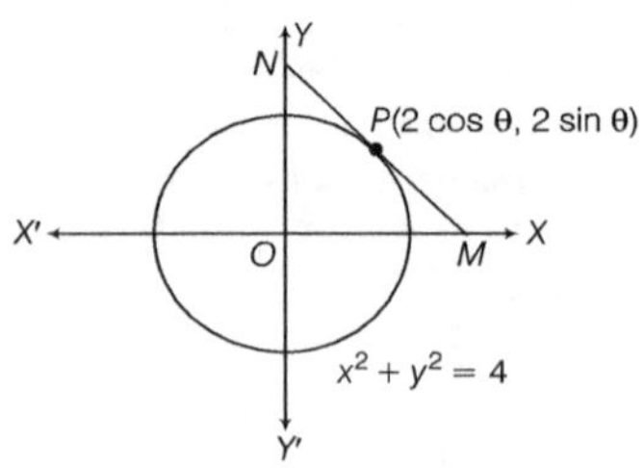

$\therefore$ The coordinates at $M\left(\dfrac{2}{\cos\theta}, 0\right)$ and $N\left(0, \dfrac{2}{\sin\theta}\right)$

Let (h, k) is mid-point of MN

$$\therefore \quad h = \frac{1}{\cos\theta} \text{ and } k = \frac{1}{\sin\theta}$$

$$\Rightarrow \quad \cos\theta = \frac{1}{h} \text{ and } \sin\theta = \frac{1}{k}$$

$$\Rightarrow \quad \cos^2\theta + \sin^2\theta = \frac{1}{h^2} + \frac{1}{k^2}$$

$$\Rightarrow \quad 1 = \frac{h^2 + k^2}{h^2 \cdot k^2}$$

$$\Rightarrow \quad h^2 + k^2 = h^2 k^2$$

$\therefore$ Mid-point of MN lie on the curve

$$x^2 + y^2 = x^2 y^2$$

7. *(a, d)* It is given that T is tangents to S_1 at P and S_2 at Q and S_1 and S_2 touch externally at M.

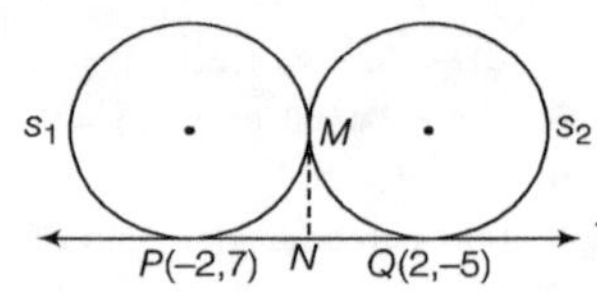

$$\therefore \quad MN = NP = NQ$$

$\therefore$ Locus of M is a circle having PQ as its diameter of circle

$\therefore$ Equation of circle

$$(x - 2)(x + 2) + (y + 5)(y - 7) = 0$$

$$\Rightarrow \quad x^2 + y^2 - 2y - 39 = 0$$

Hence, $\quad E_1 : x^2 + y^2 - 2y - 39 = 0, x \neq \pm 2$

Locus of mid-point of chord (h, k) of the circle E_1 is

$$xh + yk - (y + k) - 39 = h^2 + k^2 - 2k - 39$$

$$\Rightarrow \quad xh + yk - y - k = h^2 + k^2 - 2k$$

Since, chord is passing through (1, 1).

$\therefore$ Locus of mid-point of chord (h, k) is

$$h + k - 1 - k = h^2 + k^2 - 2k$$

$$\Rightarrow \quad h^2 + k^2 - 2k - h + 1 = 0$$

Locus is $E_2 : x^2 + y^2 - x - 2y + 1 = 0$

Now, after checking options, (a) and (d) are correct.

8. *(a, c, d)* We have,

$$sz + t\bar{z} + r = 0 \qquad \ldots\text{(i)}$$

On taking conjugate

$$\bar{s}\bar{z} + \bar{t}z + \bar{r} = 0 \qquad \ldots\text{(ii)}$$

On solving Eqs. (i) and (ii), we get

$$z = \frac{\bar{r}t - r\bar{s}}{|s|^2 - |t|^2}$$

(a) For unique solutions of z

$$|s|^2 - |t|^2 \neq 0 \Rightarrow |s| \neq |t|$$

It is true

(b) If $|s| = |t|$, then $\bar{r}t - r\bar{s}$ may or may not be zero. So, z may have no solutions.

$\therefore$ L may be an empty set.

It is false.

(c) If elements of set L represents line, then this line and given circle intersect at maximum two point.

Hence, it is true.

(d) In this case locus of z is a line, so L has infinite elements.

Hence, it is true.

www.ingramcontent.com/pod-product-compliance
Lightning Source LLC
Chambersburg PA
CBHW080210170726
48282CB00030B/1066
* 9 7 8 9 3 1 3 1 9 3 5 2 4 *